Lecture Notes in Mathematics 1543

Editors:
A. Dold, Heidelberg
B. Eckmann, Zürich
F. Takens, Groningen

A. L. Dontchev   T. Zolezzi

Well-Posed
Optimization Problems

Springer-Verlag
Berlin Heidelberg New York
London Paris Tokyo
Hong Kong Barcelona
Budapest

Authors

Asen L. Dontchev
Bulgarian Academy of Sciences
Institute of Mathematics
Acad. G. Bonchev 8
1113 Sofia, Bulgaria

Tullio Zolezzi
Università di Genova
Dipartimento di Matematica
Via L.B. Alberti 4
16132 Genova, Italy

Mathematics Subject Classification (1991): Primary: 49-02, 49K40
                                             Secondary: 41A50, 49J15, 90C31

ISBN 3-540-56737-2 Springer-Verlag Berlin Heidelberg New York
ISBN 0-387-56737-2 Springer-Verlag New York Berlin Heidelberg

Typesetting: Camera-ready by author/editor
46/3140-543210 - Printed on acid-free paper

To our parents, and
To Dora, Mira, Kiril,
Orietta and Guido

PREFACE

This book aims to present, in a unified way, some basic aspects of the mathematical theory of well-posedness in scalar optimization.

The first fundamental concept in this area is inspired by the classical idea of J. Hadamard, which goes back to the beginning of this century. It requires existence and uniqueness of the optimal solution together with continuous dependence on the problem's data.

In the early sixties A. Tykhonov introduced another concept of well-posedness imposing convergence of every minimizing sequence to the unique minimum point. Its relevance to (and motivation from) the approximate (numerical) solution of optimization problems is clear.

In the book we study both the Tykhonov and the Hadamard concepts of well-posedness, the links between them and also some extensions (e.g. relaxing the uniqueness).

Both the pure and the applied sides of our topic are presented. The first four chapters are devoted to abstract optimization problems. Applications to optimal control, calculus of variations and mathematical programming are the subject matter of the remaining five chapters.

Chapter I contains the basic facts about Tykhonov well-posedness and its generalizations. The main metric, topological and differential characterizations are discussed. The Tykhonov regularization method is outlined.

Chapter II is the key chapter (as we see from its introduction) because it is devoted to a basic issue: the relationships between Tykhonov and Hadamard well-posedness. We emphasize the fundamental links between the two concepts in the framework of best approximation problems, convex functions and variational inequalities.

Chapter III approaches the generic nature of well-posedness (or sometimes ill-posedness) within various topological settings. Parametric optimization problems which are well-posed for a dense, or generic, set of parameters are considered. The relationship with differentiability (sensitivity analysis) is pointed out.

Chapter IV establishes the links between Hadamard well-posedness and variational or epi-convergences. In this way several characterizations of Hadamard well-posedness in optimization are obtained. For convex problems the well-posedness is characterized via the Euler-Lagrange equation. An application to nonsmooth problems is presented, and the role of the convergence in the sense of Mosco is exploited, especially for quadratic problems.

Chapter V is the first one devoted to applications of the theory developed in the first four chapters. Characterizations of well-posedness in optimal control problems for ordinary (or partial) differential equations are discussed. We deal with various forms of well-posedness, including Lipschitz properties of the optimal state and control.

Chapter VI discusses the equivalence between the relaxability of optimal control problems and the continuity of the optimal value (with an abstract generalization). The link with the convergence of discrete-time approximations is presented.

Chapter VII focuses on the study of singular perturbation phenomena in optimal control from the point of view of Hadamard well-posedness. Continuity properties of various mappings appearing in singularly perturbed problems (e.g. the reachable set depending on a small parameter in the derivative) are studied.

Chapter VIII is devoted to characterizations of Tykhonov and Hadamard well-posedness for Lagrange problems with constraints in the calculus of variations, after treating integral functionals without derivatives. We also discuss the classical Ritz method, least squares, and the Lavrentiev phenomenon.

Chapter IX considers first the basic (Berge-type) well-posedness results in a topological setting, for abstract mathematical programming problems depending on a parameter. Then we characterize the stability of the feasible set defined by inequalities, via constraint qualification conditions; Lipschitz properties of solutions to generalized equations are also discussed. Hadamard well-posedness in convex mathematical programming is studied. Quantitative estimates for the optimal solutions are obtained using local Hausdorff distances. Results about Lipschitz continuity of solutions in nonlinear and linear programming end the chapter.

We have made an attempt to unify, simplify and relate many scattered results in the literature. Some new results and new proofs are included. We do not intend to deal with the theory in the most general setting; our goal is to present the main problems, ideas and results in as natural a way as possible.

Each chapter begins with an introduction devoted to examples and motivations or to a simple model problem in order to illustrate the specific topic. The formal statements are often introduced by heuristics, particular cases and examples, while the complete proofs are usually collected at the end of each section and given in full detail, even when elementary. Each chapter contains notes and bibliographical remarks.

The prerequisites for reading this book do not extend in general beyond standard real and functional analysis, general topology and basic optimization theory. Some topics occasionally require more special knowledge that is always either referenced or explicitly recalled when needed.

Some sections of this book are based in part on former lecture notes (by T. Zolezzi) under the title "Perturbations and approximations of minimum problems".

We benefited from the help of many colleagues. We would like to thank especially G. Dal Maso, I. Ekeland, P. Kenderov, D. Klatte, R. Lucchetti, F. Patrone, J. Revalski, K. Tammer, V. Veliov. The support of the Bulgarian Academy of Sciences, Consiglio Nazionale delle Ricerche, MPI and MURST is gratefully acknowledged.

We wish to thank A. Patev for drawing the figures, and C. Taverna for typing the manuscript.

June 1992   Asen L. Dontchev
                                                          Tullio Zolezzi

TABLE OF CONTENTS.

CHAPTER VII.   SINGULAR PERTURBATIONS IN OPTIMAL CONTROL.

CHAPTER VIII.   WELL-POSEDNESS IN THE CALCULUS OF VARIATIONS.

CHAPTER IX.   HADAMARD WELL-POSEDNESS IN MATHEMATICAL PROGRAMMING.

XII

Chapter I.

TYKHONOV WELL-POSEDNESS

Section 1. Definition and Examples.

Let X be a set endowed with either a topology or a convergence structure. Let

$$I : X \to (-\infty, +\infty]$$

be a proper extended real - valued function. Consider the problem

to minimize $I(x)$ subject to $x \in X$,

which we denote by (X, I).

We are interested in the well-posedness of (X, I).

A natural well - posedness concept arises when we require the following two conditions. First, we impose existence and uniqueness of the global minimum point

$$x_0 = \arg \ \min \ (X, I).$$

Second, we require that, whenever we are able to compute approximately the optimal value

$$I(x_0) = \inf \ I(X),$$

then we automatically do approximate the optimal solution x_0. That is, every method constructing minimizing sequences for (X, I) corresponds to approximately computing x_0.

More precisely, the latter condition means that, if x_n is any sequence from X such that

$$I(x_n) \to \inf \ I(X)$$

then

$$x_n \to \arg \min \ (X, I).$$

This condition is clearly of fundamental relevance to the approximate (numerical) solving of (X, I). We shall consider first the case when X is a convergence space, as follows.

Let X be a convergence space, with convergence of sequences denotes by $\longrightarrow$, see e.g. Kuratowski [1, p. 83 - 84]. Let $I : X \to (-\infty, +\infty]$ be a proper extended real-valued function. The problem (X, I) is *Tykhonov well-posed* iff I has a unique global minimum point on X towards which every minimizing sequence converges.

An equivalent definition is the following: there exists exactly one $x_0 \in X$ such that $I(x_0) \leq I(x)$ for all $x \in X$, and

$$I(x_n) \to I(x_0) \text{ implies } x_n \to x_0.$$

Notice that the existence of some x_0 as above implies its uniqueness (if x_0, y_0 do then take the minimizing sequence $x_0, y_0, x_0, y_0, ...$).

Tykhonov well - posedness of (X, I) is often stated equivalently as *strong uniqueness* of arg min (X, I), or *strong solvability* of (X, I). Sometimes well - posedness is translated literally as *correctness*. Problems which are not well - posed will be called *ill - posed*. Sometimes they are referred to as *improperly posed*.

1 Example. Let $X = R^n$ and $I(x) = |x|$ (taking any norm).

Then $0 = $ arg min (X, I) and clearly (X, I) is Tykhonov well - posed.

2 Example. Let $X = R$ and

$$I(x) = x \text{ if } x > 0, \ = |x + 1| \text{ if } x \leq 0.$$

Then the only minimum point is $x_0 = -1$ but (X, I) is Tykhonov ill - posed since the minimizing sequence $x_n = 1/n$ does not converge to x_0.

Remark. Let A be an open set in R^n, let $x_0 \in A$ and $f \in C^2(A)$. Suppose that $\nabla f(x_0) = 0$ and the Hessian matrix of f at x_0 is positive definite. Then x_0 is a local minimizer of f. By Taylor's formula, there exist some $\alpha > 0$ and a ball $B \subset A$ centered at x_0 such that

$$f(x) \geq f(x_0) + \alpha |x - x_0|^2, x \in B,$$

so that (B, f) is well - posed.

3 Example. Let X be the unit ball in $L^\infty(0, 1)$ equipped with the strong convergence. Given $u \in X$ let $x(u)$ be the only absolutely continuous solution to

$$\dot{x} = u \text{ a.e. in } (0, 1), x(0) = 0,$$

and let

$$I(u) = \int_0^1 x(u)^2 \, dt.$$

Then arg min (X, I) reduces to the single function $u = 0$ and the optimal value is 0. This (optimal control) problem is not Tykhonov well - posed since

$$u_n(t) = \sin nt$$

is a minimizing sequence because

$$x(u_n)(t) = \int_0^t u_n \, dt \ \to 0 \text{ uniformly on } [0, 1],$$

therefore $I(x_n) \to 0$. But u_n does not converge to 0 in X since $\|u_n\| = 1$ for any $n \geq 2$.

4 Example. Everything is as in example 3, except that X is now equipped with $L^1 -$ convergence . Then X is a compact metric space (by Alaoglu's theorem and separability of $L^1(0, 1)$: see Dunford-Schwartz [1, V.4.2 and V.5.1]). Let u_n be any minimizing sequence. Fix any subsequence of u_n, then a further subsequence $v_n \to u_0 \in X$ and $I(v_n) \to 0$. Then $x(v_n) \to x(u_0)$ uniformly in $[0, 1]$. So $I(u_0) = 0$, hence $u_0 = 0$. This shows that the original sequence u_n converges towards 0 in X, thus (X, I) is Tykhonov well - posed.

5 Example. Let X be the unit ball of $L^2(0,1)$ equipped with the strong convergence, and let $x(u)$ be as in example 3. Put

$$I(u) = \int_0^1 x(u)^2 \, dt + \varepsilon \int_0^1 u^2 \, dt, \quad \varepsilon \geq 0.$$

If $\varepsilon > 0$ then (X, I) is Tykhonov well - posed, but it is not when $\varepsilon = 0$.

6 Example. Let X be a sequentially compact convergence space, let I be a proper and sequentially lower semicontinuous function on X. Suppose that (X, I) has a unique global minimum point. Then (X, I) is Tykhonov well - posed. So, assuming uniqueness of the minimizer, well - posedness obtains with respect to the natural convergence associated to the direct method (compactness and lower semicontinuity) of the calculus of variations. Another case: $X = R^n$, $I : R^n \to (-\infty, +\infty)$ is strictly convex, and $I(x) \to +\infty$ as $x \to \infty$. Then I is continuous and (X, I) is Tykhonov well - posed. More generally, (X, I) is Tykhonov well - posed with respect to the strong convergence whenever X is a compact convex subset of some Banach space and I is finite-valued, strictly convex and strongly lower semicontinuous on X.

7 Example. The problem (R^k, I) is Tykhonov well - posed if I is convex, lower semicontinuous and there exists an unique $x_0 = \arg\min(R^k, I)$. To see this, replace I by $x \to I(x_0 + x) - I(x_0)$. Then we can assume without loss of generality that

$$I(0) = 0 < I(x) \text{ if } x \neq 0.$$

Let x_n be any minimizing sequence. If $|x_n| \to +\infty$ for some subsequence, then by convexity

$$0 \leq I\left(\frac{x_n}{|x_n|}\right) \leq \frac{1}{|x_n|} I(x_n) \to 0 \text{ as } n \to +\infty.$$

On the other hand, for a further subsequence

$$\frac{x_n}{|x_n|} \to y \text{ with } |y| = 1.$$

By lower semicontinuity we get $I(y) = 0$, a contradiction. So any cluster point u of x_n fulfills (for some subsequence) $I(x_n) \to I(u) = 0$, yielding $u = 0$ by uniqueness. Then $x_n \to 0$ for the original sequence.

8 Example. Let X be a convergence space and $I : X \to (-\infty, +\infty)$ be lower semi-continuous and coercive, i.e. $y_n \in X$ and $\sup I(y_n) < +\infty$ imply that some subsequence of y_n converges. Then (X, I) is Tykhonov well - posed if it has a unique global minimum point. If I is lower semicontinuous and bounded from below, then (X, I) is Tykhonov well - posed iff every minimizing sequence converges.

9 Remark. If X is a Banach space, I is convex on X and (X, I) is Tykhonov well - posed with respect to the strong convergence on X, then $I(x) \to +\infty$ as $\|x\| \to +\infty$. Indeed, let $x_0 = \arg \min (X, I)$ and arguing by contradiction, assume

$$L = \lim \inf I(x) < +\infty \text{ as } \|x\| \to +\infty.$$

By well - posedness, $I(x_0) < L$. For some sequence x_n with $\|x_n\| \to +\infty$ we have $I(x_n) \to L$.

Put $2y_n = x_n + x_0$. Then $\|y_n\| \to +\infty$. By convexity, for some $\varepsilon > 0$

$$I(y_n) \leq \frac{1}{2}I(x_0) + \frac{1}{2}I(x_n) \leq \frac{1}{2}(L - \varepsilon) + \frac{1}{2}(L + 2\frac{\varepsilon}{3})$$

thus $\lim \sup I(y_n) < L$, a contradiction.

We shall see that the conditions

$$I(x) \to +\infty \text{ as } \|x\| \to +\infty \text{ and } \arg \min (X, I) \text{ is a singleton,}$$

do not imply Tykhonov well - posedness in the infinite-dimensional setting (modify I in example 18 by imposing $I(x) \to +\infty$ as $\|x\| \to +\infty$: for example

$$I(x) = \sum_{n=1}^{\infty} \frac{< x, e_n >^2}{n^2} + (\|x\| - 1)^2 \text{ if } \|x\| > 1).$$

10 Example. Let X be a nonempty closed convex subset of the real Hilbert space H. Let $u \in H$ be fixed and let

$$I(x) = \|u - x\|, x \in X.$$

By (the proof of) the Riesz projection theorem we see that (X, I) is Tykhonov well - posed with respect to the strong convergence. In fact, let x_n be any minimizing sequence. By the parallelogram law, for every n and k

$$\|u - \frac{x_n + x_k}{2}\|^2 + \|\frac{x_n - x_k}{2}\|^2 = \frac{1}{2}(\|u - x_k\|^2 + \|u - x_n\|^2).$$

Since $(x_n + x_k)/2 \in X$, we get

$$\|u - \frac{x_n + x_k}{2}\|^2 \geq m^2$$

where $m = \text{dist}(u, X)$. Therefore

$$\|\frac{x_n - x_k}{2}\|^2 \leq \frac{1}{2}(\|u - x_n\|^2 + \|u - x_k\|^2) - m^2,$$

hence $\|x_n - x_k\| \to 0$ as $n, k \to +\infty$. Therefore there exists some $z \in X$ such that

$$x_n \to z \text{ and } \|u - x_n\| \to \|u - z\| = m.$$

Uniform convexity of H implies uniqueness of z, whence well - posedness.

Example 10 will be generalized, in a very significant way, in section II.1.

As a final example, let X be a real normed space and $A \subset X$. By the very definition (see Giles [1 p.195]), a point $x \in A$ is *strongly exposed* iff there exists $u \in X^*$ such that (A, u) is Tykhonov well-posed with respect to strong convergence and $x = \arg \min (A, u)$.

Section 2.   Metric Characterizations.

The weaker the convergence on X is, the easier Tykhonov well-posedness obtains. As a matter of fact, as we saw in example 6, under uniqueness of the minimizer Tykhonov well-posedness of (X, I) is a fairly common property, since it can be obtained by the direct method in the calculus of variations. However, in the applications we often need sufficiently strong convergence of minimizing sequences. Therefore the following setting is of interest.

Standing assumptions: X is a metric space with metric d, and

$$I : X \to (-\infty, +\infty).$$

Remark. We can assume I real-valued without loss of generality, since (X, I) is Tykhonov well-posed iff (K, I) is, where

$$K = \text{ effective domain of } I = \{x \in X : I(x) < +\infty\}.$$

We shall consider X as a convergence space equipped by the (natural) convergence structure induced by the metric.

In the sequel we shall use the following conditions:

(1) I is sequentially lower semicontinuous and bounded from below.

(2) X is complete.

The basic idea behind the next fundamental theorem can be roughly explained as follows. If (X, I) is Tykhonov well - posed then $\varepsilon-$ arg min (X, I) shrinks to the unique optimal solution as $\varepsilon \to 0$. Conversely, if diam $[\varepsilon - $ arg min $(X, I)] \to 0$ then every minimizing sequence is Cauchy, therefore it will converge to the unique solution of (X, I), provided that (1) and (2) hold.

11 Theorem *If (X, I) is Tykhonov well-posed then*

(3)   *diam $[\varepsilon - $ arg min $(X, I)] \to 0$ as $\varepsilon \to 0$.*

Conversely, (3) implies Tykhonov well-posedness under (1) and (2).

Since Tykhonov well - posedness of (X, I) amounts to the existence of some $x_0 \in$ arg min (X, I) such that

$$I(x_n) \to I(x_0) \Rightarrow d(x_n, x_0) \to 0,$$

then it is reasonable to try to find some estimate from below for $I(x) - I(x_0)$ in terms of $d(x, x_0)$. This aims to quantitative results about Tykhonov well-posedness (e.g. rates of convergence of minimizing sequences).

A function

$$c : D \to [0, +\infty)$$

is called a *forcing function* iff

$$0 \in D \subset [0, +\infty), c(0) = 0 \text{ and } a_n \in D, c(a_n) \to 0 \Rightarrow a_n \to 0.$$

An (obvious) example of forcing function is

$$c(t) = t^k, t \geq 0, k > 0.$$

Notice that (D, c) is Tykhonov well-posed if c is a forcing function.

12 Theorem (X, I) *is Tykhonov well-posed iff there exists a forcing function c and a point x_0 such that*

(4) $$I(x) \geq I(x_0) + c[d(x, x_0)] \text{ for every } x \in X.$$

Example (uniqueness implies well-posedness). Let X be a real Hilbert space (identified with its dual space) equipped with the strong convergence. Let

$$A : X \to X$$

be a linear, continuous, self-adjoint, nonnegative operator. For any fixed $y \in X$ consider

$$I(x) = <Ax, x> -2 <y, x>, x \in X.$$

If arg min (X, I) is a singleton for every $y \in X$, then (X, I) is Tykhonov well-posed. To see this, notice that for any $y \in X$ the corresponding I is a convex differentiable functional. Hence $x_0 = $ arg min (X, I) iff

(5) $$Ax_0 = y.$$

Then A is bijective, hence an isomorphism. Thus 0 belongs to the resolvent of A, which is open. If follows that the spectrum of A is contained in some half-line $[\alpha, +\infty), \alpha > 0$. Hence (Brezis [2, prop. VI.9])

$$0 < \alpha \leq \inf \{<Ax, x>: x \in X, \|x\| = 1\}$$

giving coercivity of A. Then by (5)

$$I(x) - I(x_0) = <A(x - x_0), x - x_0> \geq \alpha \|x - x_0\|^2,$$

proving well-posedness by theorem 12.

The link between well-posedness of (X, I) and unique solvability, with continuous dependence of x_0 upon y, of its Euler-Lagrange equation (5), a truly basic issue of the theory, will be developed in section II.5, IV.7 and IV.9.

If we consider

(6) $$q(s) = \sup \{t \in D : c(t) \leq s\}, s \geq 0,$$

then by (4) we obtain the following estimate.

13 Corollary (X, I) *is Tykhonov well-posed iff there exist an increasing function*

$$q : [0, +\infty) \to [0, +\infty), q(t) \to 0 \text{ as } t \to 0,$$

and a point $x_0 \in X$ *such that*

(7) $$d(x, x_0) \leq q[I(x) - I(x_0)] \text{ for every } x \in X.$$

14 Corollary (X, I) *is Tykhonov well-posed with solution* x_0 *iff for some* $x_0 \in X$ *and for every* $\varepsilon > 0$ *one can find* $\delta > 0$ *such that*

$$I(x) - I(x_0) < \delta \text{ implies } d(x, x_0) < \varepsilon.$$

From Theorem 12 we see that Tykhonov well-posedness of (X, I) may be described in a quantitative way by the biggest forcing function c fulfilling (4), which may be called *modulus of Tykhonov well- posedness* of (X, I), defined by

(8) $$c(t) = \inf \{I(x) - I(x_0) : d(x, x_0) = t\}, t \geq 0,$$

where $x_0 = \arg \min (X, I)$.

15 Proposition *Let* X *be a nonempty convex subset of a real normed space,* I *be convex on* X *and* (X, I) *be Tykhonov well-posed. Then the function* c *defined by (8) is strictly increasing and forcing.*

By taking some smaller c in (4) we may always assume that c is continuous and strictly increasing near 0 (as shown later in the proofs).

Among the above characterizations, Theorem 11 uses the set of $\varepsilon-$ optimal solutions, while theorem 12 requires the exact optimal solution x_0 . The next theorem requires the knowledge of the optimal value.

16 Theorem *If* (X, I) *is Tykhonov well-posed then there exists a forcing function* c *such that*

(9) $$-\infty < \inf I(X) \leq \max \{I(x), I(y)\} - c[d(x, y)]$$

for every $x, y \in X$. *Conversely, (9) with some forcing function* c *implies Tykhonov well-posedness of* (X, I) *under (1) and (2).*

Let K be a nonempty convex subset of a real Banach space X. Then $I : K \to (-\infty, +\infty)$ is *uniformly quasiconvex* iff there exists an increasing function

$$c : [0, +\infty) \to [0, +\infty)$$

such that $c(0) = 0, c(t) > 0$ whenever $t > 0$ and

(10) $$I(\frac{x + y}{2}) \leq \max\{I(x), I(y)\} - c(\|x - y\|)$$

for every $x, y \in K$. As easily checked, c is necessarily forcing. From theorem 16 we get

17 Corollary *If K is closed, I is lower semicontinuous, bounded from below and uniformly quasiconvex on K, then (K, I) is Tykhonov well-posed.*

By (10) we see that c is an estimate from below of the modulus of well-posedness of (K, I).

18 Example. Let X be the closed unit ball of a real separable Hilbert space H, equipped with the strong convergence. Let e_n be an orthonormal basis for H, and

$$I(x) = \sum_{n=1}^{\infty} \frac{< x, e_n >^2}{n^2}, \ x \in H.$$

Then I is strictly convex, continuous, and arg min $(X, I) = 0$. But (X, I) is not Tykhonov well-posed since e_n is a minimizing sequence. Moreover $\sqrt{n} e_n$ is an unbounded minimizing sequence for the ill- posed problem (H, I).

Now let X be a bounded subset of the real Banach space E. Then X is *dentable* iff for every $a > 0$ there exists $y \in X$ such that

$$y \notin \text{clco} \ [X \setminus \text{int} \ B(y, a)]$$

This notion plays a significant role in the geometric theory of Banach spaces (see e.g. Bourgin [1]). Dentability is related to Tykhonov well-posedness for linear functionals. In fact we have

19 Proposition *Let X be closed in E. Then X is dentable if there exists $u \in E^*$ such that (X, u) is strongly Tykhonov well-posed.*

(Here "strongly" refers to the strong convergence on X).

PROOFS.

We need

20 Lemma. Given $D \subset [0, +\infty)$ with $0 \in D$, and $c : D \to [0, +\infty)$ such that $c(0) = 0$, let q be defined by (6). Then

(11) $\qquad\qquad q \text{ is increasing }, q(t) \geq 0, t \leq q[c(t)] \ \forall t;$

(12) $\qquad\qquad q(t) \to 0 \text{ as } t \to 0 \text{ if } c \text{ is a forcing function.}$

Given $E \subset [0 + \infty), 0 \in E, q : E \to [0, +\infty)$, let

(13) $\qquad\qquad c(t) = \inf \ \{s \geq 0 : q(s) \geq t\}, t \geq 0.$

Then

(14) c is increasing, $c[q(t)] \leq t \ \forall t$;

(15) c is forcing if $q(t) \to 0$ as $t \to 0$;

(16) $s \geq 0, 0 \leq t \leq q(s) \Rightarrow c(t) \leq s$.

Proof. Of course q given by (6) and c given by (13) are extended real-valued functions. (11) is obvious from (6). Let $s_n \geq 0, s_n \to 0$. Let us show that $q(s_n) \to 0$. There exist points $t_n \geq 0$ such that $c(t_n) \leq s_n, q(s_n) \leq 1/n + t_n$. Then $c(t_n) \to 0$ so $t_n \to 0$, hence $q(s_n) \to 0$. This shows (12). (14) and (16) are obvious from the definition of c. Clearly $c(0) = 0, c(t) \geq 0 \ \forall t$. If $a_n \geq 0$ and $c(a_n) \to 0$ there exists a sequence s_n such that

$$a_n \leq q(s_n), 0 \leq s_n < c(a_n) + \frac{1}{n}.$$

Hence $s_n \to 0$, thus $q(s_n) \to 0$ and $a_n \to 0$ as required to prove (15). $\qquad\square$

Theorem 11. Suppose that (X, I) is Tykhonov well-posed. If (3) fails, there exist $\varepsilon_k \to 0$ and $a > 0$ such that

$$\text{diam } [\varepsilon_k - \text{ arg min } (X, I]) \geq 2a \ \forall k.$$

Then one can find points $u_k, v_k \in \varepsilon_k - \text{ arg min } (X, I)$ such that

(17) $d(u_k, v_k) \geq a \ \forall k.$

But u_k, v_k are both minimizing sequences, therefore (by assumption) converging to the same point, against (17). Conversely assume (3). Given $a > 0$ there exists $b > 0$ such that

(18) $\text{diam } [\varepsilon - \text{ arg min } (X, I)] < a$ if $0 < \varepsilon < b.$

Let u_n be a minimizing sequence. Given ε as in (18) we get $u_n \in \varepsilon - \text{ arg min } (X, I)$ for large n, then again by (18) u_n is Cauchy. Let $u_n \to x_0$ (by (2)). From (1) we see that $x_0 \in \text{ arg min } (X, I)$. Every minimizing sequence converges, therefore arg min (X, I) is a singleton. $\qquad\square$

Remark. The above proof shows that semicontinuity of I can be omitted in Theorem 11 if we know that arg min $(X, I) \neq \emptyset$.

Counterexample. Without (2), condition (3) is no longer sufficient for Tykhonov well-posedness. Take $I(x) = |x - \sqrt{2}|, X =$ rational numbers in $[1, 2]$

Theorem 12. Assume (4). Then clearly x_0 is the only solution. Let x_n be a minimizing sequence. Then $c[d(x_n, x_0)] \to 0$ implying $x_n \to x_0$. Conversely, let (X, I) be Tykhonov well-posed with solution x_0, c be the modulus of well-posedness given by (8). We show that c is forcing. Clearly, $c(0) = 0, c(t) \geq 0 \ \forall t$. If $t_n \geq 0$ are such that $c(t_n) \to 0$, then there exists a sequence $u_n \in X$ such that $I(u_n) \to I(x_0), d(u_n, x_0) = t_n \ \forall n$. Then u_n is minimizing, therefore $t_n \to 0$. $\qquad\square$

Corollary 13. If (X, I) is Tykhonov well-posed with solution x_0, then (4) holds. Thus by (11) we get (7). Conversely, assume (7). Let c be defined by (13). Then c is a forcing function by (15). Remembering (16), then (4) follows from (7). $\quad\square$

Corollary 14. Tykhonov well-posedness implies (7). Thus, given $\varepsilon > 0$ there exists $\delta > 0$ such that

$$q(t) < \varepsilon \text{ whenever } 0 < t < \delta,$$

whence the conclusion. Conversely, notice that if $I(x) < I(x_0)$ for some x, then $d(x, x_0) < \varepsilon$ for every $\varepsilon > 0$, a contradiction. Therefore $I(x_0) = \inf I(X)$. Now set

$$c(t) = \inf \{I(x) - I(x_0) : d(x, x_0) = t\} \geq 0.$$

Let $a_n \geq 0$ be such that $c(a_n) \to 0$. We can find points $x_n \in X$ such that

$$d(x_n, x_0) = a_n, I(x_n) - I(x_0) < c(a_n) + \frac{1}{n}.$$

Given $\varepsilon > 0$ let δ be as in the assumption. Then $I(x_n) - I(x_0) < \delta$ for n sufficiently large, therefore $a_n < \varepsilon$. This shows that c is a forcing function. Moreover (4) holds, so that the conclusion comes from theorem 12. $\quad\square$

Proposition 15. As shown in the proof of theorem 12, c is forcing. Let $0 < a < b$ be such that X contains points u_1, u_2 with $\|u_1 - x_0\| = a, \|u_2 - x_0\| = b$. Given $x \in X$ such that $\|x - x_0\| = b$, put $z = sx + (1-s)x_0 \in X$ where $s = a/b$. Then $\|z - x_0\| = a$. By convexity of I

$$I(z) \leq s\, I(x) + (1 - s)I(x_0),$$

therefore

$$c(a) \leq I(z) - I(x_0) \leq s \inf \{I(x) - I(x_0) : \|x - x_0\| = b, x \in X\} < c(b). \quad\square$$

Now we show that (in the setting of theorem 12) the function c in (4) *may be taken continuous and stricly increasing* near 0 if X contains at least two points. Put

$$c_1(t) = \inf \{c(s), s \geq t\}, t \geq 0;$$

$$c_2(t) = \frac{2}{\pi t} \int_0^t c_1(s) \, arctg\, s \, d\, s, \ t > 0; c_2(0) = 0.$$

Let X contain some point at distance $T > 0$ from x_0. Then $c_1(T) < +\infty$, therefore c_2 is continuous in $[0, T]$. Moreover $c_2(t) \leq c_1(t) \leq c(t)$ since c_1 is increasing. Finally,

$$\frac{\pi}{2}t^2 \frac{d}{dt}c_2(t) = tc_1(t)\, arctg\, t - \int_0^t c_1(s)\, arctg\, s\, d\, s$$

for a.e. $t \in (0, T)$, therefore c_2 is strictly increasing there (since $s > 0 \Rightarrow c_1(s) > 0$).

Theorem 16. Let (X, I) be Tykhonov well-posed. Put

$$q(\varepsilon) = \text{diam} \; [\varepsilon - \arg \; \min \; (X, I)], \varepsilon \geq 0,$$

and let c be defined by (13). Given $x, y \in X$ consider

$$k = \max \; \{I(x), I(y)\} - \inf \; I(X).$$

Then by (14), since $x, y \in k - \arg \; \min \; (X, I)$ we get

$$c[d(x, y)] \leq c[q(k)] \leq k,$$

giving (9). By (15), c is forcing. Conversely, assume (9). Given $\varepsilon > 0$ let $x, y \in \varepsilon - \arg \; \min \; (X, I)$. By (9)

(19) $\inf \; I(X) \leq \inf \; I(X) + \varepsilon - c[d(x, y)].$

Let q be given by (6). Then, by (11) and (19), $d(x, y) \leq q(\varepsilon)$, thus diam $[\varepsilon - \arg \; \min \; (X, I)] \leq q(\varepsilon)$.
By (12), the conclusion follows from theorem 11.   □

A direct proof of corollary 17. Let x_n be a minimizing sequence to (K, I). Then by (10), for all n and p,

$$\inf \; I(K) \leq I(\frac{x_n + x_p}{2}) \leq \inf \; I(K) + a_{np} - c(\|x_n - x_p\|)$$

where $a_{np} \to 0$ as $n, p \to +\infty$. Thus $c(\|x_n - x_p\|) \to 0$, hence x_n is a Cauchy sequence in K. By closedness and semicontinuity, x_n converges to arg $\min \; (K, I)$.   □

Proposition 19. By theorem 11, it suffices to show that dentability follows from

(20)   diam $[\varepsilon - \arg \; \min \; (X, u)] \to 0$ as $\varepsilon \to 0$ for some $u \in E^*$.

By (20), assume that $u \in E^*$ and for every $a > 0$ there exists $\varepsilon > 0$ such that

$$\text{diam} \; [\varepsilon - \arg \; \min \; (X, u)] < a.$$

Pick $x \in X$ with

(21) $u(x) < \inf u(X) + \varepsilon.$

Then

$$\varepsilon - \arg \; \min \; (X, u) \subset \text{int} \; B(x, a)$$

therefore

$$\text{cl} \; \text{co} \; [X \setminus \text{int} \; B(x, a)] \subset \text{cl} \; \text{co} \; [X \setminus \varepsilon - \arg \; \min \; (X, u)].$$

Hence $y \in$ cl co $[X \setminus \text{int} \; B(x, a)]$ implies $\varepsilon + \inf \; u(X) \leq u(y)$. By (21) we get

$$x \notin \text{cl} \; \text{co} \; [X \setminus \text{int} \; B(x, a)].$$

□

Section 3.   Topological Setting.

In this section X will denote a Hausdorff topological space, equipped with the convergence structure inherited by the topology. Therefore the sequence $x_n \to x_0$ in X iff for every neighbourhood A of x_0 there exists p such that $x_n \in A$ when $n \geq p$. Suppose

$$I : X \to (-\infty, +\infty).$$

Generalizing corollary 14, we get

21 Proposition (X, I) *is Tykhonov well-posed iff there exists* $x_0 \in X$ *such that*

(22)   *for every neighbourhood A of x_0 there exists $\delta > 0$ such that*

$$I(x) - I(x_0) < \delta \Rightarrow x \in A.$$

Moreover, (22) $\Rightarrow x_0 = arg\ min\ (X, I)$.

The characterization afforded by proposition 21 may be expressed making use of the level set multifunction

$$t \rightrightarrows \mathrm{lev}\ (I, t) = \{x \in X : I(x) = t\}, t \in R$$

with (possibly empty) values in X. In fact, (22) amounts to the following : $x_0 = arg\ min\ (X, I)$ and for every neighbourhood A of x_0 there exists $\delta > 0$ such that

$$\mathrm{lev}\ (I, t) \subset A\ \text{if}\ I(x_0) < t < I(x_0) + \delta.$$

The same is obviously true if we replace lev (I, t) by

$$\mathrm{sub\ lev}\ (I, t) = \{x \in X : I(x) \leq t\}.$$

Therefore we have proved

22 Proposition (X, I) *is Tykhonov well-posed iff arg min (X, I) is a singleton and* $(sub)lev(I, \cdot)$ *is upper semicontinuous at inf $I(X)$.*

The following (formally stronger) definition of *nets Tykhonov well-posedness* is natural when dealing with a general topological space X. We require existence and uniqueness of $x_0 = arg\ min\ (X, I)$ and convergence to x_0 of every minimizing net (see Kelley [1, ch. 2]). Of course, nets Tykhonov well-posedness entails well-posedness in the usual (sequential) sense. Conversely, let (X, I) be Tykhonov well-posed and $x_0 = arg\ min\ (X, I)$. If (X, I) fails to be nets well-posed, then there exists some minimizing net x_n such that $x_n \not\to x_0$. Therefore we can find some neighbourhood A of x_0 such that for every m there exists $n \geq m$ with $x_n \notin A$. On the other hand, by proposition 21 there exists p such that $x_n \in A$ wherever $n \geq p$, a contradiction. Hence *the sequential and nets definitions of Tykhonov well-posedness are equivalent.*

Now let X be a compact topological space, and suppose

$$I : X \to (-\infty, +\infty)$$

to be lower semicontinuous, with an unique

$$x_0 = \arg\ \min\ (X, I).$$

Let x_n be any minimizing net for (X, I). Any subnet of x_n is minimizing and, by compactness, a further subnet $y_n \to z$ say. Thus

$$I(x_0) = \lim\ \inf\ I(y_n) \geq I(z)$$

yielding $z = x_0$ by uniqueness. Then (X, I) is nets Tykhonov well-posed, therefore Tykhonov well-posed in the usual (sequential) sense. Summarizing, we have

23 Theorem *Let X be a compact Hausdorff topological space and $I : X \to (-\infty, +\infty)$ lower semicontinuous. Then (X, I) is Tykhonov well-posed iff $arg\ min(X, I)$ is a singleton.*

A different approach to well-posedness of (X, I) can be based on consideration of the relevant preorder represented on X by I. A *total preorder* on X is a reflexive, transitive and total binary relation on X. The real-valued function I on X *represents* the preorder $\leq$ iff

$$x \leq y \text{ if and only if } I(x) \leq I(y).$$

In a sense precisely defined in the next proposition, it is the preorder represented by I on X which matters for Tykhonov well-posedness of (X, I), since well - posedness is then shared by all functions representing that preorder. Such objective functions are then equivalent each other up to composition with some increasing homeorphism (i. e. rescaling). Therefore well - posedness can be considered as an ordinal property.

24 Proposition *Let X be a (nonempty) connected topological space. Let*

$$I, J : X \to (-\infty, +\infty)$$

be continuous and representing the same preorder on X. Then (X, I) is Tykhonov well-posed iff (X, J) is. In this case $arg\ min\ (X, I) = arg\ min\ (X, J)$ and moreover

$$J = h[I]$$

for some strictly increasing continuous surjective map

$$h : I(X) \to J(X).$$

PROOFS.

Proposition 21. Assume well-posedness with solution x_0. Suppose there exists some neighbourhood A of x_0 and a sequence x_n such that

$$x_n \notin A, I(x_n) - I(x_0) < \frac{1}{n} \text{ for every } n.$$

Then x_n would be a minimizing sequence, hence $x_n \to x_0$ which is a contradiction. Conversely assume (22). If there exists some $u \in X, u \neq x_0$, such that $I(u) \leq I(x_0)$, then by (18) u belongs to every neighbourhood of x_0, a contradiction since X is Hausdorff. Therefore $x_0 = \arg \min (X, I)$. Now let x_n be any minimizing sequence. Fix any neighbourhood A of x_0. Given $\delta > 0$ we get $I(x_n) - I(x_0) < \delta$ for sufficiently large n, hence $x_n \in A$ by (22), so that $x_n \to x_0$. □

Proposition 24. Denote by $\leq$ the given preorder in X. By assumption

(23) $$x \leq y \Leftrightarrow I(x) \leq I(y) \Leftrightarrow J(x) \leq J(y).$$

Given $t \in I(X)$, let $t = I(x), x \in X$ and put

$$h(t) = J(x).$$

By (23) this defines $h(t)$ unambigously, h is strictly increasing, and of course $J = h[I]$. Arguing by contradiction, suppose h is not continuous at some $t_0 \in I(X)$. Without loss of generality we can therefore assume that, for some $t_n \uparrow t_0$, some $x_n \in X$ and some real p, we have

$$J(x_n) = h(t_n) \uparrow p < J(x_0) = h(t_0).$$

Since $J(X)$ is connected, there exists some $z \in X$ such that $p < J(z) < J(x_0)$. Hence by (23)

$$t_n = I(x_n) < I(z) < I(x_0) = t_0,$$

contradicting $t_n \uparrow t_0$. □

Section 4.   Differential Characterizations.

In this section K is a nonempty convex subset of a real Banach space X, and

$$I : K \to (-\infty, +\infty).$$

We recall that $u \in X^*$ is a subgradient of I at x (relatively to K) iff

(24) $$I(y) \geq I(x) + <u, y - x> \quad \forall y \in K.$$

I is subdifferentiable on K iff $\partial I(x) \neq \emptyset$ for every $x \in K$, where $\partial I(x)$ denotes the set of all subgradients of I at x (called subdifferential of I at x relatively to K).

We shall consider K as a convergence space, equipped with the restriction of either the strong or the weak convergence in X. Accordingly, we shall consider either a *strongly* or *weakly Tykhonov well-posed* problem (K, I).

25 Theorem *Let I be subdifferentiable on K. Then (K, I) is strongly Tykhonov well-posed iff $\inf I(K) > -\infty$ and there exist $x_0 \in K$ and a forcing function c such that*

(25) $$<u, x - x_0> \geq c(\|x - x_0\|)$$

$\forall x \in K$ and $\forall u \in \partial I(x)$. Moreover (25) implies

$$(26) \qquad\qquad\qquad I(x) \geq I(x_0) + c(\frac{\|x - x_0\|}{2}).$$

By using duality and differentiability theory for convex functions we get characterizations of Tykhonov well-posedness in terms of the conjugate functional.

Throughout, X is a real Banach space with dual X^*, and

$$I : X \to (-\infty, +\infty]$$

is proper. A significant role will be played by the conclusion of proposition 21, which we isolate in the following definition.

Given $x_0 \in X$ and $y \in X^*$, I is *strongly (weakly) rotund at x_0 with slope y* iff $I(x_0)$ is finite and for every strong (weak) neighbourhood A of 0 in X there exists $\delta > 0$ such that

$$I(x_0 + u) - I(x_0) - < y, u > \leq \delta \Rightarrow u \in A.$$

From proposition 21 we immediately obtain

26 Proposition *I is strongly (weakly) rotund at x_0 with slope y iff $(X, I(\cdot) - < y, \cdot >)$ is strongly (weakly) Tykhonov well-posed with solution x_0.*

For (proper and lower semicontinuous) convex functions, strong rotundity turns out to be dual to Fréchet differentiability, while weak rotundity is a dual notion to Gâteaux differentiability. This deep and fundamental fact allows us to get an important differential characterization of Tykhonov well-posedness for convex optimization problems.

We shall denote by

$$I^*(z) = \sup\{< z, x > - I(x) : x \in X\}, z \in X^*$$

the Fenchel conjugate function to I. The (Gâteaux or Fréchet) gradient of I^* at z will be denoted by $\nabla I^*(z)$.

27 Theorem *Let I be proper, convex and lower semicontinuous, y a given point in X^*, and*

$$J(u) = I(u) - < y, u >, u \in X.$$

Then (X, J) is strongly (weakly) Tykhonov well-posed with solution x_0 iff I^* is Fréchet (Gâteaux) differentiable at y with $x_0 = \nabla I^*(y)$.

28 Corollary *Let the assumptions of theorem 27 hold. Then (X, I) is strongly (weakly) Tykhonov well-posed with solution x_0 iff I^* is Fréchet (Gâteaux) differentiable at 0 with $x_0 = \nabla I^*(0)$.*

PROOFS.

Theorem 25. Assume (25). Let x_n be a minimizing sequence. Then $I(x_n) \to \inf I(K)$. Take $u_n \in \partial I((x_n + x_0)/2)$. Then

$$I(x_n) \geq I(\frac{x_n + x_0}{2}) + < u_n, \frac{x_n + x_0}{2} - x_0 > \geq$$

$$\geq I(\frac{x_n + x_0}{2}) + c(\frac{\|x_n - x_0\|}{2}) \geq \inf \ I(K) + c(\frac{\|x_n - x_0\|}{2}),$$

hence

$$0 \leq c(\|\frac{x_n - x_0}{2}\|) \leq I(x_n) - \inf I(K),$$

whence $x_n \to x_0$. Now we show that $x_0 = \arg \ \min \ (K, I)$. Suppose $y \in K$ and $I(y) < I(x_0)$. Put

$$y_1 = \frac{y + x_0}{2} \text{ and take } u \in \partial I(\frac{y + x_0}{2}).$$

Then

$$I(x_0) > I(y) \geq I(y_1) + < u, \frac{y - x_0}{2} > = I(y_1) + < u, \frac{y + x_0}{2} - x_0 > \ \geq$$

$$\geq I(y_1) + c(\|y_1 - x_0\|) > I(y_1).$$

By considering

$$y_2 = \frac{y_1 + x_0}{2}, \cdots, y_{n+1} = \frac{y_n + x_0}{2}, \cdots$$

we see that $y_n \in K$ for each n and

$$I(x_0) > I(y) > I(y_1) > \cdots > I(y_n) > I(y_{n+1}) > \cdots,$$

hence

$$\lim \ \sup \ I(y_n) < I(y).$$

If $u \in \partial I(x_0)$ then

$$I(y_n) \geq I(x_0) + < u, y_n - x_0 >$$

which gives

$$\lim \ \inf \ I(y_n) \geq I(x_0)$$

since $y_n \to x_0$. Summarizing

$$I(y) < I(x_0) \leq \lim \ I(y_n) < I(y),$$

a contradiction. Thus $x_0 = \arg \ \min \ (K, I)$ and every minimizing sequence is strongly convergent towards some minimum point. Then we have strong Tykhonov well-posedness with solution x_0.

Mimicking the beginning of the above proof, fix $x \in K$ and $u \in \partial I((x_0 + x)/2)$. Then

$$I(x) \geq I(\frac{x_0 + x}{2}) + < u, \frac{x_0 + x}{2} - x_0 > \geq I(x_0) + c(\|\frac{x - x_0}{2}\|)$$

thereby proving (26).

Conversely, assume Tykhonov well-posedness of (K, I) with solution x_0 and set

$$(27) \qquad c(t) = \inf\{< u, x - x_0 >: x \in K, \|x - x_0\| = t, u \in \partial I(x)\}.$$

Then $< u, x - x_0 > \geq I(x) - I(x_0) \geq 0$, so by (27) we have $c(t) \geq 0$ for every $t \geq 0, c(0) = 0$.

Let $a_n \geq 0, c(a_n) \to 0$. Then for every n there exist $x_n \in K$ and $u_n \in \partial I(x_n)$ such that

$$\|x_n - x_0\| = a_n, \quad 0 \leq \; < u_n, x_n - x_0 > \; \leq c(a_n) + \frac{1}{n}.$$

Since

$$I(x_0) \geq I(x_n) + < u_n, x_0 - x_n >$$

it follows that

$$I(x_0) \geq \lim \sup \; I(x_n) \geq \lim \inf \; I(x_n) \geq I(x_0)$$

therefore x_n is a minimizing sequence. By well-posedness and (27) we have $a_n \to 0$. $\quad\square$

Now we prove that the forcing function c *given by (27) is increasing*, provided (K, I) is Tykhonov well-posed with solution x_0. Given $a, b, 0 < a < b$ let $s = a/b, x \in K, u \in \partial I(x), \omega \in \partial I[x_0 + s(x - x_0)]$. By the monotonicity of the subdifferential

$$< u - \omega, (1 - s)(x - x_0) > \; \geq 0$$

moreover

$$I[x_0 + s(x - x_0)] \geq I(x_0) \geq I[x_0 + s(x - x_0)] + < \omega, s(x_0 - x) >$$

implying $< \omega, s(x - x_0) > \; \geq 0$. Thus

$$< u, x - x_0 > - < \omega, s(x - x_0) > \; = \; < u - \omega, (1 - s)(x - x_0) >$$
$$+ < u - \omega, s(x - x_0) > + < \omega, (1 - s)(x - x_0) >$$
$$\geq [s/(1 - s)] < u - \omega, (1 - s)(x - x_0) > + [(1 - s)/s] < \omega, s(x - x_0) > \; \geq 0.$$

Therefore

$$c(b) \geq \inf \; \{< \omega, (a/b)(x - x_0) >: \omega \in \partial I[x_0 + (a/b) \; (x - x_0)], x \in K, \|x - x_0\| = b\} \geq$$
$$\inf \; \{< \omega, z - x_0 >: \omega \in \partial I(z), z \in K, \|z - x_0\| = a\} = c(a). \qquad \square$$

Theorem 27. We shall regard X^* and X as real vector spaces paired by the bilinear form $< \cdot, \cdot >$. We need some preliminary definitions. Given a nonempty subset $A \subset X^*$, we denote by

$$A^0 = \{x \in X :< u, x > \; \leq 1 \text{ for every } u \in A\}$$

the *polar* of A . Given $A \subset X^*$ and $B \subset X$ we shall say that

x_0 is an $A-$ *gradient of* I^* *at* y iff $-\infty < I^*(y) < +\infty, x_0 \in X$ and

$$\sup\{(1/t)[I^*(y + tu) - I^* (y)] - < u, x_0 >: u \in A\} \to 0 \text{ as } t \downarrow 0;$$

I   is $B-$ rotund at x_0 with slope   y
iff $-\infty < I(x_0) < +\infty, y \in X^*$ and for every $\varepsilon > 0$ there exists $\delta > 0$ such that

$$\{v : I(x_0 + v) - I(x_0) - <y,v> \le \delta\} \subset \varepsilon B.$$

We shall need the following

29 Lemma Let $\delta > 0, y \in X^*$ and $x_0 \in X$ be such that

(28) $$<y,x_0> = I(x_0) + I^*(y).$$

Then

$$\{v : I(x_0 + v) - I(x_0) - <y,v> \le \delta\}^0 \subset$$

$$\subset \frac{1}{\delta}\{u : I^*(y + u) - I^*(y) - <u,x_0> \le \delta\} \subset$$

$$\subset 2\{v : I(x_0 + v) - I(x_0) - <y,v> \le \delta\}^0.$$

Taking this lemma for granted, we prove

Step 1 . Given $A \subset X^*$, let $x_0 \in \partial I^*(y)$. Then x_0 is an $A-$ gradient of I^* at y iff I is A^0- rotund at x_0 with slope y.

Proof of Step 1. Since $x_0 \in \partial I^*(y)$, we have

$$I^*(y + tu) \ge I^*(y) + t <u,x_0>, u \in X^*, t > 0$$

hence

$$\frac{1}{t}[I^*(y + tu) - I^*(y)] \ge <u,x_0> \text{ for every } u \in X^*.$$

Therefore x_0 is an $A-$ gradient of I^* at y iff for every $\varepsilon > 0$ there exists $T > 0$ such that

(29) $$\sup\{\frac{1}{t}[I^*(y + tu) - I^*(y)] - <u,x_0> : u \in A\} \le \varepsilon \text{ if } 0 < t \le T.$$

Since $t \to (1/t)[I^*(y + tu) - I^*(y)]$ is increasing on $(0, +\infty)$, (29) is equivalent to

(30) $$\sup\{\frac{1}{T}[I^*(y + Tu) - I^*(y)] - <u,x_0> : u \in A\} \le \varepsilon$$

which amounts to

(31) $$TA \subset \{u : I^*(y + u) - I^*(y) - <u,x_0> \le T\varepsilon\}.$$

By lemma 29, from (31) we see that for every $\varepsilon > 0$ there exists $\delta > 0$ such that

(32) $$\frac{A}{\varepsilon} \subset \{v : I(x_0 + v) - I(x_0) - <y,v> \le \delta\}^0.$$

Taking polars in (32), again by lemma 29 we see that x_0 is an $A-$ gradient of I^* at y iff for every $\varepsilon > 0$ there exists $\delta > 0$ such that

$$\varepsilon A^0 \supset \{v : I(x_0 + v) - I(x_0) - < y, v > \leq \delta\}$$

(Barbu - Precupanu [1, th. 1.5 p. 99]), i.e. I is A^0- rotund at x_0 with slope y. This ends the proof of step 1.

Let us denote now by E either the collection of all singleton subsets of X^*, or the collection made of just the unit ball of X^*. Then denote by F the topology induced on X by uniform convergence of the linear functionals $< \cdot, z >, z \in X$ on all subsets in E. Of course F is either the weak or the strong topology on X. We shall say that
I^* is $E-$ differentiable at y with $x_0 = \nabla I^*(y)$
iff x_0 is an $A-$ gradient of I^* at y for every $A \in E$. Then of course x_0 will be uniquely determined as either the Gateaux or Fréchet gradient of I^* at y. Finally we shall say that
I is $F-$ rotund at x_0 with slope y
iff I is $B-$ rotund at x_0 with slope y for every $F-$ neighbourhood B of the origin of X.

Step 2. I^* is $E-$ differentiable at y with $x_0 = \nabla I^*(y)$ iff I is $F-$ rotund at x_0 with slope y.

Proof of step 2 . Let I^* be $E-$ differentiable at y with $x_0 = \nabla I^*(y)$. Then of course $x_0 \in \partial I^*(y)$. Let I be $F-$ rotund at x_0 with slope y. Then for every $F-$ neighbourhood B of the origin, given $\varepsilon > 0$ there exists $\delta > 0$ such that

$$\{v : I(x_0 + v) - I(x_0) - < y, v > \leq \delta\} \subset \varepsilon B.$$

Given $v \neq 0$, then some $F-$ neighbourhood of v does not intersect some neighbourhood of 0 (since F is Hausdorff). It follows that

$$I(x_0 + v) > I(x_0) + < y, v >, v \neq 0$$

hence $y \in \partial I(x_0)$, yielding again $x_0 \in \partial I^*(y)$. Therefore in either case step 1 is applicable. Thus I^* is $E-$ differentiable at y with $x_0 = \nabla I^*(y)$ iff I is A^0- rotund at x_0 with slope y for every $A \in E$. If E consists of just the unit ball A of X^*, then A^0 is the unit ball of X (by a corollary to the Hahn - Banach theorem) and the conclusion of step 2 holds. The same is true if E is the collection of all singleton subsets of X^*. Indeed, in this case a local basis of the weak topology in X is obtained by taking all finite intersections of sets of the form

$$\varepsilon A^0 = \{x \in X :< u, x > \leq \varepsilon\},$$

where $\varepsilon > 0$ and $A = \{u\}, u \in X^*$. Finally notice that if $B = B_1 \cap ... \cap B_m$, then I is $B-$ rotund iff I is B_i- rotund, $i = 1, ..., m$. This ends the proof of step 2. In view of proposition 26, the proof of theorem 27 is ended, provided we prove lemma 29.
Proof of lemma 29. The proper convex functions

$$f(z) = \frac{1}{\delta}[I^*(y + \delta z) - I^*(y) - \delta < z, x_0 > -\delta], z \in X^*,$$

$$g(v) = \frac{1}{\delta}[I(x_0 + v) - I(x_0) - < y, v > +\delta], v \in X,$$

are conjugate to each other, as it can be verified by direct calculation using (28). Since

$$g(0) = 1 \geq \inf\ g(X)$$

and

$$I^*(y) \geq\ <y, x_0 + v> -I(x_0 + v) \text{ for every } v,$$

by (28) we get

$$I(x_0 + v) - I(x_0) - <y, v> \geq 0$$

yielding

(33) $$1 = \inf\ g(X) = g(0).$$

Hence

$$f(0) = -1 = -g(0) = \inf\ f(X^*).$$

Now the conclusion of the lemma can be rewritten as

$$\{v : g(v) \leq 2\}^0 \subset C \subset 2\{v : g(v) \leq 2\}^0,$$

where

$$C = \{z : f(z) \leq 0\}.$$

Since C is closed convex and $0 \in C$, if follows that $C^{00} = C$ (Barbu - Precupanu [1, th.1.5 p.99]). Thus the conclusion will follow if we show that

(34) $$\{v : g(v) \leq 2\} \supset C^0 \supset (1/2)\{v : g(v) \leq 2\}.$$

Let h denote the support function of C on X, i.e.

$$h(v) = \sup\{<z, v> : z \in C\}.$$

According to Rockafellar [1 , cor.4B],

(35) $$h(v) = \inf\{tg(v/t) : t > 0\}.$$

In particular $h \leq g$, so that

$$\{v : g(v) \leq 2\} \subset \{v : h(v) \leq 2\} = 2\{v : h(v) \leq 1\} = 2C^0.$$

This establishes the right half of (34). To establish the left half, it suffices to show that

(36) $$\{v : h(v) < 1\} \subset \{v : g(v) \leq 2\},$$

since the closure of the set on the left in (36) is C^0, whereas the set on the right is closed convex. Given any v such that $h(v) < 1$, there exists by (35) some $t > 0$ such that $tg(v/t) < 1$. Then $0 < t < 1$ by (33). Convexity of g implies then that

$$g(v) \leq (1 - t)g(0) + tg(v/t) \leq 1 - t + 1 < 2.$$

Thus (36) holds, and the proof of lemma 29 is completed. □

Section 5.   Generalized Minimizing Sequences, Levitin-Polyak Well-Posedness.

Some numerical optimization methods for constrained problems (as for example exterior penalty techniques) produce a sequence of points which fail to be feasible, but tend asymptotically to fulfill the constraints, while the corresponding values approximate the optimal one. To take care of such cases one can strengthen the Tykhonov well-posedness concept, as follows.

Let X be a metric space, K a nonempty subset thereof and let $I : X \to (-\infty, +\infty)$ be a given function.   The sequence x_n is a *generalized minimizing sequence* for (K, I) iff

$$x_n \in X, I(x_n) \to \inf\ I(K),\ \text{dist}\ (x_n, K) \to 0.$$

We say that the problem (K, I) is *Levitin - Polyak well-posed* iff there exists an unique

$$x_0 = \arg\ \min\ (K, I)$$

and for every generalized minimizing sequence x_n to (K, I) one has

$$x_n \to x_0.$$

Of course, Levitin - Polyak well-posedness implies Tykhonov's. The converse is true provided that I is uniformly continuous, as easily checked, but not necessarily true if I is (only) continuous: consider $X = R^2, K = R \times \{0\}, I(x, y) = x^2 - (x^4 + x)y^2$ and the generalized minimizing sequence $(n, 1/n)$.

We know from theorem 11 that Tykhonov well-posedness can be characterized by the behaviour of diam $[\varepsilon - \arg\ \min\ (K, I)]$ as $\varepsilon \to 0$. Now define, for $\varepsilon > 0$ and I bounded from below,

$$L(\varepsilon) = \{x \in X :\ \text{dist}\ (x, K) \le \varepsilon\ \text{and}\ I(x) \le \inf\ I(K) + \varepsilon\}.$$

The behavior of $L(\varepsilon)$ takes care not only of the minimizing sequences to (K, I), which (by definition) lie in K, but also of the generalized minimizing sequences to (K, I) (which may violate the constraint K up to a small extent). Remembering theorem 11 we get in a similar fashion

30 Theorem *If X is a complete metric space, K is closed, I is lower semicontinuous and bounded from below on K, then*

(37)   *diam $L(\varepsilon) \to 0$ as $\varepsilon \to 0$*

implies Levitin - Polyak well - posedness of (K, I).

We observe that (37) is not a necessary condition to Levitin - Polyak well- posedness. Take $X = R^2, K = \{(x, y) \in R^2 : y = 0\}$ and

$$I(x, y) = \begin{cases} 0 & ,\ \text{if}\ x = n, y = 1/n, n = 1, 2, 3, ..., \\ x^2 + y^2 + 1 & ,\ \text{otherwise.} \end{cases}$$

Then (K, I) is Levitin - Polyak well-posed, but for every $\varepsilon > 0$ we have $(n, 1/n) \in L(\varepsilon)$ if n is sufficiently large, contradicting (37).

31 Example. Let X be a real reflexive Banach space, K a nonempty bounded closed convex subset in X and $I : X \to (-\infty, +\infty)$ an uniformly quasiconvex functional which is lower semicontinuous. We show that (K, I) is Levitin - Polyak well-posed with respect to the strong convergence in X. This fact extends corollary 17. To verify well-posedness, notice that by quasiconvexity and semicontinuity every sublevel set of I is weakly closed, therefore I is bounded from below over K. Now consider (corollary 17)

$$x_0 = \arg \min (K, I).$$

Let x_n be a generalized minimizing sequence for (K, I). Since dist $(x_n, K) \to 0$, there exists some sequence $u_n \in K$ such that $x_n - u_n \to 0$. For some subsequence $u_n \rightharpoonup u \in K$, therefore $x_n \rightharpoonup u$ as well, and

$$I(x_0) = \lim \inf \ I(x_n) \geq I(u)$$

yielding $x_0 = u$. Hence $x_n \rightharpoonup x_0$ for the original sequence. Then

$$I(x_0) \leq \liminf \ I[(x_n + x_0)/2] \leq \lim \sup \ I[(x_n + x_0)/2]$$
$$\leq \lim \sup \ \max \{I(x_n), I(x_0)\} = I(x_0),$$

so that

$$I(\frac{x_n + x_0}{2}) \to I(x_0),$$

therefore

$$c(\|x_n - x_0\|) \leq \ \max\{I(x_n), I(x_0)\} - I(\frac{x_n + x_0}{2}) \to 0$$

since $I(x_n) \to I(x_0)$. This shows that $x_n \to x_0$.

32 Example. Let (K, I) be Tykhonov well-posed with $K \in$ Conv (R^p) and $I \in C_0(R^p)$. Then (K, I) is Levitin - Polyak well-posed. For, let $x = \arg \min (K, I)$. Replacing I by $u \to I(u + x) - I(x)$, we may assume without loss of generality that

$$I(0) = 0 = \min \ I(K).$$

Suppose (K, I) fails to be Levitin - Polyak well-posed. Then there exists a generalized minimizing sequence u_n such that $|u_n| \geq a > 0$ for some $a \in (0, 1)$ and every n. If u_n is bounded it has some cluster point y, with $y \in \arg \min (K, I)$ by lower semicontinuity and dist $(u_n, K) \to 0$. Hence $I(y) = 0$, contradicting well-posedness since $|y| \geq a$. If $|u_n| \to +\infty$, by convexity of dist $(\cdot, K)$ we get

$$\text{dist} \ (\frac{u_n}{|u_n|}, A) \to 0$$

and by convexity of I we see that $u_n/|u_n|$ is a bounded generalized minimizing sequence. Hence it has some cluster point y with $|y| = 1$, again contradicting well-posedness.

Remark. A significant extension of this example to infinite - dimensional Banach spaces will be considered in corollary II.18.

33 Example. Let X be a real reflexive Banach space, $g : X \to R$ be a continuous uniformly convex function, that is

(38) $$g(tx + (1-t)y) \leq tg(x) + (1-t)g(y) - t(1-t)c(\|x - y\|)$$

for every $x, y \in X, 0 \leq t \leq 1$ and for some forcing function c. Put

$$K = \{x \in X : g(x) \leq 0\}.$$

Let $I : X \to (-\infty, +\infty)$ be coercive and weakly sequentially lower semicontinuous. If there exists an unique $x_0 = \arg \min (K, I)$ with $g(x_0) = 0$ then (K, I) is Levitin - Polyak well-posed. Indeed, let x_n be a generalized minimizing sequence for (K, I). Then $\|x_n - z_n\| \to 0$ for some $z_n \in K$. By (38) one has

$$\frac{1}{t}[g(tx + (1-t)y) - g(y)] \leq g(x) - g(y) - (1-t)c(\|x - y\|).$$

Letting $t \to 0^+$ and remembering prop. 2.3, p.106 of Barbu - Precupanu [1], we get

$$< u, x - y > \leq g(x) - g(y) - c(\|x - y\|) \text{ if } u \in \partial g(y).$$

Then

(39) $$0 \geq g(z_n) \geq g(x_0) + < u, z_n - x_0 > + c(\|z_n - x_0\|)$$

for some $u \in \partial g(x_0)$ and some forcing function c. By coercivity, semicontinuity and uniqueness of x_0 we get (in a standard way) $x_n \rightharpoonup x_0$ and therefore $z_n \rightharpoonup x_0$. By (39)

$$0 \leq c(\|z_n - x_0\|) \leq < u, x_0 - z_n > \to 0.$$

Thus $z_n \to x_0$, therefore $x_n \to x_0$ as required.

   Remark. An interesting connection between Levitin - Polyak well-posedness of (K, I) and the continuous dependence of $\arg \min (K, I)$ from the data K, I will be shown in theorem III. 17.

PROOF.

   Theorem 30. Assume (37). Then given $a > 0$ there exists $b > 0$ such that

(40) $$\text{diam } L\ (\varepsilon) < a \text{ if } 0 < \varepsilon < b.$$

Let u_n be any generalized minimizing sequence. Then for every ε as in (40) we have $u_n \in L(\varepsilon)$ for n sufficiently large. Thus by (40) u_n is a Cauchy sequence. Therefore $u_n \to x_0$ for some $x_0 \in K$ since K is closed. By semicontinuity, $x_0 \in \arg \min (K, I)$ which turns out to be a singleton, since every minimizing sequence converges . $\square$

Section 6.   Well-Posedness in the Generalized Sense.

The concept of Tykhonov well-posedness can be extended to minimum problems without uniqueness of the optimal solution. We consider a convergence space X and a proper extended real-valued function

$$I : X \rightarrow (-\infty, +\infty].$$

We shall require existence of solutions to (X, I) (but not uniqueness) and, for every minimizing sequence, convergence of some subsequence towards some optimal solution. In a sense, arg min (X, I) will attract all minimizing sequences.

More precisely, the problem (X, I) will be called *well-posed in the generalized sense* iff arg min $(X, I) \neq \emptyset$, and every sequence $u_n \in X$ such that $I(u_n) \rightarrow \inf I(X)$ has some subsequence $v_n \rightarrow u$ with $u \in$ arg min (X, I).

Of course, (X, I) is Tykhonov well-posed iff it is well-posed in the generalized sense and arg min (X, I) is a singleton. Thus generalized well-posedness is really a generalization of Tykhonov well-posedness.

34 Examples. If $X = (-\infty, +\infty)$ and $I(x) = |\ |x| - 1\ |$, then obviously (X, I) is well-posed in the generalized sense, without being Tykhonov well-posed.

Suppose that I fulfills (1), and $\varepsilon -$ arg min (X, I) is sequentially compact for some $\varepsilon > 0$. Then every minimizing sequence has some cluster point which of course belongs to arg min (X, I). So (X, I) is well-posed in the generalized sense whenever we can apply the direct method in the calculus of variations.

Remark. By replacing the minimum problem (X, I) by a relaxed version of it (as for example in Ekeland - Temam [1, ch.IX, sec. 4] or Buttazzo [1]) we can obtain a weak form of generalized well-posedness, i.e. the cluster points of every minimizing sequence solve the *relaxed problem*. However, we are interested in *not* changing the original problem (X, I) (and getting strong convergence for the minimizing sequences).

A completely different link between (Hadamard) well - posedness and relaxation is presented in chapter VI.

35 Example. Let X be a real normed space. A subset K thereof is called *approximatively compact* (a key notion in best approximation theory) iff for every $x \in X$, every sequence $u_n \in K$ such that

$$\|u_n - x\| \rightarrow \ \text{dist}\ (x, K)$$

has subsequences strongly convergent to some element of K. Then K is approximatively compact iff for every fixed $x \in X$ the problem (K, I) is (strongly) well - posed in the generalized sense, where $I(u) = \|x - u\|$.

If X is a metric space and I fulfills (1), then generalized well-posedness of (X, I) implies sequential compactness of every minimizing sequence (by definition). Moreover arg min (X, I) is nonempty and compact, being closed from (1) and sequentially compact. We have therefore proved

36 Proposition *Suppose that X is a metric space and I fulfills (1). Then (X, I) is well posed in the generalized sense iff every minimizing sequence is sequentially compact. In such a case, arg min (X, I) is compact nonempty and coincides with the set of cluster points of some minimizing sequence to (X, I). Moreover, well-posedness in the generalized sense is equivalent to compactness of arg min (X, I) and upper semicontinuity of*

$$\varepsilon \rightrightarrows \varepsilon - arg\ min\ (X, I)\ at\ \varepsilon = 0.$$

Let X be a metric space and let

$$I : X \rightarrow (-\infty, +\infty]$$

be proper and lower semicontinuous. If sub lev (I, t) is compact and nonempty for some t, then of course inf $I(X) > -\infty$ and (X, I) is well-posed in the generalized sense. For suitably restricted problems we have more precisely

37 Proposition *Let X be a locally compact metric space. Assume that every sublevel set of I is connected. Then the following are equivalent facts:*

(41)   sub lev (I, t) *is compact for some* $t > inf\ I(X)$;

(42) (X, I) *is well-posed in the generalized sense;*

(43)   arg min (X, I) *is nonempty and compact.*

If X and I fulfill (1) and (2), then well-posedness in the generalized sense of (X, I) is, roughly speaking, equivalent to the following: as $\varepsilon \rightarrow 0$ then $\varepsilon -$ arg min (X, I) tends to become compact, as we see from the definition and proposition 36. So it is natural to extend the basic theorem 11 to the present setting by using the Kuratowski number α (measure of noncompactness), defined for every bounded set $A \subset X$ as

$\alpha(A) = \inf\{d > 0 : A$ has a finite cover consisting of sets with diameter $< d\}$.

38 Theorem *Suppose that (1) and (2) hold. Then (X, I) is well-posed in the generalized sense iff*

(44) $\alpha[\varepsilon - arg\ min\ (X, I)] \rightarrow 0\ as\ \varepsilon \rightarrow 0$.

The metric characterization of Tykhonov well-posedness afforded by theorem 12 can be similarly extended, as follows.

39 Theorem *Suppose that (1) and (2) hold, and let X be bounded. Then (X, I) is well-posed in the generalized sense iff there exists a forcing function c such that*

$$(45) \qquad\qquad inf\ I(X) \leq sup\ I(A) - c[\alpha(A)]$$

for every $A \subset X$ such that $sup\ I(A) < +\infty$.

40 Example. We present a problem (X, I) which is well-posed in the generalized sense and

$$\alpha[\varepsilon - \arg\ \min\ (X, I)] > 0 \text{ if } \varepsilon > 0.$$

Let X be the closed unit ball of $C^0([0, 1])$ with the uniform norm, and denote by $I(u)$ the oscillation of u on $[0, 1]$, i.e.

$$I(u) = \max\{u(t) : 0 \leq t \leq 1\} - \ \min\ \{u(t) : 0 \leq t \leq 1\}, u \in X.$$

Then $M = \arg\ \min\ (X, I)$ is the set of constant functions u such that $|u(t)| \leq 1$. Moreover, if $0 < \varepsilon < 1$, then the sequence $u_k(t) = \varepsilon t^k$ belongs to $M(\varepsilon) = \varepsilon - \arg\ \min\ (X, I)$ so that $\alpha[M(\varepsilon)] > 0$, $\varepsilon > 0$. To show well - posedness in the generalized sense, let $u \in M(\varepsilon)$ and put $2c = \max\ u + \min\ u$. Then $|u(t) - c| \leq \varepsilon$ for each t. Therefore $M(\varepsilon) \subset {}^\varepsilon M$, hence $\alpha[M(\varepsilon)] \leq \alpha(M) + \varepsilon = \varepsilon$, showing well-posedness in the generalized sense by theorem 38. Of course diam $M(\varepsilon) \nrightarrow 0$ as $\varepsilon \rightarrow 0$.

A further extension of the Tykhonov well - posedness concept for optimization problems without uniqueness, in a metric space X, is defined as follows. The problem (X, I) is *stable* iff arg min $(X, I) \neq \emptyset$ and for every minimizing sequence x_n we have

$$\text{dist } [x_n, \arg\ \min\ (X, I)] \rightarrow 0.$$

The (obvious) link between stability and well-posedness in the generalized sense follows.

41 Proposition *If (X, I) is well-posed in the generalized sense, then (X, I) is stable. Conversely, if (X, I) is stable, arg $min(X, I)$ is closed and sequentially compact, then (X, I) is well-posed in the generalized sense.*

A characterization of stability may be obtained in terms of upper semicontinuity of the level sets of I at inf $I(X)$.

A multifunction

$$t \rightrightarrows S(t)$$

acting between metric spaces Y and X is called *upper Hausdorff semicontinuous* at $t_0 \in Y$ iff for every $\varepsilon > 0$ there exists a neighbourhood V of t_0 such that

$$S(t) \subset S(t_0)^\varepsilon \text{ whenever } t \in V.$$

42 Theorem *Let $I : X \rightarrow (-\infty, +\infty)$. Then the following are equivalent facts:*

$$(46) \qquad\qquad (X, I) \text{ is stable};$$

(47) *inf $I(X)$ is finite and $lev(I, \cdot)$ is upper Hausdorff semicontinuous at $inf\ I(X)$.*

43 Example. Let $X = \{(x_1, x_2) \in R^2 : 0 \leq x_1 \leq 1, x_2 \geq 0\}$ and $I(x_1, x_2) = x_1$. Then (X, I) is stable by theorem 42, since

$$t \rightrightarrows \text{lev}\ (I, t) = \{(t, x_2) : x_2 \geq 0\}$$

is upper Hausdorff semicontinuous at $t = 0$. The sequence $(1/n, n)$ is minimizing for (X, I) and has no cluster points. Therefore (X, I) is not well-posed in the generalized sense.

PROOFS.

Proposition 36. Only the last statement needs proof. Assume well-posedness in the generalized sense. Suppose, on the contrary, that for some open set V and some sequence $\varepsilon_n \to 0$

$$\text{arg}\ \min\ (X, I) \subset V, \varepsilon_n - \text{arg}\ \min\ (X, I) \not\subset V.$$

Then there is a minimizing sequence without cluster points inside arg min (X, I), a contradiction. Compactness has been already checked. Conversely, let x_n be any minimizing sequence. Since $\varepsilon_n = I(x_n) - \inf\ I(X) \to 0$, given $a > 0$ we get

$$x_n \in [\ \text{arg}\ \min\ (X, I)]^a$$

for every sufficiently large n. By compactness of arg min $(X, I), x_n$ has a cluster point there, proving generalized well-posedness.

Proposition 37. As already remarked, $(41) \Rightarrow (42)$ and $(42) \Rightarrow (43)$. We need only to show that $(43) \Rightarrow (41)$. By local compactness there exists $\varepsilon > 0$ such that

$$B = [\ \text{arg}\ \min\ (X, I)]^\varepsilon$$

has compact closure. Arguing by contradiction, suppose that for every n

$$A_n = \text{sub}\ \text{lev}\ [I, 1/n + \inf I(X)]$$

is not compact. Being closed, it cannot be contained in cl B. Therefore A_n meets $X \setminus B$ for every n. On the other hand, for any fixed $x_0 \in$ arg min (X, I) one has $x_0 \in A_n \cap B$ for every n. Since A_n is connected, there exists some $u_n \in A_n \cap \{x \in X : \text{dist}\ [x, \text{arg}\ \min\ (X, I)] = \varepsilon\}$. Let u be a cluster point of u_n. Then, by lower semicontinuity, $u \in$ arg min (X, I), moreover

$$\text{dist}\ [u, \text{arg}\ \min\ (X, I)] = \varepsilon,$$

a contradiction. $\square$

Theorem 38. Put

$$M(\varepsilon) = \varepsilon - \arg \min (X, I), M = \arg \min (X, I).$$

Assume (44). Since $M(\cdot)$ is decreasing and every $M(\varepsilon)$ is closed and nonempty, by the generalized Cantor theorem (see Kuratowski [1, th. p.318]) if follows that

(48) $\qquad$ haus $[M(\varepsilon), M] \to 0$ as $\varepsilon \to 0$ and $M = \cap\{M(\varepsilon) : \varepsilon > 0\}$

is nonempty and compact.

Let x_n be a minimizing sequence. Then for every $\varepsilon > 0$ we get $x_n \in M(\varepsilon)$ for every n sufficiently large. Then dist $(x_n, M) \to 0$ by (48). By compactness of M we get sequential compactness of x_n. Hence (X, I) is well-posed in the generalized sense by proposition 36.

Conversely, assume well-posedness in the generalized sense for (X, I) and put

$$p(\varepsilon) = \text{haus} \ (M(\varepsilon), M), \varepsilon > 0.$$

Arguing by contradiction, suppose that $p(\varepsilon) \not\to 0$ as $\varepsilon \to 0$. Then we can find $c > 0$ and a sequence $\varepsilon_j \to 0$ such that

$$\text{haus} \ (M(\varepsilon_j), M) \geq 2c.$$

Then for every j there exists $x_j \in M(\varepsilon_j)$ such that

(49) $\qquad$ dist $(x_j, M) \geq c.$

The sequence x_j is minimizing and has a cluster point y. Hence dist $(y, M) \geq c$ by (49). But $y \in M$ by well-posedness, a contradiction. Therefore

$$p(\varepsilon) \to 0 \text{ as } \varepsilon \to 0.$$

Since

$$M(\varepsilon) \subset {}^{p(\varepsilon)}M$$

we get (using an easily checked property of the Kuratowski number)

$$\alpha[M(\varepsilon)] \leq \alpha(M) + 2p(\varepsilon) = 2p(\varepsilon).$$

$\square$

Theorem 39. Taking into account theorem 38 we show that (45) is equivalent to (44). Assume (45). Then for every $\varepsilon > 0$ (notation of above proof)

$$\inf I(X) \leq \sup \ I[M(\varepsilon)] - c[\alpha(M(\varepsilon))] \leq \inf \ I(X) + \varepsilon - c[\alpha(M(\varepsilon))].$$

Then by lemma 20

$$\alpha[M(\varepsilon)] \leq q(\varepsilon) \to 0 \text{ as } \varepsilon \to 0,$$

thus proving (44). Conversely, assume (44) and put

$$q(\varepsilon) = \alpha[M(\varepsilon)], \varepsilon > 0.$$

Let the forcing function c be defined by (13). Given $A \subset X$ such that $\sup I(A) < +\infty$ let

(50) $\qquad p = \sup \ I(A) - \inf \ I \ (X).$

By (14)

$$c[\alpha(A)] \leq c[\alpha(M(p))] = c[q(p)] \leq p.$$

Then (45) follows from (50) and (15). $\square$

Theorem 42. Put

$$M = \arg\ \min\ (X, I) = \text{lev}\ [I, \inf I(X)].$$

Assume (46). Arguing by contradiction, we can find $\varepsilon > 0$, a sequence

$$t_n \to \inf\ I(X)$$

and points $u_n \in \text{lev}\ (I, t_n)$ such that

$$u_n \notin M^\varepsilon.$$

This contradicts stability since $I(u_n) = t_n$ and therefore u_n is a minimizing sequence for (X, I).

Conversely, assume (47). If $M = \emptyset$ then by semicontinuity lev $(I, t) = \emptyset$ for any t near inf $I(X)$, a contradiction. Hence $M \neq \emptyset$. Now let u_n be any minimizing sequence for (X, I). Given $\varepsilon > 0$, from semicontinuity we have

$$\text{lev}\ (I, t) \subset M^\varepsilon$$

for all t sufficiently near to inf $I(X)$. Therefore

$$u_n \in M^\varepsilon$$

for n sufficiently large, thus showing stability.   □

Section 7.   Tykhonov Regularization.

We consider a minimum problem (Q, f), where Q is a fixed subset of a given Banach space X, and

$$f : X \to (-\infty, +\infty)$$

with arg min $(Q, f) \neq \emptyset$. Since (Q, f) may be not Tykhonov well-posed with respect to the strong convergence, not all minimizing sequences will converge. We shall obtain a strongly convergent minimizing sequence of (Q, f) by approximatively solving appropriate perturbations to (Q, f) obtained by adding to f a small regularizing term. These perturbations (regularizations) partially restore well-posedness. Moreover, such a procedure is robust, since it works assuming only an approximate knowledge of f. We fix sequences $a_n > 0, \varepsilon_n > 0$ such that

$$a_n \to 0, \varepsilon_n \to 0.$$

In the following theorem we regularize (Q, f) by adding to f a small nonnegative uniformly convex term $a_n\ g$ defined on the whole space X. So g satisfies

$$g(\frac{x + y}{2}) \le (1/2)g(x) + (1/2)g(y) - c(\|x - y\|)$$

for all x, y and for some forcing function c.

44 Theorem *Let X be reflexive, f weakly sequentially lower semicontinuous and Q weakly compact. Let*

$$g : X \to [0, +\infty)$$

be strongly lower semicontinuous and uniformly convex. Then the following conclusions hold.

(51) *If* f *and* Q *are convex, then* $(Q, f + ag)$ *is strongly Tykhonov well-posed for every* $a > 0$;

(52) *if* $x_n \in \varepsilon_n - arg\ min\ (Q, f + a_n\ g)$, *then* x_n *is a minimizing sequence for* (Q, f).

Let $\varepsilon_n / a_n \to 0$, *then*

(53) $\emptyset \neq strong\ lim\ sup\ [\varepsilon_n - arg\ min(Q, f + a_n g)] \subset arg\ min\ (arg\ min\ (Q, f), g)$;

(54) *if moreover* Q *and* f *are convex, then* $x_n \in \varepsilon_n - arg\ min\ (Q, f + a_n\ g)$ *implies that*

$$x_n \to arg\ min\ (arg\ min\ (Q, f), g).$$

Finally, let $f_n : X \to (-\infty, +\infty)$ *be such that*

$$sup\ \{|f_n(x) - f(x)| : x \in Q\} \leq c_n$$

for some c_n *with* $c_n / a_n \to 0$. *Then if* $\varepsilon_n / a_n \to 0$, (53) *and* (54) *hold with* $f + a_n\ g$ *replaced by* $f_n + a_n\ g$.

In the next theorem we regularize (X, f) by adding to some approximation of f a small densely defined term with strongly relatively compact sublevel sets, provided some solution to (X, f) belongs to the domain of the regularizing functional.

We shall therefore consider
a further Banach space Y, which is densely and continuously embedded in X;
a (regularizing) strongly continuous function

$$g : Y \to [0, +\infty);$$

a sequence of functions

$$f_n : X \to (-\infty, +\infty)$$

that approximate f. Put

$$I_n = f_n + a_n\ g.$$

45 Theorem *Assume that* f *is strongly lower semicontinuous,* g *is strongly continuous in* Y *and*

(55)   *the set* $\{u \in Y : g(u) \leq t\}$ *is strongly relatively compact in* X *for every* t;

(56)   *there exists some* $z \in Y \cap arg\ min\ (X, f)$;

(57)   *for some* $c_n \to 0$ *and every* $x \in Y$

$$|f_n(x) - f(x)| \leq c_n[1 + g(x)].$$

Then the following conclusions hold.

If $lim\ sup\ c_n / a_n < 1$ *and* $lim\ sup\ \varepsilon_n / a_n < +\infty$ *then*

(58) $val(Y, I_n) \to val(X, f)$;

(59) $\emptyset \neq strong \; lim \; sup \; [\varepsilon_n - arg \; min \; (Y, I_n)] \subset arg \; min \; (X, f) \; within \; X.$

If $c_n/a_n \to 0$, $\varepsilon_n/a_n \to 0$ and moreover every sublevel set of g is weakly sequentially compact in Y, and g is weakly lower semicontinuous on Y, then

(60) $\emptyset \neq strong \; lim \; sup \; [\varepsilon_n - arg \; min \; (Y, I_n)] \subset Y \cap arg \; min \; (X, f) \; within \; X.$

In particular, if an unique minimum point u of g on $Y \cap arg \; min(X, f)$ exists, then

$$u_n \in \varepsilon_n - arg \; min \; (Y, I_n) \Rightarrow u_n \to u \; in \; X.$$

46 Example. Let $X = L^2(a, b)$ and $Y = H^1(a, b)$. Then

$$g(u) = \int_a^b q(t) \; \dot{u}^2(t) \; dt$$

is a regularizing functional which meets all requirements of theorem 45, provided $q \in L^\infty(a, b)$ and $q(t) \geq \alpha > 0$ a.e. (since the embedding of $H^1(a, b)$ into $L^2(a, b)$ is compact: see Brezis [2, th. VIII.7]).

Application: *regularization of ill-posed first kind linear equations.* Let X, Y be real Hibert spaces and

$$A : X \to Y$$

a linear bounded operator. We consider the equation (with unknown $x \in X$)

(61) $Ax = y$

for a given $y \in Y$. The following facts are known (see Groetsch [1]). Solutions to (61) may not exist, or be not unique, or depend discontinuously upon the data y. A least - squares solution to (61) is any $u \in arg \; min \; (X, f)$, where

$$f(x) = \|Ax - y\|^2,$$

which exists iff $y \in A(X) + A(X)^\perp$. In such a case, there exists an unique point $A^+ y$ of least norm among the least squares solutions, which may be considered as a kind of generalized solution to (61). Therefore

$$A^+ y = arg \; min \; [arg \; min \; (X, f), g],$$

where $g = \|\cdot\|^2$. If A is compact, except in the (relatively trivial) case when $A(X)$ is finite dimensional, A^+ is a linear unbounded operator. Therefore, due to such (Hadamard) ill-posedness, it is not possible (in general) to get approximately the generalized solution $A^+ y$ by approximating the data y through a given sequence $y_n \to y$. To overcome this difficulty, Tykhonov regularization can be applied by fixing a suitable sequence $a_n > 0$, $a_n \to 0$ and minimizing (even approximately) over X

$$I_n(x) = \|Ax - y_n\|^2 + a_n \; \|x\|^2.$$

Here we assume that

$$y \in A(X) + A(X)^{\perp}$$

and y_n is a fixed sequence such that

$$y_n \to y.$$

Then (as claimed) arg min $(X, f) \neq \emptyset$. Now take

(62) $$x_n \in \varepsilon_n - \text{arg min } (X, I_n).$$

Then we get

$$\|Ax_n - y_n\|^2 \leq I_n(x_n) \leq \varepsilon_n + I_n(0) = \varepsilon_n + \|y_n\|^2,$$

hence Ax_n is bounded. Putting

$$f_n(x) = \|Ax - y_n\|^2$$

we see that for some constant $k > 0$ and for all n

$$|f_n(x) - f(x)| \leq \|y_n - y\|(k + 2\|Ax\|).$$

Therefore for any $z \in$ arg min (X, f) and some constant h,

$$I_n(x_n) \leq \varepsilon_n + \|y_n - y\|(k + 2\|Az\|) + f(z) + a_n\|z\|^2 \leq$$

$$\leq \varepsilon_n + \|y_n - y\|(k + 2\|Az\|) + \|Ax_n - y\|^2 + a_n\|z\|^2 \leq$$

$$\leq \varepsilon_n + \|y_n - y\|(h + 2\|Ax_n\|) + \|Ax_n - y_n\|^2 + a_n\|z\|^2.$$

Hence

$$a_n\|x_n\|^2 \leq \varepsilon_n + \|y_n - y\| \text{ (constant) } + a_n\|z\|^2,$$

yielding boundedness of x_n, provided ε_n/a_n and $\|y_n - y\|/a_n$ are bounded. This shows that there exists some ball Q in X, containing A^+y, such that for all sufficiently large n

$$\text{arg min } (Q, I_n) = \text{arg min } (X, I_n).$$

Then theorem 44 applies, yielding

47 Corollary *If $y \in A(X) + A(X)^{\perp}$, $y_n \to y$ and x_n is any sequence given by (62), then*

$$x_n \to A^+y$$

provided $\varepsilon_n/a_n \to 0$ and $\|y_n - y\|/a_n \to 0$.

PROOFS.

Theorem 44. (51) follows directly from corollary 17. To prove (52), let $y \in$ arg min (Q, f). Then

(63) $f(x_n) + a_n \ g(x_n) \leq f(y) + a_n \ g(y) + \varepsilon_n \leq f(x_n) + a_n \ g(y) + \varepsilon_n$

yielding

(64) $g(x_n) \leq g(y) + \varepsilon_n / a_n.$

By the first inequality in (63), since $g \geq 0$,
$$f(y) \leq f(x_n) \leq f(y) + a_n \ [g(y) - g(x_n)] + \varepsilon_n \leq f(y) + a_n g(y) + \varepsilon_n, \text{ thus}$$
$$f(x_n) \rightarrow \inf \ f(Q),$$

thereby proving (52). By compactness, $x_n \rightharpoonup z \in Q$ for some subsequence. Then $z \in$ arg min (Q, f) by semicontinuity of f. From (64)

$$g(z) \leq \lim \inf \ g(x_n) \leq g(y),$$

hence z minimizes g on arg min (Q, f). Moreover, by the uniform convexity

$$c(\|x_n - z\|) \leq \frac{1}{2}g(x_n) + \frac{1}{2}g(z) - g(\frac{x_n + z}{2}).$$

By (64), for the previous subsequence,

$$\lim \sup \ c \ (\|x_n - z\|) \leq g(z) - \lim \inf \ g(\frac{x_n + z}{2}) \leq 0,$$

so that $x_n \rightarrow z$ since c is forcing. This proves (53). Then (54) follows (assuming convexity) by uniqueness of the minimizer of g on arg min (Q, f), towards which every subsequence from x_n will therefore converge.

The last conclusion with $f_n + a_n g$ instead of $f + a_n g$ will be proved by using some results from chapter IV. Put
$$I_n = f_n + a_n \ g.$$
Since $I_n(x) \rightarrow f(x)$ for every $x \in X$, and $u_n \rightharpoonup u$ in Q implies

$$\lim \inf \ I_n(u_n) \geq \lim \inf \ f_n(u_n) \geq f(u)$$

we see that
$$\text{weak seq. epi} - \lim \ I_n = f.$$
Then, by theorem IV.5 it follows that

$$\text{val} \ (Q, I_n) \rightarrow \text{val} \ (Q, f).$$

Now fix any $x_n \in \varepsilon_n -$ arg min (Q, I_n). Then $x_n \rightharpoonup u$ for some subsequence. Thus, for any $y \in$ arg min (Q, f)

$$I_n \ (x_n) \leq \varepsilon_n + I_n \ (y) \leq \varepsilon_n + c_n + f(y) + a_n \ g(y)$$
$$\leq \varepsilon_n + c_n + f(x_n) + a_n \ g(y) \leq \varepsilon_n + 2c_n + f_n(x_n) + a_n g(y)$$

whence
$$g(x_n) \leq \varepsilon_n / a_n + 2c_n / a_n + g(y).$$
Then the conclusion can be obtained as in the preceding proof (which has $f_n = f$). $\square$

Theorem 45. We shall rely on some results from Chapter IV. Let us show that

(65) $$\text{strong epi} - \lim I_n = f \text{ in } Y.$$

If $x_n \to x$ in y, then from (57)

$$I_n(x_n) \geq f_n(x_n) \geq f(x_n) - c_n \left[1 + g(x_n)\right],$$

hence

$$\lim \inf I_n(x_n) \geq f(x),$$

since g is continuous. Moreover

$$I_n(y) \to f(y), y \in Y,$$

thereby proving (65). Now take any $u_n \in \varepsilon_n - \arg \min (Y, I_n)$. Then by (56) and (57)

$$\begin{aligned}
a_n \, g(u_n) &= I_n(u_n) - f_n(u_n) \leq I_n(z) + \varepsilon_n - f_n(u_n) \\
&= f_n(z) + a_n \, g(z) + \varepsilon_n - f_n(u_n) = f_n(z) - f(z) \\
&\quad + f(z) - f(u_n) + f(u_n) - f_n(u_n) + a_n \, g(z) + \varepsilon_n \\
&\leq c_n \left[1 + g(z)\right] + c_n \left[1 + g(u_n)\right] + a_n \, g(z) + \varepsilon_n.
\end{aligned}$$

Hence

$$(a_n - c_n) \, g(u_n) \leq c_n \left[1 + g(z)\right] + c_n + a_n \, g(z) + \varepsilon_n,$$

whence

(66) $$g(u_n) \leq \frac{a_n + c_n}{a_n - c_n} \, g(z) + \frac{2c_n + \varepsilon_n}{a_n - c_n}.$$

For sufficiently large n we have

$$c_n/a_n \leq q < 1, \varepsilon_n/a_n \leq M < +\infty.$$

Then by (66)

$$g(u_n) \leq \frac{2}{1-q} \, g(z) + \frac{2q + M}{1-q} \leq \text{constant},$$

so that, by (55)

(67) $$u_n \to u \text{ in } X \text{ for some subsequence and some } u \in X.$$

By theorem IV.5 and (65) we know that

(68) $$\lim \sup \text{val} \, (Y, I_n) \leq \text{val} \, (Y, f) = \text{val} \, (X, f)$$

since $z \in Y$. On the other hand, for the subsequence in (67)

$$\lim \inf I_n(u_n) \geq f(u),$$

hence

$$\lim \inf \text{val} \, (Y, I_n) \geq f(u) \geq \text{val} \, (X, f).$$

Together with (68) this yields

$$\text{val}\,(Y, I_n) \to \text{val}\,(X, f).$$

Since this happens to every subsequence of f_n, (58) is proved. Then (59) comes from theorem IV.5, taking into account (56) and (65). Finally, assuming weak compactness, boundedness of $g(u_n)$ and (67) yield $u \in Y$ and $u_n \rightharpoonup u$ in Y. If $c_n/a_n \to 0$ and $\varepsilon_n/a_n \to 0$, then by (66)

$$g(u) \le g(z)$$

which proves (60). $\square$

Section 8.   Notes and Bibliographical Remarks.

To section 1. The fundamental definition of Tykhonov well-posedness goes back to Tykhonov [5]. Example 7 may be found in Rockafellar [2, p. 266] where (section 27) further well-posedness criteria can be found. Remark 9 is due to Poracká - Diviš [1].

To section 2. The basic theorem 11 is due to Furi - Vignoli [3]. Theorem 12 and proposition 15 are due to Zolezzi [8], extending previous results of Vajnberg [1], where further sufficient conditions to well-posedness may be found. A characterization of Piontkowski [1] follows easily from theorem 12.

A specialized notion of Tykhonov well-posedness for best approximation problems, called *strong uniqueness* (see e.g. Braess [1. p.13]), corresponds to (4) with a linear forcing function. See Polyak [2, p. 136 and 205] for the corresponding notion (sharp minimum point) in convex optimization, where (th. 6 p. 137) it is shown that it implies superstability, i. e. invariance of arg min under small perturbations. Characterizations of strong uniqueness of best approximation problems from subspaces are in Bartelt - Mc Laughlin [1].

We remark that strong uniqueness cannot hold for the (well - posed) best approximation problems from linear subspaces of Hilbert spaces (see Braess [1, p.15]), or of smooth spaces.

Corollary 13 (as a necessary condition to well - posedness) and theorem 16 are due to Furi - Vignoli [2], where some examples are given. Corollary 17 is due to Polyak [1], where further applications are presented. Example 18 is due to Vajnberg [1], and in a sense it is the typical example of proper convex lower semicontinuous functional in a Hilbert space which is not strongly Tykhonov well-posed, as shown by Patrone [2]. Proposition 19 is a particular case of Bourgin [1, prop. 2.3.2] (see also theorem 2.3.6 there for further characterizations of the Radon-Nikodym property via well-posedness). Under suitable conditions, Tykhonov well-posedness with solution x_0 is equivalent to dentability at x_0, see Looney [1].

To section 3. See Fiorenza [1] for a forerunner of proposition 21 (in the particular case where x_0 is a fixed point of the mapping T on the metric space X and $I(x) = $ dist $[x, T(x)]$). For related results see section 1 of Čoban - Kenderov - Revalski [1]. Proposition 22 extends prop. IV.1.1 of Bednarczuk [2] and it is due to Patrone [3]. Proposition 24 is due again to Patrone [3], where further applications of this approach to well-posedness of saddle point problems and Nash equilibria may be found.

To section 4. Theorem 25 is due to Zolezzi [8], extending previous results of Poracká-Diviš [1]. Theorem 27 is a particular case of the basic results in Asplund -Rockafellar [1], where an extension to nonconvex problems may be found. Extensions of theorem 27 to sequences of not necessarily convex problems may be found in Zolezzi [8]. For a sufficient condition of differential character see also Smarzewski [1]. Related results are in Kosmol - Wriedt [1].

To section 5. The definition of generalized minimizing sequence and the corresponding well-posedness concept were introduced by Levitin - Polyak [1], to whom example 31 is due (together with applications to optimal control problems). The sets $L(\varepsilon)$ were introduced by Revalski [1]. Theorem 30 was remarked by Georgiev [1]. Further criteria of Levitin-Polyak well - posedness are contained in Cavazzuti - Morgan [1] (to whom example 33 is due) and Beer - Lucchetti [4] (to whom example 32 belongs).

To section 6. The definition of well - posedness in the generalized sense, theorem 38, 39 and example 40 are due to Furi - Vignoli [3]. Proposition 37 is due to Beer - Lucchetti [1]. Theorem 42 is due to Bednarczuk [2]. Well-posedness in the generalized sense can be characterized, as in proposition 22 and theorem 42, via upper semicontinuity of the $\varepsilon -$ arg min set, see Čoban - Kenderov - Revalski [1, prop.1.8]. As remarked in Rolewicz [1], the norm of a Banach space X has the drop property iff (B, u) is well-posed in the generalized sense for every $u \in X^*$ with $\|u\| = 1, B$ being the unit ball in X. For characterizations of well-posedness in the generalized sense for convex problems, see section II. 3. A definition of generalized Levitin - Polyak well-posedness (corresponding to well-posedness in the generalized sense) is introduced and characterized by Revalski - Zhivkov [1]. A criterion of well-posedness in the generalized sense (which extends proposition 21) is obtained in Čoban - Kenderov - Revalski [1]. A characterization involving Čech compactification is due to Revalski [4].

The concept of stability is sometimes used in the russian literature. A strengthened notion of stability for the metric projection within $C^0(T)$, which is related to well-posedness for best approximation problems, is considered in Wu Li [1] (under the name of Hausdorff strong uniqueness), and in Park [1] for general normed spaces, where it is related to Lipschitz continuity of the metric projection.

To section 7. For surveys about Tykhonov regularization applied to ill - posed problems and extensions (a method formulated in Tykhonov [2]), see Tikhonov - Arsénine [1], [2], Groetsch [1], Bertero [1], [2] and Liskovitz [1]. Theorem 44 is an extension of th. 5 in Levitin - Polyak [1]. Particular cases of theorem 44 may be found in Budak - Berkovich - Gaponenko [1], and further results are in Leonov [1]. Results related to theorem 44 about regularization of ill-posed equations in Hilbert spaces were given by Ribière [1]. Corollary 47 is a partial extension of a result due to Groetsch (see Bertero [2, th.5.1]). See also Seidman [2]. Regularization is applied to mathematical programming problems by Vasil'ev [1].

Extensions. Notions of equi well-posedness and uniform well-posedness have been introduced by Cavazzuti - Pacchiarotti [1], Zolezzi [8] and Cavazzuti - Morgan [1], where applications to penalty methods and perturbations of linear - quadratic problems may be found. See section IV.4 for one of such definitions. A generalization of the ordinal setting as related to well-posedness (section 3) is introduced in Patrone [5]. Well-posed preorders are defined and studied in Patrone - Revalski [1], [2].

A definition of well-posed saddle-point problems and related results have been given by Cavazzuti - Morgan [2]. Characterizations of well-posedness for such problems are obtained by Cavazzuti - Morgan [1], Lucchetti [6] and Patrone - Torre [1]. For two - level problems see Morgan [1]. Attempts to define well-posedness in vector optimization problems are considered in Lucchetti [6].

Chapter II.

HADAMARD AND TYKHONOV WELL-POSEDNESS

Introduction

(for a key chapter). Let X be a fixed topological (or convergence) space, and

$$I : X \to (-\infty, +\infty]$$

be extended real-valued functions belonging to some given family F endowed with a fixed topology or convergence. Roughly speaking, we are interested in the continuous dependence of

$$\arg \min (X, I) \text{ or val } (X, I)$$

(or both) upon $I \in F$.

This is a well-posedness property of the optimization problem at hand. As in the case of Tykhonov well-posedness (a quite distinct concept by now) the simplest case arises when for every $I \in F$ there exists exactly one minimum point

$$u_0 = \arg \ \min (X, I),$$

so that we are interested in the continuity on F of

$$I \to \arg \ \min (X, I), I \to \text{val} (X, I).$$

The requirement of existence, uniqueness and continuous dependence of the (optimal) solution u_0 from problem's data $I \in F$ extends to optimization problems the classical idea of well - posedness for problems in mathematical physics, due to Hadamard.

By the way, problems in mathematical physics (e.g. boundary value problems for partial differential equations) often have a variational origin, thereby raising a very fundamental issue about the relationships between the two basic well-posedness concepts: well-posedness of the boundary value problem and of the corresponding minimum problem. (This topic will be pursued further in section IV.7.)

Assuming existence and uniqueness of $u_0 = \arg \ \min (X, I)$, the precise meaning of continuous dependence of u_0 from I is far from being obvious. The appropriate notion of continuous dependence may change significantly for different classes of optimization problems. For this reason we will not consider a fixed notion of *Hadamard well-posedness* in optimization (an appropriate term by now), but we will define it in a convenient way from time to time.

In section 2 we fix I_0, we consider various constraints defined by subsets $K \subset X$, and minimize on X

$$I = I_0 + \text{ind } K.$$

Then we consider the dependence upon K of the minimum point, or of the optimal value.

When the underlying space X is normed, the constraint K is convex and

$$I_0(x) = \|x - x_0\| \text{ for some fixed } x_0 \in X,$$

we shall get some basic results both for the theory and the applications. It turns out that, for significant classes of convex optimization problems including this (best approximation) particular case, Hadamard and Tykhonov well-posedness are in a sense equivalent. This is one of the fundamental results of the whole theory. Moreover, a precise relation to (classical) Hadamard well-posedness of operator equations (or variational inequalities) can be obtained. This will fully justify our terminology about Hadamard well-posedness.

Section 1.   Well-Posed Best Approximation Problems, E-Spaces.

In this section X denotes a real Banach space. We denote by Conv (X) the collection of all nonempty closed convex subsets of X. A *convex best approximation problem* in X is defined as (X, I) where

(1) $$I(x) = \|x - x_0\| + \text{ind } (K, x),$$

$x_0 \in X$ and $K \in \text{Conv}(X)$. If the solution to this problem is unique, then arg min(X, I) is the best approximation to x_0 from K.

Motivations for investigating Tykhonov well-posedness of convex best approximation problems come from the following facts.

(a) Many optimization problems from applications may be casted in the above setting. Examples are least squares (section I.7), optimal control problems of the linear regulator type (section V.1), convex quadratic mathematical programming, minimum principles related to partial differential equations or inequalities of mathematical physics (e.g. Dirichlet principle. See sections 4 and 5).

(b) The geometric structure of those Banach spaces, in which every convex best approximation problem is Tykhonov well-posed, is especially significant. There are interesting links with well-posedness of more general convex optimization problems.

(c) Tykhonov well-posedness of such problems is related in a very significant way to continuity properties of the solutions with respect to changes in the data x_0, K, i.e. Hadamard well-posedness.

(d) Some problems in classical approximation theory (e.g. spline approximation) fit into this setting.

Recall that the Banach space X is *rotund* iff

$$x, y \in X, \|x\| = \|y\| = 1, x \neq y \text{ and } 0 < t < 1 \Rightarrow \|tx + (1 - t)y\| < 1.$$

If X is rotund, every convex best approximation problem in X has at most one solution.

Example. Let X be a real Hilbert space. From example I.10 we see that (X, I) defined by (1) is strongly Tykhonov well-posed for every x_0 and every $K \in \text{Conv}(X)$.

The Banach space X is an E - *space* iff X is reflexive, rotund and

(2) $$x_n \rightharpoonup x, \|x_n\| \to \|x\| \Rightarrow x_n \to x.$$

1 Examples of E - spaces. If X is finite - dimensional, then it is an E - space iff it is rotund. Any Hilbert space is an E - space. This latter fact may be, in a sense, generalized as follows. X is *uniformly convex* iff

$$\|x_n\| = \|y_n\| = 1, \|x_n + y_n\| \to 2 \Rightarrow \|x_n - y_n\| \to 0.$$

Then *every uniformly convex Banach space is an E - space.* For, it is well known that X is rotund and reflexive (see Diestel [1, th.2 p. 37]). If x_n, x are as in (2), and $\|x_n\| = \|x\| = 1$, let $u \in X^*$ be such that

(3) $$\|u\| = 1, \; < u, x > = 1$$

(see Brezis [2, cor. I. 3 p. 3]). Then

$$2 \ge \|x_n + x_j\| \ge \; < u, x_n + x_j > \to 2 \text{ as } n, j \to +\infty.$$

Then $x_n - x_j \to 0$, therefore x_n converges strongly, so $x_n \to x$, thus showing that X is an E - space. Therefore every $L^p(\Omega, \Sigma, \mu)$ with (Ω, Σ, μ) countably finite and $1 < p < \infty$, every Sobolev space $W^{k,p}(\Omega)$ where $1 < p < \infty$, k a positive integer and Ω open in R^n, are examples of E - spaces (being uniformly convex as well known). However, there exist E - spaces which are not uniformly convex.

Every reflexive space is potentially an E - space, since by a theorem of Troyanski, it can be renormed with an equivalent locally uniformly convex norm (Diestel [1, th.1 p. 164]). For, this new equivalent norm fulfills the following property:

$$\|y_0\| = 1 = \|y_n\|, \|y_n + y_0\| \to 2 \Rightarrow y_n \to y_0.$$

Then if $\|x_n\| = \|x\| = 1$ and $x_n \rightharpoonup x$ as in (2), we pick u as in (3). We get

$$2 = \lim \; < u, x_n + x_0 > \; \le \liminf \|x_n + x_0\| \le \limsup \|x_n + x_0\| \le 2$$

yielding $x_n \to x_0$.

It turns out that the notion of E - space is closely related to *strong* Tykhonov well-posedness (i.e., with respect to strong convergence) of convex best approximation problems within such a space. Let indeed $K \in \text{Conv}(X)$, where X is a real reflexive, rotund Banach space. Consider the minimization of $\| \cdot \|$ over K. This problem has an unique solution x_0. Let x_n be any minimizing sequence. Since x_n is bounded, some subsequence $y_n \rightharpoonup y \in K$. Then

$$\|x_0\| \le \|y\| \le \liminf \|y_n\| = \limsup \|y_n\| = \|x_0\|.$$

By uniqueness, $y = x_0$, and the original sequence $x_n \rightharpoonup x_0, \|x_n\| \to \|x_0\|$. Assuming (2) we get strong Tykhonov well-posedness of $(K, \| \cdot \|)$. Since K is arbitrary, one may guess that the conditions on X are also necessary for Tykhonov well-posedness. As a matter of fact we have

2 Theorem X *is an E - space iff every convex best approximation problem on X is strongly Tykhonov well-posed.*

We pause to observe that the conclusion in theorem 2 can be strengthened to Levitin - Polyak well-posedness (section I.5) of every convex best approximation problem, since it amounts to Tykhonov well-posedness. To this aim, let $(K, \|\cdot\|)$ be Tykhonov well-posed. If u_n is any generalized minimizing sequence, then

$$u_n \in X, \text{dist} \,(u_n, K) \to 0, \|u_n\| \to \text{dist} \,(0, K).$$

There exists a sequence $x_n \in K$ such that

$$\|u_n - x_n\| \le \text{dist} \,(u_n, K) + 1/n.$$

Then

$$\text{dist} \,(0, K) \le \|x_n\| \le \text{dist} \,(u_n, K) + 1/n + \|u_n\|$$

hence x_n is a minimizing sequence. By well - posedness, $x_n \to x_0$, the point of least norm in K, whence $u_n \to x_0$, proving Levitin - Polyak well-posedness.

A dual characterization of E - spaces comes from section I.4. Let $y \in X^*, y \ne 0$. By corollary I. 28, strong Tykhonov well-posedness of the problem, to minimize on the closed unit ball B the linear form $x \to - < y, x >$, amounts to Fréchet differentiability at 0 of the conjugate function to

$$x \to - < y, x > +\text{ind} \,(B, x),$$

that is of the norm of X^* at y. Then the following definition is appropriate. A Banach space is *strongly smooth* iff its norm is Fréchet differentiable out of 0.

3 Theorem X *is an E - space iff X^* is strongly smooth.*

Moreover, strong Tykhonov well-posedness of every convex best approximation problem obtains iff the same holds for all problems, to minimize nontrivial continuous linear forms on B.

4 Theorem *The following are equivalent facts:*
(4) *X is an E -space;*
(5) *for every $y \in X^*, y \ne 0$, the problem of minimizing*

$$x \to < y, x > \quad subject \ to \ \|x\| \le 1$$

is strongly Tykhonov well-posed;
(6) *for every $u \in X^*, u \ne 0$, the problem of minimizing*

$$x \to \|x\| \ subject \ to \ < u, x >= 1$$

is strongly Tykhonov well-posed.

From example III. 4 we see that the conclusion of theorem 4 fails if B is replaced by a bounded closed convex set.

PROOFS.

5 Lemma Let $c \in R, u \in X^*, u \neq 0$,

$$H = \{y \in X :< u, y >= c\}.$$

Then for every $x \in X$

$$\text{dist } (x, H) = |< u, x > -c|/\|u\|.$$

Proof. If $y \in H$

$$|< u, x > -c| \leq |< u, x - y >| \leq \|u\| \, \|x - y\|$$

giving

$$\text{dist } (x, H) \geq |< u, x > -c|/\|u\|.$$

We show now the converse inequality. Given $\varepsilon > 0$, there exists $z \in X$ such that $\|z\| = 1$ and $\|u\| - \varepsilon < \, < u, z >$. Then

$$y = x - \frac{< u, x > -c}{< u, z >} z \in H,$$

moreover

$$\|x - y\| = \|\frac{< u, x > -c}{< u, z >} z\| < \frac{|< u, x > -c|}{\|u\| - \varepsilon},$$

so by letting $\varepsilon \to 0$

$$\text{dist } (x, H) \leq |< u, x > -c|/\|u\|.$$

$\square$

We prove *theorems 2, 3 and 4* simultaneously by showing that the various statements there are equivalent, as follows.

Step 1. *If X is an E - space, then every convex best approximation problem is strongly Tykhonov well-posed.* Let X be an E - space, $K \in \text{Conv } (X), x_0 \in X$, and put

$$I(x) = \|x - x_0\| + \text{ ind } (K, x).$$

(X, I) *has at least one solution.* If

$$x_n \in K, \|x_n - x_0\| \longrightarrow \text{dist } (x_0, K)$$

then (by reflexivity) some subsequence $y_n \to y_0 \in K$. Then

$$\|x_0 - y_0\| \leq \, \liminf \|y_n - x_0\| \leq \lim \|x_n - x_0\| = \, \text{dist } (x_0, K) \leq \|x_0 - y_0\|$$

thus giving $y_0 \in \text{arg min } (X, I)$.

(X, I) *has at most one solution.* If $x_1, x_2 \in \text{arg min}(X, I)$ then $x_1, x_2 \in K, \|x_0 - x_1\| = \|x_0 - x_2\| = \, \text{dist } (x_0, K)$. Thus if $0 \leq t \leq 1$

$$(7) \quad \text{dist } (x_0, K) \leq \|tx_1 + (1-t)x_2 - x_0\| \leq t\|x_1 - x_0\| + (1-t)\|x_2 - x_0\| = \, \text{dist } (x_0, K).$$

Uniqueness is trivial if dist $(x_0, K) = 0$. Let $d = \, \text{dist } (x_0, K) > 0$. By (7) the line segment between $(x_1 - x_0)/d, (x_2 - x_0)/d$ belongs to the unit sphere of X, hence $x_1 = x_2$ by rotundity.

Any minimizing sequence is strongly convergent. Let x_n be such a sequence. Then

$$x_n \in K, \ \|x_n - x_0\| \to \ \text{dist} \ (x_0, K).$$

Fix any subsequence of x_n. Then a further subsequence $y_n \rightharpoonup y_0 = \arg \min (X, I)$ by uniqueness. By (2) this shows Tykhonov well-posedness of (X, I).

Step 2. Obviously, *Tykhonov well-posedness of every convex best approximation problem implies (6)*.

Step 3. *(6)* $\Rightarrow$ *(4)*. To show reflexivity, let $y \in X^*$ be such that $\|y\| = 1$. Then the least norm on the hyperplane $< y, x >= 1$ is attained at z say, $< y, z >= 1$, and using lemma 5 we see that $\|z\| = 1$. This means that y attains its norm, so reflexivity follows by James theorem (Diestel [1, p. 12]). Now let x_1, x_2 be in X and

$$\|x_1\| = \|x_2\| = 1, \ \|x_1 + x_2\| = 2.$$

Then we can find $u \in X^*$ such that

$$\|u\| = 1, < u, x_1 + x_2 >= 2 =< u, x_1 > + < u, x_2 >$$

(Brezis [2, cor. I.2 p. 3]). Therefore

$$1 \geq < u, x_2 >= 2- < u, x_1 > \geq 1.$$

Then $< u, x_2 >= 1 = \|x_2\| = \|u\|$. This implies that x_2 is a point of least norm on the hyperplane $< u, x >= 1$, since for every such x

$$< u, x >= 1 \leq \|u\| \ \|x\| = \|x\|.$$

The same is true for x_1. By well - posedness $x_1 = x_2$, hence X is rotund. Let us show (2). This will end the proof. If $x_n \rightharpoonup x, \|x_n\| = \|x\| = 1$, let $u \in X^*$ be such that $\|u\| = 1 =< u, x >$. Then $x_n/ < u, x_n >$ is a minimizing sequence for the problem of minimizing $\|y\|$ on the hyperplane $< u, y >= 1$, whose unique solution is x. By well - posedness we see that $x_n \to x$. $\square$

Step 4. *(4)* $\Rightarrow$ *(5)*. Given $y \in X^*, y \neq 0$, by reflexivity $< y, \cdot >$ has at least one minimum point on B. Let u_1, u_2, be minimum points. Then $< y, u_1 >=< y, u_2 >= -\|y\|$, hence for $z = -y/\|y\|$ and $v = (1/2)(u_1 + u_2)$,

$$< z, u_1 >=< z, u_2 >=< z, v >= 1$$

whence

$$\|u_1\| = \|u_2\| = \|v\| = 1$$

(Brezis [1, cor. I.4 p. 4]). Rotundity implies $u_1 = u_2$, hence uniqueness of the minimum point. If x_n is any minimizing sequence, then $\|x_n\| \leq 1, < y, x_n > \to -\|y\|$, but since

$$- < y, x_n > \leq \|y\| \ \|x_n\| \leq \|y\|$$

it follows that $\|x_n\| \to 1$. Given any subsequence of x_n, a further subsequence $z_n \rightharpoonup x_0, \|x_0\| \leq 1$, hence

(8) $$< y, z_n > \to < y, x_0 >= -\|y\|.$$

Thus x_0 is the optimal solution, $\|x_0\| = 1$ by (8) and the original sequence $x_n \rightharpoonup x_0$, hence $x_n \to x_0$, showing well - posedness. $\qquad\square$

Step 5. *(5) implies that X^* is strongly smooth.*

Take $y \in X^*, \|y\| = 1$, and consider

$$I(x) = - < y, x > + \text{ind } (B, x), x \in X,$$

where B is the closed unit ball in X. I is proper, convex and lower semicontinuous on X. By (5), (X, I) is Tykhonov well -posed. For every $z \in X^*$

(9) $\quad I^*(z) = \sup\{< z, x > -I(x) : \|x\| \leq 1\} = \sup\{< z + y, x >: \|x\| \leq 1\} = \|z + y\|.$

Therefore, by corollary I.28, I^* is Fréchet differentiable at 0, thus X^* is strongly smooth. $\qquad\square$

Step 6. X^* *strongly smooth implies (4).*

X *is reflexive.* Take $y \in X^*, \|y\| = 1$ and consider again I defined as in step 5. Then by (9) I^* is Fréchet differentiable at 0. Thus by corollary I. 28, (X, I) is strongly Tykhonov well - posed. Given any minimizing sequence x_n, we get

$$< y, x_n > \longrightarrow \sup \{< y, x >: \|x\| \leq 1\} = \|y\|,$$

then by well - posedness

$$x_n \longrightarrow u = \arg \min (X, I)$$

hence y attains the norm (at u). Reflexivity then follows from James theorem (Diestel [1, p. 12]).

X *is rotund.* Let the unit sphere S of X contain a nondegenerate segment of ends x, y. Then $(x + y)/2 \in S$. So there exists $u \in X^*$ such that

$$\|u\| = 1, < u, x + y >= 2,$$

hence

$$2 =< u, x + y >=< u, x > + < u, y > \leq 2$$

implying

$$< u, x >=< u, y >= 1 = \|u\|.$$

For every $\omega \in X^*$

$$< \omega, x > - < u, x > \leq \|\omega\| - \|u\|, < \omega, y > - < u, y > \leq \|\omega\| - \|u\|.$$

Denote by j the canonical embedding of X into X^{**}. Then

$$\|\omega\| \geq \|u\| + < j(x), \omega - u >, \|\omega\| \geq \|u\| + < j(y), \omega - u >$$

which means that $j(x), j(y)$ are distinct subgradients of the norm of X^* at u, contradicting strong smoothness.

(2) holds. Given x_n, x as therein, let $\|x\| > 0, \|x_n\| > 0$ and consider $y_n = x_n/\|x_n\|, y = x/\|x\|$. Then $y_n \rightharpoonup y, \|y_n\| = \|y\| = 1$. There exists $u \in X^*$ such that $\|u\| = 1 =< u, y >$ (Brezis [2, cor. I.3 p. 3]). Then $< u, y_n > \longrightarrow 1$, so that y_n is a minimizing sequence for (X, I), where

$$I(x) = - < u, x > + \text{ind } (B, x).$$

As shown in the beginning of step 6, (X, I) is strongly Tykhonov well - posed, hence $y_n \to y$, thus $x_n \to x$. In conclusion, X is an E - space. $\qquad\square$

Section 2.   Hadamard Well-Posedness of Best Approximation Convex Problems.

Let X be a real normed space, and consider a point $x \in X$ and a nonempty subset K thereof. Under existence and uniqueness conditions, we shall denote by

$$p(x, K)$$

the best approximation to x from K, i.e. the unique solution to the problem of minimizing over X

(10) $$I(u) = \|x - u\| + \text{ind } (K, u).$$

Then the optimal value of (X, I) is dist (x, K). We are interested in strong continuity of $p(\cdot, K)$ and dist $(\cdot, K)$ (or both) for any fixed K, a basic property of Hadamard well-posedness for the best approximation problem defined by (10).

Strong continuity (even uniform Lipschitz continuity) of dist $(\cdot, K)$ obtains in general, since for every $x_1, x_2 \in X$ and $y \in K$

$$\|x_1 - y\| \leq \|x_1 - x_2\| + \|x_2 - y\|$$

giving

$$\text{dist } (x_1, K) \leq \|x_1 - x_2\| + \text{ dist } (x_2, K)$$

so that

(11) $$|\text{dist } (x_1, K) - \text{dist } (x_2, K)| \leq \|x_1 - x_2\|.$$

Here (of course) arg $\min(X, I)$ may well be empty.

Now let X be a real Hilbert space and $K \in$ Conv (X). Then $p(\cdot, K)$ is a strongly continuous function, even uniformly Lipschitz, since for every $x_1, x_2 \in X$ we have

$$\|p(x_1, K) - p(x_2, K)\| \leq \|x_1 - x_2\|$$

(for a proof see Brezis [2, prop. V.3 p.80]).

On the other hand, continuity of $p(\cdot, K)$ fails in general, as shown by

6 Example. Let $K \subset C^0([0, 1])$ be the subset of all rational functions $t \to a/(bt + c), 0 \leq t \leq 1$, where a, b, c are real constants and $bt + c > 0$ on $[0, 1]$. Every function in $C^0([0, 1])$ has exactly one best approximation from K (see Braess [1, th. 2.2 p.114]). Consider the sequence $u_n(t) = 1/(1 + nt), n = 1, 2, 3, ...,$ and let $x_n \in C^0([0, 1])$ be affine on $[0, 1/2]$ and on $[1/2, 1]$, such that

$$x_n(0) = u_n(0) - 3/4, x_n(1/2) = u_n(1/2) + 3/4, x_n(1) = u_n(1) - 3/4.$$

Then by the alternation theorem (Braess [1, p.114]) we see that

$$p(x_n, K) = u_n \text{ for all } n.$$

Moreover x_n converges to some x_0 in $C^0([0, 1])$. Put $u_0(t) = 0, 0 \leq t \leq 1$, then again by the alternation theorem we get

$$p(x_0, K) = u_0,$$

but $u_n \not\to u_0$ since $\|u_n - u_0\| = 1$ for all n.

There even exist examples of a discontinuous $p(\cdot, K)$, K a linear subspace of a reflexive rotund Banach space (see Giles [1, p. 246]).

The above facts exhibit striking differences about the strength of possible notions of Hadamard well - posedness for (X, I) with any fixed K. *Solution* or *value Hadamard well-posedness* (so to speak) may result in deeply different notions.

We shall pursue continuous dependence properties of solutions to (X, I). A significant link between Tykhonov and Hadamard well - posedness is given by

7 Proposition *Let X be an E - space and $K \in$ Conv (X). Then $p(\cdot, K)$ is strongly continuous on X.*

Proof. Let $x_n \to x$ in X and put

$$y_n = p(x_n, K), \ y = p(x, K).$$

Then

dist $(x, K) \leq \|x - y_n\| \leq \|x - x_n\| + \|x_n - y_n\| \leq \|x - x_n\| + \|x_n - y\| \leq 2\|x - x_n\| + \|x - y\| = 2\|x - x_n\| +$ dist (x, K),

hence $\|x - y_n\| \to$ dist (x, K). It follows that y_n is a minimizing sequence to (X, I), where I is defined by (10). By theorem 2, $y_n \to$ arg min$(X, I) = y$. □

Proposition 7 admits the following partial converse. If the underlying space is Hilbert and for each point x there is an unique nearest point $p(x, K)$ in K, then strong continuity of $p(\cdot, K)$ forces convexity of K. This is a necessary condition for Hadamard well - posedness. It will play a key role in characterizing well - posed control systems (see theorem V.4).

A nonempty subset K of X is a *Chebychev set* in X iff for every $x \in X$ there exists exactly one nearest point to x in K.

8 Proposition *Let X be a Hilbert space and K a Chebychev set in X. If $p(\cdot, K)$ is strongly continuous on X, then K is convex.*

Under the assumptions of proposition 8, K is convex iff for every x, y

$$\|p(x, K) - p(y, K)\| \leq \|x - y\|$$

since the metric projection does not increase distances (Brezis [2, prop. V.3 p.80]). In the Hilbert space setting, solution Hadamard well-posedness, i.e. continuity of $p(\cdot, K)$, is equivalent to Lipschitz continuity (with constant 1), a stronger form of Hadamard well - posedness in general (see chapter IX).

Proposition 7 shows the following. Given K, under Tykhonov well - posedness of every $(X, I), I$ as in (10), we get continuous dependence of arg min (X, I) upon x.

Strong continuity of the unique optimal solution $p(\cdot, K)$ to $(X, I), I$ given by (10), may be interpreted by saying that (X, I) is *solution Hadamard well -posed* (in the strong topology, as a function of x). As we shall see later in section 5, this definition of Hadamard well - posedness for optimization problems contains as a particular case the classical one (given by Hadamard) for a particular class of problems in partial differential equations. This is one of the most significant results in the theory we are developing.

We go back to (X, I) where I is defined by (10). It is of great interest to analyze the behaviour of optimal solutions under perturbations acting on the constraint K (for fixed x). In which sense is the optimal solution $p(x, K)$ and the value val (X, I) a continuous function of K ? It is significant to find the weakest convergence (if any) on Conv (X) giving strong continuity of $p(x, \cdot)$ for all $x \in X$. This is a second form of solution Hadamard well - posedness.

Let X be a real reflexive rotund Banach space. Let K_0 and $K_n, n = 1, 2, 3, ...,$ be in Conv (X). If for every x we have

$$p(x, K_n) \to p(x, K_0)$$

then in particular for each $x \in K_0$ there exists a selection from K_n, strongly convergent to x. Therefore

(12) $K_0 \subset$ strong lim inf K_n.

On the other hand, put $x_n = p(x, K_n)$ and assume (12). Fix $u \in K_0$ and let $u_n \in K_n$ be such that $u_n \to u$. Since u_n is bounded,

$$\|x_n\| \le \|x_n - x\| + \|x\| \le \|x - u_n\| + \|x\| \le \text{ constant } < +\infty.$$

Then x_n is weakly sequentially compact. Fix any subsequence of x_n. A further subsequence $y_j = x_{n_j}$ say, will converge weakly to some x_0. If we assume that

(13)   weak seq. lim sup $K_n \subset K_0$

then $x_0 \in K_0$. Now consider $v_j = u_{n_j}$. Then

$$v_j \in K_{n_j}, \|x - y_j\| \le \|x - v_j\| \text{ for every } j, v_j \to u.$$

Therefore

$$\|x - x_0\| \le \liminf \|x - y_j\| \le \lim \|x - v_j\| = \|x - u\|, u \in K_0,$$

hence $x_0 = p(x, K_0)$. By uniqueness, the original sequence

$$x_n = p(x, K_n) \to x_0 = p(x, K_0).$$

Hence

$$\|x - x_0\| \le \lim \inf \|x - x_n\| \le \limsup \|x - x_n\| \le \lim \|x - u_n\| = \|x - u\|$$

for every $u \in K_0$. Since u is arbitrary

$$\|x - p(x, K_n)\| \to \|x - p(x, K_0)\|$$

i.e.

(14)
$$\text{dist }(x, K_n) \to \text{dist }(x, K_0).$$

If we assume that X is an E - space, then

(15)
$$p(x, K_n) \to p(x, K_0).$$

The above calculations show that a convergence for closed convex sets fulfilling (12), (13) is relevant to well - posedness of the convex best approximation problems in the E - space X, i.e. to obtain (14), (15). Since all of this works within the class of E - spaces, from theorem 2 we may guess the existence of a strong link between Hadamard and Tykhonov well - posedness (at least for best approximation convex problems). Such issues are basic to the whole theory.

It is therefore natural to introduce the following definition. Let X be a real normed space, K_n a sequence of nonempty subsets of X, K_0 a nonempty subset of X. The sequence K_n *converges in the sense of Mosco* to K_0, written

$$M - \lim K_n = K_0,$$

iff (12) and (13) hold, or equivalently iff

(16) for every $u \in K_0$ there exists $u_n \in K_n$, for every n sufficiently large, such that $u_n \to u$;

(17) for every subsequence n_j of the natural numbers, if $x_j \in K_{n_j}$ for every j and $x_j \rightharpoonup x$, then $x \in K_0$.

Now let X be reflexive and rotund. Given $x \in X$ and $K \in$ Conv (X), consider I given by (10). Of course

$$\text{val }(X, I) = \text{ dist }(x, K), \text{ arg min }(X, I) = p(x, K).$$

Let $\to$ be a given convergence structure on Conv(X). Denote by

$$(Conv(X), I)$$

the set of all problems (K, I) where I is given by (10) and $K \in$ Conv (X), $x \in X$. We shall say that (Conv $(X), I$) is

solution Hadamard well − posed with respect to $\to$

iff $p(x, \cdot)$ is strongly continuous on Conv (X) equipped by $\to$, for every $x \in X$;

value Hadamard well − posed with respect to $\to$

iff dist $(x, \cdot)$ is continuous on Conv (X) equipped by $\to$, for every $x \in X$.

We shall now consider three main problems. The first, to characterize the weakest convergence on Conv (X) (if any), with respect to which there is solution or value Hadamard well - posedness. The second, to relate each other solution and value Hadamard well -

posedness. The third, to establish a link between Tykhonov and Hadamard well - posedness.

About the first and second problem we have

9 Theorem *Let X and X^* be E - spaces, and K_n, K_0 be a sequence in Conv (X). The following are equivalent facts:*

$$(18) \qquad\qquad M - \lim K_n = K_0;$$

$$(19) \qquad\qquad p(x, K_n) \to p(x, K_0) \text{ for every } x \in X;$$

$$(20) \qquad\qquad \text{dist } (x, K_n) \to \text{ dist } (x, K_0) \text{ for every } x \in X.$$

The assumption about X in theorem 9 amounts to requiring that X is a strongly smooth E - space (by theorem 3) or that, equivalently, both norms of X and X^* are Fréchet differentiable out of the origin.

Theorem 9 shows that Mosco convergence on Conv(X) is the weakest one guaranteeing solution, or equivalently, value Hadamard well - posedness for (Conv $(X), I)$, assuming strong Tykhonov well - posedness for all best approximation convex problems both in X and X^*. In the proof we shall see that some of the equivalences among (18), (19), (20) require less restrictive assumptions on X.

We consider now the (third) problem, i.e. which relations exist between Hadamard and Tykhonov well - posedness for best approximation convex problems. Let I be given by (10). Theorem 9 relates strong Tykhonov well - posedness of each problem (X, I) with Hadamard's one for (Conv $(X), I)$. The next theorem shows their equivalence, and the special role best approximation convex problems play in this setting. It is a particular instance of the equivalence between Tykhonov and Hadamard well - posedness in convex optimization, one of the main results of the theory, which we shall pursue further in the next section.

10 Theorem *Let X be reflexive and rotund. Then (Conv $(X), I)$ is solution Hadamard well - posed with respect to Mosco convergence iff every problem (X, I) is strongly Tykhonov well - posed. In other words, the following are equivalent facts:*

$$(21) \qquad\qquad X \text{ is an } E \text{ - space;}$$

$$(22) \qquad \begin{aligned} &\text{for every } x \in X \text{ and every sequence } K_n \text{ in Conv } (X) \\ &\text{with } M - \lim K_n = K_0 \text{ we have } p(x, K_n) \to p(x, K_0). \end{aligned}$$

$$(23) \qquad \begin{aligned} &x_n \to x_0 \text{ in } X, K_n \text{ and } K_0 \text{ in Conv } (X), M - \lim K_n = K_0 \\ &\Rightarrow p(x_n, K_n) \to p(x_0, K_0). \end{aligned}$$

In which spaces X is theorem 9 valid ?

Following Sonntag [2], we shall say that X is a $C-space$ (after Cudia [1]) iff X, X^* are both E - spaces. So X is a C - space iff X fulfills the assumptions in theorem 9.

From example 1 we see that theorem 9 applies if X is a Hilbert space or $X = L^p(\Omega, \Sigma, M)$ with $1 < p < \infty$ and (Ω, Σ, M) countably finite, or $X = W^{k,p}(\Omega)$ any Sobolev space with k a positive integer, $1 < p < \infty$ and Ω an open subset of R^n. But there are C - spaces not isomorphic to any uniformly convex space. See Sonntag [2, sec. 7] for a counterexample as well as further examples of C - spaces. If X is finite - dimensional, then X is a C - space iff both X and X^* are rotund. C - spaces are a very stable class. As a matter of fact, by theorem 3, X is a C - space iff X^* does. Moreover it can be proved that closed subspaces of C - spaces, and quotient spaces by closed subspaces, are again C - spaces; if X_n is a sequence of C - spaces and $1 < p < \infty$, then their l_p - product is.

Well-posedness and renorming.

Given a real reflexive Banach space X there exists an equivalent norm on X such that X is a C - space (Diestel [1, cor. 3 p.167]). This property is not particularly relevant as far as theorem 9 is concerned. Changing the norm (although in an equivalent way) may change the original optimization problem (we are interested in) to an unacceptable way. Essential features of the given problem may be lost. Renorming cannot obtain strong convergence of minimizing sequences (since the new norm is equivalent to the old one). But changing the given norm to a suitable parameterized family of new norms which are near to the original one gives often well - posedness from an ill - posed problem, without losing contact with the original one, since we deal with approximation in a variational sense. This happens e.g. in Tykhonov regularization procedures (section I.7).

Example. Abusing notation, let us denote again by $p(x, K)$ the set (not necessarily a singleton) of all best approximations to x from K. Without rotundity, Hausdorff convergence of K_n to K_0 does not imply Hausdorff convergence of $p(x, K_n)$ to $p(x, K_0)$. Take $X = R^2$ equipped with the norm $\|x\| = |x_1| + |x_2|$, and $x = 0, K_0 =$ segment with ends $(0, 1)$ and $(1, 0), K_n =$ segment with ends $(0, 1), (1 + 1/n, 1)$. See the figure of the next page. In this example however we see that the following weaker conclusion obtains: $\sup\{\text{dist } [u, p(x, K_0)] : u \in p(x, K_n)\} \to 0$ (i.e. Hadamard well - posedness in the sense we shall use in theorem 14).

PROOFS.

Proposition 8. Write simply

$$p(x) = p(x, K)$$

and consider

$$I(y) = (1/2)\|y\|^2 + \text{ind } (K, y), y \in X.$$

Then for every x, identifying X and X^*,

$$I^*(x) = \sup\{< x, y > -(1/2)\|y\|^2 : y \in K\} = (1/2)\sup\{\|x\|^2 - \|x - y\|^2 : y \in K\}$$

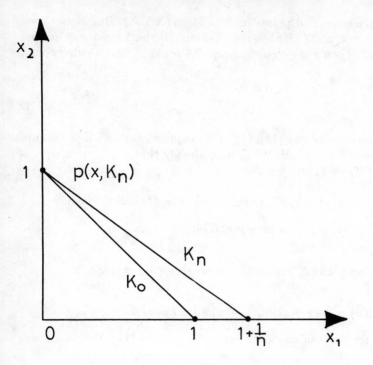

$$=< x, p(x) > -(1/2)\|p(x)\|^2.$$

Thus I^* is convex continuous on X, since $I^*(x) \leq \|x\|^2/2$ everywhere (th.1.3 p.90 of Barbu - Precupanu [1]). Now for all x, h, $\|x + h - p(x+h)\|^2 \leq \|x + h - p(x)\|^2$ yielding

$$0 \leq I^*(x + h) - I^*(x) - < p(x), h > = < x + h, p(x+h) > -$$

$$-(1/2)\|p(x+h)\|^2 - < x, p(x) > +(1/2)\|p(x)\|^2 - < p(x), h >$$

$$=< h, p(x+h) - p(x) > +(1/2)\|x - p(x)\|^2 - (1/2)\|p(x+h) - x\|^2$$

$$\leq < h, p(x+h) - p(x) >= o(\|h\|),$$

by continuity of p. Hence

$$\partial I^*(x) = p(x) \text{ for all } x \in X.$$

Now consider the second conjugate

$$I^{**}(y) = (\|y\|^2)/2 \text{ if } y \in K, I^{**}(y) = +\infty \text{ if } y \notin \text{ cl co } K.$$

This formula follows since I^{**} is the biggest lower semicontinuous convex function majorized by I (Ekeland - Temam [1, prop. 4.1 p.17]). Therefore $K \subset \text{ dom } I^{**}$ which is convex. If K is not convex, we can find points in dom $I^{**} \cap (X \setminus K)$. But K is closed being Chebychev, so by cor. 6.2 p. 32 of Ekeland - Temam [1] there exist $x \notin K$ and $u \in X$ such that $u \in \partial I^{**}(x)$. Then $x \in \partial I^*(u)$ by prop. 2.1 p.103 of Barbu - Precupanu [1], yielding

$$x = p(u) \notin K,$$

a contradiction. □

Theorem 9. Some of relations among (18), (19), (20) require less restrictive assumptions about X, as we shall see in the proof. We showed already that
if X is an E - space, then (18) $\Rightarrow$ (19).
Moreover we see that
if X is reflexive and rotund (so that $p(x, K)$ makes sense) *then (19) $\Rightarrow$ (20),* since

$$\text{dist } (x, K) = \|x - p(x, K)\|.$$

Now we prove that
(24) *if X is reflexive and rotund, and X^* is an E - space, then (20) $\Rightarrow$ (18).*
For every $x \in K_0$

$$\text{dist}(x, K_n) = \|p(x, K_n) - x\| \to \text{ dist } (x, K_0) = 0$$

yielding (16). Let $x_n \in K_n$ for some subsequence and $x_n \rightharpoonup x$, as in (17). We shall prove that

(25) $$x \in K_0$$

By theorem 3, X is strongly smooth. Let

$$J : X \to X^*$$

be the Fréchet differential of $(1/2)\|\cdot\|^2$ on X (i.e. the duality mapping of X). To prove (25) we shall need the following
11 Lemma If $K \in \text{Conv } (X)$, the function

$$x \to (1/2)[\text{dist } (x, K)]^2$$

is Fréchet differentiable on X, with derivative at x equal to $J[x - p(x, K)]$.

Taking lemma 11 for granted, consider

$$g_n(x) = (1/2)[\text{dist } (x, K_n)]^2, u_n = p(x, K_n).$$

By convexity of g_n and lemma 11

(26) $$g_n(\omega) \geq g_n(x) + <J(x - u_n), \omega - x> \quad \forall \omega \in X, \forall n.$$

Moreover $\|J(x - u_n)\| = \|x - u_n\| = \text{ dist } (x, K_n)$. By reflexivity of X^*, for some subsequence

$$J(x - u_n) \rightharpoonup y \in X^*.$$

Letting $n \to +\infty$ in (26) we have

$$g_0(\omega) \geq g_0(x) + <y, \omega - x> \quad \forall \omega \in X.$$

By lemma 11 we see that $y = J(x - u_0)$. Uniqueness of the subgradient y implies that the original sequence $J(x - u_n) \rightharpoonup J(x - u_0)$. Moreover, by assumption
$$\|J(x - u_n)\| = \|x - u_n\| = \text{ dist } (x, K_n) \to \text{ dist } (x, K_0) = \|x - u_0\|.$$
Since X^* is an E - space, $J(x - u_n) \to J(x - u_0)$. Putting $\omega = x_n$ in (26)

$$0 = [\text{dist } (x_n, K_n)]^2 \geq [\text{dist } (x, K_n)]^2 + 2 < J(x - u_n), x_n - x >$$

and letting $n \to +\infty$ we get dist $(x, K_0) \leq 0$. $\qquad \square$

Proof of lemma 11. Put $y = p(x, K)$. Given $u \in K, s = 0$ minimizes $\|su + (1 - s)y - x\|^2$ as $0 \leq s \leq 1$, therefore

$$0 \leq ((d/ds)\|su + (1 - s)y - x\|^2)_{s=0} = 2 < J(y - x), u - y > .$$

It follows that $p(x, K)$ solves the variational inequality

(27) $\qquad < J[p(x, K) - x], u - p(x, K) > \geq 0 \; \forall u \in K.$

Write $p(x) = p(x, K)$ and let $h \in X$. By Fréchet differentiability of $(1/2)\| \cdot \|^2$ we have for any x

(28) $\qquad (1/2)\|x + h - p(x + h)\|^2 \geq (1/2)\|x - p(x)\|^2 + < J[x - p(x)], h >$
$\qquad + < J[x - p(x)], p(x) - p(x + h) > \geq < J[x - p(x)], h >$

(the last inequality follows from (27)). On the other hand

(29) $\qquad (1/2)\|x + h - p(x + h)\|^2 - (1/2)\|x - p(x)\|^2 \leq (1/2)\|x + h - p(x)\|^2$
$\qquad - (1/2)\|x - p(x)\|^2 = < J[x - p(x)], h > + o(\|h\|).$

Combining (28) and (29) we see that $J[x - p(x)]$ is the Fréchet derivative, as required. $\quad \square$
This completes the proof of theorem 9. $\qquad \square$

Remark. A direct proof that $(19) \Rightarrow (18)$ goes as follows. Of course (16) is fulfilled. We need only to prove (17). Let x_i and x be as there and consider

$$y_i = p(x, K_{n_i}), \; y = p(x, K_0).$$

By (27) $< J(y_i - x), x_i - y_i > \geq 0$ for every i. Since J is strongly continuous (Giles [1, lemma p.147]), as $i \to +\infty$ we get

$$< J(y - x), x - y >= -\|y - x\|^2 \geq 0,$$

thus $x = y \in K_0$. We proved directly (before stating theorem 9) that $(18) \Rightarrow (20)$. This fact holds assuming only reflexivity of X, as follows. Let $u \in K_0$ and $u_n \in K_n$ be as in (16). Then
$$\limsup \text{ dist } (x, K_n) \leq \lim \|x - u_n\| = \|x - u\|.$$
Since $u \in K_0$ is arbitrary

(30) $$\limsup \ \text{dist} \ (x, K_n) \le \ \text{dist} \ (x, K_0).$$

Pick a subsequence of K_n such that

$$\text{dist} \ (x, K_n) \to \liminf \ \text{dist} \ (x, K_n).$$

For every n we find $u_n \in K_n$ such that

$$\|x - u_n\| \le \ \text{dist} \ (x, K_n) + 1/n.$$

Let $u \in K_0, v_n \in K_n$ be such that $v_n \to u$. Then

$$\|u_n\| \le \|x - u_n\| + \|x\| \le \|x\| + 1 + \|x - v_n\| \le \ \text{constant},$$

therefore for some subsequence n_j of the positive integers we have $x_j \rightharpoonup x_0$ where $x_j = u_{n_j}$. Then $x_0 \in K_0$ by (17). Thus

$$\|x - x_0\| \le \liminf \|x - x_j\| \le \liminf \ \text{dist} \ (x, K_n),$$

giving

(31) $$\text{dist} \ (x, K_0) \le \liminf \ \text{dist} \ (x, K_n).$$

The conclusion follows from (30) and (31). $\qquad\square$

Another proof of (18) $\Rightarrow$ (20) follows from (12) of theorem IV.5, which implies

$$\limsup \ \text{dist} \ (x, K_n) \le \ \text{dist} \ (x, K_0).$$

Hence $p(x, K_n)$ is a bounded sequence. If $p(x, K_n) \rightharpoonup u$ for some subsequence, then $u \in K_0$ and

$$\liminf \|x - p(x, K_n)\| \ge \|x - u\|$$

yielding for the original sequence

$$\liminf \ \text{dist} \ (x, K_n) \ge \ \text{dist} \ (x, K_0).$$

Theorem 10. The equivalence between (21) and (22) is a corollary to theorem 13 in the next section. Since obviously (23) $\Rightarrow$ (22), we need only to show that (21) $\Rightarrow$ (23). To this aim put

$$p_n(\cdot) = p(\cdot, K_n).$$

Let $z \in K$ and $z_n \in K_n$ be such that $z_n \to z$. Then

$$\|p_n(x_n)\| \le \|p_n(x_n) - x_n\| + \|x_n\| \le \|x_n - z_n\| + \|x_n\|,$$

hence for some subsequence

(32) $$p_n(x_n) \rightharpoonup y \in K_0$$

by Mosco convergence. Moreover

(33) $$\liminf \|x_n - p_n(x_n)\| \ge \|y - x_0\| \ge \|p_0(x_0) - x_0\|.$$

Pick $y_n \in K_n$ such that $y_n \to p_0(x_0)$. Then

$$\|x_n - p_n(x_n)\| \leq \|x_n - y_n\|,$$

hence

(34) $$\limsup \|x_n - p_n(x_n)\| \leq \|x_0 - p_0(x_0)\|.$$

By (33) and (34) we get

$$\|x_n - p_n(x_n)\| \to \|x_0 - p_0(x_0)\|.$$

By (32), $x_n - p_n(x_n) \rightharpoonup x_0 - y$, so that

$$\|x_0 - y\| \leq \liminf \|x_n - p_n(x_n)\| = \|x_0 - p_0(x_0)\|.$$

By (32) $y \in K_0$, hence $y = p_0(x_0)$. Thus $p_n(x_n) \to p_0(x_0)$.   □

Section 3.   Equivalence between Tykhonov and Hadamard Well-Posedness in Convex Optimization.

Throughout this section, X is a fixed real Banach space. Let $I : X \to (-\infty, +\infty)$ be fixed. Given a convergence structure $\to$ on Conv (X), we shall say that
(Conv $(X), I$) is solution Hadamard well - posed with respect to $\to$ iff

(35) $\arg \min (K, I)$ is a singleton for every $K \in$ Conv (X),

(36) $\arg \min (K_n, I) \to \arg \min (K_0, I)$ whenever $K_n \to K_0$ in Conv (X).

12 Remark. Suppose that X is reflexive and I is lower semicontinuous and convex. Then I fulfills (35) iff it has bounded sublevel sets and is strictly quasiconvex, i.e.

$$0 < t < 1, x \neq y \Rightarrow I(tx + (1-t)y) < \max\{I(x), I(y)\}.$$

We shall say that

(Conv $(X), I$) is strongly Tykhonov well - posed

iff (K, I) is for every $K \in$ Conv (X).

Recall that the Hausdorff distance between two nonempty subsets A, B of X is defined as

$$\text{haus} (A, B) = \max\{e(A, B), e(B, A)\},$$

where

$$e(A, B) = \sup\{\text{dist} (a, B) : a \in A\}.$$

13 Theorem *Let X be reflexive, I continuous and convex.*
37) *If I is uniformly continuous on bounded sets, then (Conv $(X), I$) strongly Tykhonov well - posed $\Rightarrow$ (Conv $(X), I$) solution Hadamard well - posed with respect to Mosco convergence.*

(38) $(Conv\ (X), I)$ *solution Hadamard well - posed with respect to Hausdorff convergence of bounded sets* $\Rightarrow (Conv\ (X), I)$ *strongly Tykhonov well - posed.*

The equivalence between (21) and (22) in theorem 10 is now a corollary to theorem 13 (remembering theorem 2), since Hausdorff convergence is stronger than Mosco on Conv (X). To prove this, let $K_n \to K$ in Conv (X) in the sense of Hausdorff. Given $x \in K$ let $y_n \in K_n$ be such that

$$\|x - y_n\| \leq \ \text{dist}\ (x, K_n) + 1/n.$$

Since dist $(x, K_n) \leq$ haus (K, K_n) we see that $K \subset$ strong lim inf K_n. Now let $x_p \in K_{n_p}$ be such that $x_p \to x$. Pick $y_p \in K$ such that

$$\|x_p - y_p\| \leq \ \text{dist}\ (x_p, K) + 1/p.$$

Since dist $(x_p, K) \leq$ haus (K_{n_p}, K), we get $y_p \to x \in K$.

Moreover, (37) generalizes (18) $\Rightarrow$ (19) of theorem 9.

Tykhonov well - posedness is equivalent to (strong) solution Hadamard well - posedness when the latter is taken either with respect to constraint perturbations, as in theorem 13, or when perturbations of the functionals are taken into account. In this latter case the equivalence may be extended to characterize well - posedness in the generalized sense with respect to strong convergence (section I.6), provided Hadamard well - posedness is meant in a sufficiently weak sense, as follows.

For sequence $I_n, I \in C_0(X)$ (i.e. extended real - valued functions on X which are proper, convex and lower semicontinuous) let us say that

$$I_n \to I\ uniformly\ on\ bounded\ sets$$

iff dom $I_n =$ dom I for n sufficiently large and

$$\sup\{|I_n(x) - I(x)| : x \in K \cap \ \text{dom}\ I\} \to 0$$

for every bounded $K \subset X$ such that $K \cap$ dom $I \neq \emptyset$.

Given $I \in C_0(X)$ with a nonempty arg min(X, I), let us say that (X, I) is *solution Hadamard well - posed* iff $I_n \in C_0(X)$ with a nonempty arg min(X, I_n), $I_n \to I$ uniformly on bounded sets imply

$$\sup\{\text{dist}[u, \text{arg min}(X, I)] : u \in \text{arg min}(X, I_n)\} \to 0 \text{ as } n \to +\infty,$$

or equivalently

(39) for every $\varepsilon > 0$, arg min$(X, I_n) \subset$ [arg min$(X, I)]^\varepsilon$ for sufficiently large n.

14 Theorem *Let X be reflexive and $I \in C_0(X)$. Then the following are equivalent facts:*

(40) (X, I) *is well - posed in the generalized sense;*

(41) *arg min(X, I) is compact nonempty and (X, I) is solution Hadamard well-posed.*

In the particular case of Tykhonov well - posedness we get directly

15 Corollary *Let I, X be as in theorem 14. Then the following are equivalent facts:*
(X, I) is strongly Tykhonov well - posed;
arg $min(X, I)$ is a singleton and $I_n \in C_0(X), I_n \to I$ uniformly on bounded sets,
$x_n \in arg\ min(X, I_n) \Rightarrow x_n \to arg\ min(X, I)$.

In corollary 15, uniform convergence cannot be weakened to Mosco's (defined later in section IV.8), as shown by
16 Example. Let X be a separable Hilbert space with basis e_n. Consider

$$I_n(x) = \sum_{i \neq n} <x, e_i>^2 + (1/n) <x - e_n, e_n>^2, I(x) = \|x\|^2.$$

Here (X, I) is strongly Tykhonov well - posed,

$$arg\ min\ (X, I_n) = e_n \not\to 0 = arg\ min(X, I),$$

although $M - \lim I_n = I$. To see this, notice that $I_n(x) \to I(x)$ for every $x \in X$. We claim that

$$x_n \to x \Rightarrow \liminf I_n(x_n) \geq I(x).$$

In fact let $y_n = x_n - <x_n, e_n> e_n$. Then $y_n \overset{\cdot}{\to} x$ and

$$I_n(y_n) = \|y_n\|^2 + (1/n) <y_n - e_n, e_n>^2 \geq \|y_n\|^2;$$

$$I_n(x_n) - I_n(y_n) = (1/n)(-2 <x_n, e_n> + <x_n, e_n>^2) \to 0,$$

proving the above claim.

If we strengthen the topology which defines Hadamard well - posedness, from Mosco to local Hausdorff (in a suitable sense), then Tykhonov well - posedness of a fixed convex problem (A, I) can be characterized again through solution and value Hadamard well - posedness. This sharpens in part corollary 15 (since we deal with constrained problems under a less stringent mode of convergence than locally uniform). It follows that asymptotically minimizing sequences of approximating (or perturbed) problems automatically converge to the unique minimizer of (A, I), provided that the convergence of minimum problems is defined as follows.
For nonempty subsets C, D of X and sufficiently large $r > 0$ put

$$h(r, C, D) = \max\{e(C \cap B(r), D), e(D \cap B(r), C)\},$$

where $B(r)$ denotes the closed ball of radius r around the origin in X. For a sequence A_n, A of nonempty subsets of X, we write

$$loc\ H - lim\ A_n = A$$

iff $h(r, A_n, A) \to 0$ as $n \to +\infty$ for every sufficiently large $r > 0$.
We shall consider the epigraph of any proper function $f : X \to (-\infty, +\infty]$ as a subset of the normed space $X \times R$ equipped with the box norm $\|(x, t)\| = \max\{\|x\|, |t|\}$.

We claim that if $A_n, A \in$ Conv (X),

$$\text{loc } H - \lim A_n = A \Rightarrow M - \lim A_n = A.$$

Indeed, if $v \in A$ with $\|v\| \leq r$, there exists $v_n \in A_n$ such that

$$\|v - v_n\| \leq 1/n + \text{ dist } (v, A_n) \to 0.$$

Arguing by contradiction, let (for some subsequence n_k) $u_k \in A_{n_k}$ be such that

$$u_k \to u \text{ and } u \notin A.$$

By the strict separation theorem (Brezis [2, th. I.7 p.7])

$$(42) \qquad\qquad < y, x > +\varepsilon \leq < y, u >$$

for some $y \in X^*, \varepsilon > 0$, all $x \in A$. Since u_k is bounded, dist $(u_k, A) \to 0$, hence $\|u_k - x_k\| \to 0$ for a suitable sequence $x_k \in A$. Then $x_k \to u$, and (42) with $x = x_k$ yields $< y, u > +\varepsilon \leq < y, u >$, a contradiction, proving our claim.

17 Theorem *Let $A \in$ Conv (X) and $I \in C_0(X)$ be such that I is continuous at some point of A. If (A, I) is Tykhonov well - posed, then*
(43) *for every sequence $A_n \in$ Conv (X), $I_n \in C_0(X)$ such that*

$$\text{loc } H - \lim A_n = A, \text{loc } H - \lim \text{ epi } I_n = \text{epi } I$$

we have

$$\text{val } (A_n, I_n) \to \text{ val } (A, I),$$

and

$$\varepsilon_n \to 0, x_n \in \varepsilon_n - \text{arg min } (A_n, I_n) \Rightarrow x_n \to \text{arg min } (A, I).$$

Of course the converse holds (without imposing continuity of I.)

18 Corollary *Let $A \in$ Conv (X) and $I \in C_0(X)$ be such that (A, I) is Tykhonov well - posed. If I is continuous at some point of A, then (A, I) is Levitin - Polyak well - posed.*

Local Hausdorff convergence will be used in section IX.5 to obtain quantitative estimates about the dependence of $\varepsilon - \text{arg min}(X, I)$ and val (X, I) upon I.

Further results about equivalence between Tykhonov and Hadamard well - posedness, in a nonconvex setting, are given in corollary III.18.

PROOFS.

Theorem 13. Let us prove (37). Let $M - \lim K_n = K_0$ in Conv (X). Write

$$x_0 = \arg \; \min(X, I), y_n = \arg \; \min(K_n, I), n = 0, 1, 2, \dots.$$

First case: $x_0 \in K_0$. Then by Tykhonov well -posedness $x_0 = y_0$. It suffices to prove that

$$(44) \hspace{4cm} I(y_n) \to I(y_0)$$

since then $y_n \to y_0$, as a minimizing sequence for (X, I).

Since $I(y_n) \geq I(y_0), \liminf I(y_n) \geq I(y_0)$. By (16) there exists $u_n \in K_n, u_n \to y_0$, so $I(y_n) \leq I(u_n)$, giving $\limsup I(y_n) \leq I(y_0)$, which entails (44).

Second case: $x_0 \notin K_0$. Put

$$\text{sub} \;\; \text{lev} \; (I, a) = \{x \in K : I(x) \leq a\}, a \in R.$$

By Tykhonov well - posedness of (X, I) we have

$$(45) \hspace{4cm} v = I(y_0) > I(x_0).$$

Since I is continuous, $x_0 \in$ int sub lev (I, v) which is an open convex set. So if $x \in$ int sub lev (I, v) there exist $h \in (0, 1), x_1 \in$ sub lev (I, v) such that

$$x = hx_0 + (1 - h)x_1.$$

Then by (45)

$$I(x) \leq hI(x_0) + (1 - h)I(x_1) < v,$$

so we see that int sub lev (I, v) and K_0 are disjoint. By the separation theorem (Braess [1, p.63]) we can find $u \in X^*, u \neq 0$ and $c \in R$ such that $< u, x > \geq c$ for all $x \in K_0$, and

$$(46) \hspace{3cm} < u, \omega > < c \text{ for all } \omega \in \text{ int sub lev } (I, v).$$

Thus $< u, y_0 > = c$ since $y_0 \in K_0 \cap$ sub lev (I, v). By (16) let $x_n \in K_n$ be such that $x_n \to y_0$. Since $I(x_n) \geq I(y_n)$ we have

$$(47) \hspace{4cm} \limsup I(y_n) \leq I(y_0),$$

thus $y_n \in$ sub lev (I, a) for some a. We shall rely on the following

19 Lemma. Let $I : X \to (-\infty, +\infty]$ be proper, convex and lower semicontinuous. Let $K \in$ Conv (X) be such that (K, I) is strongly Tykhonov well - posed. Then $K \cap$ sub lev (I, a) is bounded for every real a.

End of proof of (37). Since y_n is bounded by lemma 19, for a subsequence $y_n \to \bar{y} \in K_0$ (from (17)). Thus

$$(48) \hspace{4cm} \liminf I(y_n) \geq I(\bar{y}) \geq I(y_0).$$

The behavior of each subsequence of y_n is as above, thus by (47) we see that for the original sequence

(49)
$$I(y_n) \to I(y_0).$$

Then $I(y_0) = I(\bar{y})$ by (48). Tykhonov well-posedness gives $y_0 = \bar{y}$, so for the original sequence

(50)
$$y_n \rightharpoonup y_0.$$

Now write
$$H = \{x \in X :< u, x >= c\}, H_0 = \{x \in X :< u, x >= 0\}.$$

Since H_0 has codimension 1, there exists $z \notin H_0$ such that for every n we can find $z_n \in H_0, \alpha_n \in R$ satisfying

(51)
$$y_n - y_0 = z_n + \alpha_n z.$$

Then $< u, y_n - y_0 >= \alpha_n < u, z >$, implying by (50) that $\alpha_n \to 0$, therefore $y_n - (y_0 + z_n) \to 0$ by (51). From the uniform continuity of I and boundedness of y_n, z_n it follows that
$$I(y_n) - I(y_0 + z_n) \to 0,$$

and by (49)

(52)
$$I(y_0 + z_n) \to I(y_0).$$

By (46) we have $y_0 \in H$. The existence of $\omega \in H$ such that $I(\omega) < I(y_0)$ implies $\omega \in$ int sub lev (I, v) and this by (46) gives $\omega \notin H$, a contradiction. Hence $y_0 = \arg \min(H, I)$. Then by (52) and Tykhonov well-posedness of (H, I) we have $z_n \to 0$, hence $y_n \to y_0$ by (51).

Proof of lemma 19. The conclusion is obvious if K is bounded, or $a \leq \inf I(K)$. Let now K be unbounded. From unboundedness of $K \cap$ sub lev (I, a) for every $a > \inf I(K)$ we find some sequence $x_n \in K$ such that
$$\|x_n\| \geq n, \ I(x_n) \leq \inf I(K) + 1/n,$$

against Tykhonov well-posedness. Then $K \cap$ sub lev (I, a) is bounded for at least some $a > \inf I(K)$. Since
$$K \cap \text{ sub lev } (I, a) = \text{ sub lev } (I + \text{ ind } (K, \cdot), a)$$

we can apply cor. 4D p.56 of Rockafellar [1], obtaining boundedness for every a. □

Proof of (38). By contradiction, there exists $K \in$ Conv (X) such that (K, I) is not strongly Tykhonov well-posed. Now (K, I) is strongly Tykhonov well-posed with solution x_0 iff $(K - x_0, J)$ does with solution 0, where
$$J(u) = I(u + x_0) - I(x_0).$$

Then, without loss of generality, by (35) we can assume that

(53)
$$\arg \min(K, I) = 0, \inf I(K) = 0.$$

By ill - posedness, there exists $y \in K$ such that $y \neq 0$. Let $u \in X^*$ be such that $\|u\| = 1, < u, y >= \|y\|$. Set

$$A_j = \{x \in K :< u, x >\geq 1/j^2\},$$

a nonempty closed convex set, since $y/\|y\|j^2$ belongs to it. By (53)

(54) $$v_j = \inf I(A_j) \geq 0,$$

and $v_j = 0$ implies $0 \in A_j$ by (35) and (53), a contradiction. Therefore $v_j > 0$ for all j. Ill - posedness gives some minimizing sequence x_j for (K, I) which is bounded and does not converge strongly to 0 (if x_j is unbounded, then by (53) $x_j/\|x_j\| + (1 - 1/\|x_j\|)0$ is a bounded minimizing sequence). Thus $x_j \rightharpoonup x \in K$ for some subsequence. Therefore

$$0 = \liminf I(x_j) \geq I(x), \text{ hence } x = 0 \text{ by (53)}.$$

Then by (35), for the original sequence

$$x_j \rightharpoonup 0, \|x_j\| \not\rightarrow 0,$$

so for a subsequence $\|x_j\| \geq a > 0$. By replacing x_j by $ax_j/\|x_j\| \in K$, and (if necessary) y by a suitable point between 0 and y itself (since K is convex and contains 0) we find a subsequence y_n of x_j such that y_n is a minimizing sequence for (K, I), moreover for every n

(55) $$| < u, y_n > | \leq 1/n^2,$$

(56) $$\|y\| = \|y_n\| = a > 0,$$

(57) $$I(y_n) < v_n.$$

Now pick $u_n \in X^*$ such that

(58) $$\|u_n\| = 1, \ < u_n, y_n >= a,$$

and consider

$$T_n = \{x \in K : \|x\| \leq a, < u, x > +1/n < u_n, x >\geq a/n+ < u, y_n >\},$$
$$n = 1, 2, 3, \cdots, \ T_0 = \{x \in K : \|x\| \leq a, < u, x >\geq 0\}.$$

Then (56) and (58) yield $y_n \in T_n$ for every $n \geq 1$. For the time being, assume the following

20 Lemma $H - \lim T_n = T_0$.

End of proof of (38). Let $z_n = \arg \min(T_n, I)$. Since $0 \in T_0, z_0 = 0$. Moreover by (54) $I(z_n) \leq I(y_n) < v_n$, so $z_n \notin A_n$ by (54) again. Then

(59) $$< u, z_n >< 1/n^2.$$

Since $z_n \in T_n$, by (55) and (59)

$$< u_n, z_n >\geq a + n < u, y_n > -n < u, z_n >\geq a - 2/n,$$

yielding $z_n \not\to 0$, against Hadamard well - posedness. □

Proof of Lemma 20. By replacing y_n by y_n/a and K by K/a, we can assume $a = 1$. Now we show that for every $\varepsilon \in (0, 2)$ and every n sufficiently large, for every $x \in T_0$ there exists $\omega \in T_n$ such that

$$\|x - \omega\| \leq \varepsilon.$$

To this aim take $y \in K, y \neq 0$, whose existence is assured by ill - posedness of (X, I) and (53). We will obtain

$$\omega = (\bar{x} + b_n y)/(1 + b_n)$$

for suitable $\bar{x} \in K$ and $b_n > 0$. Then $\omega \in K$ by convexity. We take

$$\bar{x} = x \text{ if } \|x\| \leq 1 - \varepsilon/2, \bar{x} = (1 - \varepsilon/2)x \text{ if } \|x\| > 1 - \varepsilon/2.$$

It follows that $\bar{x} \in K$, since $0 \in K$. Remembering that $\|y\| = 1$ we have

$$\|x - \omega\| \leq (b_n + \varepsilon/2)/(1 + b_n) + b_n/(1 + b_n) \leq \varepsilon$$

if $b_n \leq \varepsilon/4$. Then $< u, \omega > +(1/n) < u_n, \omega > \geq 1/n+ < u, y_n >$ provided

$$(< u_n, \bar{x} > /n + b_n + b_n < u_n, y > /n)/(1 + b_n) \geq 1/n+ < u, y_n >$$

(given that $< u, y > = 1$ and $< u, \bar{x} > \geq 0$). Thus $\omega \in T_n$ provided

$$\frac{b_n}{1 + b_n} \geq [1/n+ < u, y_n > -(1/n)(\frac{< u_n, \bar{x} >}{1 + b_n} + \frac{b_n}{1 + b_n} < u_n, y >)].$$

Remembering (58) and that $\|\bar{x}\| \leq 1$, we see that $\omega \in T_n$ provided we choose $b_n > 0$ such that

$$< u, y_n > +3/n \leq b_n/(1 + b_n) \leq \varepsilon/4$$

which is feasible for all n sufficiently large due to (55). To end the proof, let us show that for every $\varepsilon \in (0, 2)$ and every n sufficiently large, given $\omega \in T_n$ there exists $x \in T_0$ such that

$$\|x - \omega\| \leq \varepsilon.$$

Take $x = \omega$ if $< u, \omega > \geq 0$. When $< u, \omega > < 0$ we have

$$- < u, \omega > \leq (1/n) < u_n, \omega > -1/n- < u, y_n > \leq \varepsilon/2$$

for n sufficiently large, independently of ω, because $\|u_n\| = 1$ and (55). Then take

$$x = \frac{\omega- < u, \omega > y}{1- < u, \omega >} \in K.$$

Then $\|x\| \leq 1, < u, x > = 0$, hence $x \in T_0$, moreover

$$\|x - \omega\| = \|\frac{< u, \omega > \omega- < u, \omega > y}{1- < u, \omega >}\| \leq \varepsilon.$$

□

Theorem 14. Write $I_n \to I$ iff we have uniform convergence on bounded sets. We need the following lemmas.

21 Lemma Let $I \in C_0(X)$ be bounded from below. If for some $c > 0$

(60) $c - $ arg min (X, I) is bounded

then $I_n \to I$ implies that for every $b > 0$ and for n sufficiently large, $b - $ arg min(X, I_n) are uniformly bounded and

$$b - \text{arg min } (X, I_n) \subset 2b - \text{arg min } (X, I).$$

22 Lemma Let Y be a metric space and $I : Y \to (-\infty, +\infty]$ a proper function. Then (Y, I) is well - posed in the generalized sense iff arg min (Y, I) is nonempty and compact, and

(61) for every $\varepsilon > 0$ there exists $\delta > 0$ such that $\delta - $ arg min $(Y, I) \subset [$ arg min $(Y, I)]^\varepsilon$.

Taking the above lemmas for granted, assume (40). From well - posedness there exists $b > 0$ (arbitrarily small) such that (60) obtains, otherwise some minimizing sequence for (X, I) without cluster points would exist. By lemma 21 and reflexivity of X we get arg min$(X, I_n) \neq \emptyset$ for large n. The conclusion then follows from lemmas 21 and 22, since (39) holds.

Conversely, assume (41). Arguing by contradiction, from lemma 22 there exist $\varepsilon > 0$ and some sequence x_n such that

$$I(x_n) \leq \inf I(X) + 1/n^2, \text{ dist } [x_n, \text{arg min}(X, I)] \geq 2\varepsilon.$$

We can assume that x_n is bounded (if $x_0 \in $ arg min(X, I) and $\|x_n - x_0\| \to +\infty$, then the sequence $x_n/\|x_n - x_0\| + (1 - 1/\|x_n - x_0\|)x_0$ is bounded, and minimizing by the convexity of I). Put

$$I_n(x) = I(x) + (1/n)\|x - x_n\|^2, u_n = \text{arg min}(X, I_n).$$

Then $I_n \to I$ and

$$(1/n)\|u_n - x_n\|^2 \leq I_n(x_n) - I(u_n) \leq \inf I(X) + 1/n^2 - I(u_n) \leq 1/n^2,$$

whence $u_n - x_n \to 0$. This yields

$$\text{dist } [u_n, \text{arg min}(X, I)] \geq \varepsilon$$

for sufficiently large n, contradicting (39).   □

Proof of lemma 21. From Rockafellar [1, cor. 4 D p. 56], we know that $\varepsilon - $ arg min (X, I) is bounded for every $\varepsilon > 0$. Take $x_0 \in $ arg min(X, I). Replacing $I(x)$ by $I(x + x_0) - I(x_0)$ we can assume, without loss of generality, that $I(0) = 0 = \inf I(X)$. Let for some $h > 0$

(62) $1 - \text{arg min}(X, I) \subset B(0, h/2)$.

Step 1. *Inf $I_n(X) > -\infty$ for all large n.*
Arguing by contradiction, let x_n be such that

$$I_n(x_n) \leq -n.$$

If x_n is bounded, then $I_n(x_n) - I(x_n) \to 0$, whence $I(x_n) \to -\infty$, contradicting weak sequential lower semicontinuity of I. If $\|x_n\| \to +\infty$ for some subsequence, let $y_n = hx_n/\|x_n\|$. Then by convexity $I_n(y_n) \le (1 - h/\|x_n\|)I_n(0)$, hence

$$\limsup I_n(y_n) \le I(0) = 0$$

and by the boundedness of y_n, $\limsup I(y_n) \le 0$. But this is impossible since $\|y_n\| = h$ and $I(y_n) > 1$ by (62).

Step 2. *For every $\varepsilon > 0$ there exist h and p such that*

$$\varepsilon - arg\ min\ (X, I_n) \subset B(0, h), n \ge p.$$

If $\varepsilon - arg\ min(X, I_n)$ were not equibounded for large n and some $\varepsilon > 0$, then (for some subsequence of I_n) there would exist x_n such that

(63) $$\|x_n\| \ge n,\ I_n(x_n) \le \inf I_n(X) + \varepsilon.$$

Given $a > 0$ let $y \in X$ be such that

$$I(y) \le \inf I(X) + a.$$

Then for n sufficiently large

$$\inf I_n(X) \le \inf I(X) + 2a$$

hence

$$\limsup \inf I_n(X) \le \inf I(X),$$

whence by (63), for sufficiently large n

(64) $$I_n(x_n) \le \inf I(X) + 2\varepsilon.$$

Put

$$y_n = hx_n/\|x_n\|.$$

Then by (62) and convexity, $1 < I(y_n) < +\infty$, hence by uniform convergence $I_n(y_n) < +\infty$ for n sufficiently large, and $\liminf I_n(y_n) \ge 1$. Thus by convexity

$$I_n(y_n) \le (h/\|x_n\|)I_n(x_n) + (1 - h/\|x_n\|)I_n(0)$$

hence by (64)

$$I_n(0) \ge (\|x_n\|/(\|x_n\| - h))[I_n(y_n) - (h/\|x_n\|)I_n(x_n)] \ge (\|x_n\|/(\|x_n\| - h))[I_n(y_n)$$
$$- (h/\|x_n\|)(\inf I(X) + 2\varepsilon)],$$

whence $\liminf I_n(0) \ge 1$, contradicting $I_n(0) \to I(0) = 0$.

End of the proof of lemma 21. Given $b > 0$, let h be as in step 2 for $\varepsilon = b$. Then for sufficiently large n we have

$$\sup\{|I_n(x) - I(x)| : x \in B(0, h) \cap \ \mathrm{dom}\ I\} \le b.$$

Now let $x \in b - arg\ min\ (X, I_n)$ for such n. Then by step 2

$$I(x) \leq I_n(x) + b \leq \inf I_n(X) + 2b.$$

$\square$

Proof of lemma 22. Assume (61) and let x_n be a minimizing sequence for (Y, I). Then for every $\varepsilon > 0$ and n sufficiently large

$$x_n \in [\arg \min (Y, I)]^\varepsilon.$$

There exist $u_n \in \arg \min (Y, I)$ such that $d(x_n, u_n) \leq \varepsilon$, where d is the distance in Y. By compactness, x_n has subsequences converging to some $x \in \arg \min(Y, I)$. If conversely (Y, I) is well - posed in the generalized sense, then $\arg \min (Y, I)$ is compact nonempty by proposition I.36. If for some $\varepsilon > 0$ and some minimizing sequence x_n to (Y, I)

$$\mathrm{dist} \, [x_n, \arg \min (Y, I)] \geq \varepsilon,$$

this would contradict the existence of some subsequence of x_n which converges towards some point in $\arg \min (Y, I)$. $\square$

From the proof of theorem 13 we may guess that in (37) it suffices to assume strong Tykhonov well - posedness of (K, I) only for every closed affine halfspace K. In the light of theorem 4, this is not too surprising. The subclass of Conv (X) made of all closed halfspaces determines strong Tykhonov well - posedness, as shown by

23 Proposition *Let X be reflexive and $I : X \to (-\infty, +\infty)$ be convex and continuous. If (H, I) is strongly Tykhonov well - posed for every closed affine halfspace H, then $(Conv \, (X), I)$ is strongly Tykhonov well - posed.*

Proof. We start showing that (X, I) is Tykhonov well - posed. Fix $u \neq 0$ in X^* and set

$$H^+ = \{x \in X :< u, x > \geq 0\}, H^- = \{x \in H :< u, x > \leq 0\},$$
$$x^+ = \arg \min (H^+, I), x^- = \arg \min (H^-, I).$$

Case 1: $x^- = x^+$. Then such a point equals $\arg \min(X, I)$ and well -posedness obtains.

Case 2: $x^- \neq x^+$. If $I(x^-) = I(x^+)$ then I is constant on the segment joining x^- and x^+, contradicting well - posedness. If e.g. $I(x^+) < I(x^-)$, then $x^+ = \arg \min (X, I)$, and we obtain again well - posedness. So (X, I) is strongly Tykhonov well - posed. Now fix $K \in$ Conv (X). Let us prove Tykhonov well - posedness of (K, I). Put

$$x_0 = \arg \min (X, I).$$

The conclusion is obvious if $x_0 \in K$. Let $x_0 \notin K$. By lemma 19, every minimizing sequence to (K, I) is bounded, therefore $\arg \min (K, I) \neq \emptyset$. Take any $y \in \arg \min (K, I)$. By well - posedness $I(y) > I(x_0)$. Set $v = I(y)$. By continuity, $x_0 \in$ int sub lev (I, v). As in the proof of theorem 13, we can separate K from sub lev (I, v) by a hyperplane $< u, x >= c$, so that

$$\mathrm{sub} \ \mathrm{lev} \, (I, v) \subset H^- = \{x \in X :< u, x > \leq c\},$$

$$K \subset H^+ = \{x \in X :< u,\ x > \geq c\},$$

$$< u, \omega > < c \text{ whenever } \omega \in \text{int sub lev } (I, v).$$

Therefore y minimizes I on H^+, and well - posedness of (H^+, I) implies well - posedness of (K, I). $\qquad\qquad\qquad\qquad\qquad\qquad\qquad\qquad\qquad\qquad\qquad\qquad\qquad$ $\square$

Theorem 17. For subsets A_n, A of X or $X \times R$ write

$$A_n \to A$$

iff loc.$H - \lim A_n = A$, and for extended real - valued functions f_n on X write

$$f_n \to f$$

iff loc. $H - \lim \text{epi } f_n = \text{epi} f$.

It is obvious that (43) implies Tykhonov well - posedness. Conversely assume well - posedness of (A, I). We shall admit

Step 1: $I_n + indA_n \to I + indA$. For a proof see Beer - Lucchetti [2, th. 3.6].

Now let $f, f_n \in C_0(X)$ be such that (X, f) is Tykhonov well - posed and $f_n \to f$.

Step 2: $val(X, f_n) \to val(X, f)$. Since $f_n \to f$ implies $M - \lim f_n = f$, from IV. (12) we get

$$\limsup \text{ val } (X, f_n) \leq_j \text{val } (X, f) = v.$$

Arguing by contradiction, let $\varepsilon > 0$ be such that for some subsequence

$$\text{val } (X, f_n) < v - 3\varepsilon.$$

Pick $u_n \in X$ such that

(65) $$f_n(u_n) < v - 2\varepsilon.$$

We need the following

Sublevel lemma. For every $t > v$ we have

$$\text{sub lev } (f_n, t) \to \text{ sub lev } (f, t).$$

For a proof see Beer - Lucchetti [3, th.3.6]. By the sublevel lemma

(66) $$\text{sub lev } (f_n, v + 2\varepsilon) \to \text{ sub lev } (f, v + 2\varepsilon).$$

By well - posedness and theorem I. 11 we see that sub lev $(f, v + 2\varepsilon)$ is bounded (for a possibly smaller ε). We shall prove that

(67) for some $r > |v| + 3\varepsilon$ and all sufficiently large n

$$\text{sub lev } (f_n, v + 2\varepsilon) \subset rB.$$

Now

(68) $$h(r, \text{ epi } f, \text{ epi } f_n) < \varepsilon \text{ for all } n \text{ sufficiently large.}$$

Taking (67) for granted we have from (65)

$$(u_n, v - 2\varepsilon) \in (rB \times [-r, r]) \cap \text{ epi } f_n.$$

It follows from (68) that there exists $(x, t) \in \text{ epi } f$ with $\|(x, t) - (u_n, v - 2\varepsilon)\| < \varepsilon$, hence $f(x) \leq t \leq v - \varepsilon$, a contradiction. The proof of step 2 is ended by proving (67). Put

$$C_n = \text{ sub lev } (I_n, v + 2\varepsilon), \quad C = \text{ sub lev } (I, v + 2\varepsilon).$$

Then $C_n \to C$ by (66) and C is bounded, hence for some $r > 0$

$$C \subset rB.$$

Fix $u_0 \in C$. Convergence of C_n yields

(69) $\quad \sup\{| \text{ dist } (x, C_n) - \text{ dist } (x, C)| : \|x\| \leq r + 3\} < 1$ for all sufficiently large n.

We claim that for such n
$$C_n \subset (r + 2)B.$$
Suppose not: then there exists $u_n \in C_n$ with $\|u_n\| > r + 2$. By (69), $\text{dist}(u_0, C_n) < 1$, hence there exists $z_n \in C_n$ with $\|z_n\| < 1 + r$. As a result, some convex combination y_n of u_n and z_n has norm $r + 2$, hence

$$| \text{ dist } (y_n, C_n) - \text{ dist } (y_n, C)| = \text{ dist } (y_n, C) \geq 2$$

contradicting (69). The proof of step 2 is now complete.

Step 3: $\varepsilon_n \geq 0, x_n \in \varepsilon_n - arg \ min \ (X, f_n), \varepsilon_n \to 0 \Rightarrow x_n \to arg \ min \ (X, f)$.

Let $x_0 = arg \ min \ (X, f)$ and $\varepsilon > 0$. By theorem I.11 there exists some $b > 0$ such that

(70) $$\text{diam } [b - arg \ min \ (X, f)] < \varepsilon/2.$$

From the sublevel lemma and (67) we get Hausdorff convergence of the sublevel sets, hence for all n sufficiently large

(71) $$\text{haus } [\text{sub } \text{ lev } (f_n, b + \hat{v}), b - arg \ min \ (X, f)] < \varepsilon/2.$$

Moreover from step 2 and n sufficiently large we have

$$\varepsilon_n < b/2, \quad \text{val } (X, f_n) \leq v + b/2.$$

Hence

$$f_n(x_n) \leq \text{val } (X, f_n) + \varepsilon_n \leq v + b.$$

Put

$$D_n = \text{ sub lev } (f_n, v + b), D = b - arg \ min \ (X, f).$$

Given $a > 0$, by definition of Hausdorff distance we get for all n, $a + \text{ haus } (D_n, D) \geq a + \text{ dist } (x_n, D) \geq \|x_n - u_n\|$ for some $u_n \in D$, hence

$$\|x_n - x_0\| \leq \|x_n - u_n\| + \|u_n - x_0\| \leq \text{ haus } (D_n, D) + \text{ diam } D$$

and finally, remembering (70) and (71),

$$\|x_n - x_0\| \leq \varepsilon \text{ for all sufficiently large } n.$$

Steps 2 and 3 establish a particular case of theorem 17 (i.e. $A_n = X$ for all n). From step 1 we can take $f_n = I_n + \text{ ind } A_n$, and get the required conclusions from steps 2 and 3. $\square$

Corollary 18. Let $x_0 = \arg \min(A, I)$ and consider any generalized minimizing sequence x_n. Thus dist $(x_n, A) \leq c_n$ where $c_n \to 0$. Clearly loc. $H - \lim(A + c_n B) = A$. We apply theorem 17 with $I_n = I$ and $A_n = A + c_n B$, obtaining by (43)

$$\text{val } (A + c_n B, I) \to \text{ val } (A, I) = I(x_0).$$

Thus (again by (43)) $x_n \to x_0$ since $x_n \in A_n$ and $I(x_n) \to I(x_0)$. $\qquad\qquad\square$

Section 4.   Convex Problems and Variational Inequalities.

Standing assumptions: X is a real Banach space, $K \in$ Conv $(X), I : X \to (-\infty, +\infty]$ is strongly lower semicontinuous on X and proper on K.

From theorem I. 11 we know that strong Tykhonov well - posedness of (K, I) amounts to

$$\text{diam } [\varepsilon - \arg \ \min(K, I)] \to 0 \text{ as } \varepsilon \to 0,$$

provided I is bounded from below on K. From cor. 11 of Ekeland [1], applied to $I(\cdot) +$ ind $(K, \cdot)$, for every $\varepsilon > 0$ and $x \in \varepsilon - \arg \ \min(K, I)$ there exists $\bar{x}$ there such that

(72) $$\|\bar{x} - x\| \leq \sqrt{\varepsilon}, I(\bar{x}) \leq I(y) + \sqrt{\varepsilon}\|\bar{x} - y\| \text{ for every } y \in K.$$

Therefore it is natural to introduce the set

$$Z(\varepsilon) = \{x \in K : I(x) \leq I(y) + \varepsilon\|x - y\| \text{ for every } y \in K\}.$$

Then by (72)

(73) $$\varepsilon^2 - \arg \ \min \ (K, I) \subset \ ^\varepsilon Z(\varepsilon), \varepsilon > 0.$$

If diam $Z(\varepsilon) \to 0$ as $\varepsilon \to 0$, by (73) we get strong Tykhonov well - posedness of (K, I).

Now let I be Gâteaux differentiable and bounded from below on K. If we apply again cor. 11 of Ekeland [1], we get the following. For every $u \in \varepsilon^2 - \arg \ \min \ (K, I)$ there exists some x there such that $\|u - x\| \leq \varepsilon$ and

$$I[x + t(y - x)] \geq I(x) - \varepsilon t \|y - x\|, 0 < t < 1, y \in K.$$

Hence $x \in T(\varepsilon)$ and

(74) $$\varepsilon^2 - \arg \ \min \ (K, I) \subset \ ^\varepsilon T(\varepsilon), \varepsilon > 0,$$

where

(75) $$T(\varepsilon) = \{x \in K :< \nabla I(x), x - y > \leq \varepsilon\|x - y\| \text{ for every } y \in K\}.$$

Then a further sufficient condition to strong Tykhonov well - posedness of (K, I) is that diam $T(\varepsilon) \to 0$ as $\varepsilon \to 0$. Thus the sets $Z(\varepsilon), T(\varepsilon)$ play a role in investigating Tykhonov well - posedness. Moreover, under convexity of I, the above conditions are necessary to well - posedness, as shown by

24 Theorem *If $Z(\varepsilon) \neq \emptyset$ for all $\varepsilon > 0$ and*

$$diam\ Z(\varepsilon) \to 0\ as\ \varepsilon \to 0$$

then (K, I) is strongly Tykhonov well - posed. The converse holds provided I is convex on X.

25 Theorem *Let I be bounded from below and Gâteaux differentiable on K. Then $T(\varepsilon) \neq \emptyset$ for every $\varepsilon > 0$. If*

$$diam\ T(\varepsilon) \to 0\ as\ \varepsilon \to 0$$

then (K, I) is strongly Tykhonov well - posed. The converse holds if moreover I is convex on X.

A natural generalization of the preceding theorems leads to

26 Theorem *Let $I, J : X \to (-\infty, +\infty]$ be convex and lower semicontinuous, J Gâteaux differentiable on K. Set*

(76) $\quad H(\varepsilon) = \{x \in K :< \bigtriangledown J(x), x - y > +I(x) \leq I(y) + \varepsilon\|x - y\|\ for\ all\ y \in K\}.$

If $(K, J + I)$ is strongly Tykhonov well - posed on K, then

$$H(\varepsilon) \neq \emptyset\ for\ all\ \varepsilon;\ diam\ H(\varepsilon) \to 0\ as\ \varepsilon \to 0.$$

The converse holds if moreover $I + J$ is bounded from below on K.

Theorem 26 contains as a particular case both Theorem 24 (when $J = 0$) and 25 (when $I = 0$) restricted to convex problems.

As well known, minimizing $J + I$ on K under the assumptions of theorem 26 is equivalent to solve the variational inequality

(77) $\qquad\qquad < \bigtriangledown J(x), x - y > +I(x) \leq I(y)\ for\ every\ y \in K$

in the unknown $x \in K$ (prop. 2.2 p. 37 of Ekeland - Temam [1]). It is therefore quite natural to define a suitable notion of well - posedness for (77) which is equivalent to strong Tykhonov well - posedness of $(K, I + J)$, as suggested by theorem 26.

Given $u \in X^*, A : X \to X^*, I$ and K as before, we consider the following variational inequality:

(78) $\qquad\qquad\qquad$ find $x \in K$ such that

$$< Ax - u, x - y > +I(x) \leq I(y)\ for\ every\ y \in K.$$

We shall say that *the variational inequality (78) is well - posed* iff

$$G(\varepsilon) \neq \emptyset\ for\ all\ \varepsilon > 0\ and\ diam\ G(\varepsilon) \to 0\ as\ \varepsilon \to 0,$$

where

(79) $G(\varepsilon) = \{x \in K :< Ax - u, x - y > +I(x) \le I(y) + \varepsilon\|x - y\|$ for all $y \in K\}$.

The above definition gives immediately

27 Theorem *Under the assumptions of theorem 24, the variational inequality (78) with $A = \bigtriangledown J$ and $u = 0$ is well - posed iff $(K, I + J)$ is strongly Tykhonov well - posed.*

PROOFS.

Theorem 24. Suppose $Z(\varepsilon) \neq \emptyset$ and diam $Z(\varepsilon) \to 0$ as $\varepsilon \to 0$. From semicontinuity, $Z(\varepsilon)$ is closed for every ε. Therefore there exists some $x_0 \in \cap\{Z(\varepsilon) : \varepsilon > 0\}$ and then

$$I(x_0) \le I(y) + \varepsilon\|x_0 - y\|, y \in K, \varepsilon > 0.$$

Thus $x_0 \in$ arg min (K, I) and I is bounded from below. Then the conclusion follows from theorem I.11 and (73). Conversely assume (K, I) strongly Tykhonov well - posed. Then $x_0 =$ arg min $(K, I) \in Z(\varepsilon)$ $\forall \varepsilon > 0$. By replacing K by $K - x_0$ and $I(u)$ by $I(u + x_0) - I(x_0)$ we can assume $x_0 = 0, I(x_0) = 0$. Arguing by contradiction, let us assume that there exist $a > 0$ and $\varepsilon_n \to 0$ such that diam $Z(\varepsilon_n) > 2a$ for every n. Then we can find $x_n \in Z(\varepsilon_n)$ such that $\|x_n\| \ge a$. Put $m = \inf\{I(x) : x \in K, \|x\| \ge a\}$. Then $m > 0$ by well - posedness. The existence of $z \in K, \|z\| \ge a$ such that $I(z) < m\|z\|/a$ implies by convexity

$$m \le I[az/\|z\| + (1 - (a/\|z\|))0] \le (a/\|z\|)I(z) < m$$

which is a contradiction. Hence $I(x_n) \ge m\|x_n\|/a$, moreover $I(x_n) \le \varepsilon_n\|x_n\|$ since $x_n \in Z(\varepsilon_n)$, yielding $\varepsilon_n \ge m/a$ for every n, a contradiction.   □

Theorem 25. Well - posedness follows from (74), as already shown. Conversely, assume well -posedness. Then $x_0 =$ arg min (K, I) solves the variational inequality $< \bigtriangledown I(x_0), y - x_0 > \ge 0, y \in K$ (prop. 2.2 p.37 of Ekeland - Temam [1]), hence $x_0 \in T(\varepsilon)$ for all ε (a further proof that $T(\varepsilon) \neq \emptyset$). As in the previous proof we may well assume $x_0 = 0, I(x_0) = 0$. Arguing by contradiction, suppose there exist $a > 0$ and $\varepsilon_n \to 0$ with diam $T(\varepsilon_n) > 2a$. Then some $x_n \in T(\varepsilon_n)$ can be found with $\|x_n\| \ge a$. Now

$$< \bigtriangledown I(ax_n/\|x_n\|), x_n > \le < \bigtriangledown I(x_n), x_n >$$

since by monotonicity of $\bigtriangledown I$

$$0 \le < \bigtriangledown I(x_n) - \bigtriangledown I(ax_n/\|x_n\|), x_n > (1 - a/\|x_n\|).$$

Thus by convexity

$$I(ax_n/\|x_n\|) - I(0) \le < \bigtriangledown I(ax_n/\|x_n\|), ax_n/\|x_n\| >$$
$$\le (a/\|x_n\|) < \bigtriangledown I(x_n), x_n > \le a\varepsilon_n$$

because of (75) with $y = 0$. It follows that $ax_n/\|x_n\| \in a\varepsilon_n -$ arg min (K, I), whose diameter tends to 0 according to theorem I.11, a contradiction.   □

Theorem 26. Assume strong Tykhonov well - posedness of $(K, I + J)$ and set $x_0 = $ arg min $(K, I + J)$. By prop. 2.2 p. 37 of Ekeland - Temam [1], $x_0 \in H(\varepsilon)$ for every $\varepsilon > 0$. By convexity of J, for all x and y in K

$$J(y) \geq J(x) - < \triangledown J(x), x - y >$$

therefore $H(\varepsilon) \subset Z(\varepsilon)$ (relative to $I + J$). Then theorem 24 gives the conclusion. Conversely assume $H(\varepsilon) \neq \emptyset$ and diam $H(\varepsilon) \to 0$ as $\varepsilon \to 0$. Then cor. 11 of Ekeland [1] can be applied, since $I + J$ is proper on K if I does. (Properness of I on K is not explicitly needed here, since if K contains at least two points and $I(x) = +\infty$ identically on K, then $K = H(\varepsilon)$ which gives a contradiction). So given $x \in \varepsilon^2 - $ arg min $(K, I + J)$ there exists some $z \in K \cap$ dom I such that $\|x - z\| \leq \varepsilon$, moreover if $0 < t < 1$ and $y \in K$

$$J(z) + I(z) \leq J[ty + (1 - t)z] + I[ty + (1 - t)z] + t\varepsilon \|y - z\|.$$

By convexity of I it follows that

$$(1/t)\{J(z) - J[z + t(y - z)]\} + I(z) \leq \varepsilon \|y - z\| + I(y).$$

Letting $t \to 0$ we have

$$< \triangledown J(z), z - y > + I(z) \leq I(y) + \varepsilon \|z - y\|$$

hence $z \in H(\varepsilon)$. This shows that

$$\varepsilon^2 - \text{arg min } (K, I + J) \subset {}^\varepsilon H(\varepsilon).$$

The conclusion then follows from theorem I.11.   □

Section 5.   Well-Posed Variational Inequalities.

In this section we consider the variational inequality (78) with $I = 0$; we will denote it as

$$(K, A, u).$$

Standing assumptions: $A : X \to X^*$ is a monotone mapping, $K \in $ Conv $(X), u \in X^*$.

As we shall see now, the notion of well - posed variational inequality (78), defined in section 4, links in a very significant way Tykhonov well - posedness for minimum problems to the classical notion of Hadamard well - posedness for linear equations involving monotone operators. This connection is one of the basic results of the theory.

28 Example. Suppose X is reflexive. Let A be hemicontinuous and strongly monotone. Then there is some $c > 0$ such that

$$< Ax - Ay, x - y > \geq c\|x - y\|^2 \text{ for every } x, \ y \in X.$$

Thus (K, A, u) has exactly one solution belonging to $G(\varepsilon)$ for every $\varepsilon > 0$ (see Baiocchi - Capelo [1, th.10.4 p. 234]). Therefore $G(\varepsilon) \neq \emptyset$. Given $x, z \in G(\varepsilon)$, by (79)

$$< Ax - u, x - z >\leq \varepsilon \|x - z\|, < Az - u, z - x >\leq \varepsilon \|x - z\|$$

hence

$$c\|x - z\|^2 \leq 2\varepsilon \|x - z\|$$

yielding diam $G(\varepsilon) \leq 2\varepsilon/c$. Therefore (K, A, u) is well - posed for every K and u.

Example. Let $A = \nabla J$, where J is as in theorem 24. Consider $H(\varepsilon)$ defined by (76) with J replaced by $J(\cdot)- < u, \cdot >$ and $I = 0$. Then $G(\varepsilon) = H(\varepsilon)$. Therefore (K, A, u) is well - posed provided that the assumptions of theorem 26 are fulfilled. Of course the variational inequality (78) is equivalent to the problem $(K, J(\cdot)- < u, \cdot >)$, which turns out to be strongly Tykhonov well - posed.

If A is hemicontinuous, then

$$G(\varepsilon) = \{x \in K :< Ay - u, x - y >\leq \varepsilon \|x - y\| \text{ for all } y \in K\}$$

as in Minty's lemma (Baiocchi - Capelo [1, th. 7.14 p.177]). By this representation we see that $G(\varepsilon)$ is closed for every ε. On the other hand the set of solutions to (K, A, u) is exactly $\cap\{G(\varepsilon) : \varepsilon > 0\}$. So if the variational inequality is well - posed, then the intersection of $G(\varepsilon)$ shrinks to a single point. Therefore

29 Theorem *If A is hemicontinuous and (K, A, u) is well - posed, then it has an unique solution.*

The converse is false, even if $G(\varepsilon)$ is nonempty and bounded. This may be seen from example I.18, remembering theorem 27. Consider X a separable Hilbert space with orthonormal basis e_n, and let
$u = 0, I(x) = \sum_{n=1}^{\infty} < x, e_n >^2 /n^2, K = $ closed unit ball of $X, A = \nabla I$.
However the converse of theorem 29 is true in the finite - dimensional setting.

30 Theorem *Let X be finite - dimensional and A hemicontinuous. If (K, A, u) has exactly one solution then it is well-posed.*

We digress briefly from our main purpose by remarking that boundedness of $G(\varepsilon)$ gives an existence theorem for variational inequalities.

31 Proposition *Let X be reflexive, A hemicontinuous on $K, G(\varepsilon) \neq \emptyset$ for all ε and bounded for at least some ε. Then (K, A, u) has solutions.*

We come back to the relationships between well - posedness of (K, A, u) and Hadamard well - posedness. A particularly significant special case obtains when X is a Hilbert space, $K = X$ and A is a continuous linear operator. Then (if $I = 0$) (78) is equivalent to the linear operator equation

$$(80) \qquad\qquad\qquad Ax = u.$$

Among other applications, (80) abstractly models many classical linear boundary - value problems of the mathematical physics. In this setting Hadamard firstly introduced his classical definition of well - posedness, which amounts to existence, uniqueness and continuous dependence of the solution x to (80) upon the datum u (see e.g. Courant - Hilbert [1, p.227]).

As far as (80) is concerned, the main result is a partial extension of theorem 13: well - posedness of (80) is equivalent to well - posedness in the classical sense of Hadamard.

32 Theorem *Let X be a real Hilbert space, A a bounded linear operator. The following are equivalent facts:*

(81) (K, A, u) *is well - posed for every* $K \in \text{Conv}(X)$;

(82) (K, A, u) *has exactly one solution for every* $K \in \text{Conv}(X)$;

(83) A *is coercive, i.e.* $< Ax, x > \geq c\|x\|^2$ *for all x and some constant $c > 0$.*

Consider now the particular instance when A is symmetric, i.e. $A = A^*$. Then (A, K, u) is equivalent to the minimum problem (K, I), where

$$I(x) = (1/2) < Ax, x > - < u, x >$$

(see Ekeland - Temam [1, prop. 2.1 p.35]). From theorem 32, for the above quadratic functional I on the Hilbert space X *the following are equivalent:*

(K, I) *is strongly Tykhonov well $-$ posed for all* $K \in \text{Conv}(X)$;

(K, I) *has exactly one solution for all* $K \in \text{Conv}(X)$;

A *is coercive.*

Since (X, A, u) amounts to (80) if A is bounded linear, its well - posedness is then equivalent to the isomorphic character of the operator A, i.e. well - posedness in the classical sense of Hadamard.

33 Theorem *Let X be a reflexive Banach space and A bounded linear. Well - posedness of (X, A, u_0) for some $u_0 \in X^*$ implies well - posedness of (X, A, u) for every $u \in X^*$, moreover A is an isomorphisms. Therefore (80) is well - posed in the classical sense of Hadamard. The converse is also true.*

A counterexample. Let $X = R^2$ with points $x = (x_1, x_2)$, and $A(x_1, x_2) = (-x_2, x_1)$. Then $(X, A, 0)$ is well - posed. But (K, A, u) is not if $u = (2, 0)$ and $K = \{x \in R^2 : x_2 \geq -1\}$. In fact (K, A, u) has no solution. Therefore well - posedness of a variational inequality on some given convex set does not entail existence of solutions on the convex subsets thereof. This is in striking contrast with the behavior of strongly Tykhonov well - posed convex minimum problems, defined by lower semicontinuous functions on reflexive spaces. In such a case existence of optimal solutions on subsets comes from boundedness of the minimizing sequences, by lemma 19.

Let X be a Hilbert space. A direct connection between Hadamard well - posedness of (80) and Tykhonov's may be obtained by considering

$$I(x) = < Ax, x >, x \in X.$$

If A is linear, bounded and one - to - one, then (80) is Hadamard well - posed iff (X, I) is strongly Tykhonov well - posed.
For, A has no continuous inverse iff there exists a sequence x_n such that

$$\|x_n\| = 1 \text{ and } Ax_n \to 0,$$

but then x_n is a minimizing sequence to (X, I), which does not converge strongly to $0 = \arg \min(X, I)$. Conversely, ill - posedness of (X, I) implies existence of a minimizing sequence u_n with $\|u_n\| \geq a > 0$. Thus

$$< A(u_n/\|u_n\|), u_n/\|u_n\| > \leq (1/a^2) \, I(u_n) \to 0$$

yielding

$$\inf\{< Au, u >: \|u\| = 1\} = 0,$$

hence no continuous inverse to A can exist. No symmetry assumption is imposed on A therefore (80) is not necessarily the Euler - Lagrange equation for (X, I). If A is not necessarily monotone, the same result holds by considering

$$I(x) = \|Ax\|^2, x \in X.$$

PROOFS.

Theorem 29. As we saw before, it suffices to show that $B(\varepsilon) = G(\varepsilon), \varepsilon > 0$, where

$$B(\varepsilon) = \{x \in K :< Ay - u, x - y > \leq \varepsilon\|x - y\| \text{ for every } y \in K\}.$$

Let $x \in G(\varepsilon)$. Then since A is monotone

$$< Ay - u, x - y > \leq < Ax - u, x - y > \leq \varepsilon\|x - y\|, y \in K$$

hence $x \in B(\varepsilon)$. If conversely $x \in B(\varepsilon)$, let $y \in K$ and $0 < t \leq 1$. Then

$$< A[x + t(y - x)] - u, x - y > \leq \varepsilon\|x - y\|.$$

Letting $t \to 0$, by hemicontinuity we get $x \in G(\varepsilon)$. $\qquad\qquad\square$

Remark. Let I be convex on X and Gâteaux differentiable on K. Then by the above proof we see that every $T(\varepsilon)$ given by (75) is closed, since

$$T(\varepsilon) = G(\varepsilon) = B(\varepsilon) \text{ if } A = \nabla I, u = 0.$$

If $T(\varepsilon) \neq \emptyset$ and diam $T(\varepsilon) \to 0$, it follows that there exists some point in $\cap T(\varepsilon)$. Such a point minimizes I on K. Therefore, adding in theorem 25 the assumptions of convexity and $T(\varepsilon) \neq \emptyset$, allows us to drop the boundedness assumption about I (∇I is then hemicontinuous since it is demicontinuous: see Asplund - Rockafellar [1, cor. 1 p.460]).

Theorem 30. Let x_0 be the unique solution to (K, A, u). Arguing by contradiction, assume that there exist $a > 0, \varepsilon_n \downarrow 0$ and points $x_n \in G(\varepsilon_n)$ such that $|x_n - x_0| \geq a$. If x_n were unbounded, then for some subsequence $|x_n| \to +\infty$. Put

$$\alpha_n = 1/|x_n - x_0|, z_n = x_0 + \alpha_n(x_n - x_0).$$

Then $z_n \in K$ for all n sufficiently large. For a further subsequence $z_n \to z \in K$ and $z \neq x_0$. If $y \in K$,

$$(84) \qquad < Ay - u, z - y > \; \leq \; |Ay - u||z - z_n| + < Ay - u, z_n - y > .$$

The last term in the right - hand side equals $(1 - \alpha_n) < Ay - u, x_0 - y >$ $+\alpha_n < Ay - u, x_n - y > .$ Thus, for n sufficiently large, (75) yields

$$(85) \qquad < Ay - u, z - y > \; \leq \; |Ay - u||z - z_n| + \alpha_n \varepsilon_n |x_n - y|,$$

since $< Ay - u, x_0 - y > \; \leq \; 0$ by Minty's lemma (Baiocchi - Capelo [1, th. 7.14]), and $x_n \in G(\varepsilon_n) = B(\varepsilon_n)$ as shown in the proof of theorem 29. Letting $n \to +\infty$ in (85) we see (again by Minty's lemma) that z is a solution, contradicting uniqueness. So x_n is bounded. Then $x_n \to \omega$ for some subsequence, and $\omega \in G(\varepsilon) = B(\varepsilon)$ for every $\varepsilon > 0$. Then ω solves the variational inequality. This is a contradiction since $|\omega - x_0| \geq a$. $\square$

Proposition 31. Assume that $G(\varepsilon)$ is bounded. Since $G(\cdot)$ decreases, given $x_n \in G(1/n)$ then for some subsequence $x_n \to x_0 \in K$. Given any $y \in K$, by monotonicity

$$< Ay - u, y - x_n > \; \geq \; < Ax_n - u, y - x_n >$$
$$\geq (-1/n)\|y - x_n\| \geq -(1/n)(\|y\| + c) \text{ for a suitable } c > 0.$$

Letting $n \to +\infty$ we get

$$< Ay - u, y - x_0 > \; \geq \; 0.$$

Minty's lemma (Baiocchi - Capelo [1, th. 7.14 p.177]) gives the conclusion. $\square$

Theorem 32. Obviously (81) $\Rightarrow$ (82) by theorem 29. We saw in example 28 that (83) $\Rightarrow$ (81). Thus we need only to prove that (82) $\Rightarrow$ (83). We shall use the following
34 Lemma Assume A positive, Y a finite - dimensional subspace of X, A not coercive on X. Then A is not coercive on $Y^\perp$.
Proof of lemma 34. There exists $x_n, \|x_n\| = 1$, such that

$$(86) \qquad\qquad < Ax_n, x_n > \; \to \; 0.$$

Given any subsequence of x_n, some further subsequence satisfies $x_n \to x_0$. Thus

$$0 \; \leq \; < Ax_0, x_0 > \; \leq \; \liminf < Ax_n, x_n > \; = \; 0$$

which entails $x_0 = 0$ by positivity. Then $x_n \to 0$ for the original sequence. Write

$$x_n = u_n + v_n, u_n \in Y, v_n \in Y^\perp;$$

then $u_n \to 0, v_n \to 0$, since orthogonal projections are linear and continuous. It follows that $\|v_n\| \to 1$ since $1 = \|x_n\|^2 = \|u_n\|^2 + \|v_n\|^2$. Then $\|v_n\| \geq 1/2$ for all large n. The inequality

$$< Az, z > \geq a\|z\|^2 \text{ for all } z \in Y^\perp, \text{ some } a > 0,$$

implies

$$< Ax_n, x_n > \geq a\|v_n\|^2 - \|Au_n\|\|v_n\| - \|Av_n\|\|u_n\| \geq a/8$$

for all large n, contradicting (86). $\qquad\qquad\qquad\qquad\qquad\square$

End of proof of theorem 32. Let us show that A is positive. Assume that, for some $z \neq 0, < Az, z >= 0$. Let $K = $ sp $\{z\}$ and denote by tz the solution to (K, A, u). Then for every real numbers

$$< A(tz) - u, (t - s)z > \leq 0$$

yielding $< u, z >= 0$. So every point in K solves the variational inequality, contradicting (82). Therefore A is positive. Arguing by contradiction, suppose A not coercive. Let $Y_1 = $ sp $\{u\}$. Then, by lemma 34, A is not coercive on $Y_1^\perp$. Hence there exists $\omega_1 \in Y_1^\perp$ such that $\|\omega_1\| = 1$ and $0 < < A\omega_1, \omega_1 > < 1$. Again by lemma 34 we see that A is not coercive on $Y_2^\perp$ where $Y_2 = $ sp $\{u, \omega_1, A\omega_1, A^*\omega_1\}$. So we find $\omega_2 \in Y_2^\perp$ such that $\|\omega_2\| = 1$ and

$$0 < < A\omega_2, \omega_2 > < 1/2.$$

In this way we consider

$$Y_n = \text{ sp } \{u, \omega_1, ..., \omega_{n-1}, A\omega_1, ..., A\omega_{n-1}, A^*\omega_1, ..., A^*\omega_{n-1}\}.$$

There exists some $\omega_n \in Y_n^\perp, \|\omega_n\| = 1$ and

$$(87) \qquad\qquad 0 < t_n =< A\omega_n, \omega_n > < 1/n.$$

Of course this procedure ends after a finite number of steps if X is finite - dimensional. Now set

$$Y = \text{ cl sp } \{\omega_n : n = 1, 2, ...\}.$$

By construction, ω_n is a basis in Y. Then define $B : Y \to Y$ as the linear bounded map such that

$$B\omega_n = t_n\omega_n, t_n \text{ given by (78)}.$$

Then by (87) we see that B is well defined and nonnegative. By (87) $\sum_{n=1}^\infty t_n\omega_n$ converges (in X). Now consider

$$K = \{x \in Y :< x, \sum_{n=1}^\infty t_n\omega_n >= 1\}.$$

Then the variational inequality (K, A, u) is equivalent to $(K, A, 0)$, since $< \omega_n, u >= 0$ for every n, whence $< u, y >= 0$ for every $y \in Y$. On the other hand, $x \in Y$ implies that for every j

$$(88) \qquad\qquad < Ax, \omega_j >= \sum_{n=1}^\infty c_n < A\omega_n, \omega_j >= c_j t_j =< Bx, \omega_j >$$

provided $x = \sum_{n=1}^\infty c_n\omega_n$. So $(K, B, 0)$ has some solution $\bar{x}$ by (82). Then

$$(89) \qquad\qquad < B\bar{x}, \bar{x} - y > \leq 0, y \in K.$$

Taking $y = \omega_n/t_n$ we get

$$< B\bar{x}, \bar{x} > \leq (1/t_n) < B\bar{x}, \omega_n > .$$

Writing $\bar{x} = \sum_{n=1}^{\infty} c_j\omega_j$ we obtain

$$< B\bar{x}, \omega_n > = c_n t_n,$$

thus by (88) and (89)

$$0 \leq < A\bar{x}, \bar{x} > = < B\bar{x}, \bar{x} > \leq c_n \text{ for every } n.$$

Since $c_n \to 0$ we obtain a contradiction because $0 \notin K$. □

Remark. If we replace Conv (X) by the set of all closed affine subspaces of X in theorem 32, then its conclusion holds true (indeed, in the above proof K is an affine subspace). Compare with proposition 23, whose conclusion follows from such an extension of theorem 32 in the particular case of a symmetric operator A for

$$I(x) = (1/2) < Ax, x > - < u, x > .$$

Theorem 33. By theorem 29, there exists an unique solution x_0 to (X, A, u_0). We verify that

(90) $$\|Ax\| \geq c\|x\| \text{ for all } x \in X \text{ and some } c > 0,$$

or equivalently that

$$\|Ax - u_0\| \geq c \text{ for all } x \text{ such that } \|x - x_0\| = 1.$$

If not, there exists a sequence x_n such that

$$\|Ax_n - u_0\| \leq 1/n, \ \|x_n - x_0\| = 1.$$

Therefore $x_0, x_n \in G(1/n)$, against well - posedness. Thus (90) is proved, hence A is one - to - one. We need the following

Lemma. The range of A is dense.

Assuming this for a moment, we fix $u \in X^*$ and consider

$$G(\varepsilon) = \{x \in X : \|Ax - u\| \leq \varepsilon\}$$

which is nonempty for all ε. If $x_1, x_2 \in G(\varepsilon)$, from (90) one obtains $c\|x_1 - x_2\| \leq 2\varepsilon$, hence diam $G(\varepsilon) \to 0$ as $\varepsilon \to 0$. This shows well -posedness of (X, A, u) for every u, hence A is an isomorphism and (80) has an unique solution (for every u) which depends continuously upon u. The converse is obvious. □

Proof of the lemma. Suppose there exist $u \in X^*, \varepsilon > 0$ such that the open ball V of center u and radius ε does not meet $A(X)$. By the Hahn - Banach theorem (Brezis [2, th.I.6 p.5]) we find some $y \in X, y \neq 0$, such that for a suitable constant k

$$< Ax, y > \leq k \leq < v, y >$$

for every $x \in X, v \in V$. Then $< Ax, y > = 0, x \in X$. Hence given $t \in R$

$$0 \leq < A(ty + x), ty + x > = t < Ay, x > + < Ax, x >$$

which yields $< Ay, x > = 0, x \in X$. Then $Ay = 0$, yielding $y = 0$, a contradiction. □

Section 6.   Notes and bibliographical remarks.

Both concepts of well - posedness isolate properties, which are basic to (and, up to some extend, were motivated by) the numerical computation of optimal solutions to minimization problems: see e.g. Daniel [1] for early results. In a broader setting, see the (Hadamard - like) definition of well -posed computing problems in Isaacson - Keller [1, p.22].

To section 1. E - spaces (somewhere called strongly convex spaces) were introduced and characterized by Fan - Glicksberg [1]. Condition (2) was introduced by Kadec [1], and generalized to a (not topological) convergence setting by Deutsch [1] for best approximation problems. Theorem 2 and related properties of E - spaces may be found in Holmes [1]. Theorem 2, strengthened to Levitin - Polyak well - posedness, is due to Šolohovič [2], where E - spaces are characterized using Levitin - Polyak well - posedness of some classes of semicontinuous functionals.

E - spaces were characterized by Šolohovič [1] in terms of convergence as $\varepsilon \to 0$ of points of least norm in $\varepsilon - \arg \min (K, I)$ (a regularization procedure) for convex problems (K, I). Related results can be found in Konyagin - Tsar 'kov [1]. Lemma 5 goes back to Ascoli [1, p.46]. Sufficient conditions to Tykhonov well -posedness of best approximation in suns are given in Smarzewski [2].

To section 2. About the classical notion of Hadamard well - posedness (mostly within the framework of problems from mathematical physics) see Courant - Hilbert [1 (p. 227)], Hadamard [1] and [2], Talenti [1], Bertero [1], Tikhonov - Arsénine [1] and [2]. Proposition 7 is a particular case of the results in Singer [1] about the continuity of the metric projection onto approximatively compact Chebyshev sets. This fact means that, for (not necessarily convex) best approximation problems, their Tykhonov well - posedness implies solution Hadamard well - posedness. The proof is exactly the same as in proposition 7. Moreover, generalized well - posedness of best approximation problems defined by K (see example I.35) implies upper semicontinuity of the multifunction $p(\cdot, K)$, as shown by Singer [1], and in a more general (not necessarily topological) framework by Deutsch [1]. (A similar (not topological) framework in the sequential setting will be adopted in chapter IV.)

In Singer [1] E - spaces are characterized as rotund real Banach spaces in which every weakly sequentially closed set is approximatively compact, see Efimov - Stečkin [1]. Example 6 is from Cheney [1, p.167] (see also Maehly - Witzgall [1]). Further examples of discontinuous metric projections can be found in Braess [1, p. 42], Holmes [1 p. 169], and Brown [1]. Well - posedness criteria for best approximation by rational functions are given in Braess [1, ch.V, sec. 2.B].

In addition to proposition 8, proposition 7 admits the following partial converse. Let X be reflexive. Then X is an E - space if for every closed convex K the metric projection onto K is single - valued and strongly continuous (a further example of equivalence between Tykhonov and solution Hadamard well - posedness: compare with theorem 2).

See Vlasov [2, ch.II], [3] and Oshman [1], [2] for this and further results about continuity of the metric projections, as well as Beer [1]. The uniform continuity of the metric projection is studied in Berdyshev [2] and [1].

For nonconvex sets K, the equivalence between Tykhonov well - posedness of best approximation problems for K, continuity of the metric projection and differentiability of dist $(\cdot, K)$ is discussed in Fitzpatrick [1], extending a former result of Asplund [2], where proposition 8 can be found. In particular, it follows that, if $K \in Cl\ (X)$ and $x \notin K$, Tykhonov and Hadamard well - posedness (i.e. continuity of $p(\cdot, K)$ at x) for the problem of best approximation to x from K, are equivalent properties (and equivalent to Fréchet differentiability of dist $(\cdot, K)$ at x), provided X and X^* are E - spaces. (See also Coyette [1]). An interesting link between differentiability of the optimal value and Tykhonov well - posedness in the calculus of variations will be presented in section VIII.2. Compare with theorem III. 35.

The conclusion of proposition 8 holds in Banach spaces X with a rotund dual space (see Giles [1, sec.4.2] for a proof), as shown by Vlasov [1]. Convexity of K is also a necessary condition for Tykhonov well - posedness of every best approximation problem from K (see Asplund [2, cor.3] from a remark of Klee).

Another example of equivalence between Tykhonov and solution Hadamard well - posedness (local Lipschitz continuity) can be found in best appoximation problems from finite - dimensional subspaces of $C^0(T)$: see Wu Li [1, th. A p.165]. For related results see Park [1].

In the russian literature, value (solution) Hadamard well - posedness is often called *stability with respect to the functional (the solution)*.

The definition of Mosco convergence (sometimes called *Kuratowski - Mosco convergence*) was firstly introduced in the framework of variational problems by Mosco [1]. Theorem 9 is due to Sonntag [1], [2]. Particular cases of parts of it were found by Mosco [2] and Pieri [1]. See also Attouch [3, prop.3.34 p.322] and [2], and Zalinescu [1, p. 285] for other proofs. Spaces in which (18) $\Leftrightarrow_r (20)$ are characterized by Borwein - Fitzpatrick [1]. A comparison among various notions of set convergence, with some bearing on theorem 9 and related results, may be found in Sonntag [4]. Applications of theorem 9 and of results in this section to optimal control will be presented in sections V.2 and V.3. Results related to theorem 9 can be found in Tsukada [1] and Lucchetti [5].

The weakest topology on Conv (X) such that both

$$K \to \ \text{dist}\ (x, K), K \to \inf\{< y, u >: u \in K\}$$

are continuous for all $x \in X, y \in X^*$, has been characterized by Beer [4]. The Mosco topology on Conv (X) (see Beer [1]) is characterized in Shunmugaraj - Pai [1] as the weakest one such that $K \to \inf f(K)$ is continuous for every continuous function f which has weakly compact sublevel sets; moreover, $K \rightrightarrows \varepsilon - \text{arg min}(K, f)$ is continuous.

For special classes of problems without uniqueness (for example, metric projections) the existence of a continuous selection of the multifunction "data $\rightrightarrows$ set of optimal solutions" can be considered as a weak form of Hadamard well - posedness.

To section 3. A characterization of strong Tykhonov well - posedness on (Conv $(X), I$) in terms of Fréchet differentiability of I^* (compare theorem I.27) is given in Kosmol - Wriedt [1]. The characterization afforded by remark 12 follows from theorem 2 of Polyak [1] and lemma 2.5 of Lucchetti - Patrone [3]. The basic theorem 13 and proposition 21 are due to Lucchetti - Patrone [3]. Particular cases of theorem 13 are in Berdyšev [4] (but see Lucchetti - Patrone [3, p.205]). A related result is considered in Zolezzi [13, p.309]. Theorem 14, corollary 15 and example 16 are due to Lucchetti [1]. Theorem 14 generalizes results of Berdyšev [6]: related results are in Bednarczuk - Penot [1].

A stability property of local minimizers for uniformly small local perturbations of Tykhonov well - posed problems is considered in Hyers [1] and [2].

Theorem 17 and corollary 18 are particular cases of Beer - Lucchetti [2], see also [3]. The significance of the local Hausdorff topology in well - posedness was firstly shown by Attouch - Wets [4], [5] and [6] (hence the equivalent name *"a w - topology"* used in the literature; *bounded Hausdorff* is also used).

To section 4. Theorems 24, 25, 26 and 27 are due to Lucchetti - Patrone [2], where further characterizations of well - posedness are given, together with related estimates for suitable forcing functions.

To section 5. The definition of well - posed variational inequality and all results in this section are due to Lucchetti - Patrone [4]. An apparently different definition of well - posedness for (K, A, u) is due to Revalski [2], where it is shown to agree with that of section 4 provided A is hemicontinuous and K is bounded.

Chapter III.

GENERIC WELL-POSEDNESS

Section 1.   Example and Motivations.

Many optimization problems involve data, which can be considered as parameters. For instance, in the setting of section II. 1, x_0 and K are natural parameters in the problem of the minimum distance from x_0 to K.

1 Example (optimal control). In a linear regulator problem, we minimize

$$\int_0^T [(x - x_0)'P(x - x_0) + (u - u_0)'Q(u - u_0)]dt$$

where the control variable u and the state variable x are connected by

$$\dot{x}(t) = A(t)x(t) + B(t)u(t), \quad \text{a.e.} \quad t \in (0, T).$$

Thus the matrices P, Q, A, B and the desired trajectory u_0, x_0 are (physical) parameters in the problem.

Therefore, we are often interested in solving a whole family $F(p)$ of optimization problems depending on a parameter p, in the form

$$\inf \{I(p, x) : x \in K(p)\}.$$

We want either to solve $F(p)$ for all prescribed p, or to solve it for most p. Sometimes the parameter models (data) perturbations acting on a fixed problem, which corresponds to a given parameter p_0. Cases of interest are, among many, the following:

(a) We are able to solve $F(p_0)$, and we want to know solvability and significant properties of optimal solutions of $F(p)$ for p near p_0; so we are interested in the local properties of the solution dependence on p (small perturbations around p_0);

(b) We are unable to solve the original problem $F(p_0)$, but the perturbed problems $F(p)$ are simpler than the original one (e.g. $F(p_0)$ is an infinite-dimensional optimization problem, while $F(p)$ are finite - dimensional if $p \neq p_0$). We may simply choose p near enough p_0 for the perturbation to be insignificant for all practical purposes, and deal with $F(p)$ instead of $F(p_0)$;

(c) We are interested in the global properties of the problems $F(p)$ when p ranges on all possible values. More specifically we would like to describe the subset of those p for which $F(p)$ is well-posed. Our goal is to find conditions or classes of problems for which (may be not all but) most of the problems are well-posed. In other words we want to know when and in what sense well-posedness is a generic property. This means that the set of those p, corresponding to an ill-posed problem $F(p)$, is small in a suitable sense: e.g., well- posedness obtains for a dense set of parameters, even residual (i.e., it contains a countable intersection of open dense sets in a complete metric space). This happens in example 2 as far as Tykhonov well-posedness is concerned.

2 Example. Minimizing $px^2 + x$ (p parameter ≥ 0) on $(-\infty, +\infty)$ gives a Tykhonov well-posed problem if $p > 0$, while $p = 0$ defines a problem without solutions. Minimizing px, $|p| \leq 1$, for $0 \leq x \leq 1$ gives an unique solution if $p \neq 0$, and infinitely many for $p = 0$.

An approach like (b) is not always feasible. As far as well-posedness is concerned, it may be grossly wrong that if the perturbation associated with p is small, that is p is near to p_0, then the behavior of $F(p)$ is nearly that of $F(p_0)$. Example I. 5 shows that well-posed problems (with parameter ε) give raise to an ill-posed problem (with $\varepsilon = 0$). In this sense, well-posedness is an unstable property: small perturbations may destroy it.

3 Example (unstable well-posedness). Minimizing $x \to \|x\|$ on a given Banach space defines a strongly Tykhonov well-posed problem, while minimizing

$$x \to \max\{\|x\|, 1/n\}, n = 1, 2, 3, ...,$$

does not (even in the generalized sense in the infinite-dimensional setting). Here we are considering uniformly small perturbations.

We note that the highly nonconstructive nature of the dense (or generic) well- posedness theorems may be a serious shortcoming. We may well know a priori the existence of a residual set of parameters, for which well-posedness obtains, but nothing may be known about a concrete characterization of such a set. What is lacking are the equations (so to speak) which define the set of well-posedness. On the other hand, generic well-posedness results are useful and significant when dealing with optimization problems involving measured data or uncertain parameters.

Functional - Analytic Aspects

Generic well-posedness of minimum problems for continuous linear forms is strictly related to the geometric theory of Banach spaces.

Let X be a real Banach space and K a closed bounded subset thereof. Given any $p \in X^*$ we wish to minimize $x \to <p, x>$ on K. From theorem II.4 we know that if X is not an E - space, and K is its unit ball, then there exists some non-zero $p \in X^*$ such that $(K, <p, \cdot>)$ is not Tykhonov well-posed. In general, there exists $p \in X^*$ such that $(K, <p, \cdot>)$ is Tykhonov well-posed iff there exists $x_0 \in K$ which is strongly exposed (by definition), and

$$x_0 = \arg \min (K, <p, \cdot>) :$$

see Giles [1, p.195].

The following exhibits K and p such that $(K, <p, \cdot>)$ is not strongly Tykhonov well-posed in a Hilbert space (so the conclusion of theorem II.4 may fail if K is not the unit ball).

4 Example. Let $X = l^2, x_n = (0, ..., 0, 1, 0, ...)$ with 1 in the n - th place, $n = 1, 2, 3, ...,$

$$K = \text{cl co } \{x_n : n = 1, 2, 3, ...\}, p = (1, 1/2, 1/3, ..., 1/n, ...) \in X^*.$$

Then $0 = \arg \min (K, <p, \cdot>)$, x_n is a minimizing sequence and $x_n \rightharpoonup 0$, but $x_n \not\to 0$.

The following known theorem shows relevant links between the geometric theory of Banach spaces and strong Tykhonov well-posedness of constrained minimum problems for continuous linear forms. Terminology is that of Giles [1, sec. 3.5] where definitions and proofs may be found.

5 Theorem $(K, <p, \cdot>)$ *is strongly Tykhonov well-posed for every p in a dense set of* X^* *if K is convex and weakly compact, or if K is bounded, closed and convex and X has the Radon - Nikodým property.*

Notice that

(i) the Radon - Nikodým property for X is related to deep geometrical properties of the space: see Giles [1], Diestel [1], Diestel - Uhl [1], Bourgin [1];

(ii) a relevant property in the geometric theory of Banach spaces, for a given bounded subset A of X, is the existence of slices of arbitrarily small diameter, i.e. the behavior of diam $[\varepsilon - \arg \min (A, p)]$ for some $p \in X^*$ is involved. If moreover A is closed, this is equivalent by theorem I.11, to strong Tykhonov well-posedness of $(A, <p, \cdot>)$ for some $p \in X^*$, i.e. p strongly exposes A (compare with proposition I.19).

Section 2.   Generic Tykhonov Well-Posedness for Convex Problems.

Let X be a Banach space and

$$I : X \to (-\infty, +\infty].$$

We get a generic well-posedness criterion by applying theorem I.27 to the minimization on X of

$$x \to I(x) + <p, x>, p \in X^*,$$

where I is proper, convex and lower semicontinuous. A Banach space X is called a *(weak) Asplund space* iff every continuous convex function on an open convex subset of X is Fréchet (Gâteaux) differentiable on a dense G_δ subset of its domain.

Suppose moreover that

(1) I^* is everywhere finite - valued.

Then I^* is continuous on X^* (Barbu - Precupanu [1, prop. 1.6 p. 91]). Given a topological space P, a subset K of X and

$$J : P \times X \to (-\infty, +\infty]$$

we shall say that $[K, J(p, \cdot)]$ *is densely (generically) Tykhonov well - posed on P* iff there exists a dense (residual) set D in P such that $[K, J(p, \cdot)]$ is Tykhonov well-posed for every $p \in D$. By theorem I.27 we get

6 Theorem *Let* $I \in C_0(X)$ *and assume* (1). *If* X^* *is a (weak) Asplund space then* $(X, I(\cdot) + <p, \cdot>)$ *is strongly (weakly) Tykhonov well- posed, generically on* X^*.

Therefore, generic strong Tykhonov well-posedness obtains if X is reflexive and $I(x)/\|x\| \to +\infty$ as $\|x\| \to +\infty$, or if (1) holds and X^* has an equivalent Fréchet differentiable norm (out of the origin). These conclusions come from known criteria for Asplund spaces (Giles [1, sec. 3.3 and 3.4]). In section 6 we shall extend theorem 6 to nonconvex problems with a less restricted dependence upon the parameter p.

If P is a given topological or convergence space of functions defined on X, we say that (X, f) *is generically (densely) Tykhonov well-posed in* P iff there exists a residual (dense) set D in P such that (X, f) is Tykhonov well-posed for every $f \in D$.

Given a Banach space X, we shall denote by $LSC(X)$ the convergence space of all functions

$$f : X \to (-\infty, +\infty)$$

which are lower semicontinuous and convex, equipped with the uniform convergence on bounded sets in X.

Let X be a reflexive space. Then we can find an equivalent locally uniformly convex norm $\| \cdot \|$ on X, so that (as already noted in section II.1) X becomes an E - space (Diestel [1, th.1 p. 164]). Take any $f \in LSC(X)$, then every Tykhonov regularization of f,

$$(2) \qquad\qquad I(x) = f(x) + c\|x\|^2, \quad c > 0,$$

gives raise to (X, I) which is strongly Tykhonov well-posed. To this end, we know that f has continuous affine minorants, therefore I fulfills (1) and it is strictly convex. So we have exactly one point $x_0 = \arg \min (X, I)$ towards which every minimizing sequence x_n converges weakly. Then

$$\lim \inf \|x_n\|^2 \geq \|x_0\|^2.$$

Moreover $I(x_n) \to I(x_0)$, so that

$$\lim \sup c\|x_n\|^2 \leq I(x_0) - \lim \inf f(x_n) \leq c\|x_0\|^2,$$

yielding $x_n \to x_0$, as X is an E - space. We get

7 Lemma Let X be reflexive and $f \in LSC(X)$ with I given by (2), and $\| \cdot \|$ an equivalent norm on X such that X is an E - space. Then for every $c > 0, (X, I)$ is strongly Tykhonov well-posed.

By taking $c = 1/n, n = 1, 2, 3, ...$, we see by lemma 7 that (X, f) *is densely Tykhonov well-posed in* $LSC(X)$.

In the rest of this section we shall use some definitions and results borrowed from chapter IV.

By the above reasoning, every function in $C_0(X)$ is an uniform limit on bounded sets of convex functions which generate strongly Tykhonov well-posed problems. By weakening the convergence we can approximate every element of $C_0(X)$ (with respect to Mosco convergence on the reflexive space X) by Fréchet differentiable functions on X corresponding to strongly Tykhonov well-posed problems. This allows us to get approximate necessary optimality conditions. Such a result is of interest due to the strong variational properties of the Mosco convergence.

Fix $f \in C_0(X)$. Let $\| \cdot \|$ be an equivalent norm on X such that both X and X^* are locally uniformly convex and the new norms are Fréchet differentiable out of the origin on X and X^* (Diestel [1, cor.3, p. 167]). Since $f^* \in C_0(X^*)$ (Barbu - Precupanu [1, cor. 1.4 p. 97]), from lemma 7 it follows that

$$(X^*, f^*(\cdot) + \frac{1}{2n} \| \cdot \|^2), n = 1, 2, 3, ...,$$

is strongly Tykhonov well-posed. Moreover

(3)
$$M - \lim \left(f^* + \frac{1}{2n} \| \cdot \|^2 \right) = f^*.$$

Similarly to the proof of lemma 7, we get strong Tykhonov well-posedness for each $x \in X$ of

$$(X^*, f^*(\cdot) + (1/2n)\| \cdot \|^2 + < \cdot, x >).$$

Now consider

(4)
$$I_n = (f^*(\cdot) + (1/2n)\| \cdot \|^2)^* + (1/2n)\| \cdot \|^2.$$

From theorem I.27 we get Fréchet differentiability of I_n on X, moreover, by lemma 7, (X, I_n) is strongly Tykhonov well-posed for every n (as a Tykhonov regularization of the Yosida regularization of f). Thus by th. 2.3 p. 121 of Barbu - Precupanu [1] we get $I_n \to f$ pointwise on X. Moreover, by the continuity of the conjugacy (Attouch [3, th. 3.18 p. 295]) from (3) we deduce

$$M - \lim (f^*(\cdot) + (1/2n)\| \cdot \|^2)^* = f,$$

so by adding $(1/2n)\| \cdot \|^2$ which goes to 0 uniformly on bounded sets we finally obtain

$$M - \lim I_n = f,$$

and therefore

8 Proposition *Let X be reflexive and $f \in C_0(X)$. If I_n is given by (4), then : $I_n \in C_0(X), (X, I_n)$ is strongly Tykhonov well-posed, I_n is Fréchet differentiable on X, $I_n \to f$ both pointwise and in the sense of Mosco.*

Strong Tykhonov well-posedness is indeed sequentially generic in $LSC(X)$. This stronger fact will be proved by a deeper analysis, introducing a method which will be useful in many instances.

We put

(5)
$$W_n = \{ f \in LSC(X) : f \text{ is bounded from below on } X \text{ and}$$
$$\text{diam } [\varepsilon - \text{arg } \min(X, f)] \geq 1/n \text{ for every } \varepsilon > 0 \}, n = 1, 2, 3, ...;$$

(6)
$$W_0 = \{ f \in LSC(X) : \text{diam } [\varepsilon - \text{arg} \min(X, f)] = +\infty \text{ for every } \varepsilon > 0 \}.$$

Now from theorem I.11 and lemma 14 in the sequel we see that the set of Tykhonov ill-posed problems is exactly $\cup_{n=0}^{\infty} W_n$. Sequentially generic Tykhonov well-posedne

(X, f) in $LSC(X)$ will follow (since density has already been proved) if we show that every W_n is closed. We shall show that

$$f \to \text{diam} \left[\varepsilon - \arg\min(X, f) \right]$$

is upper semicontinuous in the following sense. If $f_k \to f$ in $LSC(X)$, $a > 0$ and diam $[\varepsilon - \arg\min(X, f_k)] \geq a$ for every $\varepsilon > 0$, then the same inequality is true for f, yielding closedness of each W_n. Thus we get

9 Theorem *Let X be reflexive. Then (X, f) is strongly Tykhonov well-posed for every f in a sequentially dense G_δ subset of $LSC(X)$.*

It is of interest to weaken the convergence on $C_0(X)$ by retaining significant variational properties, and see if Tykhonov well-posedness is still generic. A natural choice is then Mosco convergence, taking into account proposition 8.

Abusing notation, let us now denote by W_0 the set defined by (6) with $LSC(X)$ replaced by $C_0(X)$.

10 Example. Let X be a separable Hilbert space of infinite dimension, with orthonormal basis e_n. Let

$$I_n(x) = \|x\|^2 - \langle x, e_n \rangle^2; \quad I(x) = \|x\|^2.$$

Of course (X, I) is strongly Tykhonov well-posed. On the other hand $I_n \in W_0$, since $\arg\min (X, I_n)$ contains sp $\{e_n\}$. We check now that

$$M - \lim I_n = I$$

and this will show that W_0 is not closed in $C_0(X)$ equipped with Mosco convergence (so that the reasoning leading to theorem 9 does not work in this setting). Since $\langle x, e_n \rangle \to 0$, then $I_n(x) \to I(x)$ for every x. Now let $x_n \rightharpoonup x$, then

$$\liminf I_n(x_n) \geq I(x) \text{ if } x = 0.$$

$x \neq 0$ and put $y_n = x_n - x \rightharpoonup 0$. Expanding squares

$$) - I(x) = \|y_n\|^2 + 2\langle x, y_n \rangle - \langle x, e_n \rangle^2 - \langle y_n, e_n \rangle^2 - 2\langle x, e_n \rangle\langle y_n, e_n \rangle,$$

$$\liminf [I_n(x_n) - I(x)] = \lim \inf (\|y_n\|^2 - \langle y_n, e_n \rangle^2) \geq 0,$$

d to show convergence in the sense of Mosco.
le 10 is (in a sense) fairly typical, since we have

ition *Let X be an infinite - dimensional separable Hilbert space. Then W_0 (X) equipped with the Mosco convergence.*

nov   - dimensional setting, ill-posed convex problems are a first category set
ss of   pi-convergence (i.e. Mosco convergence).

12 Theorem (R^P, f) *is generically Tykhonov well-posed in* $C_0(R^P)$ *equipped with epi - convergence.*

Similar methods show the following. Denote by D the set of real-valued convex differentiable functions on R^P, such that $f(\cdot)- < u, \cdot >$ is Tykhonov well-posed for each $u \in R^P$.

13 Theorem D *is a residual in the space of extended real-valued functions which are convex and lower semicontinuous on* R^P, *equipped by the metric of epi - convergence.*

For the proof of theorem 13 see Patrone [4].

PROOFS.

Theorem 9. We shall need

14 Lemma Let $f \in LSC(X)$ with X reflexive. Then $f \in W_0$ whenever $\arg\min(X, f) = \emptyset$.

Proof of lemma 14 . Suppose that $\varepsilon - \arg\min (X, f)$ has a finite diameter for some $\varepsilon > 0$. Fix a minimizing sequence x_n to (X, f). Then

$$x_n \in \varepsilon - \arg\min(X, f)$$

for all large n, hence $x_n \rightharpoonup y$ for a subsequence. Thus

$$f(y) \leq \liminf f(x_n) = \inf f(X),$$

but this is a contradiction, either if f is unbounded from below or not. $\qquad\square$

Proof of theorem 9. It suffices to prove that $f_k \to f$ in $LSC(X), a > 0$, diam $[\varepsilon - \arg\min(X, f_k)] > a$ for every k and every $\varepsilon > 0$ imply

$$\text{diam } [\varepsilon - \arg\min(X, f)] > a \text{ for every } \varepsilon > 0.$$

The only case of interest is when diam $[\varepsilon - \arg\min(X, f)] < +\infty$. Lemma 14 then gives lower boundedness of f. Let $u \in X$ be such that

$$(7) \qquad\qquad f(u) \leq \inf f(X) + \varepsilon/3.$$

Then for all k sufficiently large

$$f_k(u) \leq f(u) + \varepsilon/4.$$

Put for such k

$$B_k = \{x \in X : f_k(x) \leq f(u) + \varepsilon/3\}.$$

If f_k is bounded from below, then for large k

(8) $\qquad \alpha - \arg\min (X, f_k) \subset B_k$ when α is sufficiently small,

since $f_k(x) \leq \inf f_k(X) + \alpha$ implies $f_k(x) \leq f_k(u) + \alpha \leq f(u) + \varepsilon/3$ whenever $\alpha \leq \varepsilon/12$. If f_k is unbounded from below, then (8) is true if $-1/\alpha \leq f(u) + \varepsilon/3$. So (8) gives diam $B_k > a$ for sufficiently large k. By convexity of B_k and the definition of diameter, there exist $T > 0$ and points x, y in B_k such that

(9) $\qquad\qquad\qquad\qquad a < \|x - y\| \leq T.$

Set now

$$C_k = \{w \in B_k : \|w - u\| \leq T\}.$$

Of course $u \in C_k$. If C_k contains some segment with an end at u and length $> a$ (i.e. either x or y may be chosen as u), then diam $C_k > a$. If otherwise $\|u - z\| \leq a$ for all $z \in B_k$, then

$$\|u - x\| \leq T, \quad \|u - y\| \leq T,$$

so that x and y belong to C_k. Then by (9) diam $C_k > a$ again. Now set

$$C = \bigcup C_k,$$

union taken on all k sufficiently large. Since C is bounded, for $x \in C$ and k sufficiently large (remembering (7))

$$f(x) \leq f_k(x) + \varepsilon/3 \leq f(u) + 2\varepsilon/3 \leq \inf f(X) + \varepsilon/3$$

giving $C \subset \varepsilon - \arg\min (X, f)$, so that diam $[\varepsilon - \arg\min (X, f)] > a$. $\qquad\square$

Proposition 11. If f is unbounded from below, or lacks minimum points, then $f \in W_0$ by lemma 14. So it suffices to approximate every $f \in C_0(X)$ which possesses minimum points. We can assume without restriction that

(10) $\qquad\qquad\qquad\qquad f(0) = 0 = \inf f(X).$

Let indeed $f(x_0) = \inf f(X)$, and set

$$g(x) = f(x + x_0) - \inf f(X).$$

Then g fulfills (10), while $M - \lim g_n = g$ implies $M - \lim f_n = f$, and $f_n \in W_0$ if $g_n \in W_0$, where

$$f_n(x) = g_n(x - x_0) + \inf f(X).$$

Proposition 8 gives a sequence I_n which converges towards f in the Mosco sense, and (4) implies that for each n

(11) $\qquad\qquad\qquad\qquad I_n(x)/\|x\| \to +\infty$ as $\|x\| \to +\infty$

since $(f^* + \|\cdot\|^2/2n)^*$ has continuous affine minorants. On the other side, Mosco's convergence on the set of functions $f \in C_0(X)$ fulfilling (8) is induced by a metric (see Attouch [3, th. 3.36 p. 324]). Then we need only to approximate f by a sequence in W_0 with terms fulfilling (10), for each f satisfying (11). Let now e_n be an orthonormal basis in X, and consider

$$g_n(x) = \begin{cases} 0, & \text{if } x \in \text{sp } (\{e_n\}), \\ f(x), & \text{otherwise;} \end{cases}$$

$$f_n = g_n^{**}$$

(having identified X with X^{**}).

We shall use freely in the proof some elements of convex analysis which may be found in ch. 2 of Barbu - Precupanu [1]. Of course $f_n \in C_0(X)$. We shall show that f_n fulfills (10), $f_n \in W_0$, and $M - \lim f_n = f$. By (10), $0 \le g_n \le f$, giving $g_n^* \ge f^*$. Therefore $f_n(0) \le 0$, but $f_n \ge 0$ since f_n is the upper envelope of the continuous affine minorants to g_n, so f_n fulfills (10), moroever $f_n \in W_0$, since arg min (X, f_n) contains sp $(\{e_n\})$. Since inf $f^*(X) = -f(0) = 0$, then $f^* \ge 0$. By definition

$$g_n^*(x) = \max\{\sup [t < x, e_n >: t \in R], \sup[< x, y > -f(y) : y \notin \text{sp } (\{e_n\})]\}.$$

Then $g_n^*(x) \le f^*(x)$ if $< x, e_n >= 0$. So we get

$$g_n^*(x) = \begin{cases} +\infty, & \text{if } < x, e_n > \ne 0, \\ f^*(x), & \text{if } < x, e_n >= 0. \end{cases}$$

The proof is ended by showing that $M - \lim f_n = f$, or equivalently $M - \lim g_n^* = f^*$ (Attouch [3, th. 3.18 p. 295]). Since $g_n^* \ge f^*$ we need only to show that

for every x there exists $x_n \to x$ such that $g_n^*(x_n) \to f^*(x)$.

To this end take

$$x_n = x - < x, e_n > e_n.$$

Then $x_n \to x, < x_n, e_n > = 0$ therefore $g_n^*(x_n) = f^*(x_n)$. Coercivity of f implies dom $f^* = X$, therefore f^* is everywhere continuous, yielding $g_n^*(x_n) \to f^*(x)$, as required. □

Theorem 12. Consider W_n as in (5), (6) with $LSC(X)$ replaced by $C_0(R^P)$. Now $C_0(R^P)$ equipped with epi-convergence is a complete metric space (Attouch [3, th. 3.36 p. 324]). It suffices to prove that each W_n is closed, since density follows by proposition 8. Therefore we need only to show that

(12) f and f_n in $C^0(R^P), M - \lim f_n = f, a > 0$, diam $[\varepsilon - \text{arg min } (R^P, f_n)] \ge a$ for every n and $\varepsilon > 0$ imply diam $[\varepsilon - \text{arg min}(R^P, f)] \ge a$ for every $\varepsilon > 0$.

The only case of interest occurs when

$$\text{diam}[\varepsilon - \text{arg min}(R^P, f)] < +\infty \text{ for some } \varepsilon > 0.$$

Then inf $f(R^P) > -\infty$, and of course

(13) sub lev (f, t) is nonempty and compact for every $t > \text{inf } f(R^P)$ and sufficiently near to it.

We shall show that

(14) $K - \lim \sup \mathrm{lev}\,(f_n, t) = \mathrm{sub\ lev}\,(f, t)$ for every $t > \inf f(R^P)$.

Taking (14) for granted, we see by (13) that the convergence in (14) takes place with respect to the Hausdorff distance for t as in (13) (see Hausdorff [1, p. 172]). Therefore, for such t, the diameters of the sublevel sets converge. Since

$$\lim \sup \ \inf f_n(R^P) \leq \inf f(R^P)$$

(by theorem IV.5) we see that

$$s > \inf f(R^P) \text{ implies } s > \inf f_n(R^P) \text{ for } n \text{ sufficiently large,}$$

whence for such $n, s,$ and $\varepsilon > 0$

$$\varepsilon - \arg \min(R^P, f_n) \subset \mathrm{sub\ lev}\,(f_n, s + \varepsilon),$$

thus

$$a \leq \mathrm{diam\ sub\ lev}\,(f_n, s + \varepsilon).$$

By the convergence of the diameters we get

$$a \leq \mathrm{diam\ sub\ lev}\,(f, t)$$

for t as in (13). This proves (12).

Proof of (14). If n_j is any subsequence from the positive integers and

$$x_j \in \mathrm{sub\ lev}\,(f_{n_j}, t), \quad x_j \to x,$$

then by Mosco convergence

$$t \geq \lim \inf f_{n_j}(x_j) \geq f(x),$$

hence $x \in \mathrm{sub\ lev}\,(f, t)$. Conversely, let $x \in \mathrm{sub\ lev}\,(f, t)$. If $f(x) < t$, there exists (by Mosco convergence) some $x_n \to x$ such that $f_n(x_n) \to f(x)$, hence $x_n \in \mathrm{sub\ lev}\,(f_n, t)$. If $f(x) = t > \inf f(R^P)$, there exists some y such that $f(y) < t$, therefore putting

$$z = (1 - s)x + sy, \quad 0 < s < 1,$$

we get

(15) $$|z - x| = s|y - x|, \quad f(z) < t.$$

Now take $s = 1/(n|y-x|)$ for n sufficiently large, and consider the corresponding sequence z_n. Then by (15), $f(z_n) < t$. By what we have just proved, there exists x_n such that

$$x_n - z_n \to 0, f_n(x_n) < t,$$

hence $x_n \to x$ by (15) and $x_n \in \mathrm{sub\ lev}\,(f_n, t)$. $\square$

Section 3.   Lower Semicontinuous Functions.

Let X be a complete metric space with metric d. Consider $BLS(X)$, the set of all real - valued lower semicontinuous functions that are bounded from below on X, equipped with the metric of uniform convergence on X. Consider also $BC^0(X)$, the set of all real - valued continuous functions that are bounded from below on X with the same metric. Then, as well known, $BC^0(X)$ and $BLS(X)$ are complete metric spaces. Generic well - posedness of (X, f) in $BLS(X)$ comes (as already seen in section 2) from closedness and rareness of each

$$W_n = \{f \in BLS(X) : \text{ diam } [\varepsilon - \text{arg min } (X, f)] \geq 1/n \text{ for every } \varepsilon > 0\}, n = 1, 2, 3, \ldots.$$

By theorem I.11, Tykhonov ill - posed problems coincide with $\bigcup_{n=1}^{\infty} W_n$. In this way we get

15 Theorem (X, f) *is generically Tykhonov well - posed both in* $BLS(X)$ *and in* $BC^0(X)$.

Proof. Let us show closedness of each W_n in $BLS(X)$. Given $f \notin W_n$, there exists $b > 0$ such that

$$\text{diam } [\varepsilon - \text{arg min}(X, f)] < 1/n \text{ if } 0 < \varepsilon < b.$$

It suffices to show that

(16) $g \in BLS(X)$ and $|g(x) - f(x)| < b/3$ imply $g \notin W_n$.

Now $|\inf g(X) - \inf f(X)| \leq b/3$, so that if $x \in b/3 - \text{arg min}(X, g)$, then

$$f(x) \leq \inf f(X) + b.$$

Thus

$$\text{diam } [b/3 - \text{arg min}(X, g)] \leq \text{ diam } [b - \text{arg min}(X, f)] < 1/n$$

and (16) is proved. We show rareness of W_n in $BLS(X)$ by proving that

$$BLS(X) \setminus \bigcup_{n=1}^{\infty} W_n \text{ is dense.}$$

Given $f \in BLS(X)$ and $\varepsilon > 0$, pick $y \in \varepsilon/3 - \text{arg min}(X, f)$ and put

$$g(x) = \begin{cases} f(x), & \text{if } x \neq y, \\ f(y) - 2\varepsilon/3, & \text{if } x = y. \end{cases}$$

Then $|f(x) - g(x)| < \varepsilon$ for all $x, g \in BLS(X)$ and diam $[b - \text{arg min } (X, g)] = 0$ if $b < \varepsilon/3$ (since it reduces to the singleton y). Thus (X, g) is Tykhonov well - posed. This proves the conclusion in the $BLS(X)$ case. Now we turn to the $BC^0(X)$ setting, and (abusing notation) we consider

$$W_n = \{f \in BC^0(X) : \text{ diam } [\varepsilon - \text{arg min}(X, f)] \geq 1/n \text{ for every } \varepsilon > 0\}.$$

Closedness of W_n in $BC^0(X)$ follows as in the proof for the lower semicontinuous setting. We show now that $BC^0(X) \setminus W_n$ is dense. Given $\varepsilon > 0$ and $f \in BC^0(X)$ let $y \in \varepsilon/3 - \arg\,\min\,(X, f), 0 < \delta < 1/2n$ be such that

$$d(x, y) < \delta \quad \text{implies} \quad |f(x) - f(y)| < \varepsilon/3.$$

Now set
$$A = \{x \in X : d(x, y) < \delta\}, B = \{x \in X : d(x, y) \leq \delta/2\}.$$

Since B and $X \setminus A$ are closed disjoint sets, by Urysohn's lemma (Kelley [1, p. 115]) there exists a continuous h such that

$$h(x) = 0 \quad \text{if} \quad x \notin A, h(x) = -2\varepsilon/3 \quad \text{if} \quad x \in B, -2\varepsilon/3 \leq h \leq 0.$$

Setting $g = f + h$ we get $g \in BC^0(X)$ and $|g(x) - f(x)| < \varepsilon$ for all x. It remains only to show that $g \notin W_n$. If $x \in B \cap \varepsilon/3 - \arg\,\min(X, f)$, then

$$g(x) \leq \inf f(X) - \varepsilon/3, \quad \text{giving} \quad \inf g(X) \leq \inf f(X) - \varepsilon/3.$$

If $x \notin A$ then (remembering the previous inequality) $g(x) \geq \inf g(X) + \varepsilon/3$. So if $x \in b - \arg\,\min\,(X, g)$ with $b < \varepsilon/3$ we see that $x \in A$, implying diam $[b - \arg\,\min\,(X, g)]$ $\leq 2\,\delta < 1/n$. $\qquad\square$

Section 4. Dense Solution Hadamard Well-Posedness of Constrained Minimum Problems for Continuous Functions.

In this section
(X, d) is a given metric space which is complete and bounded;
$BC^0(X)$ is equipped with the metric

$$(17) \qquad \rho(f, g) = \sup\{|f(x) - g(\dot{x})|/(1 + |f(x) - g(x)|) : x \in X\}$$

(which induces on $BC^0(X)$ the topology of uniform convergence); $Cl(X)$ is equipped with the Hausdorff distance. Then $BC^0(X)$ and $Cl(X)$ are complete metric spaces (Kuratowski [1, p. 314]). We denote by

$$P = Cl(X) \times BC^0(X)$$

the product space equipped by a product metric, say

$$[\text{haus}^2(A, B) + \rho^2(f, g)]^{1/2}.$$

We shall consider generic well - posedness properties for the optimization problems $(A, f) \in P$, and we begin with a Tykhonov well - posedness result (an extension of theorem 15).

16 Theorem (A, f) *is generically Tykhonov well - posed in* P.

We shall consider generic Hadamard well - posedness properties for the optimization problem (A, f) within P. Given $(A_0, f_0) \in P$ we shall say that (A_0, f_0) is (solution) *Hadamard well - posed* iff arg min (A_0, f_0) is a singleton x_0 and

$(A_n, f_n) \to (A_0, f_0)$ in P, arg min $(A_n, f_n) \neq \emptyset$ for every n and

$$x_n \in \text{arg min } (A_n, f_n) \Rightarrow x_n \to x_0.$$

In other words, Hadamard well - posedness of (A_0, f_0) means here (upper semi) continuity and single - valuedness of the multifunction

$$\text{arg min } : P \rightrightarrows X,$$

both at (A_0, f_0). Of course solution Hadamard well - posedness, as defined here, implies value Hadamard well - posedness.

We need to characterize Hadamard well - posedness from a metric point of view. If A is fixed we know from theorem I.11 that the behaviour of diam $[\varepsilon - \text{arg min } (A, f)]$ as $\varepsilon \to 0$ is relevant as far as Tykhonov well - posedness of (A, f) is concerned. Now define, for $\varepsilon > 0$ and $(A, f) \in P$,

$$L(A, f, \varepsilon) = \{x \in X : \text{dist } (x, A) \leq \varepsilon, f(x) \leq \varepsilon + \text{val } (A, f)\}.$$

The behaviour of $L(A, f, \varepsilon)$ is relevant to Levitin - Polyak well - posedness (as we saw in theorem I.30), which takes care of the generalized minimizing sequences. The interesting fact is that Hadamard well - posedness of (A, f) is related to Levitin - Polyak well - posedness implied by the shrinkening of $L(A, f, \varepsilon)$ as $\varepsilon \to 0$ to a single point.

17 Theorem *Let* $(A, f) \in P$. *Then the following are equivalent facts, each implying Tykhonov well - posedness of* (A, f) :

(18) (A, f) *is Hadamard well - posed;*

(19) $inf \ \{diam \ L(A, f, \varepsilon) : \varepsilon > 0\} = 0.$

As a particular case, since

$$L(X, f, \varepsilon) = \varepsilon - \text{arg min}(X, f)$$

by theorems I.11 and 17 we get

18 Corollary *Let* $f \in BC^0(X)$. *Then* (X, f) *is Hadamard well - posed iff it is Tykhonov well - posed.*

The following counterexample shows that (19) no longer implies Hadamard well - posedness if f is only lower semicontinuous.

19 Example. Let $X = A = [0, 1]$,

$$f(x) = 2 - x \quad \text{if} \quad 0 < x \leq 1, = 0 \quad \text{if} \quad x = 0.$$

Then for small ε

$$L(A, f, \varepsilon) = \{0\} = \text{arg min}(f, A)$$

but arg min $([1/n, 1], f) = \{1\}$.

Now we come to generic well - posedness results within P.

20 Theorem (A, f) *is generically Hadamard well - posed in P.*

A similar result holds for convex problems. Let us denote by E a fixed Banach space, by X a nonempty bounded, closed, convex subset thereof, and by $CC(X)$ the set of all functions $f : X \to (-\infty, +\infty)$ which are continuous and convex (hence bounded from below, as shown later), equipped with the (uniform convergence) metric (17). Let $\text{Conv}(X)$ be equipped with the Hausdorff distance. Then both $CC(X)$ and $\text{Conv}(X)$ are complete metric spaces, so is

$$T = \text{Conv}(X) \times CC(X)$$

equipped with the above product metric.

21 Theorem (A, f) *is generically Hadamard well - posed in T.*

If X is unbounded, dense (Tykhonov) well - posedness fails in general under the metric (17). Take $X = R$ and $f = 0$. Then (R, g) is ill - posed for every $g \in CC(R)$ such that $\varrho(f, g) \leq \varepsilon < 1$, since g is constant.

Now we fix f and ask for well - posedness of (A, f) for most constraint sets A. Denote by Q the set of all nonempty closed convex bounded subsets of E, equipped with the Hausdorff distance.

22 Theorem *Let* $f : E \to (-\infty, +\infty)$ *be convex and continuous. Assume that* (E, f) *is Tykhonov well - posed. Then* (A, f) *is generically Hadamard well - posed in Q.*

The conclusion of theorem 22 means that for every A in some residual subset of $Q, f_n \to f$ uniformly on $E, A_n \to A$ in Q, f_n continuous on E and $x_n \in \text{arg min}(A_n, f_n)$ imply $x_n \to \text{arg min}(A, f)$.

By fixing $y \in E$, taking

$$f(x) = \|x - y\|$$

in theorem 22 and remembering theorem 17, we get the following
Corollary. *For every* $y \in E, (A, \|y - \cdot\|)$ *is generically Tykhonov well - posed in Q.*

PROOFS.

Theorem 16. We need the following
23 Lemma For any $f \in BC^0(X), A \in Cl(X)$ and $\varepsilon > 0$ there exists a point $y \in A$ such that

$$[A, f(\cdot) + \varepsilon d(y, \cdot)]$$

is Tykhonov well - posed with solution y.
Proof of lemma 23. Let h be the restriction of f to A. From Ekeland's theorem (see Ekeland [1, th. 1 p.444]) we know that given $\varepsilon > 0$ there exists $y \in A$ such that

$$(20) \qquad h(y) < h(x) + (\varepsilon/2)d(x, y), \quad x \in A, \ x \neq y,$$

yielding $y = \arg \min [A, f(\cdot) + \varepsilon d(y, \cdot)]$. Let x_n be any minimizing sequence for $[A, h(\cdot) + \varepsilon d(y, \cdot)]$. Then $x_n \in A$, moreover

$$h(y) \leq h(x_n) + \varepsilon d(x_n, y) = h(y) + \varepsilon_n \text{ for some } \varepsilon_n \to 0.$$

Then by (20), $d(x_n, y) \leq 2\varepsilon_n / \varepsilon \to 0$.   □

Proof of theorem 16. Given $(A, f) \in P$ and $\varepsilon > 0$, let y be as in lemma 23 and consider $g(\cdot) = f(\cdot) + \varepsilon d(y, \cdot)$. Then

$$\sup\{|g(x) - f(x)|/(1 + |g(x) - f(x)|) : x \in X\} \leq \varepsilon \text{ diam } X.$$

This proves the density part. To prove the G_δ part we need the following

24 Lemma $(A, f) \to \text{val}\,(A, f)$ is upper semicontinuous in P.

Remark. The above lemma is a particular case of theorem IV.5.

Proof of lemma 24. Let $(A_n, f_n) \to (A_0, f_0)$ in P. Given $u \in A$ there exists $u_n \in A_n$ such that $u_n \to u$. Therefore $f_n(u_n) \to f_0(u)$ since uniform convergence and continuity of f_0 imply continuous convergence. Since

$$\text{val}\,(A_n, f_n) \leq f_n(u_n)$$

we get

$$\lim \sup \ \text{val}\,(A_n, f_n) \leq f_0(u).$$

Arbitrariness of u gives the conclusion.   □

End of proof of theorem 16. Consider

$$V_n = \{(A, f) \in P : \inf\{\ \text{diam } [\varepsilon - \arg \ \min (A, f)] : \varepsilon > 0\} < 1/n\}, n = 1, 2, 3, \dots.$$

We show that every V_n is open. Let $(A_0, f_0) \in V_n$, then for some $\varepsilon > 0$, diam $[\varepsilon - \arg \min (A_0, f_0)] < 1/n$. By lemma 24, for $\delta > 0$ sufficiently small, if (A, f) lies at distance $\leq \delta$ from (A_0, f_0) in P then

(21) $\text{val}\,(A, f) \leq \text{val}\,(A_0, f_0) + \varepsilon/3.$

Moreover, for every $x \in X$

$$|f(x) - f_0(x)|/(1 + |f(x) - f_0(x)|) \leq \sup\{|f(z) - f_0(z)|/(1 + |f(z) - f_0(z)|) : z \in X\} \leq \delta$$

and for δ sufficiently small

$$|f(x) - f_0(x)| \leq \delta/(1 - \delta) < \varepsilon/3, x \in X.$$

It follows that if $x \in \varepsilon/3 - \arg \min (A, f)$, by (21)

$$f_0(x) \leq f(x) + \varepsilon/3 \leq 2\varepsilon/3 + \text{val}\,(A, f) \leq \text{val}\,(A_0, f_0) + \varepsilon,$$

whence

$$\varepsilon/3 - \arg \ \min (A, f) \subset \varepsilon - \arg \ \min (A_0, f_0),$$

yielding $(A, f) \in V_n$. This shows openness of V_n. The conclusion then comes from theorem I. 11 since (A, f) is Tykhonov well - posed iff $(A, f) \in \bigcap_{n=1}^{\infty} V_n$.   □

Theorem 17. Since

$$\varepsilon - \arg \min (A, f) \subset L(A, f, \varepsilon),$$

by theorem I.11, (19) implies Tykhonov well - posedness. It suffices to show equivalence between (18) and (19).

(18) $\Rightarrow$ **(19)**. Arguing by contradiction, suppose that for some δ

$$\inf_n \text{ diam } L(A, f, 1/n) > \delta > 0.$$

Put $x_0 = \arg \min (A, f)$. Then we can find points $x_n \in X$ such that

$$\text{dist } (x_n, A) \le 1/n, f(x_n) \le \inf f(A) + 1/n,$$

$$(22) \qquad\qquad d(x_n, x_0) > \delta/2.$$

We apply cor. 11 of Ekeland [1] to f on

$$A_n = A \cup \{x_n\},$$

with $\varepsilon = 1/n$. For every n there exists $a_n \in A_n$ such that $f(a_n) \le f(x_n)$,

$$(23) \qquad\qquad d(x_n, a_n) \le 1/\sqrt{n},$$

and $f(a_n) < f(x) + (1/\sqrt{n})d(x, a_n)$ for every $x \in A_n, x \ne a_n$. Setting $f_n(\cdot) = f(\cdot) + (1/\sqrt{n})d(\cdot, a_n)$, we get

$$|f_n(x) - f(x)| \le (1/\sqrt{n}) \text{ diam } X,$$

so that $(A_n, f_n) \to (A, f)$ in P. Then

$$a_n = \arg \min (A_n, f_n) \to x_0,$$

whence by (23) $x_n \to x_0$, contradicting (22).

(19) $\Rightarrow$ **(18)**. Let $(A_n, f_n) \to (A_0, f_0)$ and fix $x_n \in \arg \min (A_n, f_n), n = 0, 1, 2, \ldots.$ Given $\delta > 0$ there exists $\varepsilon > 0$ such that

$$L(A_0, f_0, \varepsilon) \subset B(x_0, \delta).$$

For n sufficiently large

$$(24) \qquad\qquad \text{haus } (A_n, A_0) \le \varepsilon, \rho(f_n, f_0) < a,$$

where $a/(1 - a) < \varepsilon/2$. Thus

$$|f_n(x) - f_0(x)| < a/(1 - a) < \varepsilon/2, x \in X.$$

Moreover, for n sufficiently large, thanks to lemma 24,

$$f_n(x_n) = \inf f_n(A_n) \le \inf f_0(A_0) + \varepsilon/2.$$

By (24), dist $(x_n, A_0) \le \varepsilon$ and

$$f_0(x_n) \le f_n(x_n) + \varepsilon/2 \le \varepsilon + \inf f_0(A_0),$$

thus $x_n \in L(A_0, f_0, \varepsilon) \subset B(x_0, \delta)$ for sufficiently large n. This shows that $x_n \to x$. $\qquad \square$

Theorem 20. The G_δ part is completely analogous to that of theorem 16, taking into account theorem 17. In fact, remembering that P is a complete metric space, we consider

$$Z_n = \{(A, f) \in P : \inf_{\varepsilon > 0} \text{ diam } L(A, f, \varepsilon) < 1/n\}$$

and show openness of every Z_n in a completely analogous fashion as in the proof of theorem 16. Now we prove density. Fix $(A, f) \in P$ and $\varepsilon > 0$. We apply lemma 23. For every fixed $q, 0 < q < \varepsilon/\text{ diam } X$, we find $a \in {}^\varepsilon A$ such that

(25) $\qquad [{}^\varepsilon A, f(\cdot) + qd(a, \cdot)]$ is Tykhonov well - posed with solution a.

Set

$$f_\varepsilon(\cdot) = f(\cdot) + q\ d(a, \cdot), B_\varepsilon = A \cup \{a\}.$$

Then the distance in P between (A, f) and $(B_\varepsilon, f_\varepsilon)$ is $\leq \varepsilon$. We show Hadamard well - posedness of $(B_\varepsilon, f_\varepsilon)$. By theorem 17 it suffices to show that

$$\inf\{ \text{ diam } L(B_\varepsilon, f_\varepsilon, c) : c > 0\} = 0.$$

This will follow by proving that

(26) $\qquad x_n \in L(B_\varepsilon, f_\varepsilon, 1/n)$ for every $n \Rightarrow x_n \to a$.

Case 1: dist $(a, A) < \varepsilon$. Since dist $(x_n, B_\varepsilon) \leq 1/n$, then $x_n \in A^\varepsilon$ for all n sufficiently large. Since

(27) $\qquad\qquad\qquad f_\varepsilon(x_n) \leq 1/n + \inf f_\varepsilon(B_\varepsilon)$

then $x_n \to a$ by (25).

Case 2: dist$(a, A) = \varepsilon$. If $x_n \nrightarrow a$ then for some subsequence and some $c, d(x_n, a) \geq c > 0$. Since dist $(x_n, B_\varepsilon) \to 0$, it follows that dist $(x_n, A) \to 0$. So $x_n \in A^\varepsilon$ for large n, and as in case 1 we get $x_n \to a$, a contradiction. So (26) is proved, and this ends the proof. $\qquad\qquad\qquad\qquad\qquad\qquad\qquad\qquad\qquad\qquad\qquad\qquad\qquad\qquad$ □

Before proving theorem 21, we notice that *every $f \in CC(X)$ is bounded from below.* If not, $f(x_n) \to -\infty$ for some $x_n \in X$. Let $y \in X$ be fixed. By (lower semi) continuity of f at y we have for some $a > 0$

$$f(x) \geq f(y) - 1 \text{ if } \|x - y\| \leq a,$$

hence $\|x_n - y\| > a$ for all n sufficiently large. Let $t_n \in [0, 1]$ be such that

$$\|z_n - y\| = a, \text{ where } z_n = t_n y + (1 - t_n)x_n.$$

Then

$$f(y) - 1 \leq f(z_n) \leq t_n f(y) + (1 - t_n)f(x_n) \leq |f(y)| + (a/\text{diam}X)f(x_n) \to -\infty,$$

a contradiction.

Theorem 21. The G_δ part is similar to that in the proof of theorem 16 (remembering that T is a complete metric space). The proof of the density part is similar to that of theorem 20 up to (25) and the definition of f_ε. Now consider

$$C_\varepsilon = \text{co } (A \cup \{a\}).$$

Again we show (26) with B_ε replaced by C_ε, which is closed. Let $x \in X$ be such that

(28) $$\text{dist } (x_n, C_\varepsilon) \to 0, f_\varepsilon(x_n) \to f_\varepsilon(a).$$

If dist $(a, A) < \varepsilon$ we conclude that $x_n \to a$ as in the proof of theorem 20. Now let

$$\text{dist } (a, A) = \varepsilon \text{ and } \|x_n - a\| > h \text{ for some } h > 0 \text{ and for some subsequence.}$$

Let $z \in C_\varepsilon$ be any point such that $\|z - a\| > h/2$. There exists $b \in A$ such that

$$z = ta + (1 - t)b \text{ for some } t \in [0, 1).$$

Then

$$h/2 < \|z - a\| = (1 - t)\|a - b\|$$

giving

$$t < 1 - h/2 \text{ diam } C_\varepsilon.$$

But dist $(\cdot, A)$ is convex, therefore

(29) $$d(z, A) \leq t \text{ dist } (a, A) < \varepsilon - \frac{h\varepsilon}{2 \text{ diam } C_\varepsilon}.$$

There exists $y_n \in C_\varepsilon$ such that $\|x_n - y_n\| \to 0$, so for n sufficiently large

$$\|y_n - a\| \geq h/2, \|x_n - y_n\| < \frac{h\varepsilon}{2 \text{ diam } C_\varepsilon}.$$

Thus by (29)

$$\text{dist } (x_n, A) \leq \|x_n - y_n\| + \text{ dist } (y_n, A) < \varepsilon,$$

whence $x_n \in A^\varepsilon$, entailing $x_n \to a$ by (25) and (28). This is a contradiction, therefore $x_n \to a$ as required. □

Theorem 22. Q is a complete metric space. The G_δ part is similar to that in the proof of theorem 16. Now let us show density. Fix $\varepsilon > 0, X \in Q$ and set

$$m = \text{ val } (^\varepsilon X, f), x_0 = \arg \min (E, f).$$

Case 1: $x_0 \in {}^\varepsilon X$. Let $f_n \to f$ uniformly on $E, A_n \to {}^\varepsilon X$ in Q and $z_n \in \arg \min (A_n, f_n)$. Since $f_n(z_n) - f(z_n) \to 0$ and $x_0 = \arg \min (^\varepsilon X, f)$, by lemma 24

$$\lim \sup f_n(z_n) = \lim \sup f(z_n) \leq m = f(x_0).$$

But $f(z_n) \geq f(x_0)$, whence $f(z_n) \to f(x_0)$. Then $z_n \to x_0$ by well - posedness, and Hadamard well - posedness is obtained in this case.

Case 2: $x_0 \notin {}^\varepsilon X$. Then $f(x_0) < m$, otherwise by well - posedness some sequence in the closed set $^\varepsilon X$ would converge to x_0. Set

$$A = \{x \in E : f(x) \leq m\}.$$

We shall need the following

25 Lemma For every $\alpha > 0$ there exists $\delta > 0$ such that dist $(x, A) < \alpha$ whenever $x \in X$ and $f(x) < m + \delta$.

Assume the conclusion of the lemma, and set for $U, V \subset E$

$$d(U, V) = \inf\{\|u - v\| : u \in U, v \in V\}.$$

From lemma 25, given $\alpha > 0$ let $\delta > 0$ correspond to α and let $x \in \ {}^{\varepsilon}X$ be such that $f(x) < m + \delta$. Then dist $(x, A) < \alpha$, therefore $d(A, \ {}^{\varepsilon}X) = 0$, whence $d(X, A) \leq \varepsilon$. Suppose now that $d(X, A) < \varepsilon$. Then there exist some $z \in A$ and $\delta > 0$ such that

$$B(z, \delta) \subset \ {}^{\varepsilon}X.$$

Let $y = tz + (1 - t)x_0, 0 < t < 1$. For t sufficiently near to 1

$$\|y - z\| = (1 - t)\|z - x_0\| < \delta,$$

showing that $y \in \ {}^{\varepsilon}X$. Therefore

$$m \leq f(y) \leq tf(z) + (1 - t)f(x_0) < m,$$

a contradiction. So $d(X, A) = \varepsilon$ and we can apply th.1.2 of Georgiev [1]. There exists $a \in A \cap \ {}^{2\varepsilon}X$ such that the so - called drop

$$D = \text{co }(\{a\} \cup X)$$

fulfills

(30) $$a = A \cap D;$$

(31) $$x_n \in D, \quad \text{dist } (x_n, A) \to 0 \Rightarrow x_n \to a$$

(i.e. Tykhonov well - posedness for the minimization of dist (x, A) subject to the constraint $x \in D$). By theorem 17, Hadamard well - posedness of (D, f) will be proved by showing that

(32) $$\text{diam } L(D, f, \alpha) \to 0 \text{ as } \alpha \to 0.$$

To this end, let $y_n \in E$ be such that

$$\text{dist } (y_n, D) \to 0, f(y_n) \to \inf f(D).$$

By (30), $f(x) \geq m$ for every $x \in D, x \neq a$. On the other hand $f(a) \leq m$, since $a \in A$. Fix $z \in D$, then by convexity

$$m \leq f(ta + (1 - t)z) \leq tf(a) + (1 - t)f(z), 0 \leq t < 1,$$

and letting $t \to 1$ we get

$$f(a) = m = \inf f(D).$$

Then applying lemma 25 we get dist $(y_n, A) \to 0$. There exists some $x_n \in D$ such that $\|x_n - y_n\| \to 0$. Therefore dist $(x_n, A) \to 0$ and $x_n \to a$ by (31), whence $y_n \to a$. So (32) is proved. Finally, consider any $x \in D$. Then

$$x = ta + (1 - t)b, b \in X, 0 \leq t \leq 1.$$

By convexity of dist $(\cdot, X)$,

$$\text{dist } (x, X) \leq \text{ dist } (a, X) \leq 2\varepsilon,$$

whence haus $(D, X) \leq 2\varepsilon$, ending the proof. $\qquad\square$

Proof of lemma 25. Arguing by contradiction, there exist $\alpha > 0, x_n \in X$ such that

$$f(x_n) < m + 1/n, \text{ dist } (x_n, A) \geq \alpha.$$

Now let $0 < t < 1, 0 < 1 - t < \alpha/\text{ dist } (x_0, X), 0 < \delta < (1 - t)[m - f(x_0)]/t$. Consider

$$y = tx_n + (1 - t)x_0, \quad n > 1/\delta.$$

Then

$$f(y) \leq tf(x_n) + (1 - t)f(x_0) < t(m + \delta) + (1 - t)m - \delta t < m$$

whence $y \in A$. But

$$\alpha \leq \text{ dist } (x_n, A) \leq \|x_n - y\| = (1 - t)\|x_n - x_0\| \leq (1 - t) \text{ dist } (x_0, X) < \alpha,$$

a contradiction. $\qquad\square$

Section 5.   A Topological Approach to Generic Well-Posedness.

An unifying approach to generic well - posedness in constrained optimization may be obtained by exploiting suitable generic continuity properties of multifunctions, and applying them to arg min.

We first apply this approach to obtain a different proof of theorems 16 and 20 when X is a compact space. This special setting is particularly well suited for a clear understanding of the methods involved.

A second application of this approach will show generic well - posedness within the space of lower semicontinuous functions equipped with the epi - topology (a concept borrowed from chapter IV).

We consider X a compact metric space, $Cl(X)$ equipped with the Hausdorff distance and $C^0(X)$ equipped with the topology of uniform convergence on X.

The multifunction

$$\text{arg min } : Cl(X) \times C^0(X) \rightrightarrows X$$

will play a key role. We shall work within the complete metric space

$$P = Cl(X) \times C^0(X)$$

(see Kuratowski [1, ch. III p.314]). The meaning of Hadamard well - posedness of a given $(A, f) \in P$ is the same as in section 4.

26 Lemma The multifunction arg min is upper semicontinuous.

Proof. It suffices to prove that arg min is a closed multifunction (Klein - Thompson [1, th. 7.1.16]). To this aim let $(A_n, f_n) \rightarrow (A, f)$ in $P, x_n \in \text{arg min } (A_n, f_n), x_n \rightarrow x$ in X. Since

$$f_n(x_n) \leq \text{ val } (A_n, f_n),$$

by lemma 24 and continuous convergence of f_n to f we get

$$f(x) = \lim f_n(x_n) \leq \lim \sup \text{ val } (A_n, f_n) \leq \text{ val } (A, f)$$

whence $x \in \text{arg min } (A, f)$. $\qquad\square$

Remark. Lemma 26 is a particular case of IV (13).

We need the following definition. Given two topological spaces Y, Z and a point $x_0 \in Y$, the multifunction

$$F : Y \rightrightarrows Z$$

is *lower almost continuous at x_0* iff for every open $V \subset Z$ such that

$$V \cap F(x_0) \neq \emptyset$$

there exists a neighbourhood U of x_0 in Y and some $W \subset U$ such that

$$W \text{ is dense in } U, \text{ and } V \cap F(x) \neq \emptyset \text{ for every } x \in W.$$

Lower almost continuity is a considerable weakening of lower semicontinuity, as seen from

27 Proposition *Let Z be second countable. Then every multifunction $F : Y \rightrightarrows Z$ is lower almost continuous at every point of some residual subset of Y.*

By using this result, we get

28 Proposition *Let Z be a separable metric space. If $F : Y \rightrightarrows Z$ is upper semicontinuous with nonempty values, then F is lower semicontinuous at every point of some residual subset of Y.*

Now, remembering lemma 26, we apply proposition 28 to the arg min multifunction. Therefore there exists some residual set $D \subset P$ at every point of which arg min is both upper and lower semicontinuous. We shall prove theorems 16 and 20 (in our special case), as follows.

Step 1. Arg min (A, f) is a singleton for every $(A, f) \in D$.

Proof. Fix $(A, f) \in D$. Given $x_0 \in$ arg min (A, f), for every neighbourhood U of x_0 lower semicontinuity yields

(33) $U \cap$ arg min $(B, g) \neq \emptyset$ for every (B, g) sufficiently near to (A, f) in P.

Arguing by contradiction, suppose that there exists $x_1 \in$ arg min (A, f) with $x_1 \neq x_0$. Let $c > 0$ be such that

$$x_1 \notin L = B(x_0, c).$$

We can find $h \in C^0(X)$ such that

$$0 \leq h \leq 1, h(x) = 1 \text{ if } x \in L, h(x_1) = 0.$$

Fix $\varepsilon > 0$ and put

$$g = f + \varepsilon h.$$

Then $g \in C^0(X)$. For every $x \in A$

$$g(x) \geq f(x) \geq f(x_1) = g(x_1)$$

whence $g(x_1) = \inf g(A)$. But if $x \in A \cap B(x_0, c)$

$$g(x) = f(x) + \varepsilon > f(x) \geq f(x_1) = g(x_1)$$

whence $B(x_0, c) \cap \arg \min (A, g) = \emptyset$. Moreover

$$\max\{|f(x) - g(x)| : x \in X\} \leq \varepsilon.$$

These facts contradict (33). □

Step 2. (A, f) is both Tykhonov and Hadamard well - posed for every $(A, f) \in D$.

Proof. By step 1, arg min (A, f) is a singleton, therefore (A, f) is Tykhonov well - posed by example I.6. Finally, let $(A_n, f_n) \to (A, f)$ in P, and $x_n \in \arg \min (A_n, f_n)$. By compactness of X and lemma 26 we see that $x_n \to \arg \min (A, f)$. This shows Hadamard well - posedness and concludes the proof. □

In the second application we shall employ some facts from chapter IV. We consider X a compact metric space, and $LS(X)$ the set of all proper extended real - valued functions on X which are lower semicontinuous, equipped with the epi - topology. It turns out that $LS(X)$ is a complete metrizable space (see Attouch [3, cor. 2.79 p. 255]). Moreover

$$\arg \min (X, \cdot) : LS(X) \rightrightarrows X$$

is upper semicontinuous by theorem IV.5. By proposition 28, arg min $(X, \cdot)$ is lower semicontinuous at every point of some residual subset D of $LS(X)$. We show that

(34)   arg min (X, f) is a singleton for every $f \in D$.

Mimicking the proof of step 1 above, suppose on the contrary that $x_0, x_1 \in \arg \min (X, f)$ with $x_0 \neq x_1$. For $\varepsilon > 0$ define .

$$f_\varepsilon(x) = f(x) \text{ if } x \neq x_1, f_\varepsilon(x_1) = f(x_1) - \varepsilon.$$

Then $f_\varepsilon \in LS(X)$ and epi - lim $f_\varepsilon = f$, as $\varepsilon \to 0$. Let U be a neighborhood of x_0 with $x_1 \notin U$. Then U fails to meet arg min $(X, f_\varepsilon) = \{x_1\}$ for all $\varepsilon > 0$, which contradicts lower semicontinuity. Then (34) is proved. Summarizing, we have

29 Theorem (X, f) *is generically Tykhonov and Hadamard well - posed in* $LS(X)$.

PROOFS.

Proposition 27. Let $V_n, n = 1, 2, ...$, be a countable base in Z. Directly from the definition we see that

(35) F is lower almost continuous at x_0 iff cl $F^{-1}(V_n)$ is dense in some neighborhood of x_0 for every n such that $F(x_0) \cap V_n \neq \emptyset$.

Now consider

(36) $H_j = F^{-1}(V_j) \setminus \text{int cl } F^{-1}(V_j), j = 1, 2, ...$

We prove that

(37) F is lower almost continuous at every point of $Y \setminus \bigcup_{j=1}^{\infty} H_j$.

To this aim, let $x_0 \in Y \setminus \bigcup_{j=1}^\infty H_j$ and fix any n such that $V_n \cap F(x_0) \neq \emptyset$. By (35) it suffices to show that cl $F^{-1}(V_n)$ is dense in some neighborhood of x_0. Since $x_0 \in F^{-1}(V_n)$, then $x_0 \in$ int cl $F^{-1}(V_n)$, therefore we get density as claimed and (37) is proved. But H_j given by (36) is rare, yielding the conclusion. □

Proposition 28. We show that F is lower semicontinuous at every point $x_0 \in Y$ in which F is lower almost continuous. Arguing by contradiction, there exists an open subset V of Z such that

$$F(x_0) \cap V \neq \emptyset,$$

and every neighborhood of x_0 contains some point y such that

(38) $$F(y) \cap V = \emptyset.$$

There exists an open ball $S \subset V$ such that

(39) $$\text{cl } S \subset V, \quad F(x_0) \cap S \neq \emptyset.$$

Therefore by lower almost continuity

(40) $$F^{-1}(S) \text{ is dense in some neighborhood of } x_0.$$

On the other hand

$$F(y) \cap \text{ cl } S = \emptyset$$

for each y fulfilling (38). Then by upper semicontinuity, (38) and (39), every neighborhood of x_0 contains some point y and some neighborhood W of y such that

$$F(W) \cap S = \emptyset$$

contradicting (40). □

Section 6.   Dense Well-Posedness for Problems with a Parameter.

We have already met examples of parametric minimum problems. In general we consider

(41) $$I : X \times P \rightarrow (-\infty, +\infty)$$

where the decision variable (with respect to which we minimize) belongs to X, a complete metric space, while the parameter space P is a Banach space.

We want conditions giving dense, or generic, well - posedness of $[X, I(\cdot, p)]$ for $p \in P$, either in the sense of Tykhonov or of Hadamard, by which now we mean continuous dependence on p of the optimal solution.

Examples. $I(x, p) = \|x - p\| +$ ind (K, x) (distance of p to K);
$I(x, p) = <p, x> +$ ind (K, x) (optimization of a linear form on K);
the linear regulator problem of example 1, where (for instance) $p = (u_0, x_0)$.

A theorem about dense well - posedness in P may be obtained as follows. We consider first a special case: additive perturbations by a positive multiple of $\|x - p\|^2$ to a lower semicontinuous function.

30 Proposition *Let P, P^* be uniformly convex Banach spaces, f proper, lower semi-continuous and bounded from below on $P, c > 0$ and*

$$U(x, p) = f(x) + c\|x - p\|^2.$$

Then $[P, U(\cdot, p)]$ is strongly Tykhonov well - posed generically in P.

Given I as in (41) we consider the optimal value (or marginal) function

$$V(p) = \inf\{I(x, p) : x \in X\},$$

assuming (lower semi) continuity of V on P. From proposition 30 and some results of dense approximate differentiability (in a suitably weak sense) of lower semicontinuous functions, we show the existence of a dense set T in P, such that for every $u \in T$ there exists a minorant $q \in C^1(P)$ for V which is exact at u. From this property we deduce dense strong Tykhonov well - posedness of $[X, I(\cdot, p)]$, assuming sufficient smoothness of I. The continuity of V gives strong Hadamard well - posedness, i.e. continuous dependence of the optimal solution on the parameter.

31 Theorem *Let P, P^* be uniformly convex, X a complete metric space, I as in (41). Then $[X, I(\cdot, p)]$ is densely Tykhonov well - posed in P, and $p \to arg\ min\ [X, I(\cdot, p)]$ is strongly continuous on the dense set of Tykhonov well - posedness, under the following assumptions:*

(42)   *V is everywhere continuous; moreover either V is bounded from below or P is separable :*

(43)   *$I(x, \cdot)$ is Fréchet differentiable on P for every x, $I(\cdot, p)$ is lower semicontinuous on X for every $p \in P$, ∇I is continuous on $X \times P$, and $\nabla I(\cdot, p)$ is one-to-one for every $p \in P$;*

(44)   *$p_n \to u$ in $P, x_n \in X, I(x_n, p_n) \to V(u)$ and $\nabla I(x_n, p_n)$ strongly convergent in P^* imply that x_n has strongly convergent subsequences;*

(45)   *$p_n \to u$ in P, x_n a bounded sequence and $I(x_n, p_n)$ bounded imply $I(x_n, p_n) - I(x_n, u) \to 0$.*

32 Applications of theorem 31. (a) Let X be a bounded closed set in the separable Banach space Y, let $f : Y \to (-\infty, +\infty)$ be lower semicontinuous and bounded from below,

$$I(x,p) = f(x) + <p, x>, p \in Y^* = P,$$

Y and Y^* uniformly convex. Checking assumptions is straightforward. We get a partial extension of theorem 6.

(b) Let $X = P$ and

$$I(x,p) = f(x) + c\|x - p\|^q$$

where $q > 1$ and c, f, P are as in proposition 30. Assumption (43) follows from Diestel [1, th.1 p.36], and (44) is fulfilled due to the homeomorphic character of the duality mapping of P. Continuity of V, even local Lipschitz continuity, may be checked as follows. Fix $p \in P$. Since $I(x, \cdot)$ is continuous, V is upper semicontinuous at p. Therefore there exist a real constant k and some ball S centered at p such that

$$V(u) \leq k \quad \text{if } u \in S.$$

Given $p' \in S$ let $x_n \in X$ be such that

(46) $$f(x_n) + c\|x_n - p'\|^q \leq V(p') + 1/n \leq k + 1.$$

Then $\|x_n\| \leq d$ for some constant d and every n, since f is bounded from below. For some constant L, every n and every $p', p'' \in S$ we have

$$|c\|x_n - p'\|^q - c\|x_n - p''\|^q| \leq L\|p' - p''\|,$$

hence by (46)

$$V(p') + 1/n \geq f(x_n) + c\|x_n - p''\|^q - L\|p' - p''\| \geq V(p'') - L\|p' - p''\|,$$

yielding

$$V(p'') - V(p') \leq L\|p' - p''\|,$$

hence the local Lipschitz continuity of V.

Taking $f = 0$ in 32 (b) we get a slightly weaker form of a theorem of Stečkin [1] (see also Braess [1, th.1.7] and Edelstein [1]), about density of points with a unique best approximation from a given closed set (as a matter of fact X needs only to be an E-space, moreover such points form a residual set : see Borwein - Fitzpatrick [2, th. 6.6]).

The following is an example of a point in a Hilbert space without best approximations from a closed subset thereof.

33 Example. Let X be a separable Hilbert space with orthonormal basis e_n. Let $K = \{e_1, e_2, e_3, ...\}$. Then K is closed. Given $x = \sum_{j=1}^{\infty} x_j e_j$ with every $x_j < 0$, we compute

$$\|x - e_n\|^2 = \|x\|^2 + 1 - 2x_n > \|x\|^2 + 1 = \text{dist}(x, K).$$

Thus we see that $\inf\{\|x - e_j\|^2 : j = 1, 2, 3, ...\}$ is never attained.

PROOFS.

Proposition 30. Put

$$V(p) = \inf\{f(x) + c\|x - p\|^2 : x \in P\},$$

$$S(p, a) = \{x \in P : f(x) + c\|x - p\|^2 \leq V(p) + a\},$$

$$T(\varepsilon) = \{p \in P : \text{ diam } S(p, a) \leq \varepsilon \text{ for some } a > 0\}.$$

We need to prove that

(47) $\qquad\qquad\qquad T(\varepsilon)$ is open and dense for every $\varepsilon > 0$.

In fact (47) implies $D = \cap_{n=1}^{\infty} T(1/n)$ dense in P from Baire's theorem (Kelley [1, th. 34 p.200]), moreover if $p \in D$ then given n there exists $a_n > 0$ such that

$$\text{diam } S(p, a) \leq 1/n \text{ if } 0 < a \leq a_n,$$

giving well - posedness corresponding to p by theorem I.11. Now for the proof of (47).

T (ε) **is open.** Fix $p \in T(\varepsilon)$. Then diam $S(p, a) \leq \varepsilon$ for some a. Given $q \in P$ and $x_1, x_2 \in S(q, a/3)$ then

(48) $\qquad\qquad\qquad f(x_j) + c\|x_j - q\|^2 \leq V(q) + a/3, j = 1, 2.$

Now V is upper semicontinuous, as an infimum of continuous functions, so if q is sufficiently near to p

(49) $\qquad\qquad\qquad\qquad V(q) \leq V(p) + a/3.$

Then

$$f(x_j) + c\|x_j - q\|^2 \leq V(p) + 2a/3$$

if $\|p - q\|$ is sufficiently small. By (48) and (49), for a suitable constant M and $j = 1, 2$

$$f(x_j) + c\|x_j - p\|^2 \leq f(x_j) + c(\|x_j - q\| + \|q - p\|)^2 \leq V(p) + 2a/3 + M\|p - q\|$$

since

$$c\|x_j - q\|^2 \leq V(q) + a/3 - f(x_j) \leq V(p) + 2a/3 - \inf f(P),$$

yielding uniform boundedness of $c\|x_j - q\|$. Then for q sufficiently near to p

$$f(x_j) + c\|x_j - p\|^2 \leq V(p) + a$$

hence $x_1, x_2 \in S(p, a)$, therefore

$$\|x_1 - x_2\| \leq \text{ diam } S(p, a) \leq \varepsilon$$

so that $q \in T(\varepsilon)$.

To show density of $T(\varepsilon)$, we need the following definition. Let $\varepsilon > 0$ and $p \in P$. Then V *is locally* ε - *subdifferentiable at* p iff there exists $u \in P^*$ such that for q sufficiently near to p

$$V(p) \leq V(q) + < u, p - q > + \varepsilon\|p - q\|.$$

34 Lemma V is locally ε - subdifferentiable at each point of a dense set in P.

Proof of lemma 34. V is lower semicontinuous, even locally Lipschitz, as shown in 32 (b). Fix any $p \in P$, let $\delta_1 > 0$ be such that

$$V(q) \geq V(p) - 1 \text{ when } \|q - p\| \leq \delta_1.$$

By the uniform convexity of P and P^*, the norm of P is Fréchet differentiable out of the origin (Diestel [1, th.1 p.36]). Fix any positive $\delta < \delta_1$, then

$$h(x) = \exp\left(\frac{1}{\delta^2 - \|x\|^2}\right) \text{ if } \|x\| < \delta, h(x) = +\infty \text{ if } \|x\| \geq \delta$$

is Fréchet differentiable on its effective domain. Set

$$g(q) = V(q) + h(q - p).$$

By th.1 bis p.445 of Ekeland [1] there exists p_1 such that

$$\|p_1 - p\| < \delta, g(p_1) \leq g(q) + (\varepsilon/2)\|q - p_1\| \text{ for all } q.$$

Then for $\|q - p\| < \delta$

$$V(p_1) - V(q) \leq h(q - p) - h(p_1 - p) + (\varepsilon/2)\|q - p_1\|.$$

By Fréchet differentiability of h at p_1, there exists $u \in P^*$ such that (with a perhaps smaller δ)

$$h(q - p) - h(p_1 - p) \leq \ <u, q - p_1> +(\varepsilon/2)\|q - p_1\|$$

when $\|q - p\| < \delta$. Then for such q

$$V(p_1)+ <u, p_1> \leq V(q)+ <u, q> +\varepsilon\|q - p_1\|.$$

Summarizing, V is locally ε - subdifferentiable at p_1, arbitrarily near to p. $\square$

T (ε) is dense. Since P is uniformly convex, the duality mapping of P,

$$J : P \to P^*$$

is the Fréchet differential of $(1/2)\| \cdot \|^2$. Again by uniform convexity of P and P^*, J is a homeomorphism between the strong topologies (Diestel [1, ch. 2, sec.4]), and J^{-1} is the duality mapping of P^*. From uniform convexity of P, J^{-1} is uniformly strongly continuous on bounded sets by a theorem of Smulian (prop. 3.1 p.42 of Browder [1]), moreover $\|Ju\| = \|u\|$ for all $u \in P$. Then, given $p \in P$ and a bounded set L in P, we can find $\delta > 0$ such that

(50) $$\|x_1 - x_2\| \leq \varepsilon \text{ if } x_1, x_2 \in L \text{ and } 2c\|J(p - x_1) - J(p - x_2)\| \leq \delta.$$

Now pick $p \in P$ at which V is locally $\delta/6 = b-$ subdifferentiable. By lemma 34 this happens on a dense set in P. Then for some $c_1 > 0, u \in P^*$

(51) $$V(p) \leq V(q) + b\|p - q\|+ <u, p - q> \text{ if } \|p - q\| < c_1.$$

Since J is a Fréchet differential,

(52) $$U(x,q) \leq U(x,p) + 2c < J(p-x), q-p > +b\|p-q\|$$

whenever $\|p-q\| \leq c_2$ for a suitable c_2. Let

$$c_3 = \min\{c_1, c_2\}, a = bc_3, x \in S(p,a).$$

From (52), if $\|p-q\| \leq c_3$

$$V(q) \leq V(p) + 2c < J(p-x), q-p > +2a.$$

Remembering (51) we get

$$< u - 2cJ(p-x), q-p > \quad \leq \; 3a \text{ when } \|p-q\| \leq a/b,$$

yielding

$$\|u - 2cJ(p-x)\| \leq \delta/2.$$

Then if $x_1, x_2 \in S(p,a)$

$$2c\|J(p-x_1) - J(p-x_2)\| \leq \delta.$$

Thus by (50) there exists $a > 0$ such that

$$x_1, x_2 \in S(p,a) \text{ implies } \|x_1 - x_2\| \leq \varepsilon,$$

giving $p \in T(\varepsilon)$ for each p of a dense set in P. This completes the proof of proposition 30. □

The proof of theorem 31 will require the following

Minorant lemma. Let f, P be as in proposition 30. Then there exists a dense set D in dom f, such that for every $x \in D$ we can find $q \in C^1(P)$ fulfilling

$$q \leq f, \quad q(x) = f(x).$$

Proof. By proposition 30, for every $c > 0$

$$\text{arg min } (P, f(\cdot) + c\|p - \cdot\|^2) \neq \emptyset$$

for every p in a dense set of P. Given $u \in \text{dom } f$ and $\varepsilon > 0$ set

$$c = (8/\varepsilon^2)[f(u) - \inf f(P)].$$

We show existence of q as required, q exact at some point at distance $\leq \varepsilon$ from u. If $c = 0$, the required q is the constant function $f(u)$. Otherwise $c > 0$ and we can find p such that $\|u - p\| < \varepsilon\sqrt{2}/4$, and x which minimizes $f(\cdot) + c\|p - \cdot\|^2$ on P. Then we define

(53) $$q(y) = f(x) + c\|p - x\|^2 - c\|p - y\|^2 \leq f(y), y \in P.$$

Therefore q minorizes $f, q \in C^1(P), q(x) = f(x)$. Setting $y = u$ in (53) we get

$$c\|x - p\|^2 \leq f(u) - \inf f(P) + c\|p - u\|^2 \leq \varepsilon^2 c/4$$

giving

$$\|u - x\| \leq \|u - p\| + \|p - x\| \leq \varepsilon.$$

□

Sketch of a geometrical proof of the minorant lemma.

By Stečkin's theorem (see Stečkin [1] ; see also Edelstein [1] for a weaker (density) result; a particular case of proposition 30 under our assumptions about P) there is a dense set of points in $P \times R$ with an unique nearest point from epi f. This gives a dense set of points x in dom f such that there exists a ball intersecting epi f exactly at $(x, f(x))$. Then for all y and some $u \in P^*$

$$f(y) \geq f(x) + <u, y - x> + o(\|y - x\|) = q(y), \text{ say } .$$

By a suitable smoothing of $o(t)$ we get the required minorant q.

Theorem 31. There exists a dense set D in P, for each point u of which V has a minorant in $C^1(P)$, exact at u. It suffices to apply the minorant lemma if V is bounded from below. If otherwise P is separable, let p_n be dense in P. By (42) there is a closed ball B_n around p_n such that V is bounded from below on B_n. Now using the minorant lemma for $f = V + \text{ind } B_n$, we find a dense set D_n in B_n with the required minorant property for V at each point of D_n. Then the same holds for the dense set $D = \bigcup_{n=1}^{\infty} D_n$ in P. Let $u \in D, q \in C^1(P)$ be such that

$$q \leq V, \quad q(u) = V(u).$$

We show that $[X, I(\cdot, u)]$ is Tykhonov well - posed. Let x_n be a minimizing sequence. Then for some $\varepsilon_n \neq 0$ and $\varepsilon_n \to 0$

$$I(x_n, u) - q(u) = I(x_n, u) - V(u) \leq \varepsilon_n^2.$$

We apply cor. 11 p. 456 of Ekeland [1] to

$$F_n(\omega) = I(x_n, \omega) - q(\omega), \omega \in P.$$

Then there exists $p_n \in P$ such that

$$F_n(p_n) \leq F_n(u); \|p_n - u\| \leq \varepsilon_n; \| \nabla F_n(p_n)\| \leq \varepsilon_n;$$

(54) $$F_n(p_n) \leq F_n(\omega) + \varepsilon_n \|\omega - p_n\| \text{ for all } \omega.$$

Then

$$\nabla I(x_n, p_n) - \nabla q(p_n) \to 0,$$

but $p_n \to u, q \in C^1(P)$, so that

(55) $$\nabla I(x_n, p_n) \to \nabla q(u).$$

By (54)

$$q(p_n) \leq I(x_n, p_n) \leq q(p_n) + I(x_n, u) - V(u) + \varepsilon_n \|p_n - u\| \leq q(p_n) + \varepsilon_n^2 + \varepsilon_n \|p_n - u\|,$$

and by continuity of q

$$I(x_n, p_n) \to V(u).$$

By (44) some subsequence $x_n \to x_0 \in X$, and by (45) $I(x_n, p_n) - I(x_n, u) \to 0$. Therefore

$$I(x_0, u) \leq \liminf I(x_n, u) = V(u),$$

thereby showing that x_0 minimizes $I(\cdot, u)$ on X. Continuity of the gradient gives by (55)

(56) $$\nabla I(x_n, p_n) \to \nabla q(u) = \nabla I(x_0, u).$$

Injectivity of the gradient gives uniqueness of the optimal solution to $[X, I(\cdot, u)]$ and convergence to it of all minimizing sequences, thus showing Tykhonov well - posedness for all $u \in D$. We end the proof by showing that

$$u_n \to u \text{ in } D \Rightarrow s(u_n) \to s(u)$$

where

$$s(u) = \arg \min [X, I(\cdot, u)], u \in D.$$

Let x_n be a given subsequence of $s(u_n)$. Then by (42)

$$I(x_n, u_n) = V(u_n) \to V(u).$$

Set

$$G_n(p) = I(x_n, p) - q(p), p \in P.$$

We can assume

$$G_n(u_n) = V(u_n) - q(u_n) \leq \varepsilon_n^2,$$

where $\varepsilon_n > 0$ and $\varepsilon_n \to 0$, so that

$$G_n(u_n) \leq \inf G_n(P) + \varepsilon_n^2$$

since $G_n \geq 0$. Again by Ekeland's theorem (cor. 11 of Ekeland [1]), for every n there exists $z_n \in P$ such that

$$\|z_n - u_n\| \leq \varepsilon_n, \| \nabla G_n(z_n)\| \leq \varepsilon_n, G_n(z_n) \leq \varepsilon_n^2,$$
$$G_n(z_n) \leq G_n(\omega) + \varepsilon_n \|z_n - \omega\|, \omega \in P.$$

This implies

(57) $$z_n \to u, G_n(z_n) \to 0, \nabla I(x_n, z_n) \to \nabla q(u)$$

giving

$$I(x_n, z_n) - q(z_n) \to 0$$

therefore

$$I(x_n, z_n) \to V(u),$$

thus by (44) a further subsequence $x_n \to x_0$. Then by (45)

$$I(x_n, z_n) - I(x_n, u) \to 0,$$

implying that x_0 minimizes $I(\cdot, u)$ on X. Well - posedness gives uniqueness, therefore

$$x_0 = s(u) = \lim s(u_n)$$

for the original sequence. $\square$

Remark. From the last convergence in (57) we see that for all u in a dense subset of P, $x_0 = \arg\min\,[X, I(\cdot, u)]$ satisfies the necessary condition

$$\nabla q(u) = \nabla I(x_0, u).$$

If in particular V turns out to be densely Fréchet differentiable with a continuous gradient, then we may assume $q = V$, obtaining $\nabla V(u) = \nabla I(x_0, u)$.

Section 7.   Well-Posedness and Sensitivity Analysis.

Let P be a Banach space, $p \in P$ a fixed point there, and let L be a closed ball in P around p. We consider a convergence space X and

$$I : X \times L \to (-\infty, +\infty).$$

Let

$$V : L \to [-\infty, +\infty)$$

be the value (marginal) function, defined by

$$V(q) = \text{val}\,[X, I(\cdot, q)] = \inf\{I(x, q) : x \in X\}.$$

Stability analysis for $q \to [X, I(\cdot, q)]$ deals with (semi) continuity properties of

$$q \to \varepsilon - \arg\min[X, I(\cdot, q)] \text{ and } q \to V(q),$$

while *sensitivity analysis* (roughly speaking) is concerned with differentiability properties of V and $q \to \arg\min[X, I(\cdot, q)]$.

We see from the proof of theorem 31 that differentiability properties of V at p play an important role as far as Tykhonov well - posedness of $[X, I(\cdot, p)]$ is concerned. By adapting that reasoning we get

35 Theorem $[X, I(\cdot, p)]$ *is Tykhonov well - posed, moreover*

$$(58) \qquad p_n \to p, x_n \in X, \quad I(x_n, p_n) - V(p_n) \to 0 \Rightarrow x_n \to \arg\min\,[X, I(\cdot, p)]$$

under the following assumptions:

$(59) \qquad$ *V is Fréchet differentiable on int L, with ∇V continuous at p;*

for every x, $I(x, \cdot)$ is Fréchet differentiable on int L, with ∇I continuous, and I lower semicontinuous on $X \times L$, and $\nabla I(\cdot, p)$ is one - to - one on $\arg\min[X, I(\cdot, p)]$;

$(60) \qquad p_n \to p, I(x_n, p_n) \to V(p), \nabla I(x_n, p_n)$ *strongly convergent in P^**

imply that x_n has a convergent subsequence.

Proof. Let x_n be any minimizing sequence to $[X, I(\cdot, p)]$, so that

$$I(x_n, p) \to V(p).$$

Let
$$F_n(\omega) = I(x_n, \omega) - V(\omega), \omega \in L.$$

Then there exists a sequence $\varepsilon_n \to 0$ with $\varepsilon_n > 0$ such that
$$F_n(p) \leq \inf F_n(L) + \varepsilon_n^2.$$

From cor. 11 p.456 of Ekeland [1], we find $p_n \in L$ such that
$$F_n(p_n) \leq F_n(\omega) + \varepsilon_n \|\omega - p_n\|, \omega \in L; \ \|p_n - p\| \leq \varepsilon_n; \ \|\nabla F_n(p_n)\| \leq \varepsilon_n.$$

Thus
$$V(p_n) \leq I(x_n, p_n) \leq V(p_n) + I(x_n, p) - V(p) + \varepsilon_n^2$$

yielding
$$I(x_n, p_n) \to V(p),$$

moreover
$$\nabla I(x_n, p_n) - \nabla V(p_n) \to 0.$$

Then by (59)
$$\nabla I(x_n, p_n) \to \nabla V(p).$$

Therefore, by (60), for some subsequence
$$x_n \to x_0 \in \arg \ \min \ [X, I(\cdot, p)],$$

and for the same subsequence

(61)
$$\nabla I(x_n, p_n) \to \nabla I(x_0, p) = \nabla V(p).$$

Given any $y_0 \in \arg \ \min \ [X, I(\cdot, p)]$, we apply the previous reasoning to the constant sequence $y_0, y_0, y_0, \cdots$. By injectivity of the gradient and (61)
$$x_0 = \arg \ \min \ [X, I(\cdot, p)],$$

therefore every minimizing sequence converges to x_0, giving Tykhonov well - posedness. Now let x_n, p_n be as in (58). We can assume
$$I(x_n, p_n) \leq V(p_n) + \varepsilon_n^2, \ \varepsilon_n \to 0.$$

Once more we apply cor.11 of Ekeland [1] to
$$G_n(q) = I(x_n, q) - V(q).$$

We find z_n satisfying
$$\|z_n - p_n\| \leq \varepsilon_n, \|\nabla G_n(z_n)\| \leq \varepsilon_n, G_n(z_n) \leq \varepsilon_n^2;$$
$$G_n(z_n) \leq G_n(\omega) + \varepsilon_n \|z_n - \omega\|, \omega \in L.$$

Hence
$$I(x_n, z_n) \to V(p), \nabla I(x_n, z_n) \to \nabla V(p).$$

A subsequence of x_n converges towards $\arg \ \min \ [X, I(\cdot, p)]$ by (60), and the uniqueness of the optimal solution gives the conclusion. $\square$

Another interesting link between stability and sensitivity analysis is based on the following fact. Let $f : R^n \to (-\infty, +\infty]$ be lower semicontinuous. Then f is locally Lipschitz near a given point $x \in R^n$ iff its Clarke's generalized gradient $\partial f(x)$ is nonempty and bounded (see Rockafellar [7, th.4]). By applying this theorem to the optimal value function of mathematical programming problems with data perturbations, it is possible to obtain criteria (even necessary and sufficient) for local Lipschitz behaviour of the optimal value function with respect to perturbations. The generalized gradient of such a function, however, is related to the set of (suitably defined) multipliers. Thus, sensitivity analysis in terms of the behaviour of (generalized) Lagrange multipliers implies relevant informations about Hadamard well - posedness.

Section 8.   Notes and Bibliographical Remarks.

To section 2. For an extension of theorem 6 without assuming convexity of I see Stegall [2] (compare with 32 (a)). Propositions 8, 11, example 10 and theorems 9, 12 are due to Lucchetti- Patrone [1]. The proof of theorem 12 follows Patrone [4].

Generic uniqueness for best approximation problems in uniformly convex Banach spaces was proved by Stečkin [1], Edelstein [1]. For a sharper result see De Blasi - Myjak [4]. Generic Tykhonov well -posedness for best approximation problems is obtained in De Blasi - Myjak [1], [3]. Related results for the metric projection are given in Beer [1] and Georgiev [1].

The conclusion of theorem 12 holds in the setting of quasi - convex functions on R^P, as shown in Beer -Lucchetti [1] (since in this setting uniqueness implies well - posedness, as easily shown by mimicking example I.7).

To section 3. Theorem 15 is due to Lucchetti - Patrone [1]. Topological criteria for genericity of Tykhonov, or generalized, well - posedness in $BC^0(X)$, X a topological space, which generalize theorem 15, are due to Čoban - Kenderov - Revalski [1] and Revalski [4].

Results of Beer - Lucchetti [2], [3] show that in the infinite -dimensional case, most problems (X, I) with $I \in C_0(X)$ are not Tykhonov well - posed with respect to the Mosco topology (compare with proposition 11), while most convex problems are Tykhonov well - posed for any Banach space X with respect to the local epi-distance topology (often called "aw - topology": see section II.3). Moreover, most problems are Levitin - Polyak well -posed in that topological setting.

To section 4. Theorems 16, 17, 20, 21 and corollary 18 are due to Revalski [1], [3] and [5], where the lower semicontinuous setting is considered. For related results see Lucchetti - Patrone [1], and Čoban - Kenderov - Revalski [1] for a topological approach. For single problems without uniqueness see also Bednarczuk - Penot [1], [2].

Metric spaces in which Hadamard and Levitin - Polyak well - posedness are equivalent are characterized by Revalski - Zhivkov [1], with links to a notion of strong well - posedness introduced by Beer - Lucchetti [2]. See Beer - Lucchetti [3, th. 4.2] for a characterization of well - posedness in the generalized sense (section I.6) for convex problems, analogous to corollary 18 (but using instead the local epi - distance convergence).

The counterexample after theorem 21 is due to De Blasi - Myjak [2]. Theorem 22 is due to Georgiev [1] (where related genericity results are presented), and its corollary to De Blasi - Myjak [1].

To section 5. The definition of lower almost continuity and proposition 27 are particular cases of Kenderov [1], where further results about generic uniqueness and well - posedness may be found (as a byproduct of generic continuity properties for multifunctions). Generic uniqueness (step 1) is proved by Kenderov [2] for a class of topological spaces. Related results based on Kenderov's approach are given in Revalski [3], Beer - Kenderov [2], Beer [2], Todorov [1].

In Beer [2] it is shown that generic well - posedness holds in the more appropriate setting obtained by identifying in P pairs (A, f) and (A, g) when $f = g$ on A, since then (A, f) and (A, g) define the same constrained optimization problem. Generic well - posedness is characterized by Revalski [4] in the quotient space obtained by defining equivalence between (X, f) and (X, g) whenever they have the same minimizing sequences.

Proposition 28 is related to the result of Fort [1] (originally proved for multifunctions with compact values in a metrizable space). The proof we present here is due to Kenderov.

To section 6. Theorem 31 is obtained by particularizing results of Ekeland - Lebourg [1], who unified a number of scattered results. See also Aubin - Ekeland [1, ch. 5, sec. 7]. In that paper, Lipschitz continuity of the value function and generic fulfillment of the first - order optimality conditions are also considered. Related results are surveyed in Lebourg [2]. Some ideas in the proof of theorem 31 are due to Lebourg [1]. Particular cases of theorem 31, not requiring however uniform convexity of P^*, are in Baranger [1] which contains applications to the optimal control of partial differential equations, and in De Blasi - Myjak [3]. For a stronger minorant lemma see cor. 7.1 of Borwein - Strojwas [1], where density results for nearest points to closed sets may be found. A counterexample to density for nearest points was given by Edelstein [2]. For some partial generalizations see Borwein [1].

Generic existence in a setting similar to that of theorem 31 is shown in Baranger - Temam [1]. Dense Tykhonov well - posedness in the setting of application 32 (a), if Y has the Radon - Nikodým property, is proved by Stegall [1]. See also Witomski [1], Phelps [1 p. 88]. Application 32 (a) when Y has the Radon - Nikodým property is due to Asplund [1].

Generic Hadamard well - posedness for linear semi - infinite optimization problems may be found in Todorov [1], and density results in Nürnberger [1]. Density of points in function spaces having strongly unique (hence Tykhonov well - posed) best approximation from given subspaces is obtained by Nürnberger - Singer [1] and Angelos - Schmidt [1]. For generic well - posedness of the metric projection in E - spaces see Konyagin [1].

To section 7. The link between sensitivity analysis and Lipschitz behaviour of the optimal value function in mathematical programming is exploited (via nonsmooth analysis) in Rockafellar [9] and [10]. See section VIII.2 for an interesting connection between Tykhonov well-posedness and differentiability of the optimal value function for Lagrange problems.

Chapter IV.

WELL-POSEDNESS AND VARIATIONAL, EPI- AND MOSCO CONVERGENCES

Section 1. Variational Convergences and Hadamard Well-Posedness.

We are interested here in a well - posedness analysis for general variational problems (X, I). Therefore we assume as little structure as possible. We suppose

$$X \quad \text{a given convergence space}$$

in the sense of Kuratowski [1, p. 83 - 84], and consider functions

$$I : X \to [-\infty, +\infty]$$

The sequential setting is sufficiently simple and useful for most applications in optimization.

There are many forms of Hadamard well - posedness for (X, I), as far as the dependence upon I is considered. Given a convergence for sequences of functions, denoted by $I_n \to I$, and a convergence $\to$ on subsets of X, the following are various forms of Hadamard well - posedness:

(1)
$$\text{val } (X, I_n') \to \text{val } (X, I);$$

(2)
$$\varepsilon - \arg\min (X, I_n) \to \varepsilon - \arg\min (X, I) \text{ for every fixed } \varepsilon \geq 0;$$

of particular significance is the convergence of $\arg\min(X, I_n)$ to $\arg\min (X, I)$;

(3) $\varepsilon_n \to 0, x_n \in \varepsilon_n - \arg\min (X, I_n)$ imply that every cluster point of x_n belongs to $\arg\min (X, I)$; conversely given $x \in \arg\min(X, I)$ there exists $\varepsilon_n \to 0$ such that $x \in \liminf [\varepsilon_n - \arg\min (X, I_n)]$.

We remark that if $I_n = I$ for every n, then (3) coincides with (Tykhonov) well - posedness in the generalized sense (section I.6) provided

$$\limsup[\varepsilon_n - \arg\min (X, I)] \neq \emptyset.$$

If (1) is fulfilled, then

$$I_n(x_n) \to \text{val } (X, I) \text{ whenever } x_n \in \varepsilon_n - \arg\min (X, I_n), \ \varepsilon_n \to 0.$$

Under suitable conditions, it may happen that $I(x_n) \to \text{val } (X, I)$. This means that ε_n - solutions to the approximating problems (X, I_n) can be used as approximate solutions to (X, I).

Under some regularity conditions we are also interested whether, loosely speaking,

(4)   there is some form of convergence of the corresponding necessary
optimality conditions (e.g., the set of multipliers do converge).

The main problem is of course that of finding a suitable "variational" convergence

$$I_n \to I$$

which entails some properties among (1),...,(4). Roughly speaking, we need passing in the limit under the operation of taking minimum, that is we want

$$\min(\lim I_n) = \lim(\min I_n)$$

as far as optimal values and (approximate) optimal solutions are concerned.

There is a strong link between the above issues and *approximation* and *convergence* problems in optimization. Having fixed the underlying space X (in which the optimization problems of interest take place)
  approximation means that (X, I) is given. Its solution is obtained approximately by finding a sequence $I_n \to I$ in a "variational" sense and solving (X, I_n);
  convergence means that the sequence (X, I_n) is given, and the right "variational" limit (X, I) is to be found.

As a first approach to these problems, we could guess that a reasonable answer lies in using, as "variational" convergences, classical modes of convergence (e.g. locally uniform). However, there exist very significant optimization problems in which classical convergences turn out to be either unduly restrictive or not "variational" (in the sense that none among (1),..., (4) can be obtained).

1 Example: locally uniform convergence may be not "variational".
Let $X = R, I(x) = |x|$,
  $I_n(x) = |x|$ if $|x| \le n, = 2n - |x|$ if $n < |x| \le 3n, = |x| - 4n$ if $|x| > 3n$.
Then $I_n \to I$ uniformly on compact sets, but

$$\text{val}\,(R, I_n) = -n \not\to \text{val}\,(R, I) = 0,$$

$$\arg\min\,(R, I_n) = \{\pm 3n\} \not\to \arg\min\,(R, I) = \{0\}$$

in any reasonable sense (and the same is true for $\varepsilon - \arg\min(X, I_n)$). See the figure on the next page.
  Of course, if K is any (nonempty) set and $I_n, I : K \to (-\infty, +\infty)$ are bounded from below, then uniform convergence of I_n towards I yields $\inf I_n(K) \to \inf I(K)$. Indeed, put

$$a_n = \inf I_n(K), \quad a = \inf I(K).$$

Then $a_n \le I_n(x)$ for every x and n, hence $\limsup a_n \le I(x)$, yielding $\limsup a_n \le a$. For every $\varepsilon > 0$ and n pick $u_n \in K$ such that $I_n(u_n) \le a_n + \varepsilon$. Hence $a - \varepsilon + I_n(u_n) - I(u_n) \le a_n$, and letting $n \to +\infty$ we get $a - \varepsilon \le \liminf a_n$. Thus $a_n \to a$, as claimed.

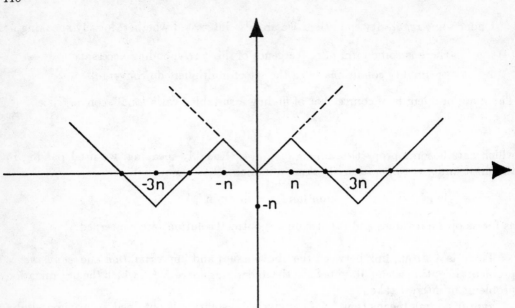

2 Example: classical convergences may be restrictive. Let X be an infinite - dimensional separable Hibert space, with orthonormal basis e_n. Fix any $x_0 \in X$ and set

$$I_n(x) = \|x - x_0\| + \text{ ind } (K_n, x), I(x) = \|x - x_0\| + \text{ ind } (K, x)$$

where

$$K = \{0\}, K_n = \text{closed segment with ends } 0 \text{ and } e_n.$$

Here I_n does not converge locally uniformly to I, but from theorem II.9

$$\arg \min (X, I_n) \to 0 = \arg \min (X, I),$$

$$\text{val } (X, I_n) \to \|x_0\| = \text{val } (X, I)$$

since $M - \lim K_n = K$.

3 Example: pointwise convergence may be inadequate, because not "variational".
Let $X = H_0^1(0, 2\pi)$ equipped with the weak convergence;

$$a_n(t) = 2 + \frac{\sin(nt)}{|\sin(nt)|}, n = 1, 2, 3, ...; f \in L^2(0, 2\pi);$$

$$I_n(x) = (1/2) \int_0^{2\pi} a_n(t)\dot{x}(t)^2 dt - \int_0^{2\pi} f(t)x(t)dt.$$

Which is the "variational" limit of I_n (if any)? Since $a_n \to b$ in $L^\infty(0, 2\pi)$, where $b(t) = 2$ a. e. in $(0, 2\pi)$, we get pointwise convergence, i.e.

$$I_n(x) \to J(x) = \int_0^{2\pi} \dot{x}^2 dt - \int_0^{2\pi} fx \, dt, x \in X.$$

On the other hand

$$x_n = \arg\min (X, I_n)$$

is the weak (distributional) solution to the Euler - Lagrange equation

(5) $$-\frac{d}{dt}(a_n \dot{x}_n) = f, \quad x_n \in H_0^1(0, 2\pi).$$

Now let $g \in H^1(0, 2\pi)$ be such that $\dot{g} = f$. We get

$$-\dot{x}_n = \frac{1}{a_n}(g + c_n)$$

where

$$c_n = -(\int_0^{2\pi} g \, a_n^{-1} dt)(\int_0^{2\pi} a_n^{-1} dt)^{-1}$$

since $\int_0^{2\pi} \dot{x}_n dt = 0$. But $a_n^{-1} \to a_0^{-1}$ in $L^\infty(0, 2\pi)$, where $a_0(t) = 3/2$ a.e. Then

$$c_n \to c_0 = -(\int_0^{2\pi} g a_0^{-1} dt)(\int_0^{2\pi} a_0^{-1} dt)^{-1}$$

thus giving $x_n \to x_0$ in $H_0^1(0, 2\pi)$, where

$$-\frac{d}{dt}((3/2)\dot{x}_0) = f, \quad x_0 \in H_0^1(0, 2\pi).$$

Summarizing, if we define

$$I(x) = (3/4)\int_0^{2\pi} \dot{x}^2 dt - \int_0^{2\pi} fx \, dt$$

we get

$$\arg\min (X, I_n) \to \arg\min (X, I) \text{ in } H_0^1(0, 2\pi).$$

Moreover, multiplying (5) by x_n and integrating by parts we get

$$\text{val } (X, I_n) = \frac{1}{2}\int_0^{2\pi} a_n \dot{x}_n^2 dt - \int_0^{2\pi} fx_n \, dt = -\frac{1}{2}\int_0^{2\pi} fx_n \, dt$$

$$\to -\frac{1}{2}\int_0^{2\pi} fx_0 \, dt = \text{ val } (X, I).$$

Therefore the "variational" limit of (X, I_n) is given by (X, I), and not by the pointwise limit (X, J) which has a different optimal value and a different minimizer. (See example VIII.16).

Under restrictive compactness conditions, local uniform convergence is "variational".

4 Proposition *Let $f_n, f : X \to (-\infty, +\infty)$ be with f lower semicontinuous and $f_n \to f$ uniformly on compact sets. If there exists a compact set K such that $\inf f_n(K) = \inf f_n(X)$ for all n, then as $n \to +\infty$*

$$\inf f_n(X) \to \inf f(X);$$

if for some subsequence $f_n(x_n) - \inf f_n(X) \to 0$ and $x_n \to x_0$, then $x_0 \in arg\ min\ (X, f)$;

$$limsup\ \varepsilon - arg\ min\ (X, f_n) \subset \varepsilon - arg\ min\ (X, f), \varepsilon \geq 0.$$

Proof. Given $\varepsilon > 0$, there exists p such that if $x \in K$ and $n \geq p$ then $f(x) - \varepsilon \leq f_n(x)$. By assumption

$$\inf f(X) \leq \inf f(K) \leq \varepsilon + \inf f_n(K) = \varepsilon + \inf f_n(X),$$

hence

$$\inf f(X) \leq \lim \inf \inf f_n(X).$$

By pointwise convergence, for any u

$$f_n(u) \to f(u), \quad f_n(u) \geq \inf f_n(X)$$

yielding

$$\lim \sup \inf f_n(X) \leq \inf f(X),$$

thereby proving value convergence. Now consider x_n as in the assumption, with $x_{n_j} \to x_0$, and the compact set

$$E = \{x_{n_j}\} \cup \{x_0\}$$

on which f_n converges uniformly toward f. Then $f_{n_j}(x_{n_j}) - f(x_{n_j}) \to 0$. By lower semicontinuity and value convergence

$$\inf f(X) = \lim f_n(x_n) = \lim f(x_{n_j}) \geq f(x_0).$$

Finally if $c \in R, y_n \in X$ with a cluster point y and $f_n(y_n) \leq c$, then as before $f(y) \leq c.\square$
The above examples show that proposition 4 is of limited value.

Section 2. Variational Convergence.

We introduce now a suitable convergence, which implies (1), (2), (3) under minimal assumptions. Let X be a given convergence space, and

$$I, I_n : X \to [-\infty, +\infty]$$

be a given sequence of extended real-valued functions. For the time being, assume that for every n sufficiently large, we have val $(X, I_n) > -\infty$. Fix $\varepsilon \geq 0$.
Suppose that

(6) $\qquad\qquad x_n \to x$ in X implies $\liminf I_n(x_n) \geq I(x)$.

Given $y_n \in \varepsilon - \arg\min(X, I_n)$ we get

$$I_n(y_n) \leq \text{val}(X, I_n) + \varepsilon.$$

Let now $y_{n_j} \to y$ for some subsequence n_j. Consider $z_n = y_{n_j}$ if $n_j \leq n < n_{j+1}$, then $z_n \to y$ and by (6)

(7) $\qquad I(y) \leq \liminf I_n(z_n) \leq \liminf I_{n_j}(y_{n_j}) \leq \liminf \text{val}(X, I_{n_j}) + \varepsilon.$

We remark explicitly here that, as seen from the above calculation yielding (7), (6) amounts to

(8) for every subsequence n_j of the positive integers,

$$\text{if } x_j \to x \text{ then } \liminf I_{n_j}(x_j) \geq I(x).$$

Moreover suppose that

(9) $\qquad$ for every $u \in X$ there exist $u_n \in X$ such that $\limsup I_n(u_n) \leq I(u)$.

Fix $u \in X$ and let u_n be as in (9). Then

(10) $\qquad \liminf \text{val}(X, I_{n_j}) \leq \limsup \text{val}(X, I_{n_j}) \leq \limsup \text{val}(X, I_n)$
$\qquad\qquad \leq \limsup I_n(u_n) \leq I(u).$

By (7), (10) and arbitrariness of u

(11) $\qquad\qquad \limsup \text{val}(X, I_n) \leq \text{val}(X, I), y \in \varepsilon - \arg\min(X, I).$

Now let $\varepsilon_n \to 0, x_n \in \varepsilon_n - \arg\min(X, I_n)$ be such that for some $x \in X$ a subsequence $x_{n_j} \to x$. Then by (8) and (11)

$$I(x) \leq \liminf I_{n_j}(x_{n_j}) \leq \limsup I_n(x_n) \leq \limsup \text{val}(X, I_n) \leq \text{val}(X, I)$$

yielding

$$x \in \arg\min(X, I).$$

Moreover, if there exists a relatively compact sequence $x_n \in \varepsilon_n - \arg\min(X, I_n)$, any subsequence of x_n has a further convergent subsequence. Fix n_j such that

$$\text{val}(X, I_{n_j}) \to \liminf \text{val}(X, I_n).$$

Then some subsequence $x_{n_{j_k}} \to y$, and by (8)

$$\text{val}(X, I) \leq I(y) \leq \liminf I_{n_{j_k}}(x_{n_{j_k}}) \leq \liminf \text{val}(X, I_{n_{j_k}}) = \liminf \text{val}(X, I_n).$$

By (11) we conclude that

$$\text{val}(X, I_n) \to \text{val}(X, I).$$

Summarizing, conditions (6) and (9) isolate a mode of convergence which entails good well - posedness properties.

We shall say that I_n *converges variationally to* I, and write

$$var - lim\, I_n = I$$

(more precisely $c\ var - \lim I_n = I$ if c is the convergence on X and there is danger of confusion), iff

(6) $x_n \to x$ implies $\liminf I_n(x_n) \geq I(x)$

and

(9)   for every $u \in X$ there exists $u_n \in X$ such that $\limsup I_n(u_n) \leq I(u)$.

The following basic theorem justifies the definition of variational convergence.

5 Theorem *Assume that var - $\lim I_n = I$. Then*

(12) $\limsup val\, (X, I_n) \leq val\, (X, I);$

(13) $\limsup [\varepsilon - arg\, min\, (X, I_n)] \subset \varepsilon - arg\, min\, (X, I)$
          for all sufficiently small $\varepsilon \geq 0;$

(14)   *if $\varepsilon_n \geq 0, \varepsilon_n \to 0$, then $\limsup [\varepsilon_n - arg\, min\, (X, I_n)] \subset arg\, min\, (X, I)$.*

Moreover if $\limsup [\varepsilon_n - arg\, min\, (X, I_n)] \neq \emptyset$ then $\limsup val\, (X, I_n) = val\, (X, I)$ and of course arg min $(X, I) \neq \emptyset$. Finally, if there exist $\varepsilon_n \to 0$ and a relatively compact sequence $x_n \in \varepsilon_n - arg\, min\, (X, I_n)$, then

(15) $val\, (X, I_n) \to val\, (X, I) = min\, I(X).$

Theorem 5 says that, with respect to variational convergence, the following forms of Hadamard well - posedness hold:

$I \to\ val\, (X, I)$ is upper semicontinuous;

$I \rightrightarrows \varepsilon - arg\, min\, (X, I)$ is upper semicontinuous by inclusion;

moreover, under compactness assumptions

$I \to\ val\, (X, I)$ is continuous;

limits of (approximate) $\varepsilon-$ solutions to (X, I_n) as $\varepsilon \to 0$ are global optimal solutions to (X, I).

Remark. Let X be a topological space and let I be any extended real - valued function on X. Then

$$(\varepsilon, J) \rightrightarrows \varepsilon - \arg\min (X, J)$$

is upper semicontinuous at $(0, I)$ in the following sense. Suppose that

$$\text{var} - \lim I_n = I$$

(convergence on X being induced by the topology) and I_n is equicoercive (see (17) below). Then for every neighborhood U of $\arg\min(X, I)$ there exist $\varepsilon > 0$ and an integer p such that

$$\varepsilon - \arg\min (X, I_n) \subset U \text{ for all } n \geq p.$$

For, suppose not. Then we can find a neighborhood V of $\arg\min(X, I)$ and points $y_n \in c_n - \arg\min (X, I_n)$ with $c_n \to 0$ and y_n relatively compact, such that $y_n \notin V$. This contradicts theorem 5.

Remark. Theorem 5 yields an existence theorem for optimal solutions to the variational limit problem of (X, I_n); by (14) it is in fact sufficient that for some $\varepsilon_n \to 0$

$$(16) \qquad\qquad \limsup [\varepsilon_n - \arg\min (X, I_n)] \neq \emptyset.$$

This approach to existence generalizes the direct method of calculus of variations, based on lower semicontinuity and compactness. Suppose that $I : X \to (-\infty, +\infty]$ is lower semicontinuous. Then the constant sequence $I_n = I$ is variationally convergent toward I. In this case, (16) amounts to the existence of a minimizing sequence with cluster points. A sufficient condition to (16) in a frequently encountered application is the following. Let X be a reflexive Banach space equipped with the weak convergence. Suppose that

$$\text{var} - \lim I_n = I, \quad I \text{ is finite somewhere}$$

and

$$u_n \in X, \sup I_n(u_n) < +\infty \Rightarrow u_n \text{ is bounded.}$$

Then by (11), for some $x_0 \in X$

$$\limsup \text{val} (X, I_n) \leq I(x_0).$$

Therefore $x_n \in \varepsilon_n - \arg\min (X, I_n)$ implies that $I_n(x_n)$ is bounded from above. Equicoercivity (see (17) below) entails weak sequential compactness of x_n.

Remarks. (a). (14) and (15) hold if $\text{var} - \lim I_n = I$ and there exists a relatively compact set K such that

$$\text{val} (X, I_n) = \text{val} (K, I_n) \text{ for each } n.$$

(b). Assume that $\text{var} - \lim I_n = I$, $\varepsilon_n \to 0$ and $x_n \in \varepsilon_n - \arg\min (X, I_n)$. If I is upper semicontinuous and x_n is relatively compact, then

$$I(x_n) \to \text{val} (X, I),$$

since for every subsequence, some further subsequence $x_n \to y$ say, and by (14)

$$\text{val}\,(X,I) \le \lim\inf I(x_n) \le \lim\sup I(x_n) \le I(y) = \ \text{val}\,(X,I).$$

6 Corollary *Let T be a set of extended real - valued functions on X such that each arg min $(X,I), I \in T$, is a singleton. For each sequence $I_n \in T$ let there exist $\varepsilon_n \to 0$ and some relatively compact sequence x_n such that $x_n \in \varepsilon_n - arg\ min\ (X, I_n)$. Then*

$$I \to \ val\,(X, I), I \to arg\ min\ (X, I)$$

are continuous on T equipped with variational convergence.

Example. Existence of x_n as in corollary 6 obtains under *equicoercivity* of T, i. e.

(17) $I_n \in T, \ x_n \in X, \ \sup I_n(x_n) < +\infty \Rightarrow x_n$ relatively compact.

This happens if X is a reflexive Banach space equipped with the weak convergence, and there exists

$$a\colon [0, +\infty) \to R, \ a(t) \to +\infty \text{ as } t \to +\infty$$

such that for every $I \in T$

$$I(x) \ge a(\|x\|), \ x \in X.$$

By corollary 6, under uniqueness of the minimizers and equicoercivity, we get solution and value Hadamard well - posedness with respect to the variational convergence.

Example 1 shows that (15) fails in general. Another example follows. Take $X = R, I_n(x) = x$ if $x \ge 0, = x/n$ if $-n \le x \le 0, = -1$ if $x \le -n$. Here var - lim $I_n = I$ where $I(x) = \max\{0, x\}$ (in fact $I_n \to I$ uniformly on compact sets), but

$$\arg\min (X, I_n) = -n; \quad \text{val}\,(X, I_n) = -1, \ \text{val}\,(X, I) = 0.$$

We have trivially proper inclusions in (13) since $\varepsilon - \arg\min (X, I_n) = (-\infty, n(\varepsilon - 1)]$ for $0 < \varepsilon < 1$, hence $\lim\sup [\varepsilon - \arg\min (X, I_n)] = \emptyset$.

7 Remark. If $I_n \to I$ continuously, i. e.

$$x_n \to x \Rightarrow I_n(x_n) \to I(x)$$

then obviously var $- \lim I_n = I$. Variational convergence is not a Kuratowski convergence, since uniqueness of the limit may fail. For example the constant sequence $I_n(x) = x^2, X = R$, converges variationally toward both $J_1(x) = x^2$ and $J_2(x) = 0$.

Note on terminology. *Variational convergence* is a technical term referring to the convergence defined by (6) and (9). The more vague term *variational convergences* refers to any convergence yielding at least one form of Hadamard well - posedness (1), (2), (3). (Sometimes we say that such convergences are variational.)

PROOFS.

Theorem 5. **Proof of (12).** The only case of interest arises when

$$-\infty < \limsup \operatorname{val}(X, I_n).$$

Suppose that $\operatorname{val}(X, I) > -\infty$. Take $\varepsilon > 0$ and $y \in X$ such that

$$I(y) \leq \operatorname{val}(X, I) + \varepsilon.$$

From (9) there exists y_n such that

(18) $$\limsup I_n(y_n) \leq I(y)$$

implying

$$\limsup \operatorname{val}(X, I_n) \leq \limsup I_n(y_n) \leq \varepsilon + \operatorname{val}(X, I)$$

and the conclusion follows. Suppose that $\operatorname{val}(X, I) = -\infty$. Then for every $\varepsilon > 0$ there exists y such that $I(y) \leq -1/\varepsilon$, and by (9) we can find y_n fulfilling (18), so that

$$-\infty < \limsup \operatorname{val}(X, I_n) \leq' \limsup I_n(y_n) \leq -\frac{1}{\varepsilon},$$

a contradiction, so (12) is proved.

Proof of (13). Let $x \in \limsup [\varepsilon - \arg\min (X, I_n)]$. Then there are points $x_n \in \varepsilon-\arg\min (X, I_n)$ such that a subsequence $x_{n_j} \to x$. Suppose that $\limsup \operatorname{val}(X, I_{n_j}) > -\infty$. By (12)

$$\operatorname{val}(X, I) > -\infty.$$

Moreover, for some further subsequence, $\operatorname{val}(X, I_n) > -\infty$, hence

$$I_n(x_n) \leq \operatorname{val}(X, I_n) + \varepsilon.$$

By (8)

$$I(x) \leq \liminf I_{n_j}(x_j) \leq \varepsilon + \limsup \operatorname{val}(X, I_{n_j}) \leq \operatorname{val}(X, I) + \varepsilon,$$

hence the conclusion. Suppose that $\operatorname{val}(X, I_n) \to -\infty$. For j suffiently large

$$\operatorname{val}(X, I_{n_j}) + \varepsilon < -\frac{1}{\varepsilon},$$

hence $I_{n_j}(x_{n_j}) \leq -1/\varepsilon$. Then as before

$$I(x) \leq \liminf I_{n_j}(x_{n_j}) \leq -1/\varepsilon.$$

The conclusion follows in both cases $\operatorname{val}(X, I) = -\infty$, $\operatorname{val}(X, I) > -\infty$, provided ε is sufficiently small.

Proof of (14). The only case of interest is when $\limsup\,[\varepsilon_n - \arg\,\min\,(X, I_n)] \neq \emptyset$. Let $x_n \in \varepsilon_n - \arg\,\min\,(X, I_n)$ be such that $x_n \to x$ for some subsequence. It follows that $x_n \in \varepsilon - \arg\,\min\,(X, I_n)$ for all n sufficiently large. By (13)

$$x \in \cap\{\varepsilon - \arg\,\min\,(X, I) : \varepsilon > 0\} = \arg\min\,(X, I).$$

If the subsequence $x_{n_j} \to x$, then by (8)

$$\mathrm{val}\,(X, I) \leq I(x) \leq \liminf I_{n_j}(x_{n_j}) \leq \limsup I_{n_j}(x_{n_j}) \leq \limsup \mathrm{val}\,(X, I_n)$$

both when $\mathrm{val}\,(X, I_{n_j}) > -\infty$ for infinitely many indices and when $\mathrm{val}\,(X, I_{n_j}) = -\infty$ for all j sufficiently large. The conclusion follows by (12).

Proof of (15). By (12) it suffices to prove that

$$\liminf \mathrm{val}\,(X, I_n) \geq \mathrm{val}\,(X, I).$$

The only case of interest is when $\mathrm{val}\,(X, I) > -\infty$. For a subsequence p_k we have

$$\mathrm{val}\,(X, I_{p_k}) \to \liminf \mathrm{val}\,(X, I_n).$$

For a further subsequence n_j we get $x_{n_j} \to x$. Then by (8)

$$-\infty < \mathrm{val}\,(X, I) \leq I(x) \leq \liminf I_{n_j}(x_{n_j}),$$

thus $\mathrm{val}\,(X, I_{n_j}) > -\infty$ for every j sufficiently large. Then for such j

$$I_{n_j}(x_{n_j}) \leq \varepsilon_{n_j} + \mathrm{val}\,(X, I_{n_j}),$$

hence
$$\mathrm{val}\,(X, I) \leq I(x) \leq \liminf \mathrm{val}\,(X, I_{n_j}) = \liminf \mathrm{val}\,(X, I_n). \qquad \square$$

Section 3.   Sequential Epi-Convergence and Hadamard Well-Posedness.

Variational convergence is the weakest known sequential notion which allows to obtain Hadamard well - posedness (theorem 5). Such a convergence notion may be no longer adequate when uniqueness of the variational limit problem is required (see remark 7). Moreveor stability of convergence of the minima under additive perturbations may be not preserved, since $\mathrm{var}\,\lim I_n = I$ does not imply (in general) $\mathrm{var}\,\lim\,(I_n + J) = I + J$.

We need to introduce a stronger notion of convergence in order to obtain sharper results.

Let X be a convergence space and

$$I_n, I : X \to [-\infty, +\infty].$$

We shall say that

$$seq.\ epi - \lim I_n = I$$

(or more precisely c seq. epi - lim $I_n = I$ if c is the convergence on X) iff

(6) $$x_n \to x \Rightarrow \liminf I_n(x_n) \geq I(x);$$

(19)   for every $u \in X$ there exists $u_n \to u$ such that $I_n(u_n) \to I(u)$.

Remark. By (6) and (19) we see that seq. epi - lim $I_n = I$ iff for every $x \in X$

(20) $I(x) = \min\{\liminf I_n(x_n): x_n \to x\} = \min\{\limsup I_n(x_n): x_n \to x\}.$

This in turn is equivalent to (6) and

(21)   for every u there exists $u_n \to u$ such that $\limsup I_n(u_n) \leq I(u)$.

From the above definition we get immediately

8 Proposition *Seq. epi - lim $I_n = I$ implies var - lim $I_n = I$.*

Hence epi - convergence is variational, since theorem 5 and corollary 6 apply to it. If seq. epi - lim $I_n = I$ and

$$f : X \to (-\infty, +\infty) \text{ is continuous}$$

then

$$x_n \to x \Rightarrow \liminf [I_n(x_n) + f(x_n)] \geq I(x) + f(x).$$

If u_n, u are as in (19), then

$$I_n(u_n) + f(u_n) \to I(u) + f(u).$$

Thus sequential epi - convergence is stable under continuous additive perturbations. Moreover we get a convergence in the sense of Kuratowski on $LSC(X)$ since we have the following technical facts.

9 Proposition *Assume that seq. epi - lim $I_n = I$. Then*

(22)   *seq. epi - lim $(I_n + f) = I + f$*
           for every continuous function $f : X \to (-\infty, +\infty)$;

(23)   *I is uniquely defined;*

(24)   *for every subsequence n_j, seq. epi - lim $I_{n_j} = I$;*

(25)   *if I is lower semicontinuous, then the constant sequence*
           $I_n = I$ converges to I in the sequential epi - sense.

Examples of sequential epi - convergence. Assume that each $I_n \in SC(X)$. Then if I_n is a monotone *increasing* sequence, we get

$$\text{seq. epi - lim } I_n = \sup I_n.$$

To verify this claim, we set $I = \sup I_n$, then $I_n(x) \to I(x)$ for all x. Now let $x_n \to x$. Having fixed k,

$$I_k(x_n) \leq I_n(x_n) \text{ when } n > k,$$

so by semicontinuity

$$\liminf I_n(x_n) \geq \lim_{n \to +\infty} \inf I_k(x_n) \geq I_k(x)$$

and the claim follows by taking sup with respect to k.

If I_n is a monotone *decreasing* sequence and $I = \inf I_n$ is lower semicontinuous, then

$$\text{seq. epi - lim } I_n = I = \inf I_n$$

since $I_n \to I$ pointwise, while if $x_n \to x$ then (6) obtains from $I_n(x_n) \geq I(x_n)$ and semicontinuity of I.

Let $X = R$ and

$$I_n(x) = nxe^{-n^2x^2}.$$

Then $I_n \to 0$ pointwise (even uniformly on each closed set K such that $0 \notin K$,) but not in the seq. epi - sense. Otherwise, by (15) we get the contradiction

$$\text{val } (R, I_n) = -\frac{1}{\sqrt{2e}} = I_n(\frac{1}{n\sqrt{2}}) \to 0$$

since $1/n\sqrt{2} = \arg\ \min(R, I_n)$ converges. In fact
seq. epi - lim $I_n = I$, where $I(x) = 0$ if $x \neq 0$, $I(0) = -1/\sqrt{2e}$.

10 Example. Let X be a topological space, with convergence induced by the topology. Assume that $I_n \to I$ pointwise, and let I_n be *locally equicontinuous*. This means that for every $\varepsilon > 0$ and $x \in X$ there exists a neighborhood V of x such that for all n

$$|I_n(x) - I_n(y)| < \varepsilon, y \in V.$$

(A particular case: (X, d) is a metric space and I_n is *locally equi - Lipschitz continuous*, i. e. for every $x \in X$ we can find $L > 0$ and $a > 0$ such that

$$d(x', x) \leq a,\ d(x'', x) \leq a \Rightarrow |I_n(x') - I_n(x'')| \leq L\ d(x', x'') \text{ for all\ n}).$$

Then seq. epi - lim $I_n = I$, since given $x_n \to x$ and $\varepsilon > 0$ we have for all n sufficiently large

$$I_n(x_n) \geq I_n(x) - \varepsilon$$

hence

$$\liminf I_n(x_n) \geq -\varepsilon + \liminf I_n(x) = I(x) - \varepsilon,$$

yielding (6). A particular case: let X be an open convex set in R^P (with the usual convergence), I_n finite - valued convex functions on X, pointwise converging to I. Then

$$\text{seq. epi - lim } I_n = I$$

since I_n are locally equi - Lipschitz continuous (Rockafellar [2, th. 10.8]).

Assuming seq. epi - convergence of I_n to I, by taking essential advantage from (19) we obtain equivalence between value Hadamard well - posedness and lower semicontinuity type properties of $\varepsilon - \arg \min (X, I_n)$, as follows.

11 Theorem *Assume that seq. epi - $\lim I_n = I$. Then the following facts are equivalent:*

(15) $$val\ (X, I_n) \to val\ (X, I)\ and\ arg\ min\ (X, I) \neq \emptyset;$$

(26) *for every $x_0 \in arg\ min\ (X, I)$ there exists $\varepsilon_n \to 0$ and*
      $$x_n \in \varepsilon_n - arg\ min\ (X, I_n)\ such\ that\ x_n \to x_0.$$

12 Theorem *Assume that*

$$seq.\ epi - \lim I_n = I,\ val\ (X, I_n) \to val\ (X, I).$$

Then

(27) $$arg\ min\ (X, I) = \cap \{\ lim\ inf\ [\varepsilon - arg\ min\ (X, I_n)] :\ \varepsilon > 0\}$$
      $$= \cap \{lim\ sup\ [\varepsilon - arg\ min\ (X, I_n)] :\ \varepsilon > 0\}.$$

As a partial converse, we remark that
(15), (26) and (27) are equivalent properties, provided

$$seq.\ epi - lim\ I_n = I\ and\ arg\ min\ (X, I) \neq \emptyset.$$

Indeed, if $x_0 \in arg\ min\ (X, I)$ and (27) holds, for every $\varepsilon > 0$ there exists $x_n \in \varepsilon - \arg \min (X, I_n)$ such that $x_n \to x_0$. Thus by (6) and (12) (if either $\lim \inf val\ (X, I_n) > -\infty$ or $= -\infty$)

$$I(x_0) = val\ (X, I) \leq \lim \inf I_n(x_n) \leq \lim \inf val\ (X, I_n)$$
$$\leq \lim \sup val\ (X, I_n) \leq val\ (X, I),$$

yielding (15).

The assumptions of theorem 12 do not imply solution Hadamard well - posedness in the form

$$\arg \min (X, I) \subset \lim \inf \arg \min (X, I_n)$$

as shown by taking

$$X = R,\ I(x) = \max \{0, |x| - 1\}, I_n(x) = \max \{x^2/n,\ I(x)\}.$$

In other words, $\varepsilon_n = 0$ in (26) does not work. (Compare with the next theorem 19).

Generalizing (27), we have

13 Proposition *Suppose I is bounded from below and*

$$seq.\ epi\ -\ lim\ I_n = I,\quad val\ (X, I_n) \to val\ (X, I).$$

Then for every $\varepsilon \geq 0$

$$\varepsilon -\ arg\ min\ (X, I) = \cap\ \{\ lim\ inf\ [b -\ arg\ min\ (X, I_n)] : b > \varepsilon\}$$
$$= \cap\ \{\ lim\ sup\ b -\ arg\ min\ (X, I_n) : b > \varepsilon\}.$$

The conclusion of proposition 13 is sufficient to get seq. epi - convergence from variational convergence in the metric case, and to obtain value Hadamard well - posedness (a partial converse to the above proposition).

14 Theorem *Assume that var - lim $I_n = I$, and for every $\varepsilon \geq 0$*

$$\varepsilon -\ arg\ min\ (X, I) \subset \cap\ \{lim\ inf\ [b -\ arg\ min\ (X, I_n)] : b > \varepsilon\}.$$

Then if I is bounded from below

$$val\ (X, I_n) \to val\ (X, I),$$

and if the convergence on X is induced by a metric

$$seq.\ epi\ -\ lim\ I_n = I.$$

Value Hadamard well - posedness for every compact set in X is equivalent to seq. epi - convergence.

15 Proposition *If for every compact set K in X*

$$val\ (K, I_n) \to val\ (K, I)$$

with I lower semicontinuous, then

(28) $$seq.\ epi\ -\ lim\ I_n = I.$$

Conversely, if (28) holds, then

$$lim\ inf\ val\ (K, I_n) \geq val\ (K, I)\ for\ every\ compact\ \ K\ in\ X;$$
$$lim\ sup\ val\ (A, I_n) \leq val\ (A, I)\ for\ every\ open\ set\ A\ in\ X.$$

Now assume that (X, I) is a metric space, every I_n is lower semicontinuous on X and

$$seq.\ epi\ -\ lim\ I_n = I.$$

Let $a: X \to (-\infty, +\infty)$ exist such that

(29) $a(x) \le I_n(x)$ for every n and x

and every sublevel set of a is sequentially compact. Hence (17) holds, i. e. I_n is equico-ercive. Given $u \in X$ and $c > 0$, consider

$$g_n(x) = I_n(x) + cd(x, u)^2, \ g(x) = I(x) + cd(x, u)^2.$$

By (22) we see that

$$\text{seq. epi - } \lim g_n = g.$$

On the other hand $g_n \ge I_n$, therefore by (29) $\arg\min (X, g_n) \ne \emptyset$ for all n. Thus by (15) for every c and u

$$\text{val} (X, g_n) \to \text{val} (X, g)$$

which means pointwise convergence of such (Moreau - Yosida) regularizations to I_n. So

(30) $\inf\{I_n(x) + cd(x, u)^2 \colon x \in X\} \to \inf\{I(x) + cd(x, u)^2 \colon x \in X\},$

moreover by (14)

(31) $\lim \sup \ \arg\min [X, I_n(\cdot) + cd(\cdot, u)^2] \subset \arg\min [X, I(\cdot) + cd(\cdot, u)^2],$

for all $c > 0$ and $u \in X$. Under suitable conditions on I the converse holds, so that (30) and (31) characterize seq. epi - convergence.

16 Theorem *Assume that I is proper, bounded from below and I, I_n are lower semicontinuous. If (29) holds, then seq. epi - $\lim I_n = I$ iff (30) and (31) hold for every $u \in X$ and $c > 0$.*

17 Counterexample. Without equicoercivity, (28) may not imply (30). Consider

$$a_n(t) = 2 + \sin(nt)/|\sin(nt)|; \ \ I_n(x) = \int_0^{2\pi} a_n x^2 dt,$$

$$X = L^2(0, 2\pi) \text{ equipped with the strong convergence.}$$

Then $I_n(x) \to I(x)$ for every x, where

$$I(x) = 2 \int_0^{2\pi} x^2 dt,$$

since $a_n \rightharpoonup a$ in $L^\infty(0, 2\pi)$, $a(t) = 2$ a. e. Moreover by the Cauchy - Schwartz inequality

$$|I_n(u) - I_n(v)| \le 3 \int_0^{2\pi} |u - v| \, |u + v| dt \le 3\|u - v\| \, \|u + v\|,$$

whence I_n is locally equicontinuous. Then by example 10

$$\text{seq. epi - } \lim I_n = I.$$

But

$$\inf\{I_n(x) + \|x - u\|^2 : x \in X\} = \inf \int_0^{2\pi} [a_n x^2 + (x - u)^2] dt$$

$$= \int_0^{2\pi} \inf\{a_n y^2 + (y - u)^2 : y \in R\} dt = \int_0^{2\pi} u^2 a_n/(1 + a_n) dt.$$

Similarly

$$\inf\{I(x) + \|x - u\|^2 : x \in X\} = (2/3) \int_0^{2\pi} u^2 dt,$$

therefore (30) fails since $a_n/(1 + a_n) \to 5/8$ in $L^\infty(0, 2\pi)$.

PROOFS.

Proposition 9. **Proof of (23).** Suppose seq. epi - lim $I_n = A$, seq. epi - lim $I_n = B$. Given $u \in X$, there exists $u_n \to u$ with $I_n(u_n) \to A(u)$, so by (6)

$$\liminf I_n(u_n) = A(u) \geq B(u).$$

Symmetrically, there exists $v_n \to u$ such that $I_n(v_n) \to B(u)$, therefore $B(u) \geq A(u)$, hence $A = B$.

Proof of (24). Given $x_j \to x$, let $u_n = x_j$, $n_j \leq n < n_{j+1}$. Then $u_n \to x$ and

$$I(x) \leq \liminf I_n(u_n) \leq \liminf I_{n_j}(x_j)$$

proving (6), while (19) is stable under subsequencing. Finally, (25) is immediate. □

Theorem 11. By theorem 5 we know that if seq. epi - lim $I_n = I$, then the existence of $\varepsilon_n \to 0$ and some relatively compact $x_n \in \varepsilon_n - \arg\min(X, I_n)$ implies (15). Conversely, assume (15). Pick any $x_0 \in \arg\min(X, I)$. If $I(x_0) > -\infty$ then by (15) val $(X, I_n) > -\infty$ for all n sufficiently large. From (20) there exists $x_n \to x_0$ such that $I_n(x_n) \to I(x_0)$, thus for some $c_n \to 0$ (by (15))

$I_n(x_n) = I_n(x_n) - I(x_0) + I(x_0) \leq |I_n(x_n) - I(x_0)| + c_n + \text{val}(X, I_n)$.
    Then

(32) $$x_n \in \varepsilon_n - \arg\min(X, I_n),$$

where $\varepsilon_n = c_n + |I_n(x_n) - I(x_0)|$. If $I(x_0) = -\infty = \text{val}(X, I)$, then by (19) there exist $\varepsilon_n > 0, \varepsilon_n \to 0$ and $x_n \to x_0$ such that $I_n(x_n) \leq -1/\varepsilon_n$ (just take $\varepsilon_n = -1/I_n(x_n)$) for n sufficiently large.
    Hence (31) holds in this case too, and (26) is proved. □

Theorem 12.   Assume (15) and let $x_0 \in \arg\min(X, I)$. If $\mathrm{val}(X, I) > -\infty$, by (26) there exist $\varepsilon_n \to 0$ and x_n such that

$$I_n(x_n) \leq \varepsilon_n + \mathrm{val}(X, I_n), x_n \to x_0.$$

Then $x_n \in \varepsilon - \arg\min(X, I_n)$ for any fixed $\varepsilon > 0$ and n sufficiently large. Hence

$$\arg\min(X, I) \subset \liminf[\varepsilon - \arg\min(X, I_n)]$$

whence

(33) $$\arg\min(X, I) \subset \cap\{\liminf[\varepsilon - \arg\min(X, I_n)] : \varepsilon > 0\}.$$

If $\mathrm{val}(X, I) = -\infty$, again by (26) there exist $x_n \to x_0$ and some $\varepsilon_n > 0, \varepsilon_n \to 0$ such that $I_n(x_n) \leq -1/\varepsilon_n$. Then given $\varepsilon > 0$ we get $I_n(x_n) \leq -1/\varepsilon$ for n sufficiently large, yielding (33). By (13)

$$\cap\{\limsup[\varepsilon - \arg\min(X, I_n)] : \varepsilon > 0\} \subset \cap\{\varepsilon - \arg\min(X, I) : \varepsilon > 0\}$$
$$= \arg\min(X, I)$$

even if $\arg\min(X, I) = \emptyset$, in which case (33) is trivially true.   □

Proposition 13. If $\varepsilon = 0$, theorem 12 gives the conclusion. Now let $\varepsilon > 0$ and $x_0 \in \varepsilon - \arg\min(X, I)$. By (9) there exists $x_n \to x_0$ with $I_n(x_n) \to I(x_0)$. By assumption, $\mathrm{val}(X, I) > -\infty$, then $\mathrm{val}(X, I_n) > -\infty$ for all n sufficiently large. Therefore given $a > 0$

$$I_n(x_n) \leq I(x_0) + a \leq \mathrm{val}(X, I) + \varepsilon + a \text{ for all large } n.$$

Then for fixed $b > \varepsilon$ we get $x_n \in b - \arg\min(X, I_n)$, hence $x_0 \in \liminf[b - \arg\min(X, I_n)]$, yielding

$$\varepsilon - \arg\min(X, I) \subset \cap\{\liminf[b - \arg\min(X, I_n)] : b > \varepsilon\}.$$

By (13), $\limsup[b - \arg\min(X, I_n)] \subset b - \arg\min(X, I)$, and the conclusion follows.   □

Theorem 14.   Before proving value convergence, let us verify that

(34) $$\mathrm{val}(X, I_n) > -\infty \text{ for all sufficiently large } n.$$

Arguing by contradiction, we have for some subsequence

$$\mathrm{val}(X, I_{n_j}) = -\infty \quad (\text{for all } j).$$

Let $\varepsilon \in (0, 1)$ be such that $(-1/\varepsilon) + \varepsilon < \mathrm{val}(X, I)$ and pick x_0 such that $I(x_0) \leq \varepsilon + \mathrm{val}(X, I)$. Then by assumption there exists $x_n \to x_0$ such that

$$I_{n_j}(x_{n_j}) \leq -(1/\varepsilon) + \varepsilon \text{ for all } j.$$

By (6)

$$\mathrm{val}(X, I) \leq I(x_0) \leq \liminf I_{n_j}(x_{n_j}) \leq \varepsilon - 1/\varepsilon,$$

a contradiction. Thus (34) is proved. Now fix $\varepsilon > 0$, $x_0 \in \varepsilon - \arg\min(X, I)$ and $b > \varepsilon$. There exists $x_n \to x_0$ with

$$I_n(x_n) \leq \mathrm{val}(X, I_n) + b$$

therefore by (6)

$$\text{val}\,(X, I) \leq I(x_0) \leq \liminf I_n(x_n) \leq b + \liminf \text{val}\,(X, I_n).$$

Since b and ε are arbitrary,

$$\text{val}\,(X, I) \leq \liminf \text{val}\,(X, I_n)$$

and the first conclusion follows from (11). Now we prove epi - convergence, that is (19). Fix $u \in \text{dom}\,I$ and set

$$\varepsilon = I(u) - \text{val}\,(X, I).$$

Given $k = 1, 2, 3, ...$, let $b = \varepsilon + 1/k$. Then by assumption

$$u \in \liminf [b - \arg\min(X, I_n)].$$

So we can find $x_n^k \to u$ such that

$$I_n(x_n^k) \leq \text{val}\,(X, I_n) + b.$$

By cor. 1.18 of Attouch [3] there exists $k(n) \to +\infty$ with

$$u_n = x_n^{k(n)} \to u, \; I_n(u_n) \leq \text{val}\,(X, I_n) + \varepsilon + \frac{1}{k(n)}.$$

Then by (12)

$$\limsup I_n(u_n) \leq \text{val}\,(X, I) + \varepsilon = I(u).$$

$\square$

Proposition 15.   From convergence of the infima on compact sets, by taking K to be a singleton we get pointwise convergence, hence (19). Let now $x_n \to x$. By contradiction, suppose that there exists a subsequence n_j and some $a \in R$ such that

(35) $$I_{n_j}(x_{n_j}) \leq a < I(x) \text{ for every } j.$$

Fix any j and $k > j$, then

$$\inf\{I_{n_k}(x_{n_p}) : p \geq j\} \leq I_{n_k}(x_{n_k}) \leq a.$$

Now consider the compact set

$$T_j = \{x\} \cup \{x_{n_p} : p \geq j\}.$$

Then

$$\inf I_{n_k}(T_j) \leq \inf\{I_{n_k}(x_{n_p}): p \geq j\} \leq I_{n_k}(x_{n_k}) \leq a.$$

Letting $k \to +\infty$, by assumption we get

$$\inf I(T_j) = \min\{I(x), \inf[I(x_{n_p}): p \geq j]\} \leq a.$$

Thus $\inf\{I(x_{n_p}): p \geq j\} \leq a$. Letting $j \to +\infty$, by semicontinuity of I we get

$$I(x) \leq \liminf I(x_{n_j}) \leq a$$

contradicting (35). Then seq. epi - $\lim I_n = I$.

Let now K be compact. The only case of interest is when val $(K, I) > -\infty$. Suppose that lim inf val $(K, I_n) = -\infty$, then for some subsequence n_j and points $u_j \in K$

$$I_{n_j}(u_j) \to -\infty.$$

Another subsequence $u_j \to x \in K$, therefore by (6)

$$-\infty < \text{val } (K, I) \leq I(x) \leq \text{ lim inf } I_{n_j}(u_j) = -\infty,$$

a contradiction. Hence lim inf val $(K, I_n) > -\infty$. Then val $(K, I_n) > -\infty$ for all n sufficiently large. For some subsequence

$$\text{val } (K, I_n) \to \text{ lim inf val } (K, I_n).$$

Let $u_n \in K$ be such that

$$I_n(u_n) \leq \text{val } (K, I_n) + 1/n.$$

Then for some further subsequence $u_{n_j} \to x \in K$. By (6) and (24)

$$\text{val } (K, I) \leq I(x) \leq \text{ lim inf } I_{n_j}(u_{n_j}) \leq \text{ lim inf val } (K, I_n).$$

Now let A be an open set. By (19) for any $x \in A$ there exists $x_n \to x$ such that $I_n(x_n) \to I(x)$, then $x_n \in A$ for all large n. Since $I_n(x_n) > \inf I_n(A)$, it follows that

$$\text{lim sup inf } I_n(A) \leq I(x).$$

The conclusion follows by arbitrariness of x. $\qquad\qquad\qquad\qquad\qquad\qquad$ □

Theorem 16. Assume (30) and (31).
Proof of (6). Let $x_n \to u$. Then

$$I_n(x_n) + cd(x_n, u)^2 \geq \inf\{I_n(x) + c \, d(x, u)^2 : x \in X\}$$

thereby giving from (30)

(36) lim inf $I_n(x_n) = $ lim inf $[I_n(x_n) + cd(x_n, u)^2] \geq \inf\{I(x) + cd(x, u)^2 : x \in X\}, c > 0.$

By semicontinuity of I, the Moreau - Yosida regularizations to I converge pointwise, i. e. as $c \to +\infty$

$$\inf\{I(x) + c \, d(x, u)^2 : x \in X\} \to I(u)$$

(see Attouch [1, th. 2.64 p. 228]). Hence letting $c \to +\infty$ in (36) we get

$$\text{lim inf } I_n(x_n) \geq I(u).$$

Proof of (19). Fix $u \in X$ and $\varepsilon > 0$ such that

$$I(u) \leq \text{val } (X, I) + \varepsilon.$$

We apply th. 1 p. 444 from Ekeland [1] with metric $kd, k = 1, 2, 3, \dots.$ Given k we find $y_k \in X$ such that

(37) $$d(y_k, u) \leq 1/k,$$

$$I(y_k) \leq I(u), I(y_k) < I(x) + \varepsilon k d(y_k, x) \text{ if } x \neq y_k.$$

Then $I(u) \le \liminf I(y_k) \le \limsup I(y_k) \le I(u)$, so that

(38) $$a_k = |I(y_k) - I(u)| \to 0.$$

Moreover y_k is the unique minimizer on X for $I(\cdot) + \varepsilon k d(y_k, \cdot)$. Fix k and consider

$$x_n^k \in \arg\min \ [X, I_n(\cdot) + \varepsilon \ k d(y_k, \cdot)].$$

By (31) with $u = y_k, c = \varepsilon k$, and by (29) it follows that every subsequence of x_n^k converges as $n \to +\infty$ to y_k, hence

(39) $$x_n^k \to y_k \text{ for every } k.$$

By (30), as $n \to +\infty$

$$I_n(x_n^k) + \varepsilon k \ d(y_k, x_n^k) \to I(y_k),$$

hence by (39)

(40) $$I_n(x_n^k) \to I(y_k) \text{ for every } k.$$

By (39) and (40) there exists a strictly increasing sequence n_k such that

(41) $$d(y_k, x_n^k) < 1/k, |I_n(x_n^k) - I(y_k)| < 1/k \text{ whenever } n \ge n_k.$$

Given j sufficiently large, there exists an unique $k = k(j)$ such that $n_k \le j < n_{k+1}$. Put $z_j = x_j^{k(j)}$, then by (38), (39) and (41)

$$d(z_j, u) \le \ d(x_j^{k(j)}, y_{k(j)}) + d(y_{k(j)}, u) \le \frac{2}{k(j)},$$

$$|I_j(z_j) - I(u)| \le |I_j(z_j) - I(y_{k(j)})| + |I(y_{k(j)}) - I(u)| \le 1/k(j) + a_{k(j)}.$$

Then we conclude that $z_j \to u$ and $I_j(z_j) \to I(u)$. $\qquad\Box$

Section 4. Epi-Convergence and Tykhonov Well-Posedness.

Suppose that (X, d) is a metric space, I_n and I_0 are extended real - valued functions on it with a finite val (X, I_0). We wish to find conditions under which (X, I_0) is Tykhonov well - posed, knowing that each (X, I_n) is and seq. epi - lim $I_n = I_0$. Of course, Tykhonov well - posedness is not stable under epi - convergence. Consider for example $X = R, I_0(x) = \max\{0, |x| - 1\}$ and $I_n(x) = I_0(x) + x^2/n$ (which converges even locally uniformly to I_0).

Let us consider the case of uniqueness of the minimizers for all (X, I_n), and put

$$x_n = \arg\min \ (X, I_n).$$

If I_n is equicoercive, i.e. (17) holds, then seq. epi - convergence of I_n towards I_0 implies

$$x_n \to x_0, \quad I_n(x_n) \to I_0(x_0)$$

by theorem 5. Moreover the following property holds:
(42) given $\varepsilon > 0$ there exists $\delta > 0$ such that for every $n \ge 1$,

$$I_n(x) - I_n(x_n) \leq \delta \Rightarrow d(x_n, x) < \varepsilon.$$

To prove (42) we argue by contradiction. Assume that for some $\varepsilon > 0$ and some sequence y_n we have

$$I_n(y_n) - I_n(x_n) < 1/n, \quad d(x_n, y_n) \geq \varepsilon.$$

Then $y_n \in 1/n - \arg\min(X, I_n)$. By equicoercivity and uniqueness of $\arg\min(X, I_0)$ it follows from (14) that $y_n \to x_0$, hence $d(x_n, y_n) \to 0$, a contradiction. So (42) is proved. By corollary I.14 we get Tykhonov well - posedness for each $(X, I_n), n \geq 1$. Moreover (42) is a property of *equi-well-posedness*, since we have Tykhonov well - posedness which is uniform with respect to n. Thus set

(43) $$c(t) = \inf\{I_n(x) - I_n(x_n) : d(x_n, x) = t, n = 1, 2, 3, ...\}.$$

It turns out that c is forcing. By definition of c

$$I_n(x) \geq I_n(x_n) + c[d(x_n, x)], x \in X.$$

As a matter of fact, this inequality amounts to (42). By sequential epi - convergence, we pass to the limit as $n \to +\infty$, obtaining

$$I_0(x) \geq I_0(x_0) + c[d(x, x_0)], x \in X,$$

which entails Tykhonov well - posedness by theorem I.12. We get

18 Theorem *Let $I_n, n \geq 0$, be extended real - valued functions on the metric space (X, d) with a unique $x_n = \arg\min(X, I_n), n = 0, 1, 2, ...$, and a finite val (X, I_0), fulfilling (42). Assume $I_n, n \geq 1$, equicoercive and let seq. epi- lim $I_n = I_0$. Then*
(44) *there exists a forcing function c such that for every x and $n = 0, 1, 2, ...$*

$$I_n(x) \geq I_n(x_n) + c[d(x_n, x)];$$

in particular each (X, I_n) is Tykhonov well - posed;
(45) *val $(X, I_n) \to$ val (X, I_0); if y_n is asymptotically minimizing for I_n then*

$$y_n \to \arg\min(X, I_0).$$

Summarizing, under uniqueness of the minimizers, equicoercivity and equi-well-posedness of the approximating problems, the epi - limit problem is automatically Tykhonov well - posed. Moreover we get value and solution Hadamard well - posedness.

Condition (42), generalized to the case of lack of uniqueness of minimum points, is the key to get

(46) $$\arg\min(X, I_0) \subset \liminf \arg\min(X, I_n)$$

under sequential epi - convergence. In this way we shall obtain a converse to (13) when $\varepsilon = 0$ (under restrictive conditions). Notice that (46) is stronger than (26).

19 Theorem *Assume* $val\,(X, I_n)$ *finite and* $arg\ min\ (X, I_n) \neq \emptyset, n = 0, 1, 2, ...$ *Suppose* epi - $lim\ I_n = I_0$, *and consider the following condition:*

(47) $\qquad\qquad$ *for every* $\varepsilon > 0$ *there exists* $\delta > 0$ *such that*

$$I_n(x) - val\,(X, I_n) < \delta \Rightarrow dist\,[x,\ arg\ min\ (X, I_n)] < \varepsilon.$$

Then (15) and (47) imply (46), while

$$lim\ sup\ [\varepsilon_n - arg\ min\ (X, I_n)] \neq \emptyset\ for\ some\ \varepsilon_n \to 0$$

and (46) imply (47).

PROOFS.

Theorem 18. Let us show that c defined by (43) is forcing. Of course $c(t) \geq 0$. Let $t_k \geq 0$ be such that $c(t_k) \to 0$. For each k there exists an integer n_k and a point $y_k \in X$ such that

$$d(y_k, x_{n_k}) = t_k, \quad I_{n_k}(y_k) - I_{n_k}(x_{n_k}) \to 0.$$

Fix any subsequence of t_k.

Case 1: $\sup n_k = +\infty$. We may assume (for a subsequence) n_k strictly increasing. Then by (42) it follows that $t_k \to 0$.

Case 2: $\sup n_k < +\infty$. Then some subsequence is constant, $n_k = p$ say, so that

$$I_p(y_k) - I_p(x_p) \to 0.$$

By (42) again we get $t_k \to 0$. This shows that c is forcing, and proves (44) for $n \geq 1$. Tykhonov well - posedness comes from theorem I.12. Now fix x and consider $u_n \to x$ with $I_n(u_n) \to I_0(x)$.

Then by (44)

$$I_n(u_n) \geq I_n(x_n) + c[d(x_n, u_n)].$$

We know from chapter I, p. 10, that we may assume c continuous in $[0, T]$, for some $T \geq d(x_n, u_n)$ for every n. Taking limits as $n \to +\infty$ we get Tykhonov well - posedness of (X, I_0) by theorem I.12. Finally, (45) follows from theorem 5. $\qquad\square$

Theorem 19. Assume (15) and (47). Let $u \in arg\,min\,(X, I_0)$. By (19) there exists $x_n \to u$ such that $I_n(x_n) \to I_0(u)$. If dist $[x_n, arg\,min\,(X, I_n)] \not\to 0$, then there exists $\varepsilon > 0$ such that for some subsequence

$$\varepsilon \leq dist\,[x_n, arg\,min\,(X, I_n)].$$

By (15), for all n sufficiently large

$$I_n(x_n) - val\,(X, I_n) < \delta,$$

hence by (47) dist $[x_n, arg\,min\,(X, I_n)] < \varepsilon$, a contradiction. Then

$$dist\,[x_n, arg\,min\,(X, I_n)] \to 0.$$

Now let $u_n \in arg\,min\,(X, I_n)$ be such that

$$d(x_n, u_n) \leq 1/n + \text{dist}[x_n, \arg\min(X, I_n)].$$

Then $d(u, u_n) \leq d(u, x_n) + d(x_n, u_n) \to 0$, proving (46). Assume now (46). If (47) fails, there exist $\varepsilon > 0$ and y_n such that

$$I_n(y_n) - \text{val}\,(X, I_n) < 1/n, \ \text{dist}\,[y_n, \arg\min\,(X, I_n)] \geq \varepsilon$$

Since $y_n \in 1/n - \arg\min\,(X, I_n)$, it follows from the assumption that $y_n \to y$ for a subsequence. Then $y \in \arg\min\,(X, I_0)$ by (14). From (46) we can find $u_n \to y$ such that $u_n \in \arg\min\,(X, I_n)$ and

$$0 < \varepsilon \leq \text{dist}\,[y_n, \arg\min\,(X, I_n)] \leq d(y_n, y) + \text{dist}\,[y, \arg\min\,(X, I_n)]$$
$$\leq d(y_n, y) + d(y, u_n) \to 0,$$

a contradiction. $\square$

Remark. Given $I_n, n \geq 1$, (46) amounts to the existence of a forcing function c such that for all n and x

(48) $$I_n(x) \geq \text{val}\,(X, I_n) + c(\text{dist}\,[x, \arg\min\,(X, I_n)])$$

(analogous to (44) subject to the assumptions of a finite val (X, I_n) and of course $\arg\min\,(X, I_n) \neq \emptyset$ for all n). To prove this claim, assume (47) and set

$$c(t) = \inf\{I_n(x) - \text{val}\,(X, I_n) : \text{dist}\,[x, \arg\min\,(X, I_n)] = t, \ n = 1, 2, 3, ...\}.$$

Of course $c \geq 0$. Let $t_k \geq 0$ be such that $c(t_k) \to 0$. A reasoning similar to that in the proof of theorem 18 shows that $t_k \to 0$. Conversely assume (48) with a forcing function c. Then

$$c(\text{dist}\,[x, \arg\min\,(X, I_n)]) \leq I_n(x) - \text{val}\,(X, I_n),$$

so by I.(11) we obtain

$$\text{dist}[x, \arg\min\,(X, I_n)] \leq q[I_n(x) - \text{val}\,(X, I_n)],$$

where q is given by I.(6). Then (47) follows by I.(12).

Section 5.   Topological Epi-Convergence.

It is possible to introduce, in a topological setting, a satisfying definition of epi - convergence, which retains good well - posedness properties. Topological epi - convergence agrees with the sequential one under first countability of the underlying topology. Such a property is important not only for theoretical reasons, but also from a technical point of view. In fact, given such a topology it is possible to work with topological epi - convergence making use of the characterization afforded by (6) and (21), sometimes easier to handle than the topological definition, which follows.

Throughout this section we assume that:
(X, t) is a fixed topological space,

$I, I_n : X \to [-\infty, +\infty]$, $n = 1, 2, 3, \dots$ is a given sequence.

Given $x \in X$, nbh (x) is the set of all open neighborhoods of x.

20 Definition

$$Epi - \lim I_n = I \text{ iff for each } x \in X$$

(49)
$$I(x) = \sup\{\liminf \inf I_n(V) : V \in \text{ nbh } (x)\}$$
$$= \sup \{\limsup \inf I_n(V) : V \in \text{ nbh } (x)\}.$$

21 Remark. A sequential epi - limit is not necessarily a lower semicontinuous function, but a topological epi - limit necessarily is (a technically useful property). Let indeed

$$I = \text{epi - } \lim I_n.$$

To show lower semicontinuity of I, fix $x \in X$ and set

$$f(V) = \liminf \inf I_n(V), V \in \text{ nbh } (x).$$

If $p < I(x)$, then by (49), $p < f(U)$ for some $U \in \text{nbh}(x)$. Again by (49)

$$I(y) = \sup \{f(A) : A \in \text{ nbh } (x)\} \geq f(U) > p \text{ for all } y \in U$$

implying semicontinuity.

22 Proposition *If (X, t) is first countable then seq. epi - $\lim I_n = I$ iff epi - $\lim I_n = I$.*

Remark. Epi - convergence of a sequence of functions is equivalent to Kuratowski convergence of the corresponding epigraphs (Attouch [3, th. 1.39 p. 98]). Such a geometrical interpretation of epi - convergence justifies the terminology.

Another important case when the topological and sequential notions agree is given by

23 Proposition *Let X be a real reflexive separable Banach space. Let I_n satisfy the following (equicoercivity) assumption:*

$$x_n \in X, \text{ sup } I_n(x_n) < +\infty \Rightarrow x_n \text{ is bounded.}$$

Then
$$weak \text{ seq. epi - } \lim I_n = I \text{ iff weak epi - } \lim I_n = I.$$

Topological epi - convergence is variational, since it implies all the conclusions of theorem 5. Assume that

$$\text{epi - } \lim I_n = I.$$

Then by (49)

$$\limsup \text{val } (X, I_n) \leq I(y) \text{ for all } y,$$

yielding (12). Let now $\varepsilon > 0$ be given and $x_n \in \varepsilon - \arg\min(X, I_n)$ with some subsequence $x_{n_j} \to x$ say. Assume for instance that $\limsup \operatorname{val}(X, I_{n_j}) > -\infty$. Hence $\operatorname{val}(X, I) > -\infty$ by (12). Then

$$(50) \qquad\qquad I_n(x_n) \leq \operatorname{val}(X, I_n) + \varepsilon.$$

For any $V \in \operatorname{nbh}(x)$ we have

$$I_{n_j}(x_{n_j}) \geq \inf I_{n_j}(V)$$

for some subsequence n_j, hence

$$\liminf I_{n_j}(x_{n_j}) \geq \liminf \inf I_{n_j}(V).$$

By (49) and (50), for such a subsequence

$$I(x) \leq \liminf I_{n_j}(x_{n_j}) \leq \limsup I_n(x_n) \leq \limsup \operatorname{val}(X, I_n) + \varepsilon \leq \operatorname{val}(X, I) + \varepsilon$$

which entails (13). The general case is obtained following the proof of theorem 5 as well as (14) and (15). In doing so we use (6) and (9), which are implied by topological epi - convergence (see (51) in the proof of proposition 22). Summarizing, we get

24 Theorem *If epi - $\lim I_n = I$ then the conclusions of theorem 5 hold true.*

Hence we obtain Hadamard well - posedness (compare with corollary 6). Denote by $SC(X)$ the set of all extended real - valued lower semicontinuous functions on X.

25 Corollary *Let $T \subset SC(X)$ be such that $\arg\min(X, f)$ is a singleton for all $f \in T$. Assume that for every sequence $f_n \in T$ there exists $\varepsilon_n \to 0$ such that*

$$\limsup [\varepsilon_n - \arg\min(X, f_n)] \neq \emptyset.$$

Then

$$f \to \operatorname{val}(X, f), \quad f \to \arg\min(X, f)$$

are continuous on T equipped with the epi - convergence.

The above conclusion holds in a topological setting if (X, t) is a Hausdorff locally compact topological space. Then epi - convergence is induced by a topology on $SC(X)$ (see Attouch [3, th. 2.78 p.254]). Thus, under the assumption of corollary 25

$$f \to \operatorname{val}(X, f) \quad f \to \arg\min(X, f)$$

are continuous on T with respect to such a topology.

PROOFS.

Proposition 22. Of course we consider X as a convergence space, with the convergence induced by t. Assume that

$$\text{epi - } \lim I_n = I.$$

Let us verify (6). Let $x_n \to x$ and fix any neighborhood V of x. Then $x_n \in V$ for all large n, hence

$$I_n(x_n) \geq \inf I_n(V)$$

thereby

(51) $$\lim \inf I_n(x_n) \geq \lim \inf \inf I_n(V),$$

which yields (6) by taking the supremum with respect to V because of (49), without using the countability assumption. Now we prove (19). Given $x \in X$, let U_k be a countable fundamental system of neighborhoods of x, so $x \in U_{k+1} \subset U_k$ for every k. Then by the topological definition

$$\lim \sup \inf I_n(U_k) \leq I(x) \text{ for each } k.$$

Then there exists some subsequence n_k such that for all k

$$\inf I_n(U_k) < I(x) + 1/k \text{ if } n \geq n_k.$$

Now take $y_{n_k+1}, ... y_{n_{k+1}}$ in U_k such that

$$I_n(y_n) \leq I(x) + 1/k, \ n_k + 1 \leq n < n_{k+1},$$

so that $y_n \to x$ and $\lim \sup I_n(x_n) \leq I(x)$. Therefore (19) is proved, whence

$$\text{seq. epi - } \lim I_n = I.$$

Conversely, we need to show that if seq. epi - $\lim I_n = I$ then, for any $x \in X, A = B = I(x)$, where

$$A = \sup\{\lim \inf \inf I_n(V) : V \in \text{ nbh } (x)\}, B = \sup\{\lim \sup \inf I_n(V) : V \in \text{ nbh } (x)\}.$$

We may assume without restriction I_n, I equibounded, according to the following
Lemma. Let (X, t) be a topological space. Then

$$\text{epi - } \lim I_n = I \text{ iff epi - } \lim \text{arctg } I_n = \text{arctg } I.$$

Let X be a convergence space. Then

$$\text{seq. epi - } \lim I_n = I \text{ iff seq. epi - } \lim \text{arctg } I_n = \text{arctg } I.$$

(Here arctg $(\pm\infty) = \pm\pi/2$.)

Proof of the lemma. For every nonempty subset $L \subset [-\infty, +\infty]$ we have

$$\text{arctg (inf } L) = \text{inf arctg } L; \quad \text{arctg (sup } L) = \text{sup (arctg } L).$$

The conclusion then follows since we can write

$$I(x) = \sup \{\sup_{k} \inf_{n \geq k} \inf I_n(V) : V \in \text{nbh } (x)\} = \sup \{\inf_{k} \sup_{n \geq k} \inf I_n(V) : V \in \text{nbh } (x)\}$$

iff epi - $\lim I_n = I$, while (14) shows that seq. epi - $\lim I_n = I$ iff

$$I(x) = \inf\{\sup_{k} \inf_{n \geq k} I_n(x_n) : x_n \to x\} = \inf\{\inf_{k} \sup_{n \geq k} I_n(x_n) : x_n \to x\}.$$

End of proof of proposition 22. Let us show that $A = I(x)$. From (19) and (51) we have $I(x) \geq A$. We shall prove the converse inequality. By first countability we can find a decreasing sequence V_k of neighborhoods of x such that $x = \bigcap_{k=1}^{\infty} V_k$. Then

$$A = \lim_{k \to +\infty} \liminf_{n \to +\infty} \inf I_n(V_k).$$

Let $x_{nk} \in V_k$ be such that

$$I_n(x_{nk}) \leq \inf I_n(V_k) + 1/n,$$

then

$$\liminf I_n(x_{nk}) = \liminf_{n \to +\infty} \inf I_n(V_k)$$

for every k, hence

$$A = \lim_{k \to +\infty} \liminf_{n \to +\infty} I_n(x_{nk}).$$

By Attouch [1, lemma 1.17 p. 33] there exists an increasing map $n \to k(n)$ such that $k(n) \to +\infty$ and

$$A \geq \liminf I_n(x_{nk(n)}).$$

Then $y_n = x_{nk(n)} \in V_{k(n)}$ implying $y_n \to x$. Since $A \geq \liminf I_n(y_n)$, by (6) $A \geq I(x)$, hence $A = I(x)$.

A similar reasoning gives $B = I(x)$.   □

Proposition 23.   Assume that epi - $\lim I_n = I$. If $x_n \to x$, by (51) we get $\liminf I_n(x_n) \geq I(x)$. Let us prove (19). Fix $u \in X$ such that $I(u) < +\infty$ (the only case of interest). As well known (see Dunford - Schwartz [1, th. V. 5. 2]) there exists a metric d on X such that

$$x_n \to x \text{ iff } x_n \text{ is bounded and } d(x_n, x) \to 0.$$

Therefore for every positive integer p, the set $\{x \in X : d(x, u) < 1/p\}$ contains a weak neighborhood of u. By definition 20, for each p we have

$$I(u) \geq \limsup \inf\{I_n(x) : d(x, u) < 1/p\}.$$

Then for every p there exists a sequence $v_n^p \in X$ and a positive integer n_p, such that $n_p < n_{p+1}$ and

$$d(v_n^p, u) \leq 1/p, \quad I_n(v_n^p) \leq I(u) + 1/p \text{ if } n \geq n_p.$$

Hence $u_n = v_{n_p}^p$ for $n_p \leq n < n_{p+1}$ fulfills

$$d(u_n, u) \to 0, \quad \limsup I_n(u_n) \leq I(u),$$

therefore by equicoercivity $u_n \rightharpoonup u$, yielding (21). Now *assume that seq. epi - lim $I_n = I$.* Fix $u \in X$ and put

$$A = \sup \{\liminf \inf I_n(U) : U \in \text{nbh }(u)\};$$

$$B = \sup \{\limsup \inf I_n(U) : U \in \text{nbh }(u)\}.$$

Let us show that

(52) there exists $v_n \rightharpoonup u$ such that $\liminf I_n(v_n) = A$.

If $A < +\infty$, for every p we have

$$A \geq \liminf \inf\{I_n(x) : d(x, u) < 1/p\}.$$

Therefore for some subsequence n_p and some sequence x_p

$$d(x_p, u) < 1/p, \quad I_{n_p}(x_p) < A + 1/p,$$

hence

$$d(x_p, u) \to 0, \quad \liminf I_{n_p}(x_p) \leq A.$$

Thus setting $v_n = x_p$ if $n_p \leq n < n_{p+1}$, by equicoercivity we get $v_n \rightharpoonup u$, so by (51) $\liminf I_n(v_n) \geq A$. Hence

$$A \leq \liminf I_n(v_n) \leq \liminf I_{n_p}(v_{n_p}) \leq \limsup I_{n_p}(x_p) \leq A.$$

This proves (52) when $A < +\infty$. Now let $A = +\infty$. Then (52) holds with $v_n = u$ since $I_n(u) \geq \inf I_n(U)$ for every $U \in \text{nbh }(u)$. Now we prove that

(53) $$u_n \rightharpoonup u \Rightarrow B \leq \limsup I_n(u_n).$$

Fix $U \in \text{nbh }(u)$. Noticing that

$$I_n(u_n) \geq \inf I_n(U) \text{ for all large } n$$

we get

$$\limsup I_n(u_n) \geq \limsup \inf I_n(U) \text{ if } U \in \text{nbh }(u),$$

whence (53) by taking sup with respect to U. By (19) there exists $w_n \rightharpoonup u$ such that $I_n(w_n) \to I(u)$.

Therefore by (6), (52) and (53)

$$I(u) \leq A \leq B \leq \limsup I_n(w_n) = I(u),$$

showing that $I = \text{epi - lim } I_n$. □

Section 6.   Epi-Convergence and Hadamard Well-Posedness for Convex Problems.

Standing assumptions: X is a real reflexive separable Banach space, $a : X \to (-\infty, +\infty)$ is a given proper, convex, lower semicontinuous function satisfying

(54) $$a(x)/\|x\| \to +\infty \text{ as } \|x\| \to +\infty.$$

$C(X, a)$ denotes the set of all $f \in LSC(X)$ such that

$$f(x) \geq a(x) \text{ all } x \in X.$$

We wish to characterize Hadamard well - posedness of (X, f) in $C(X, a)$.

As a preliminary remark, if f_n are convex on X, and seq. epi - lim $f_n = f$ either in the weak or strong sense, then f is convex as well. E.g. in the weak sense, given $b \in (0, 1)$ and $x', x'' \in X$, by (19) there exist $x'_n \rightharpoonup x', x''_n \rightharpoonup x''$ such that

$$f_n(x'_n) \to f(x'), f_n(x''_n) \to f(x'').$$

By convexity

$$f_n[bx'_n + (1 - b)x''_n] \leq bf_n(x'_n) + (1 - b)f_n(x''_n)$$

and by (6)

$$f(bx' + (1 - b)x'') \leq \lim \inf \ f_n(bx'_n + (1 - b)x''_n) \leq$$

$$\leq b \lim \inf \ f_n(x'_n) + (1 - b) \lim \inf \ f_n(x''_n) = bf(x') + (1 - b)f(x'').$$

Therefore convexity is stable under sequential epi - convergence (as a matter of fact, with respect to any convergence on X which is compatible with the linear structure of X.)

Hadamard well - posedness of (X, f) as $f \in C(X, a)$ is related to weak epi - convergence (sequential or topological does not matter, by proposition 23), as we know from the above results in this chapter. If moreover $f \in C(X, a)$ is Gâteaux differentiable on X, it is natural to link Hadamard well - posedness of (X, f) with Hadamard well - posedness of the Euler - Lagrange equation

(55) $$\nabla f(u) = y, y \in X^*$$

in the classical sense, i.e. existence and uniqueness of the solution u to (55) together with continuous dependence of u upon the data f and y. If f is strictly convex, then u solves (55) iff $u \in \arg \ \min(f(\cdot) - < y, \cdot >)$. It is then natural to analyze Hadamard well - posedness of the whole set of such additive linear continuous perturbations to (X, f).

146

Let $f_n \in C(X, a)$ and assume

(56)
$$\text{weak epi - lim } f_n = f.$$

Fix $y \in X^*$. From (22)

$$\text{weak epi - lim } (f_n(\cdot) - <y, \cdot>) = f(\cdot) - <y, \cdot>.$$

Then, setting

$$I_n(x) = f_n(x) - <y, x>, I(x) = f(x) - <y, x>$$

we get from theorem 5

(57)
$$\limsup \text{ val } (X, I_n) \leq \text{ val } (X, I).$$

Let f be proper (which amounts to say that f is not identically $+\infty$, since $f \geq a$). If $x_n \in X$ and

$$I_n(x_n) \leq \text{ val } (X, I_n) + 1/n$$

then by (57)

$$a(x_n) \leq f_n(x_n) \leq \|y\| \, \|x_n\| + \text{ constant,}$$

so x_n is bounded (otherwise, if for a subsequence $\|x_n\| \to +\infty$, then for n sufficiently large

$$a(x_n)/\|x_n\| \leq \|y\| + c/\|x_n\|, c \text{ a suitable constant,}$$

contradicting (54)). Since x_n is weakly sequentially compact, by (15) we get

(58)
$$\text{val } (X, f_n(\cdot) - <y, \cdot>) \to \text{ val } (X, f(\cdot) - <y, \cdot>) \text{ for all } y \in X^*.$$

Now assume again (56) and strict convexity of f. By (15), equicoercivity of f_n yields

(59)
$$\text{val } (X, f_n) \to \text{ val } (X, f).$$

What is the behaviour of arg min $(X, f_n(\cdot) - <y, \cdot>)$ as $n \to +\infty$? Fix $y \in X^*$. Then $I(\cdot) = f(\cdot) - <y, \cdot>$ has exactly one minimum point x_0 on X. By (11)

(60)
$$\sup \text{ val } (X, I_n) < +\infty.$$

Now let $x_n \in$ arg min (X, I_n) (which exists by coercivity). Then $a(x_n)$ is bounded by (60), and by (14) every weakly convergent subsequence tends to x_0, whence the original sequence $x_n \to x_0$. Then $\{x_0\}$ equals both lim sup and lim inf of arg min (X, I_n). Then by (56)

(61) arg min $(X, f(\cdot) - <y, \cdot>)$
= weak seq. lim arg min $(X, f_n(\cdot) - <y, \cdot>)$ for every $y \in X^*$.

Interesting enough, the opposite implications hold in $C(X, a)$.

26 Theorem *Let* f_n, f *be in* $C(X, a)$, *where a fulfills (54). The following are equivalent facts:*

(56)
$$\text{weak epi - lim } f_n = f;$$

(58)
$$\text{val } (X, f_n(\cdot) - <y, \cdot>) \to \text{ val } (X, f(\cdot) - <y, \cdot>) \text{ for every } y \in X^*.$$

If moreover f *is strictly convex, then (56) or (58) are equivalent to the pair of conditions*

(62) *for every* $y \in X^*$ *and every selection* $u_n \in arg\ min\ (X, f_n(\cdot) - \,< y, \cdot >)$, *one has*

$$u_n \rightharpoonup arg\ min\ (X, f(\cdot) - \,< y, \cdot >)$$

and

(63) $$val\ (X, f_n) \to val\ (X, f).$$

Theorem 26 shows that the weakest convergence on $C(X, a)$ implying value Hadamard well - posedness for each linear perturbation of any fixed problem (X, f) is given by weak epi - convergence. The equivalence between (56) and (58) provides the variational meaning of weak seq. epi - convergence on $C(X, a)$, as far as Hadamard well - posedness is concerned. If moreover f is strictly convex, weak epi - convergence on $C(X, a)$ is the weakest one giving solution Hadamard well - posedness to every linear continuous perturbation of (X, f) (under value Hadamard well - posedness (63)). Furthermore, (58) may be rephrased as pointwise convergence of the conjugates f_n^* toward f^*. In turn, this is equivalent to strong epi - convergence of the conjugates if f is proper, due to the following

27 Lemma Let X be a reflexive space (separability is not assumed). Let $f_n \in C(X, a)$, with a fulfilling (54), be such that sup val $(X, f_n) < +\infty$. Then f_n^* are locally equicontinuous.

Upper boundedness of val (X, f_n) follows from strong epi - convergence of f_n^* to f^*, since then lim inf $f_n^*(0) \geq f^*(0) > -\infty$ by (6); and from (56), by (15) and properness of f. Hence epi - convergence follows from example 10. Thus, theorem 26 shows a duality property of epi - convergence. More precisely, if f and f_n are in $C(X, a)$ and f is proper, then (56) or (58) are equivalent to

$$\text{strong epi - } \lim f_n^* = f^*.$$

28 Example. Let Ω be an open bounded set in R^p, and consider $X = H_0^1(\Omega)$. We are given a sequence of real - valued functions

$$a_{ji}^n = a_{ij}^n \in L^\infty(\Omega), n = 0, 1, 2, ...; i, j = 1, ..., p,$$

such that there exist positive constants c_1, c_2 with

$$c_1\ |t|^2 \leq \sum_{i,j=1}^{p} a_{ij}^n(x) t_i t_j \leq c_2\ |t|^2$$

for every n, a.e. $x \in \Omega$, every $t \in R^p$. Now set

$$f_n(u) = (1/2) \int_\Omega \sum_{i,j=1}^{p} a_{ij}^n \frac{\partial u}{\partial x_i} \frac{\partial u}{\partial x_j} dx, u \in H_0^1(\Omega), n = 0, 1, 2, 3, ...$$

Here $H_0^1(\Omega) \subset L^2(\Omega) = L^2(\Omega)^*$ (as usual) with continuous and dense embedding. For every $y \in L^2(\Omega)$ we consider the weak solution $u_n(y) \in H_0^1(\Omega)$ of the Dirichlet problem

$$(64) \qquad -\sum_{i,j=1}^{p} \frac{\partial}{\partial x_i}\left(a_{ij}^n \frac{\partial u}{\partial x_j}\right) = y \text{ in } \Omega, u = 0 \text{ in } \partial\Omega$$

(see e.g. Brezis [2, p. 178]). Here (64) is the Euler - Lagrange equation of $(X, f_n(\cdot) - < y, \cdot >)$. We claim that

weak epi $-$ *lim* $f_n = f_0$ *iff* $u_n(y) \to u_0(y)$ *in* $H_0^1(\Omega)$ *for every* $y \in L^2(\Omega)$, *or equivalently iff* $u_n(y) \to u_0(y)$ *in* $L^2(\Omega)$ *for every* $y \in L^2(\Omega)$.

Proof. Given $y, u_n(y) = \arg \min (X, f_n(\cdot) - < y, \cdot >)$, therefore, by theorem 26, weak epi - lim $f_n = f_0$ implies weak convergence of $u_n(y)$, strong in $L^2(\Omega)$ by compact embedding (Brezis [2, p. 193]). Conversely assume weak convergence of $u_n(y)$ for each $y \in L^2(\Omega)$. By (64), writing $u_n = u_n(y)$ and noticing that

$$< y, v >= \int_\Omega yv dx, v \in H_0^1(\Omega)$$

(Brezis [2, p.82]), we get

$$(65) \qquad \int_\Omega \sum_{i,j=1}^{p} a_{ij}^n \frac{\partial u_n}{\partial x_i}\frac{\partial u_n}{\partial x_j} dx - \int_\Omega y u_n dx = 0,$$

hence

$$\text{val } (H_0^1(\Omega), f_n(\cdot) - < y, \cdot >)$$
$$= -(1/2)\int_\Omega y u_n dx \to -(1/2)\int_\Omega y u_0 dx = \text{ val } (H_0^1(\Omega), f_0(\cdot) - < y, \cdot >)$$

for every $y \in L^2(\Omega)$. Equivalently

$$f_n^*(y) \to f_0^*(y) \text{ for every } y \in H^{-1}(\Omega)$$

by dense embedding of $L^2(\Omega)$ in $H^{-1}(\Omega)$ and local equicontinuity of f_n^*. Therefore epi - convergence of f_n to f_0 obtains from theorem 26. Finally assume $u_n(y) \to u_0(y)$ in $L^2(\Omega)$. By uniform ellipticity and Poincaré inequality (Brezis [2, p.174]) $u_n(y)$ is bounded in $H_0^1(\Omega)$. Fix any subsequence of $u_n(y)$, then a further subsequence converges weakly in $H_0^1(\Omega)$ and strongly in $L^2(\Omega)$. The limit must be $u_0(y)$, and for the original sequence $u_n(y) \to u_0(y)$. So we are back in the above case.

PROOFS.

Theorem 26. We saw that (56) implies (61) and (63). It implies (58) if f is not identically $+\infty$. If $f = +\infty$ and (56) holds, let us show that val $(X, I_n) \to +\infty, y \in X^*$, where

$$I_n(\cdot) = f_n(\cdot) - <y, \cdot >.$$

Fix y and assume that for some subsequence n_k

$$\text{val } (X, I_{n_k}) < +\infty \text{ for all } k$$

(otherwise there is nothing to prove). If lim sup val $(X, I_{n_k}) < +\infty$, then by (15) we get a contradiction, hence (56) $\Rightarrow$ (58).

Proof of (58) $\Rightarrow$ (56). Let z_k be a countable dense set in the unit sphere of X^*, and consider

$$d(u, v) = \sum_{k=0}^{\infty} 2^{-k} | <z_k, u - v > |; u, v \in X.$$

Then X equipped with the metric d is a separable metric space (since X is), hence second countable. Fix any subsequence of f_n. By the compactness of epi - convergence (Attouch [3, th. 2.22 p.152]) there exist $g : X \to [-\infty, +\infty]$ and some further subsequence f_{n_j} which converges toward g in the epi sense on (X, d), hence in the sequential epi sense by proposition 22. Moreover, a sequence $x_j \to x$ in X iff x_j is bounded and $d(x_j, x) \to 0$. Thus, $x_j \to x$ implies lim inf $f_{n_j}(x_j) \ge g(x)$. If $u \in X$ and $g(x) < +\infty$, there exists a sequence u_j such that $d(u_j, u) \to 0$ and lim sup $f_{n_j}(u_j) \le g(u)$, hence u_j is bounded by (54), yielding $u_j \to u$. It follows that weak seq. epi - lim $f_{n_j} = g$, hence weak epi - lim $f_{n_j} = g$ by proposition 23. Therefore g is convex and lower semicontinuous, hence $g \in C(X, a)$. Thus

$$f^*_{n_j}(y) \overset{c}{\to} g^*(y), y \in X^*$$

since (as already proved) (56) $\Rightarrow$ (58), yielding $f^* = g^*$ by (58), hence $f = g$ (see Barbu - Precupanu [1, th. 1.4 p.97]). Summarizing, every subsequence of f_n has some further subsequence which converges toward f in the weak epi - sense, yielding (56) (Attouch [3, prop. 2.72 p. 245]).   □

Lemma 27. If for some n and $u \in X^*$ we have $f^*_n(u) = -\infty$, then $f^*_n = -\infty$ (Ekeland - Temam [1, prop. 3.1 p.14]), hence $f_n = +\infty$, contradicting upper boundedness of val (X, f_n). Therefore each f^*_n is finite on X^*. Moreover a^* is continuous in X^* (Ekeland - Temam [1, lemma 2.1 p.11]), hence for some $\varepsilon > 0$ we have

$$a^*(u) \le 1 \text{ if } \|u\| \le \varepsilon.$$

Fix any $x \ne 0$ in X^* and put

$$z = -\varepsilon x/\|x\|, t = \varepsilon/(\varepsilon + \|x\|).$$

It follows that $tx + (1 - t)z = 0$, and for some real constant k we have

$$k \le f^*_n(0) = f^*_n(tx + (1-t)z) \le tf^*_n(x) + (1-t)a^*(z) \le tf^*_n(x) + 1$$

since $a^*(z) \leq 1$. Thus

$$f_n^*(x) \geq (k-1)/t = (\varepsilon + \|x\|) (k-1)/\varepsilon.$$

With (54), this yields equiboundedness of f_n^* from above and below by two fixed convex continuous functions on X^*. The proof is ended by mimicking that of cor. 2.4 p.12 of Ekeland - Temam [1]. Separability of X is not required. $\qquad\square$

A second proof that (58) $\Rightarrow$ (56). We want to sketch a completely different, more direct proof because it exploits an interesting equivalent definition of epi - convergence in the convex setting.

Proof of (6). The set $C(X)$ of all extended real - valued, convex and lower semicontinuous functions on X is a lattice if equipped with the partial ordering

$$f \leq g \text{ iff } f(x) \leq g(x) \text{ for every } x \in X$$

and the lattice operations

$$(f \vee g)(x) = \max \{f(x), g(x)\}, \ (f \wedge g)(x) = \sup \{h(x) : h \in C(X), h \leq f, h \leq g\}$$

(see Asplund [1, cor.2.14]). Now set

$$g_j = \wedge \{f_n : n \geq j\}, \ j = 1, 2, 3, \ldots.$$

Let $x_n \to x$ in X. Since g_j is lower semicontinuous,

$$g_j(x) \leq \lim \inf g_j(x_n) \leq \liminf f_n(x_n) \text{ for each } j.$$

Then (6) obtains if we show that

(66) $$g_j(x) \to f(x) \text{ as } j \to +\infty.$$

In the lattice $C(X)$ we define

$$\lim {}'f_n = \vee_n \wedge \{f_k : k \geq n\}; \ \lim {}''f_n = \wedge_n \vee \{f_k : k \geq n\}.$$

The conjugacy correspondence $g \to g^*$ reverses inequalities and turns out to be a lattice antiisomorphism between $C(X)$ and $C(X^*)$, yielding

(67) $$(\lim {}'f_n)^* = (\vee_n \wedge \{f_k : k \geq n\})^* = \wedge_n(\wedge\{f_k : k \geq n\})^*$$
$$= \wedge_n \vee \{f_k^* : k \geq n\} = \lim {}''f_n^*$$

(see Asplund [3, cor. 2.14]). We shall need the following

29 Lemma For every subsequence of f_n, $\lim'' f_n^* = \lim \sup f_n^*$.

Proof of lemma 29. Fix any subsequence of f_n. For every $x \in X$ we have

$$(\lim {}'' f_n^*)(x) = \sup \{h(x) : h \in C(X^*), h \leq \sup_{k \geq n} f_k^* \text{ for all } n\}.$$

Since $n \to \sup \{f_k^* : k \geq n\}$ decreases, it suffices to prove that

$$\limsup f_n^* = \lim_{n \to +\infty} \sup \{f_k^* : k \geq n\} \in C(X^*).$$

Convexity of the right - hand side comes from convexity of every $\sup \{f_k^* : k \geq n\}$. Let us show lower semicontinuity. The conclusion is obvious if $f_n^* = -\infty$ for every n sufficiently large. If otherwise f_n is proper for some subsequence (i.e. f_n is not identically $+\infty$), then the corresponding subsequence f_n^* is proper as well, hence real - valued. By (58) we get boundedness of val (X, f_n), thus, by lemma 27, given $x_0 \in X^*$ and $\varepsilon > 0$ there exists a neighborhood A of x_0 such that if $x \in A$

$$\limsup f_n^*(x_0) - \varepsilon \leq \limsup f_n^*(x).$$

Then $\limsup f_n^*$ is lower semicontinuous on X^*, and lemma 29 is thereby proved.

Proof of (6) continued. By (58), (67) and lemma 29 we have

(68) $$\lim {}'' f_n^* = f^* = (\lim {}' f_n)^* \text{ for every subsequence.}$$

Taking conjugates in (68), since f and $\lim' f_n$ belong to $C(X)$ we get

$$f = \lim {}' f_n = \vee_j \wedge \{f_n : n \geq j\} = \vee_j g_j \text{ for every subsequence, hence for every } x$$

$$f(x) = \sup_j \; g_j(x) = \lim g_j(x)$$

yielding (66), and therefore (6).

Proof of (19). We shall equivalently prove that

(69)   for every x such that $f(x) < +\infty$ there exists $y_n \to x$ such that $f_n(y_n) \to f(x)$,

since if $f(x) = +\infty$ then it suffices to take $y_n = x$ by (6). So we assume that $f(x) = +\infty$.
30 Lemma (69) is true if $\partial f(x) \neq \emptyset$.
Taking this lemma for granted, by Brondsted - Rockafellar's theorem (Ekeland - Temam [1, th. 6.2 p.31]) there exists x_k such that

$$\|x_k - x\| < 1/k, |f(x_k) - f(x)| < 1/k, \partial f(x_k) \neq \emptyset \text{ for every } k.$$

From lemma 30, for every k there exists $v_{nk} \in X$ such that

(70) $$v_{nk} \to x_k, f_n(v_{nk}) \to f(v_k) \text{ as } n \to +\infty.$$

By Dunford - Schwartz [1, th. V.5.2], the restriction of the weak topology to each bounded set of X is metrizable. Let ω_n be a dense countable set in the unit sphere of X^*, and put

$$d(u, v) = \sum_{n=1}^{\infty} (1/2)^n | < \omega_n, u - v > |.$$

Then the metric d induces the weak topology on bounded sets of X. Therefore

$$z_j \rightharpoonup z \text{ iff } z_j \text{ is bounded and } d(z_j, z) \to 0.$$

Then by (70), for every k there exists n_k such that

$$d(v_{nk}, x_k) < 1/k, |f_n(v_{nk}) - f(x_k)| < 1/k, n \geq n_k.$$

We can assume n_k strictly increasing. For every $j \geq n_1$ there exists exactly one k_j such that $n_{k_j} \leq j < n_{k_j+1}$. Setting $y_j = v_{jk_j}$ we get

$$d(y_j, x) \leq d(y_j, x_{k_j}) + d(x_{k_j}, x) \leq 1/k_j + \|x_{k_j} - x\| \leq 2/k_j,$$

$$|f_j(y_j) - f(x)| \leq |f_j(y_j) - f(x_{k_j})| + |f(x_{k_j}) - f(x)| \leq 2/k_j.$$

Hence

$$d(y_j, x) \to 0, \ f_j(y_j) \to f(x),$$

so for j sufficiently large

$$a(y_j) \leq f_j(y_j) \leq 1 + f(x).$$

Therefore y_j is bounded so that $y_j \rightharpoonup x$, and (69) is proved.

The last step is therefore

Proof of lemma 30. Assume in addition that f is strictly convex. Let $y \in \partial f(x)$ and set

$$I_n(\cdot) = f_n(\cdot) - <y, \cdot>, I(\cdot) = f(\cdot) - <y, \cdot>.$$

Let

$$x_n \in \text{arg min } [X, I_n(\cdot) + d(x, \cdot)], u_n \in \text{arg min } (X, I_n).$$

Then

(71) $$I_n(x_n) + d(x, x_n) \leq I_n(u_n) + d(x, u_n).$$

We claim that $u_n \rightharpoonup x$. Fix any subsequence of u_n. By (58)

$$a(u_n) \leq f_n(u_n) \leq I_n(u_n) + <y, u_n> \leq \|y\| \|u_n\| + \text{ constant},$$

yielding boundedness of u_n. So for some further subsequence $u_n \rightharpoonup u$ say. By (58) and (6)

$$I(u) \leq \text{lim inf } I_n(u_n) = \text{ val } (X, I) = I(x)$$

but $x = \text{arg min } (X, I)$ by strict convexity of f. It follows that $u = x$, hence the original sequence $u_n \rightharpoonup x$. Now fix any subsequence of x_n. By (71), the definition of d and (58)

$$a(x_n) \leq f_n(x_n) \leq I(x_n) + d(x, x_n) + <y, x_n> \leq \text{ constant } + \|x - x_n\| + \|y\| \|x_n\|$$

yielding boundedness of x_n. Then a subsequence $x_n \rightharpoonup x_0$ say. Moreover $d(x, u_n) \to 0$. Letting $n \to +\infty$ in (71) we get (by (6) and (58))

$$I(x_0) + d(x, x_0) \leq I(x) < +\infty$$

whence $I(x_0) < +\infty$. Since x minimizes I

$$I(x_0) + d(x, x_0) \leq I(x_0)$$

which implies $d(x, x_0) = 0$. Therefore $x = x_0$ and $x_n \rightharpoonup x$ for the original sequence, yielding

$$f(x) \leq \lim \inf f_n(x_n).$$

By (58) we know that $I_n(u_n) \to I(x)$, therefore by (71)

$$\lim \sup f_n(x_n) \leq \lim \sup [I_n(x_n) + d(x, x_n)] + \lim \sup [< y, x_n > -d(x, x_n)] \leq f(x).$$

This entails (69) and therefore (58) $\Rightarrow$ (56).

In the general case (without assuming strict convexity) we notice that $d(x, \cdot)$ is convex, weakly sequentially continuous, and $d(x, u) > 0$ iff $u \neq x$, hence $x = \arg \min [X, I(\cdot) + d(x, \cdot)]$. Fix any subsequence of x_n, and let $x_{n_p} \rightharpoonup z$ for a further subsequence. By Marcellini [2, th. 1] we obtain

$$\lim {}' [I_n(\cdot) + d(x, \cdot)] = I(\cdot) + d(x, \cdot) \text{ for every subsequence}.$$

Consider $J_n(\cdot) = I_n(\cdot) + d(x, \cdot), J(\cdot) = I(\cdot) + d(x, \cdot), h_n = \wedge\{J_k : k \geq n\}$. Then for any fixed n

$$h_n(z) \leq \lim \inf_{p \to +\infty} h_n(x_{n_p}) \leq \lim \inf \text{val } (X, J_{n_p}).$$

Since $h_n(z) \to J(z)$ we obtain

$$J(z) \leq \lim \inf \text{val } (X, J_{n_p}).$$

As we saw before

$$J^* = \lim {}'' J_n^* \text{ for every subsequence},$$

hence $J_n^*(0) \to J^*(0)$, yielding

$$\text{val } (X, J_n) \to \text{val } (X, J).$$

It follows that $J(z) = \text{val } (X, J)$, hence $x = z$ and again $x_n \rightharpoonup x$ for original sequence.

$\square$

Remark. The above proof shows that, in $C(X, a)$, weak epi - $\lim f_n = f$ iff $f = \lim' f_n$ for every subsequence of f_n, provided f is proper.

End of proof of theorem 26. Now we prove that (if f is stricly convex) *(62) and (63) imply (56)*. The converse follows from theorem 24. By assumption, f is proper. Fix any subsequence of f_n. Let ω_n be a countable dense set in X^*. Let $x_n \in 1/n - \arg \min (X, f_n)$. Then by (63), $a(x_n)$ is bounded, hence x_n is, as well as $f_n(x_n)$. Thus

$$a^*(\omega_1) \geq f_n^*(\omega_1) \geq < \omega_1, x_n > -f_n(x_n) \geq \text{constant}$$

implying that $f_n^*(\omega_1)$ has convergent subsequences, say $f_{n1}^*(\omega_1)$. Similary $f_{n1}^*(\omega_2)$ is bounded, therefore it has convergent subsequences, say $f_{n2}^*(\omega_2)$, and so on. Consider the diagonal sequence $g_n = f_{nn}^*$. Then $g_n(\omega_p)$ converges for every p as $n \to +\infty$. By lemma 27 we see that g_n converges pointwise on X^*,

$$g_n(y) \to g(y) \text{ say }, y \in X^*.$$

Setting
$$k = g^*$$
we get $g \leq a^*$ since $g_n \leq a^*$, thus $k \in C(X, a)$. By the equivalence between (56) and (58)
$$\text{weak epi - lim } f_{nn} = k.$$

Now set
$$h_n = f_n - \text{ val } (X, f_n), h = f - \text{ val } (X, f).$$
By (63)

(72) $$\text{weak epi - lim } h_{nn} = k - \text{ val } (X, f).$$

We claim that

(73) $$\partial h(u) \subset \partial k(u) \text{ for every } u \in X.$$

Given $y \in \partial h(u)$, we have
$$u \in \text{arg min } (X, h(\cdot) - < y, \cdot >),$$
so by (62) there exists $x_n \in \text{arg min } (X, f_n(\cdot) - < y, \cdot >)$ such that $x_n \rightharpoonup u$. Then
$$x_n \in \text{arg min } (X, h_n(\cdot) - < y, \cdot >).$$

By (14) and (22)
$$u \in \text{arg min } (X, k(\cdot) - < y, \cdot >)$$
yielding $y \in \partial k(u)$, hence (73) is proved. Setting
$$p = k - \text{ val } (X, f)$$
we get $\partial h \subset \partial p$ by (73). Then by Barbu - Precupanu [1, cor. 2.4 p.123] there exists some constant c such that
$$p = h + c.$$
By (72), weak epi - lim $h_{nn} = p$, and $h_n \geq a +$ constant by (63). Hence by (15)
$$0 = \text{ val } (X, h_{nn}) \rightarrow \text{ val } (X, p)$$
and this yields $0 = \text{ val } (X, h) + c = c$, whence $p = h$. By proposition 23 and the "subsequences property" inherited by epi - convergence (see Attouch [3, (iii) of th.2.8.1 p.245]), it follows that for the original sequence
$$\text{weak seq. epi - lim } h_n = h.$$

Then by (63) we obtain
$$\text{weak epi - lim } [h_n + \text{ val } (X, f_n)] = \text{weak epi - lim } f_n = h + \text{ val } (X, f) = f,$$
i.e. (56). □

31 Remark. Due to lemma 27 (which does not require separability of X), (56) is equivalent to convergence of the values for every y in any given dense subset of X^*, provided sup val $(X, f_n) < +\infty$.

Section 7.   Hadamard Well-Posedness of Euler-Lagrange Equations and Stable Behaviour of Generalized (or sub) Gradients.

Differentiable convex problems.

Suppose that $f_n, f \in C(X, a)$ (with a fulfilling (54)) are Gâteaux differentiable functions on X, and f is strictly convex.

Then, given $y \in X^*$, $u_n \in \arg\ \min\ (X, f_n\ (\cdot) - \ < y, \cdot\ >)$ iff u_n is a solution to the Euler-Lagrange equation

$$\nabla f_n(u_n) = y.$$

Similarly, the only solution $u \in X$ to

$$(74) \qquad\qquad\qquad \nabla f(u) = y$$

minimizes $f(\cdot) - \ < y, \cdot\ >$ on X, and conversely. By theorem 26, value Hadamard well - posedness for all linear bounded perturbations of (X, f) is equivalent to Hadamard well - posedness of the related Euler - Lagrange equations. This means weakly continuous dependence of their solution upon the functional, for every fixed right - hand side, together with (63) which is to be considered here as a normalization condition (after all, ∇f_n gives back f_n up to an additive constant). So we have

32 Corollary *Let X, a, f, f_n be as in theorem 26, let f and f_n be Gâteaux differentiable with f strictly convex. Then the following are equivalent facts:*

$$(56) \qquad\qquad\qquad weak\ epi - lim\ f_n = f;$$

$$(75) \qquad\qquad for\ every\ y \in X^*, \nabla f_n(u_n) = y, \nabla f(u) = y \Rightarrow u_n \rightharpoonup u$$

and val $(X, f_n) \rightarrow val\ (X, f)$.

It is also true, under the assumptions of theorem 26, that weak epi - convergence of f_n to f is equivalent to solution or value Hadamard well - posedness of all continuous linear perturbations of (X, f) which are strongly convergent. This in turn amounts to Hadamard well - posedness of the Euler - Lagrange equation (74), meaning continuous dependence of its solution upon the datum (f, y), thereby including the right hand side y.

To see this, assume (56) and suppose $y_n \to y$ in X^*. Then $f_n^* \to f^*$ pointwise by theorem 26. By lemma 27, $f_n^*(y_n) \to f^*(y)$, giving

$$val\ (X, f_n(\cdot) - \ < y_n, \cdot\ >) \rightarrow\ val\ (X, f(\cdot) - \ < y, \cdot\ >).$$

Moreover, by the strong convergence of y_n to y,

$$weak\ epi - \lim\ (f_n(\cdot) - \ < y_n, \cdot\ >) = f(\cdot) - \ < y, \cdot\ > .$$

So, as in the proof of (56) $\Rightarrow$ (61), we see that $u_n \in \arg\min\ (X, f_n(\cdot) - \ < y_n, \cdot\ >)$ and f strictly convex imply

$$u_n \rightharpoonup \arg\min \ (X, f(\cdot) - <y, \cdot>).$$

Therefore we have

33 Corollary *Let X, a, f_n and f be as in theorem 26. The following are equivalent facts:*

(56) $$\text{weak seq. epi - lim } f_n = f;$$

(76) $$y_n \rightarrow y \text{ in } X^* \text{ implies val } (X, f_n(\cdot) - <y_n, \cdot>) \ \rightarrow \ val \ (X, f(\cdot) - <y, \cdot>).$$

If moreover f is strictly convex, the above conditions are equivalent to (63) along with
(77) *if $y_n \rightarrow y$ in X^* then for every selection $u_n \in \arg\min \ (X, f_n(\cdot) - <y_n, \cdot>)$
we have $u_n \rightharpoonup \arg\min \ (X, f(\cdot) - <y, \cdot>)$.*

   If every f_n and f are Gâteaux differentiable on X and f is strictly convex, then the above conditions are equivalent to (63) together with

(78) $$y_n \rightarrow y \text{ in } X^*, \ \nabla f_n(u_n) = y_n, \nabla f(u) = y \Rightarrow u_n \rightharpoonup u.$$

Thus the weak epi - topology is the weakest on $C(X, a)$ which gives Hadamard well - posedness of the Euler - Lagrange equation (74) (in the above sense).

Nondifferentiable convex problems

   Weak sequential epi - convergence (56), under Gâteaux differentiability of f_n and f, is related not only to Kuratowski convergence of the epigraphs, but also to the convergence of the graphs of the gradients; more precisely, to the following pair of conditions:
   if $y_n = \nabla f_n(u_n)$ and for some subsequence n_k we have $u_{n_k} \rightharpoonup u, y_{n_k} \rightarrow y$, then $y = \nabla f(u)$;
   for every u there exists $u_n \rightharpoonup u$ such that $y_n = \nabla f_n(u_n) \rightarrow y = \nabla f(u)$.
   The above mode of convergence of gph ∇f_n to gph ∇f is a particular case of the following definition. Let

$$T_n : X \rightrightarrows X^*$$

be given multifunctions, $n = 0, 1, 2, \dots$. Denote by

$$\text{gph } T_n = \{(x, y) \in X \times X^* : y \in T_n(x)\}$$

the graph of the n - th multifunction. We write

(79) $$G(w, s) - \lim T_n = T_0$$

iff $K - \lim \text{gph } T_n = \text{gph } T_0$ in $(X, \text{weak}) \times (X^*, \text{strong})$.
   By definition, this means that
(80) for every subsequence n_k and every sequence x_k, y_k such that $y_k \in T_{n_k}(x_k), x_k \rightharpoonup x$ and $y_k \rightarrow y$, we have $y \in T_0(x)$;
(81) for every y, x such that $y \in T_0(x)$ there exist $x_n, y_n \in T_n(x_n)$ such that $x_n \rightharpoonup x$ and $y_n \rightarrow y$.
Then (78) is equivalent to

$$G(w,s) - \lim \text{gph } \nabla f_n = \text{gph } \nabla f.$$

From corollary 33 we may well guess that for nondifferentiable convex functionals f_n, f, the condition

(82) $$G(w,s) - \lim \partial f_n = \partial f$$

together with a normalization condition, is equivalent to Hadamard well - posedness of all linear bounded perturbations of (X, f).

Condition (82) expresses a stable behaviour of first - order necessary conditions for optimality under data perturbations.

$G(s,w) - \lim \partial f_n^* = \partial f^*$ means that (symmetrically to (82)) the Kuratowski convergence

$$K - \lim \text{gph } \partial f_n^* = \text{gph } \partial f^*$$

obtains in $(X^*, \text{strong}) \times (X, \text{weak})$.

34 Theorem *Let X be a separable and reflexive Banach space, f_n and f in $C_0(X)$. Assume that*

$$u_n \in X, \sup f_n(u_n) < +\infty \Rightarrow u_n \text{ is bounded}.$$

Then the following are equivalent facts:

(83) $$\text{weak seq. epi - } \lim f_n = f;$$

(84) $$G(w,s) - \lim \partial f_n = \partial f, \text{ and there exist } u \in X, y \in \partial f(u), u_n \rightharpoonup u,$$
$$y_n \in \partial f_n(u_n) \text{ such that } y_n \to y, f_n(u_n) \to f(u);$$

(85) $$\text{strong epi - } \lim f_n^* = f^*;$$

(86) $$G(s,w) - \lim \partial f_n^* = \partial f^*, \text{ and there exist } v \in X, u \in \partial f^*(v), v_n \to v,$$
$$u_n \in \partial f_n^*(v_n) \text{ such that } u_n \rightharpoonup u, f_n^*(v_n) \to f^*(v).$$

Now we generalize corollary 32, as follows. Take f_n and f as in theorem 34. If

$$G(w,s) - \lim \partial f_n = \partial f, \text{ val } (X, f_n) \to \text{ val } (X, f),$$

then the normalization condition in (82) holds true provided we assume uniqueness of the minimizer u of f on X. To see this, let $u_n \rightharpoonup u, y_n \to 0$ be such that $y_n \in \partial f_n(u_n)$. Pick $v_n \in 1/n - \arg \min (X, f_n)$. By equicoercivity, v_n is bounded, hence

$$< y_n, v_n - u_n > \to 0.$$

Since

$$f_n(v_n) \geq f_n(u_n) + < y_n, v_n - u_n >$$

it follows that $\lim \sup f_n(u_n) \leq f(u)$, while of course $f_n(u_n) \geq \text{ val } (X, f_n)$, hence $\lim \inf f_n(u_n) \geq f(u)$. From theorem 34 we get

35 Corollary *Let X, f_n and f be as in theorem 34. Suppose that f has exactly one minimum point on X. Then the following are equivalent facts:*

$$weak\ seq.\ epi\text{ - }lim\ f_n = f;$$

$$G(w,s) -\ lim\ \partial f_n = \partial f\ and\ val\ (X, f_n) \to val\ (X, f).$$

Stable behavior of Clarke's generalized gradients (in the finite - dimensional setting).

The convergent behavior under perturbations of subgradients of convex functions we observed in theorem 34 can be generalized to locally Lipschitz continuous functions on some euclidean space R^P, by using the generalized gradient of Clarke (see Clarke [3]).

We shall apply such a result to the stability analysis of multipliers in mathematical programming problems with nonsmooth data (section IX. 4). This may be interpreted as a Hadamard well - posedness result (in some extended form) for the corresponding constrained optimization problems.

We consider a fixed sequence

$$f_n : R^P \to (-\infty, +\infty), n = 0, 1, 2, ...,$$

each term being assumed locally Lipschitz continuous.

Notation: $\partial f_n(x)$ = generalized gradient of f_n at x (see definition in Clarke [3, p. 27, and th. 2. 5. 1];

$u \in \partial^- f_n(x)$ iff $u \in R^P$ is a *lower semi - gradient* of f_n at x. This means that there exists some real - valued function k such that

(87) $$\text{for every } y \in R^P, f(y) \geq f(x) + u'(y - x) + k(x, y),$$

where

(88) $$k(x, y)/|x - y| \to 0 \text{ as } y \to x.$$

A sequence of multifunctions $T_n : R^P \rightrightarrows R^P$ is called *locally equibounded* iff for every bounded set $L \subset R^P$ there exists a constant c such that $|u| \leq c$ for all $x \in L$, all n and $u \in T_n(x)$.

The sequence f_n is called *equi - lower semidifferentiable* iff for every ball $L \subset R^P$ there exists a continuous function k on $L \times L$ fulfilling (88), moreover for every x, y in L, any n and $u \in \partial^- f_n(x)$ we have

(89) $$f_n(y) \geq f_n(x) + u'(y - x) + k(x, y).$$

36 Theorem *Let f_n be equi - lower semidifferentiable and locally equibounded, with locally Lipschitz terms, such that epi - lim $f_n = f_0$. Then*

$$\emptyset \neq\ lim\ sup\ gph\ \partial f_n \subset gph\ \partial f_0.$$

A comparison with the convex case. Suppose that every f_n is convex. Then (89) is automatically true (up to some extent, equi - lower semidifferentiability may be considered here as a surrogate convexity), as well as local Lipschitz continuity. The conclusion of theorem 36 is, however, not so sharp as in the convex case (corollary 35).

PROOFS.

Theorem 34. We give the proof in the special case that f and $f_n \in C(X, a)$, where a fulfills (54). A proof in the general case, as stated in theorem 34, can be found in Attouch [3, sec. 3.2.2 and 3.8.3]. The equivalence between (83) and (85) follows by theorem 26 and lemma 27 (observe that if we assume (85), then by (6)

$$\liminf f_n^*(0) = -\limsup \text{val}\,(X, f_n) \geq f^*(0) \geq \text{ constant } > -\infty).$$

Proof that (83) $\Rightarrow$ (84).   Let us prove (80). We have

$$y_k \in \partial f_{n_k}(x_k), x_k \rightharpoonup x, y_k \to y.$$

Then for all z and k

$$(90) \qquad\qquad f_{n_k}(z) \geq f_{n_k}(x_k) + <y_k, z - x_k>.$$

Given $u \in X$, by (24) there exists $u_k \rightharpoonup u$ such that $f_{n_k}(u_k) \to f(u)$. By (90) with $z = u_k$, as $k \to \infty$ we get

$$f(u) \geq f(x) + <y, u - x>$$

whence $y \in \partial f(x)$, so proving (80). Let us prove (81). Given $y \in \partial f(x)$, by Barbu - Precupanu [1, prop. 2.1 p. 103] we have

$$(91) \qquad\qquad f(x) + f^*(y) = <y, x>.$$

By assumption, there exists $x_n \rightharpoonup x$ such that $f_n(x_n) \to f(x)$. Since (83) is equivalent to (85), we know that there exists $y_n \to y$ such that $f_n^*(y_n) \to f^*(y)$. By definition of conjugate,

$$(92) \qquad\qquad f_n^*(y_n) \geq <y_n, x_n> -f_n(x_n),$$

therefore by (91) and (92) there exists a sequence $c_n \geq 0$ such that $c_n \to 0$ and

$$f_n(x_n) + f_n^*(y_n) = <y_n, x_n> +c_n.$$

By Brondsted - Rockafellar's theorem (Ekeland - Temam [1, th. 6. 2 p.31]) for every n there exist u_n, v_n such that

$$(93) \qquad\qquad v_n \in \partial f_n(u_n), \quad u_n - x_n \to 0, \quad v_n - y_n \to 0,$$

hence $u_n \rightharpoonup x, v_n \to y$ and (81) is proved.

Proof of the normalization condition in (84). Having fixed any x and $y \in \partial f(x)$, it suffices to prove that u_n, v_n in (93) verify

$$f_n(u_n) \to f(x).$$

By (93)

$$f_n(x_n) \geq f_n(u_n) + < v_n, x_n - u_n >.$$

Since $u_n \to x$ we get

$$f(x) \geq \limsup f_n(u_n) \geq \liminf f_n(u_n) \geq f(x),$$

thus (83) $\Rightarrow$ (84). The **proof that (84)** $\Longleftrightarrow$ **(86)** follows by noticing that the bijective continuous mapping $(x, u) \to (u, x)$, from $(X, \text{weak}) \times (X^*, \text{strong})$ to $(X^*, \text{strong}) \times (X, \text{weak})$, sends gph ∂g onto gph ∂g^* (Barbu - Precupanu [1, prop. 2.1 p. 103]) for every $g \in C_0(X)$.

Proof that (86) $\Rightarrow$ (85). Fix any subsequence of f_n. By the compactness of the epi - convergence (Attouch [3, th. 2.22 p. 152]) we obtain a further subsequence n_k such that

(94) $$\text{strong epi-lim } f_{n_k}^* = h.$$

Then $h \in C(X^*)$ by remark 21, and of course $h \leq a^*$. Let us show that $h(v) > -\infty$, where v is given by (86). This will imply properness of h (Barbu - Precupanu [1, prop. 1.4 p. 88]). If $h(v) = -\infty$ there exists $y_k \to v$ such that

(95) $$f_{n_k}^*(y_k) \to -\infty.$$

But the normalization condition in (86) yields

$$f_{n_k}(u_{n_k}) = < v_{n_k}, u_{n_k} > - f_{n_k}^*(v_{n_k}) \to < v, u > - f^*(v) \in R,$$

while from (95)

$$f_{n_k}(u_{n_k}) \geq < y_k, u_{n_k} > - f_{n_k}^*(y_k) \to +\infty,$$

a contradiction. So h is proper, that is $h \in C_0(X^*)$. Then by (94) and since (85) $\Rightarrow$ (86) we have

$$G(s, w) - \lim \partial f_{n_k}^* = \partial h = \partial f^*.$$

By Barbu - Precupanu [1, cor 2.4 p.123] there exists a constant c such that

$$h = f^* + c.$$

We claim that $c = 0$. It will follow that $h = f^*$, and therefore (85) for the original sequence (by Attouch [1, (iii) of th 2.8.1 p. 245]). Let $z_k \to v$ be such that $f_{n_k}^*(z_k) \to f^*(v) + c$. Then by (86)

$$f_{n_k}^*(z_k) \geq f_{n_k}^*(v_{n_k}) + < u_{n_k}, z_k - v_k >,$$

and letting $k \to +\infty$

$$f^*(v) + c \geq f^*(v),$$

whence $c \geq 0$. However, by (94) and (86)

$$f^*(v) = \liminf f_{n_k}^*(v_{n_k}) \geq f^*(v) + c$$

whence $c \leq 0$. Thus $c = 0$, establishing the claim. $\square$

Theorem 36. We need some preliminary results. Let

$$T_n : R^P \rightrightarrows R^P, \quad n = 0, 1, 2, \dots$$

be a sequence of multifunctions. Set

$$Q_n(x) = \limsup_{y \to x} T_n(y).$$

37 Lemma If $\limsup \operatorname{gph} T_n \subset \operatorname{gph} T_0$, then $\limsup \operatorname{gph} Q_n \subset Q_0$.
Proof. By definition of Q_n, $\operatorname{gph} Q_n = \operatorname{cl} \operatorname{gph} T_n$. Then

$$\limsup \operatorname{gph} Q_n = \limsup \operatorname{cl} \operatorname{gph} T_n \subset \operatorname{cl} \operatorname{gph} T_0 = \operatorname{gph} Q_0.$$

$\square$

38 Lemma Assume T_n locally equibounded, and $\limsup \operatorname{gph} T_n \subset \operatorname{gph} T_0$. Then

$$\limsup \operatorname{gph} \operatorname{co} Q_n \subset \operatorname{gph} \operatorname{co} Q_0.$$

Proof. Let $(x, u) \in \limsup \operatorname{gph} \operatorname{co} Q_n$. Then for suitable sequences we have

$$x_k \to x, \ u_k \to u, \ u_k \in \operatorname{co} Q_{n_k}(x_k).$$

By Carathéodory's convexity theorem (Rockafellar [2, th. 17.1]), given k there exist points $v_{ik} \in Q_{n_k}(x_k)$ and numbers $a_{ik} \geq 0$, $i = 1, \dots, p + 1$, with $\sum a_{ik} = 1$, such that $u_k = \sum a_{ik} v_{ik}$. Here $\sum$ is short for $\sum_{i=1}^{p+1}$. By local equiboundedness of T_n we see that $\cup\{Q_{n_k}(x_k) : k = 1, 2, 3, \dots\}$ is bounded. Thus for some subsequence

$$a_{ik} \to a_i, \ a_i \geq 0, \ \sum a_i = 1, v_{ik} \to v_i \text{ say.}$$

Therefore for every i

$$(x, v_i) \in \limsup \operatorname{gph} Q_n.$$

By lemma 37, $v_i \in Q_0(x)$, so that

$$u = \sum a_i v_i \in \operatorname{co} Q_0(x).$$

$\square$

39 Lemma Lim sup $\operatorname{gph} \partial^- f_n \subset \operatorname{gph} \partial^- f_0$.
Proof. For some subsequence n_j let

$$x_j \to x, \ u_j \to u, \ u_j \in \partial^- f_{n_j}(x_j).$$

Given $z \in R^P$, let L be an open ball containing both z and x. By epi - convergence, there exists $z_j \to z$ such that $f_{n_j}(z_j) \to f_0(z)$, thus z_j and $x_j \in L$ for j sufficiently large. By (89)

$$f_{n_j}(z_j) \geq f_{n_j}(x_j) + u'_j(z_j - x_j) + k(x_j, z_j).$$

Since $\liminf f_{n_j}(x_j) \geq f_0(x)$ and k is continuous

$$f_0(z) \geq f_0(x) + u'(z - x) + k(x, z).$$

$\square$

40 Lemma $\partial^- f_n$ is locally equibounded.

Proof. Given any bounded $L \subset R^P$, $x \in L$, $u_n \in \partial^- f_n(x)$ and z such that $|z| \leq 1$, by (88) we get

$$f_n(x + z) \geq f_n(x) + u'_n z + k(x, x + z),$$

hence

(96) $$u'_n z \leq f_n(x + z) - f_n(x) - k(x, x + z).$$

By the continuity of k and local equiboundedness of f_n

$$u'_n z \leq \text{ constant if } |z| \leq 1, \text{ any } n, \ x \in L.$$

$\square$

Proof of theorem 36. Given $x \in R^P$, let $x_n \to x$ and consider $u_n \in \partial f_n(x_n)$. It follows from Rockafellar [8, cor. 4 T] that for each n

(97) $$\partial f_n(x) = \text{ cl co } \limsup_{y \to x} \partial^- f_n(y), \ x \in R^P.$$

Then u_n is bounded by lemma 40, therefore $\limsup$ gph $\partial f_n \neq \emptyset$. Again by lemma 40 we get compactness of $\limsup_{y \to x} \partial^- f_n(y)$ for every n and x, so that its closed convex envelope is the same as co $\limsup_{y \to x} \partial^- f_n(y)$. Then the conclusion follows from lemmas 38 and 39. $\square$

Section 8. Convergence in the Sense of Mosco and Well-Posedness.

As we saw in section II.2, Mosco convergence in Conv (X), X a given Banach space, plays a key role in the analysis of Hadamard well - posedness of best approximation problems.

Let us come back to the definition. Given a sequence of nonempty subsets K_n of $X, n \geq 0$, then

(98) $$M - \lim K_n = K_0$$

iff

(99) $$\text{for every subsequence } n_j, \ x_j \in K_{n_j} \text{ and } x_j \to x \Rightarrow x \in K_0;$$

(100) $$\text{for every } u \in K_0 \text{ there exists } u_n \in K_n \text{ such that } u_n \to u.$$

These conditions amount to

$$\text{weak seq. } \limsup K_n = K_0 = \text{ strong } \liminf K_n.$$

We translate (98) into a mode of convergence for the indicator functions

$$I_n(\cdot) = \text{ ind } (K_n, \cdot),$$

as follows. Assume (99). Then

(101) $\qquad\qquad\qquad x_n \rightharpoonup x$ in X implies $\liminf I_n(x_n) \geq I_0(x)$,

indeed the conclusion is obvious if $x \in K_0$, otherwise if $x \notin K_0$ and $x_n \in K_n$ for some subsequence, then a contradiction arises by (99). Thus $x_n \notin K_n$ for every n sufficiently large, hence $I_n(x_n) \to +\infty = I_0(x)$. Conversely, (101) implies (99). To see this, let $x_j \rightharpoonup x$ with $x_j \in K_{n_j}$. Put

$$y_n = x_j, \ n_j \leq n < n_{j+1}.$$

Then $y_n \rightharpoonup x$, hence

$$0 = \liminf I_{n_j}(x_j) = \liminf I_{n_j}(y_{n_j}) \geq \liminf I_n(y_n) \geq I_0(x)$$

yielding $x \in K_0$. Therefore (99) $\Leftrightarrow$ (101).

Now assume (100). Then

(102) $\qquad$ for every $u \in X$ there exists $u_n \to u$ such that $\limsup I_n(u_n) \leq I_0(u)$.

Indeed, (102) is obvious when $u \notin K_0$, while u_n as in (100) works when $u \in K_0$. If conversely we assume (102), then if $u \in K_0$ we take u_n as there, so that $u_n \in K_n$ and (100) is fulfilled. We see that Mosco convergence of K_n to K_0 amounts to (101) and (102). This fact justifies the following definition.

Let

$$f_n : X \to [-\infty, +\infty], n \geq 0$$

be a given sequence of functions on the Banach space X. We write

$$M - \lim f_n = f_0$$

and say that f_n *converges to f_0 in the sense of Mosco* iff

(103) $\qquad\qquad$ weak seq. epi - $\lim f_n = f_0 =$ strong epi - $\lim f_n$.

This means that

$\qquad\qquad$ if $x_n \rightharpoonup x$ then $\liminf f_n(x_n) \geq f_0(x)$;

$\qquad$ for every u there exists a sequence $u_n \to u$ such that $f_n(u_n) \to f_0(u)$.

Then we get

41 Proposition *Let K_n, $n \geq 0$, be nonempty subsets of X. Then*

$$M - \lim K_n = K_0 \ \text{iff} \ M - \lim \text{ind}(K_n, \cdot) = \text{ind}(K_0, \cdot).$$

Example. Let K_n be nonempty subsets of the Banach space $X, n \geq 0$. If $f : X \to (-\infty, +\infty)$ is continuous and convex, then it is easily checked that

$$M - \lim K_n = K_0 \Rightarrow M - \lim(f + \text{ind } K_n) = f + \text{ind } K_0.$$

A particular case: x is a fixed point of X and

$$f(u) = \|x - u\|,$$

so that $(X, f + \text{ind } K_n)$ defines the best approximation problem to x by elements from K_n.

Example. Let $X = R^P$. Then obviously

$$M - \lim f_n = f \text{ iff epi - } \lim f_n = f.$$

Let $f_n \in C_0(R^P)$ be such that $a \leq f_n \leq b$ for every n and some $a, b \in C_0(R^P)$, both fulfilling (54). Then

$$M - \lim f_n = f \text{ iff } f_n(x) \to f(x) \text{ for all } x \in R^P.$$

Proof. Let $f_n \to f$ pointwise. Since f_n are locally equicontinuous (Rockafellar [2, th. 10.6]), then epi - $\lim f_n = f$. Conversely assume that epi - $\lim f_n = f$. Then $f_n^*(y) \to f^*(y), y \in R^P$, by theorem 26. Since $b^* \leq f_n^* \leq a^*, f_n^*$ are locally equicontinuous (lemma 27), hence epi - $\lim f_n^* = f^*$. The conclusion follows from theorem 26 by taking conjugates. $\square$

We pursue now the question, whether one can preserve well-posedness by taking limits. Theorem 18 shows that under suitable conditions, Tykhonov well- posedness is preserved under epi - limits, moreover solution and value Hadamard well-posedness is present.

By imposing Mosco convergence, it is possible to obtain a similar result. We obtain again strong Hadamard well-posedness and (in the convex setting) Mosco convergence as a necessary condition, complementing theorem 26, and Tykhonov well-posedness of the limit problem.

42 Theorem *Let X be a reflexive Banach space, I_n a sequence of proper extended real-valued functions on X. Assume that there exists an unique $x_n = \arg \min (X, I_n), n \geq 1$, and a forcing function c such that for every x and every $n \geq 1$*

$$(104) \qquad I_n(x) \geq I_n(x_n) + c(\|x_n - x\|);$$

$$\sup I_n(u_n) < +\infty \quad \Rightarrow \quad \sup \|u_n\| < +\infty; \quad M - \lim I_n = I_0.$$

Then (X, I_0) is strongly Tykhonov well- posed; val $(X, I_n) \to$ val (X, I_0), and every asymptotically minimizing sequence y_n for (X, I_n) fulfills $y_n \to \arg \min (X, I_0)$; finally (104) is true for $n = 0$.

Conversely, if X is separable, a satisfies (54), every $I_n \in C(X, a)$ and I_0 is strictly convex, then $M - \lim I_n = I_0$ provided val $(X, I_n) \to$ val (X, I_0) and $u_n \to \arg \min (X, I_0(\cdot) - < y, \cdot >)$ for every $y \in X^*$ and every selection $u_n \in \arg \min (X, I_n(\cdot) - < y, \cdot >)$.

If Mosco convergence is modified in theorem 42 to uniform convergence on strongly compact sets, then strong solution Hadamard well- posedness may fail: consider a separable Hilbert space X with basis e_n and

$$I_0(x) = \|e_1 - x\|^2 + 1, \quad I_n(x) = \|e_1 + e_n - x\|^2$$

(notice that $\lim \inf I_n(e_n) < I_0(0)$).

43 Example. We keep the same data and notation of example 28. We claim that

$$M - \lim f_n = f_0 \text{ iff } u_n(y) \to u_0(y) \text{ in } H_0^1(\Omega) \text{ for all } y \in L^2(\Omega).$$

Proof. If $u_n(y) \to u_0(y)$, then weak epi - $\lim f_n = f_0$ by example 28. Since $y \to u_0(y)$ is an isomorphism from $H^{-1}(\Omega)$ to $H_0^1(\Omega)$ (Gilbarg - Trudinger [1, sec. 8.2]), then $\{u_0(y) : y \in L^2(\Omega)\}$ is dense in $H_0^1(\Omega)$. Then by the local equicontinuity of f_n, to prove Mosco convergence it suffices to shows that

$$(105) \qquad\qquad f_n[u_0(y)] \to f_0[u_0(y)], \ y \in L^2(\Omega).$$

Fix $y \in L^2(\Omega)$ and set $u_n = u_n(y)$. Remembering (64) and (65) we obtain

$$(106) \qquad\qquad f_n(v - u_n) = f_n(v) - \int_\Omega y \, v \, dx \ + \ f_n(u_n) =$$

$$= f_n(v) - \int_\Omega y \, v \, dx \ + \ (1/2) \int_\Omega y \, u_n \, dx, \ v \ \in H_0^1(\Omega).$$

Setting $v = u_0$ in (106) we have

$$f_n(u_0) = f_n(u_0 - u_n) + \int_\Omega y \, u_0 \, dx - (1/2) \int_\Omega y \, u_n \, dx.$$

By local equicontinuity of f_n we get $f_n(u_0 - u_n) \to 0$, hence $f_n(u_0) \to (1/2) \int_\Omega y \, u_0 \, dx = f_0(u_0)$. This proves (105), yielding Mosco convergence.

Conversely assume $M - \lim f_n = f_0$. Then by (65) and (106)

$$f_n(v) - < y, v > = \text{ val } (H_0^1(\Omega), \ f_n(\cdot) - < y, \cdot >) + f_n(v - u_n), v \in H_0^1(\Omega).$$

Now put $I_n(v) = f_n(v) - < y, v >$. Then

$$(107) \qquad\qquad I_n(v) \geq I_n(u_n) + c\|v - u_n\|^2, \ v \in H_0^1(\Omega), n = 1, 2, 3, ...,$$

for a suitable constant $c > 0$, since

$$2 f_n(w) \geq c_1 \int_\Omega |\nabla w|^2 dx$$

and $(\int_\Omega |\nabla w|^2 dx)^{1/2}$ is an equivalent norm on $H_0^1(\Omega)$ thanks to Poincaré inequality (Brezis [2, p. 174]). Thus (107) entails $u_n \to u_0$ by theorem 42.   □

Mosco convergence does not imply value convergence. Consider for example

$$X = R, \ I_0 = 0, \ I_n(x) = -1 \ \text{if} \ x \leq -n, \ = x/n \ \text{if} \ x \geq -n.$$

Here $M- \lim I_n = I_0$ but val $(X, I_n) \not\to$ val (X, I_0) (compare with theorem 5). A basic property of Mosco convergence states that if X is a reflexive Banach space and $I_n \in C_0(X)$ for $n \geq 0$, then $M- \lim I_n = I_0$ iff $M- \lim I_n^* = I_0^*$ (see Attouch [3, th. 3.18 p. 295], and Beer [3] for another proof). Hence value Hadamard well-posedness obtains under Mosco convergence up to the addition of suitable small linear perturbation to I_n due to (102), as follows.

44 Theorem *Let X be a reflexive Banach space, $I_n \in C_0(X)$ for $n \geq 0$ be such that $M- \lim I_n = I_0$. Then for every $u \in X^*$ there exists $u_n \to u$ in X^* such that*

$$val \ (X, I_n(\cdot)- < u_n, \cdot >) \to \ val \ (X, I_0(\cdot)- < u, \cdot >).$$

Mosco convergence in $C_0(X)$, for X a reflexive Banach space, implies $G(w,s)$ convergence of the subdifferentials without assuming equicoercivity. A perusal of the proof that $(83) \Rightarrow (84)$ in theorem 34 shows that equicoercivity is used there only to obtain the equivalence $(83) \Leftrightarrow (85)$, which now follows from the above mentioned continuity of the conjugacy with respect to Mosco convergence (Attouch [3, th.3.18 p.295]; for separable spaces, it is a corollary to theorem 34, since $g^{**} = g$ if $g \in C_0(X)$). So we have

45 Corollary *Let X be a reflexive Banach space, $I_n \in C_0(X)$ with $M- \lim I_n = I_0$. Then $G(w,s)- \lim \partial I_n = \partial I_0$.*

This corollary will play a role in proving the stable behavior of multipliers in convex mathematical programming (section IX.3).

Proof of theorem 42. Given $z \in$ dom I_0 there exists $u_n \to z$ such that $I_n(u_n) \to I_0(z)$. Then by (104) we get

$$I_n(x_n) \leq I_n(x_n) + c(\|x_n - u_n\|) \leq I_n(u_n) \leq \ \text{constant},$$

hence for some subsequence, by (14)

$$x_n \rightharpoonup x_0 \in \text{arg min } (X, I_0)$$

and by (15), val $(X, I_n) \to$ val (X, I_0). Moreover there exists $z_n \to x_0$ such that $I_n(z_n) \to I_0(x_0)$, whence by (104)

$$I_n(z_n) \geq I_n(x_n) + c(\|x_n - z_n\|).$$

Letting $n \to +\infty$ we see that lim sup $c(\|x_n - z_n\|) \leq 0$, whence $x_n - z_n \to 0$. This entails $x_n \to x_0$ for the original sequence. Moreover, given $x \in$ dom I_0 let $v_n \to x$ be such that $I_n(v_n) \to I_0(x)$. Then

$$I_n(v_n) \geq \ \text{val} \ (X, I_n) + c(\|x_n - v_n\|)$$

and letting $n \to +\infty$ we obtain

$$I_0(x) \geq I_0(x_0) + c(\|x_0 - x\|).$$

Thus (104) is true for $n = 0$ too, hence (X, I_0) is strongly Tykhonov well - posed. If y_n is asymptotically minimizing for (X, I_n), then

$$I_n(y_n) - \text{val } (X, I_n) \to 0.$$

By (104) with $x = y_n$ we get $x_n - y_n \to 0$, therefore $y_n \to x_0 = \text{arg min } (X, I_0)$. Conversely, let $I_n \in C(X, a)$. Then weak epi - lim $I_n = I_0$ by theorem 26. Let $u \in X$ be such that $I_0(u) < +\infty$. If there exists $y \in \partial I_0(u)$, then there is some $u_n \in$ arg min $(X, I_n(\cdot) - < y, \cdot >)$ such that $u_n \to u = \text{arg min } (X, I_0(\cdot) - < y, \cdot >)$ by assumption. By weak epi-convergence, there exists $x_n \rightharpoonup u$ with $I_n(x_n) \to I_0(u)$. Then

$$I_n(u_n) - < y, u_n > \leq I_n(x_n) - < y, x_n >,$$

hence $\limsup I_n(u_n) \leq I_0(u)$. Finally, if I_0 is not subdifferentiable at u, we proceed as in the proof of (19) in theorem 26 (p. 151). Hence there exists a sequence x_k such that

$$\|x_k - u\| < 1/k, |I_0(x_k) - I_0(u)| < 1/k, \partial I_0(x_k) \neq \emptyset.$$

For every k we find a sequence v_{nk} such that

$$v_{nk} \to x_k, I_n(v_{nk}) \to I_0(v_k) \text{ as } n \to +\infty.$$

As in that proof, for some subsequence k_j the sequence $y_j = v_{jk_j}$ fulfills $y_j \to u$ and $I_j(y_j) \to I_0(u)$. □

Section 9.   Hadamard Well - Posedness of Convex Quadratic Problems.

From example 43 we see that Mosco convergence on $H_0^1(\Omega)$ of I_n to I_0, where

$$I_n(u) = (1/2) \int_\Omega \sum_{i,j=1}^{p} a_{ij}^n \frac{\partial u}{\partial x_i} \frac{\partial u}{\partial x_j} dx$$

is the weakest one giving *strong* solution Hadamard well - posedness of $(H_0^1(\Omega), I_0(\cdot) - < y, \cdot >)$ for each fixed $y \in L^2(\Omega)$. On the other hand, example 28 tells us that the weakest convergence to obtain *weak* solution Hadamard well - posedness turns out to be weak epi - convergence. This fact will be generalized to quadratic functions which are equicoercive on a given Banach space, as follows.

Standing assumptions: X is a reflexive Banach space; D is a dense subset of X^*;

$$A_n : X \to X^*, n = 0, 1, 2, ...,$$

is a given sequence of linear bounded symmetric operators; c_1, c_2 are positive constants such that for every n and $x \in X$ we have

$$< A_n x, x > \geq c_1 \|x\|^2; \ \|A_n\| \leq c_2.$$

Free problems. Given $y \in X^*$, let us denote by $u_n = u_n(y)$ the unique solution to

$$(108) \qquad < A_n u_n, v > = < y, v > \ \text{for every } v \in X.$$

Existence and uniqueness of u_n follows from Lax - Milgram's theorem (Brezis [2, p. 88]). Given $y \in X^*$, put

$$(109) \qquad I_n(u) = (1/2) < A_n u, u >; \ J_n(u) = I_n(u) - < y, u >, \ u \in X.$$

Then u_n solves (108) iff

$$u_n = u_n(y) = \arg \ \min \ (X, J_n),$$

since (108) is the Euler - Lagrange equation of (X, J_n).

Notice that, for every fixed y, (X, J_n) is equi - well - posed, since (remembering (105))

$$(110) \qquad J_n(x) - J_n(u_n) = (1/2) < A_n x, x > - < A_n u_n, x > + (1/2) < A_n u_n, u_n > =$$

$$= (1/2) < A_n(u_n - x), u_n - x > \geq (c_1/2) \|x - u_n\|^2,$$

so that (104) is fulfilled. Then by theorem 42

$$M - \lim I_n = I_0 \Rightarrow u_n(y) \to u_0(y), y \in X^*,$$

since $M - \lim I_n = I_0$ implies $M - \lim J_{n_c} = J_0$ by (22). Conversely, assume that $u_n(y) \to u_0(y)$ for every $y \in D$. Then by (108)

$$J_n(u_n) = \text{val} \ (X, I_n(\cdot) - < y, \cdot >) = -(1/2) < y, u_n > \to -(1/2) < y, u_0 >$$
$$= \text{val} \ (X, I_0(\cdot) - < y, \cdot >).$$

Then (58) follows by density of D and remark 31, whence (56). So we shall obtain Mosco convergence by showing that $I_n(u) \to I_0(u), u \in X$. By local equicontinuity of I_n we need only to prove that

$$(111) \qquad I_n[u_0(y)] \to I_0[u_0(y)], y \in D,$$

since $y \to u_0(y)$ is an isomorphism of X^* onto X (Lax - Milgram's theorem: consider indeed the inverse mapping of A_0), hence $u_0(D)$ is dense in X. By (110) and (108)

$$I_n(u_0) = < y, u_0 > -(1/2) < y, u_n > + I_n(u_n - u_0)$$

which entails (111) due to the local equicontinuity of I_n and (108). Then $M - \lim I_n = I_0$, as desired.

We shall also show that Mosco convergence of I_n to I_0 amounts to strong operator (i.e. pointwise) convergence of A_n to A_0. This allows explicit characterizations in a number of applications.

46 Theorem *The following are equivalent facts:*

$$(112) \qquad\qquad M - \lim I_n = I_0;$$

(113) *for every* $y \in D$, $\arg\min (I_n(\cdot) - \ <y, \cdot>) \ \to \ \arg\min (I_0(\cdot) - \ <y, \cdot>)$ *in* X;

$$(114) \qquad\qquad A_n x \to A_0 x \text{ for every } x \in X.$$

Thus Mosco convergence is the weakest one for the above quadratic equipositive-definite functionals I which yields strong solution Hadamard well - posedness for every $(X, I(\cdot) - \ <y, \cdot>), y \in D$. Similarly, weak epi - convergence is the weakest one giving weak solution Hadamard well - posedness for each $(X, I(\cdot) - \ <y, \cdot>), y \in D$, according to

47 Theorem *Let X be separable. The following are equivalent facts:*

$$(115) \qquad\qquad weak\ epi - \lim I_n = I_0;$$

(116) *for every* $y \in D$, $\arg\min (I_n(\cdot) - \ <y, \cdot>) \ \to \ \arg\min (I_0(\cdot) - \ <y, \cdot>)$ *in* X;

$$(117) \qquad for\ every\ y \in D,\ val\ (X, I_n(\cdot) - \ <y, \cdot>) \to \ val\ (X, I_0(\cdot) - \ <y, \cdot>);$$

$$(118) \qquad\qquad A_n^{-1}y \to A_0^{-1}y \text{ for every } y \in D.$$

By (108) we get indeed

$$A_n^{-1}y = \ \arg\ \min\ (X, I_n(\cdot) - \ <y, \cdot>).$$

We apply theorem 26. Hence (115) implies (116), (117) and (118). Of course (116) $\Leftrightarrow$ (118). Since

$$\|A_n^{-1}\| = (\inf\{\|A_n x\| : \|x\| = 1\})^{-1} \le 1/c_1$$

(Anselone [1, p. 5]) we get (118) $\Rightarrow$ (115). Finally, by local equicontinuity of I_n^*, (117) implies (116) and (115).

Corollary 35 entails the equivalence between each of the conclusions of theorem 47 and G - *convergence* of A_n to A_0, defined as

$$G(w, s) - \lim\ gph\ A_n = \ gph\ A_0,$$

which of course is equivalent to $G(s, w)-$ convergence of gph A_n^{-1} to gph A_0^{-1} (compare with theorem 34, remembering that $I_n^*(y) = (1/2) < A_n^{-1}y, y >$).

Constrained problems. We are given a sequence K_n of nonempty closed convex subsets of X and a sequence $y_n \in X^*$. We consider the minimum problems (K_n, J_n), where J_n is defined by (109) with $y = y_n$. We look for conditions giving strong convergence of arg min (K_n, J_n) to arg min (K_0, J_0). So we are interested in solution Hadamard well - posedness with respect to all data K, A, y in the variational inequality (K, A, y), i.e. to find $\omega \in K$ such that

$$(119) \qquad < A\omega - y, \ v - \omega > \geq 0 \text{ for every } v \in K.$$

Tykhonov well - posedness of (119) (and the link with Hadamard's for fixed A when $K = X$) has been studied in section II.5.

48 Theorem *Assume that*

$$M - \lim K_n = K_0, \ M - \lim I_n = I_0, \ y_n \to y_0,$$

Then

$$arg\ min\ (K_n, J_n) \to arg\ min\ (K_0, J_0)\ and\ val\ (K_n, J_n) \to val\ (K_0, J_0).$$

Now we fix the constraint K_0, assumed to be an affine manifold given by

$$Sx = y,$$

where

$$S : X \to R^m$$

is a given linear bounded surjective operator, and $y \in R^m$ is fixed too. The finite dimensionality assumption about the range of S (as an operator acting between Banach spaces) is equivalent, taking a quotient if necessary, to compactness of S and closure of its range. Whitin this framework we characterize Hadamard well - posedness of (K_0, J_0) having fixed $K_0 = S^{-1}y$, under perturbations I_n (acting on I_0) given by (109).
Put

$$x_n = arg\ min\ (S^{-1}y, I_n).$$

Then x_n, whose existence and uniqueness is readily verified, is characterized by the optimality condition (Barbu - Precupanu [1, th.1.8 p.185])

$$(120) \qquad Sx_n = y, A_n x_n = S^* z_n \text{ for some } z_n \in R^m.$$

Here

$$S^* : R^m \to X^*$$

denotes the adjoint operator of S (identifying R^m with its dual space). Then S^* is one - to - one since S is onto. Thus the n - th *Lagrange multiplier* z_n is unique. A suitable weakening of (118) suffices for weak solution and value Hadamard well - posedness, and convergence of multipliers, as follows.

49 Theorem *(a). If $A_n^{-1} u \to A_0^{-1}u$ for every $u \in S^*(R^m)$, then for every y*

$$x_n \rightharpoonup x_0, \ z_n \to z_0, \ val \ (S^{-1}y, I_n) \to \ val \ (S^{-1}y, \ I_0).$$

$$\text{If } A_n^{-1}u \to A_0^{-1}u \text{ for every } u \in S^*(R^m), \text{ then } x_n \to x_0.$$

(b). The following are equivalent facts:

(121) $$val \ (S^{-1}y, I_n) \to \ val \ (S^{-1}y, I_0) \ for \ every \ y;$$

(122) $$z_n \to z_0 \ for \ every \ y;$$

(123) $$SA_n^{-1} \ u \to SA_0^{-1} \ u \ for \ every \ u \in S^*(R^m).$$

The equivalence between value and weak solution Hadamard well - posedness holds for free problems (theorem 47), but may fail in the constrained setting, as shown in the following

Counterexample. Let $X = R^2$, $m = 1$, $Su = (1, 0)u$,

$$A_0 = \begin{pmatrix} 1/2 & 0 \\ 0 & 2 \end{pmatrix}, \quad A_n = \begin{pmatrix} 1 & 1 \\ 1 & 2 \end{pmatrix} \text{ if } n \geq 1.$$

Here $u \in S^*(R^1)$ iff $u = (v, 0)'$ for some real v. Condition (123) holds since

$$SA_n^{-1}S^* \ v = 2v = SA_0^{-1}S^*v, \ v \in R,$$

but for every y the optimal solution turns out to be

$$x_n = (y, -y/2)' \text{ if } n \geq 1, \ x_0 = (y, 0)'.$$

PROOFS.

Theorem 46. We need only to prove that $(112) \iff (114)$. We shall need the following fact. Let $f_n \in C_0(X), n = 0, 1, 2, ..,$ be such that $M - \lim f_n = f_0$. Then

(124) for every u and $v \in \partial f_0(u)$ there exists $u_n \to u$, $v_n \to v$ such that $v_n \in \partial f_n(u_n)$.

This fact follows from the proof that $(83) \Rightarrow (84)$ of theorem 34, more precisely the step which proves (81). The existence of y_n there follows now by the continuity of the conjugacy under Mosco convergence. Assume (112). By (124), given $x \in X$ there exists $u_n \to x$ such that $A_n u_n = \partial I_n(u_n) \to A_0 x$, yielding

$$A_n x = A_n(x - u_n) + A_n u_n \to A_0 x$$

since $\|A_n(x - u_n)\| \leq c_2 \|x - u_n\| \to 0$. This yields (114). Conversely assume (114). Then obviously $I_n(x) \to I_0(x)$ for every x. Let $u_n \rightharpoonup u$. Then

$$2 \ I_n(u_n) = \ < A_n(u_n - u), \ u_n - u > + 2 < A_n u, u_n >$$
$$- < A_n u, u > \ \geq \ 2 < A_n u, u_n > \ - \ < A_n u, u >$$

and the last term converges to $< A_0 u, u >= 2I_0(u)$.   □

Theorem 48. Put

$$f_n = J_n + \text{ind } K_n, \quad w_n = \arg \min (K_n, J_n).$$

Then w_n is the only solution to (X, f_n). In order to apply theorem 42, we note that if $x \in K_n$

$$f_n(x) - f_n(w_n) = (1/2) < A_n(x - w_n), \; x - w_n >$$
$$+ < A_n w_n - y_n, \; x - w_n > \geq (c_1/2) \|x - w_n\|^2$$

thanks to (119). Thus (104) is verified. It remains to show that $M - \lim f_n = f_0$. As readily verified, $M - \lim J_n = J_0$. Since both J_n and ind K_n are $M-$ convergent (proposition 41), if follows that

$$u_n \rightharpoonup u \text{ in } X \Rightarrow \liminf f_n(u_n) \geq f_0(u).$$

Now given $x \in X$ there exists $x_n \to x$ such that

$$\text{ind } (K_n, x_n) \to \text{ ind } (K_0, x)$$

by proposition 41. By theorem 46, $A_n x_n \to A_0 x$, hence $f_n(x_n) \to f_0(x)$. □

Theorem 49. **Proof of (a).** Since for every $u \in X$

$$c_1 \|u\|^2 \leq < A_n u, u > \leq c_2 \|u\|^2$$

taking conjugates we get

$$\frac{1}{c_2} \|x\|^2 \leq < A_n^{-1} x, x > \leq \frac{1}{c_1} \|x\|^2, \; x \in X^*.$$

Hence for every $u \in R^m$

$$(125) \qquad < S A_n^{-1} S^* u, u > \geq \frac{1}{c_2} \|S^* u\|^2.$$

Since S is onto, there exists some constant $c > 0$ such that for every $u \in R^m$

$$(126) \qquad \|S^* u\| \geq c \|u\|$$

(Dunford - Schwartz [1, VI.6.1]). Therefore $S A_n^{-1} S^*$ is invertible and from the optimality condition (120) we get

$$(127) \qquad z_n = (S A_n^{-1} S^*)^{-1} y.$$

By Anselone [1, p. 5], and (125), (126)

$$\|(S A_n^{-1} S^*)\|^{-1} = \inf\{\|S A_n^{-1} S^* u\| : \|u\| = 1\} \geq \frac{c^2}{c_2},$$

thus z_n is a bounded sequence. Fix any subsequence of A_n and consider the corresponding subsequences of z_n, x_n. Let

$$z_n \to z$$

for some further subsequence. Then $S^* z_n \to S^* z$. Hence by (120) and the assumption,

(128) $x_n = A_n^{-1} S^* (z_n - z) + A_n^{-1} S^* z \rightharpoonup A_0^{-1} S^* z = x$, say,

since $\|A_n^{-1}\|$ is bounded. Thus

$$Sx_n \rightharpoonup Sx = y, \quad A_0 x = S^* z,$$

hence by uniqueness and (120)

$$x = x_0 = \arg\ \min\ (S^{-1} y, I_0), z = z_0.$$

Therefore for the original sequence

$$x_n \rightharpoonup x_0, \quad z_n \rightharpoonup z_0.$$

To prove convergence of the optimal values, write by (120)

(129) $< A_n x_n, x_n > = < S^* z_n,\ A_n^{-1} S^* z_n >$

and notice that by (128)

$$A_n^{-1} S^* z_n \rightharpoonup A_0^{-1} S^* z_0.$$

To complete the proof of (a), it suffices to notice that, due to the assumption, $\rightharpoonup$ converts to $\rightarrow$ in (128).

Proof of (b). The equivalence of (121) and (122) follows by (127), since by (129)

(130)   2 val $(S^{-1} y, I_n) = < z_n, SA_n^{-1} S^* z_n > = < (SA_n^{-1} S^*)^{-1}\ y, y > .$

Assume now the convergence of the optimal values. Then by (130)

$$(SA_n^{-1} S^*)^{-1} y \rightarrow (SA_0^{-1} S^*)^{-1} y \text{ for every } y.$$

Hence we get (123) by equiboundedness of $\|(SA_n^{-1} S^*)^{-1}\|$. Of course (123) implies (121) by (130), and this completes the proof.   □

Section 10.   Notes and Bibliographical Remarks.

To section 1. Example 3 is due to De Giorgi (see Spagnolo [1]) and was the starting point of the theory of G - convergence, which in turn originated in part epi - convergence (through its variational meaning).

To section 2. The notion of variational convergence was introduced by Zolezzi [4] where applications to penalty techniques and the Rietz method are considered. Theorem 5 can be found in Zolezzi [11] and [14]. On the space of lower semicontinuous functions, variational convergence is the weakest one giving closed graph for the level set multifunction and upper semicontinuity of the value, see Lucchetti - Malivert [1]. For extensions of theorem 5 see Dolecki [1] and [2]. Value and solution well - posedness can be obtained through discrete approximation methods, see Vasin [1]. This approach has points of contact with variational convergences (but it is formally independent of it).

To section 3. Epi - convergence was firstly introduced in the finite - dimensional setting by Wijsman [1]. It was defined and studied in a topological setting (section 5) by De Giorgi - Franzoni [1] (the detailed proofs are given in De Giorgi - Franzoni [2]). For first countable spaces, they proved that the definition agrees with the sequential one, considered in the sequential setting by Moscariello [1]. Earlier, Mosco [2] introduced his convergence mode (in a sequential fashion).

The very beginning of the theory was de Giorgi's example 3, which started G - convergence developed by Spagnolo [1], [2]. Its abstract variational meaning was discovered by Marcellini [1] in the convex setting, and generalized by Boccardo - Marcellini [1]. They showed the equivalence between the definition of Marcellini [1], which is based on lattice - theoretic properties of convex functions (as introduced in section 6) and sequential epi - convergence.

A variational counterpart of elliptic G - convergence was given by De Giorgi - Spagnolo [1] (see section VIII. 3). Let us remark that epi - convergence is called "Γ (or Γ^-) *convergence*" in most papers of Italian authors, and sometimes "*G - convergence*"in the early literature.

Stability of the epi - convergence under additive perturbations is investigated in Attouch - Sbordone [1]. As proved in Del Prete - Lignola [1] (see also Dal Maso [3, sec. 8]), given a metric space X, then epi - $\lim I_n = I$ iff for every continuous function f

$$\text{val}\,(X, I_n + f) \to \text{val}\,(X, I + f) \text{ and } \limsup \arg \min (X, I_n + f) \subset \arg \min (X, I + f)$$

(compare with (22)), provided I, I_n are equicoercive.

Convergence of arg min (R^P, f_n) under uniform convergence and strict convexity of f_n is proved in Kanniappan - Sastry [1].

Thorem 12 is due to Attouch - Wets [1]. See also Wets [1], where proposition 13 can be found. For results related to theorems 11 and 12 see Dolecki [1] and Dal Maso [3, sec. 6]. About the behaviour of $\varepsilon -$ arg min see Shunmugaraj - Paj [2, sec. 8]. The second half of proposition 15 (whose converse also holds) is due to Verwaat [1]: for a related result see Beer [5]. About theorem 16 and example 17 see Attouch [3, ch. 2, sec. 2.7].

Hadamard well - posedness for saddle problems is obtained in Attouch - Wets [2], [3], Guillerme [1] and Cavazzuti [1]. For Nash equilibria see Cavazzuti [2] and Cavazzuti - Pacchiarotti [2]. Stability results for approximate solutions to two - level optimization problems are obtained by Loridan - Morgan [1]. For multiobjective optimization see Lemaire [1]. For stability of Pareto optima under epi - convergent - like perturbations see Attouch - Riahi [1]; for min - sup problems see Lignola - Morgan [4].

For variational and epi - convergence of marginal functions, see Lignola - Morgan [1], [2]. Results related to those in this section are in Mc Linden [1].

To section 4. Equi - well - posedness in the form (42) was introduced in Cavazzuti - Pacchiarotti [1] (who generalized to minimum problems a notion of strong uniqueness for fixed points, see Fiorenza [1]), and was considered in the form (44) by Zolezzi [8].

Conditions under which asymptotically minimizing sequences are minimizing sequences to the unperturbed problem are considered in Levin [1].

Equi - well - posedness is compared with uniform well - posedness, with applications to external and internal penalty techniques by Cavazzuti - Morgan [1]. Equi well - posedness (42) was further developed via epiconvergence by Del Prete - Lignola [1], to whom theorem 19 is due .

It is easy to give conditions under which every *local* minimizers of I_0 is a limit of *local* minimizers of I_n (the local version of (46)), see Kohn - Sternberg [1, th. 4.1]. For a generalization of theorem 19 see Dolecki [1], [2]. Without equicoercivity, value and solution Hadamard well - posedness is obtained by Beer - Lucchetti [2], assuming Tykhonov well - posedness of the limit problem under local epi - distance convergence.

To section 5. The topological definition, remark 21 and proposition 22 are due to De Giorgi - Franzoni [1]. Proposition 23 is due to Ambrosetti - Sbordone [1]. Epi - topology is (nearly) the weakest one yielding graph closedness of the ε − arg min multifunction, see Beer - Kenderov [1] and Lucchetti - Malivert [1].

To section 6. Theorem 26 is due to Marcellini [1] and Boccardo - Marcellini [1]. The lattice - theoretic based definition of weak epi - convergence introduced in the proof of theorem 26 is due to Marcellini [1], where example 28 can be found. Hadamard well - posedness in the finite - dimensional quasiconvex setting is obtained in Beer - Lucchetti [1].

To section 7. Corollaries 32 and 33 can be found in Boccardo - Marcellini [1]. Theorem 34 is due to Matzeu [1], and theorem 36 to Zolezzi [12]. (For a version of the condition (89) in the infinite - dimensional setting, see Ambrosetti - Sbordone [1].) About the behavior of the ε - subdifferentials, see Meccariello [1].

To section 8. The definition of Mosco convergence was introduced (of course!) by Mosco [1], [2]. Examples 43 and at p. 164 can be found in Marcellini [1]. Theorem 42 is from Boccardo [1]; for a particular case see Mosco [2, cor. of th. B, p. 546]. For theorem 44 see (in a particular case) Salinetti - Wets [1, th. 6].

Hadamard well - posedness for convex - concave problems (based on an appropriate extension of Mosco convergence) is in Attouch - Azé - Wets [1].

To section 9. About theorems 46 and 47 see Boccardo - Marcellini [1, th. 4.1] and Marcellini [1]. In this abstract setting, G - convergence means (roughly speaking) weak convergence of the solutions to (108) for any fixed right - hand side, see Spagnolo [3]. Theorem 48 is from Boccardo - Capuzzo Dolcetta [1]. Theorem 49 is due to Pedemonte [1]. Further results are given in Pedemonte [2].

Chapter V.

WELL-POSEDNESS IN OPTIMAL CONTROL

Section 1.   Tykhonov and Hadamard Well-Posedness of Linear Regulator Problems with respect to the Desired Trajectory.

In this section we consider the well-posedness of a basic problem in the optimal control of ordinary differential equations: to minimize a quadratic performance in order to track a given desired trajectory by the solutions of a controlled ordinary differential system. We want to minimize

$$(1) \quad \int_0^T [(u - u^*)'Q(u - u^*) + (x - x^*)'P(x - x^*)]dt + [x(T) - y^*]'E[x(T) - y^*]$$

on the set of all pairs (u, x) satisfying the state equation

$$(2) \quad \begin{cases} \dot{x}(t) & = A(t)x(t) + B(t)u(t) + C(t), \text{ a.e. } t \in [0, T], \\ x(0) & = v, \end{cases}$$

subject to the constraints

$$(3) \quad (u, x) \in K.$$

Let us denote briefly by L^P the usual Lebesgue space $L^P(0, T)$ of R^r - valued functions, r unambigously defined by the context.

In the above problem (1), (2), (3), $T > 0, u \in R^k$ is the control variable, $x \in R^m$ is the state variable,

$$z^* = (y^*, u^*, x^*)$$

is the given *desired trajectory* in $R^m \times L^2 \times L^2; v \in R^m$ is the initial state, which we assume fixed.

We shall need the following assumptions:

(4) for every $t, Q(t), P(t)$ and E are symmetric positive semidefinite matrices with entries in L^∞, and there exists a constant $a > 0$ such that $p' Q(t) p \geq a|p|^2$ for a.e. t, all $p \in R^k$;

(5) $A, C \in L^1, \quad B \in L^2$;

(6) K is a given convex closed set in $L^2 \times L^2$, and there exist *admissible controls*,

i.e. functions $u \in L^2$ such that (u, x) fulfills (3), where x is the state (i.e. the unique absolutely continuous solution to (2)) corresponding to u.

We denote by U the set of the admissible controls (for any z^*) and by $I(u)$ the quadratic functional (1), with x the solution of (2) corresponding to u. Of course, U is a closed convex subset of L^2.

We shall consider (as usual) two forms of well-posedness of (1), (2), (3). The first is Tykhonov well-posedness of (U, I), where U is equipped with the strong L^2 - topology. Now given any desired trajectory z^*, the existence and uniqueness of the *optimal control* $\bar{u}$ is well-known (and easily proved by standard means). Denote by $\bar{x}$ the *optimal state*, i.e. the solution to (2) corresponding to $\bar{u} = \arg\min (U, I)$.

Hadamard well-posedness with respect to z^* means here continuity of the map

$$z^* \to \bar{u}$$

from $(R^m \times L^2 \times L^2, \text{ strong})$ to $(L^2, \text{ strong})$.

1 Theorem *Under assumptions (4), (5), (6), problem (1), (2), (3) is both Tykhonov and Hadamard well-posed for every desired trajectory. Moreover, the following maps are continuous from* $(R^m \times L^2 \times L^2, \text{ strong})$:

$$z^* \to \bar{x}, \text{ to } C^0([0, T]) \text{ with the uniform convergence,}$$

$$z^* \to I(\bar{u}) = val \ (U, I).$$

Proof. Fix z^* and denote

$$< z_1, z_2 > = \int_0^T z_1' \ z_2 \ dt.$$

Then for every $u \in L^2$

$$(7) \qquad I(u) = < u - u^*, Q(u - u^*) > + < x - x^*, P(x - x^*) > + $$
$$ + [x(T) - y^*]' E [x(T) - y^*],$$

where x is the state corresponding to u.

Now let

$$(Lu)(t) = F(t) \int_0^t F^{-1}(s) \ B(s) \ u(s) \ ds,$$

$$y(t) = F(t)[v + \int_0^t F^{-1}(s) \ C(s) \ ds],$$

where F is the fundamental matrix of $\dot{x} = A(t)x$, principal at 0. Then $x = Lu + y$ and $L : L^2 \to L^2$ is a bounded linear operator.

Tykhonov well-posedness will be obtained by theorem I.12. For every admissible control u we have from (7)

(8) $I(u) - I(\bar{u}) = <u - \bar{u},\ Q(u - \bar{u}) > + <x - \bar{x},\ P(x - \bar{x}) > +$

$$+ [x(T) - \bar{x}(T)]'E[x(T) - \bar{x}(T)] + 2 < \bar{u} - u^*,\ Q(u - \bar{u}) > +$$

$$+ 2 < P(\bar{x} - x^*),\ x - \bar{x} > +2[\bar{x}(T) - y^*]'E[x(T) - \bar{x}(T)].$$

A direct calculation gives the Gâteaux differentiability of I, and for every $u, w \in L^2$

(9) $$< \nabla I(u), w > = 2 < u - u^*, Qw > +2 < P(x - x^*), Lw > +$$

$$+ 2[x(T) - y^*]'E(Lw)(T).$$

Since $\bar{u}$ is the optimal control, by convexity (Ekeland-Temam [1, prop. 2.1, p.35])

(10) $$< \nabla I(\bar{u}),\ u - \bar{u} > \geq 0 \text{ for all } u \in U,$$

which entails (from (4), (8), (9))

$$I(u) \geq I(\bar{u}) + a\|u - \bar{u}\|^2,\ u \in U$$

where $\| \cdot \|$ is the standard norm in L^2. This yields I.(4), hence Tykhonov well-posedness.

Hadamard well-posedness is obtained as follows. Consider any sequence of desired trajectories $z_n^* \to z_0^*$ in $R^m \times L^2 \times L^2$, where $z_n^* = (y_n^*,\ u_n^*, x_n^*)$. Denote by I_n the corresponding performance (7) and let $\bar{u}_n$ be the optimal control when $z^* = z_n^*$. Then by (4)

(11) $$a\|u_n^* - \bar{u}_n\|^2 \ \leq\ I_n(\bar{u}_n) \leq I_n(\bar{u}_0) \leq \text{ constant}, n = 1, 2, 3, ...$$

Then for some $\omega \in U$ and some subsequence, $\bar{u}_n \to \omega$ in L^2. Letting $n \to +\infty$ in (11), by the compactness of $L : L^2 \to C^0([0,T])$, $I_n(\bar{u}_0) \to I_0(\bar{u}_0)$, and the weak sequential lower semicontinuity we obtain

$$\text{val } (U, I_0) \leq I_0(\omega) \leq \liminf I_n(\bar{u}_n) \leq \limsup I_n(\bar{u}_n) \leq \lim I_n(\bar{u}_0) = I_0(\bar{u}_0)$$

$$= \text{ val } (U, I_0),$$

hence we have convergence of the optimal values, moreover $\bar{u}_0 = \omega$ and $\bar{u}_n \to \bar{u}_0$ for the original sequence, by the uniqueness of $\bar{u}_0$. From (9), (10)

$$< \nabla I_n(\bar{u}_0),\ \bar{u}_0 - \bar{u}_n > \geq < \nabla I_n(\bar{u}_0) - \nabla I_n(\bar{u}_n), \bar{u}_0 - \bar{u}_n > \geq 2a\|\bar{u}_0 - \bar{u}_n\|^2.$$

On the other hand, by (10)

$$< \nabla I_0(\bar{u}_0),\ \bar{u}_n - \bar{u}_0 > \ \geq 0.$$

Adding, we get

$$2a\|\bar{u}_n - \bar{u}_0\| \ \leq\ \| \nabla I_n(\bar{u}_0) - \nabla I_0(\bar{u}_0)\|$$

hence $\bar{u}_n \to \bar{u}_0$ in L^2 since $\nabla I_n(\bar{u}_0) \to \nabla I_0(\bar{u}_0)$, therefore $\bar{x}_n = L\bar{u}_n + y \to \bar{x}_0$ uniformly on $[0,T]$.   □

Remark. The above calculations show that IV. (44) holds, yielding equi-well-posedness of (U, I_n) with respect to the strong topology of L^2.

Section 2.   Dense Well-Posedness of Nonlinear Problems and a Well-Posedness Characterization of Linear Control Systems.

Dense Tykhonov and Hadamard well-posedness.   We consider the following optimal control problem (which generalizes (1),(2),(3)): to minimize

$$(12) \qquad l[x(T)] + \int_0^T L[t, x(t), u(t)]dt$$

on the set of all pairs (u, x) subject to the constraint

$$(3) \qquad (u, x) \in K,$$

that satisfy

$$(13) \qquad \dot{x}(t) = g[t,\ x(t),\ u(t)], \text{ a.e. } t \in [0, T]; \quad x(0) = v,$$

where l, L and g are given , and $T > 0, v$ are fixed.

We shall consider admissible controls in some space $L^q = L^q([0, T]), 1 < q < \infty$, under conditions giving uniqueness in the large for the corresponding states (i.e., solutions to (13)). Then the problem of minimizing

$$(14) \qquad l[x(T)] + \int_0^T L[t,\ x(t),\ u(t)]dt + c\, \|u - u^*\|^q$$

(norm of L^q, $c > 0$, $u^* \in L^q$) subject to (13) and (3) will be called here *Hadamard well-posed in L^q* on a given subset G of L^q iff for every $u^* \in G$ there exists an unique optimal control $\bar{u}$ with optimal state $\bar{x}$, and the following maps are continuous from G equipped with the strong L^q convergence:

$$u^* \to \bar{u}, \text{ to } (L^q, \text{ strong});$$

$$u^* \to \bar{x}, \text{ to } C^0([0, T]) \text{ with the uniform convergence};$$

$$u^* \to \text{ optimal value}.$$

The choice of (14) is motivated by the results of section III.6.

We shall use the following assumptions:

(15) there exist $q > 1$ and $u \in L^q$ such that (13) has exactly one absolutely continuous solution $x(u)$ corresponding to u, such that $(u, x(u)) \in K$.

Denote by U the set of all such admissible controls.

(16) K is closed in $(L^q, \text{ strong}) \times (C^0, \text{ uniform convergence})$, and there exists some $u \in U$ such that $L(\cdot, x(u)(\cdot), u(\cdot)) \in L^1$;

(17) $l : R^m \to R$ is lower semicontinuous and bounded from below; $L : [0,T] \times R^m \times R^k \to R$ is a Carathéodory function, $L(t,x,u) \geq \varphi(t)$ for all t,x,u and some $\varphi \in L^1$;

(18) $u_n, u \in U$ and $u_n \to u$ in L^q imply $x(u_n) \to x(u)$ uniformly in $[0,T]$.

2 Theorem *Under assumptions (15), (16), (17), (18), for every $c > 0$ there exists a dense set D in L^q such that problem (14), (13), (3) is both Tykhonov well-posed for every $u^* \in D$ with respect to the strong convergence in L^q, and Hadamard well-posed in L^q on D.*

Counterexample. Strengthening the convergence (from L^2 to L^∞ or uniform) may result in lack of Tykhonov well-posedness (for many optimal control problems).

Consider the minimization of $\int_0^1 (x^2 + u^2)dt$ subject to $\dot{x} = u$, $x(0) = 0$. This is a Tykhonov well-posed problem with respect to the strong convergence in L^2. However, fix any $c > 0$ and let

$$u_n(t) = 0 \text{ if } t > 1/n, \ = c \text{ if } 0 \leq t \leq 1/n.$$

Then u_n is a minimizing sequence, but if $u_0(t) = 0$ (the optimal control), then

$$\text{ess sup } \{|u_n(t) - u_0(t)| : 0 \leq t \leq 1\} = c.$$

A characterization of well-posed regulator problems for every desired trajectory.

Consider the nonlinear regulator problem defined by minimizing (1) subject to (13) and (3). As we saw in theorem 1, whenever K is convex and the dynamics g are affine, that is

(19) $\qquad g(t,x,u) = A(t)x + B(t)u + C(t)$, a.e. t, every x and u,

then the corresponding (affine) regulator problem is both Hadamard and Tykhonov well-posed for every desired trejectory z^*. When g is not affine, Hadamard and Tykhonov well-posedness will be obtained for a dense set of desired trajectories (theorem 3).

This subsection is devoted to answering the natural question, which control systems give rise to a regulator problem which is well-posed for *all* desired trajectories?

A striking characterization of such g' s will be obtained for the class of unconstrained optimal regulator control problems, when the control is effected through both u and the initial state v. Consider the problem defined by minimizing (1) on the trajectories (v,u,x) of (13) with the constraint

(20) $\qquad\qquad\qquad (v,u,x) \in K$

for a given $K \subset R^m \times L^2 \times L^2$, assuming the existence of admissible controls. Given $D \subset R^m \times L^2 \times L^2$, problem (1), (13), (20) is *Tykhonov well-posed in D* iff for every $z^* \in D$ there exists an unique optimal control $(\bar{v}, \bar{u}) \in R^m \times L^2$ to which every minimizing sequence converges strongly; and *Hadamard well-posed in D* iff for every $z^* \in D$ there exists an unique optimal control $(\bar{v}, \bar{u})$, with corresponding optimal state $\bar{x}$, and the mapping

$$z^* \to (\bar{v}, \bar{u}), \text{ to } R^m \times (L^2, \text{strong})$$

is continuous in the strong topology of D.

We shall need the following assumptions:

(21) For every $t_0 \in [0,T], v \in R^m, u \in L^2$ there exists an unique absolutely continuous solution to the differential equation in (13) such that $x(t_0) = v$, defined in the whole interval $[0,T]$, which depends continuously on $(v,u) \in R^m \times (L^2$, strong) with respect to the uniform convergence on $[0,T]$.

(22) There exists a closed set $L \subset [0,T]$ of Lebesgue measure 0 such that for every $u \in R^k$, $g(\cdot, \cdot, u)$ is continuous at every point $(t,x) \in [0,T] \times R^m$ if $t \notin L$.

(23) There exists a positive constant a such that for every vectors b,c and a.e. $t \in (0,T)$

$$b' \, Q(t)b \geq a|b|^2, \; c'P(t)c \geq a|c|^2, \; c'Ec \geq a|c|^2.$$

Under assumption (21), Hadamard well-posedness of (1), (13), (20) implies (obviously) the following additional well-posedness property: the maps

$$z^* \to \bar{x} \text{ to } (C^0, \text{ uniform convergence);}$$

$$z^* \to \text{ optimal value,}$$

are continuous with respect to the strong topology of $R^m \times L^2 \times L^2$.

As far as dense well-posedness is concerned, we see that, under the above assumptions, solving the regulator problem (1), (13), (20) amounts to project the desired trajectory on a given closed set, so that results of section III.6 apply (in particular Stečkin's theorem (p. 105) can be used).

3 Theorem *Assume (21), (23) and let K be closed. Then there exists a dense subset D of $R^m \times L^2 \times L^2$ such that (1), (13), (20) is both Tykhonov and Hadamard well-posed in D.*

Well-posed control systems are characterized by the next theorem. We need the following condition

(24) $g(\cdot, x, 0) \in L^1$ for every $x \in R^m$ and $g(\cdot, 0, u) - g(\cdot, 0, 0) \in L^2$ for every $u \in R^k$.

4 Theorem *Assume (21), (22), (23). Then a necessary (assuming (24)) and sufficient condition such that $g(t, \cdot, \cdot)$ is affine for a. e. $t \in [0,T]$ is that the optimal control problem (1), (13) be Tykhonov or Hadamard well - posed for every desired trajectory. In this case (19) holds with A, B, C continuous in $[0,T] \setminus L$.*

Comment. The meaning of the above results is the following. For many nonlinear control systems (13), given any desired trajectory, we can modify it by an arbitrarily small amount (in the L^2 sense) to obtain a well-posed problem (1), (13) (theorem 3). But if no changes of the desired trajectory are allowed, and well-posedness is required on the whole space of the desired trajectories, then the non affine control systems (13) are necessarily ruled out (theorem 4).

This is a further instance of the restrictive nature of well-posedness, when required for a sufficiently broad set of natural parameters in the problem.

PROOFS.

Theorem 2. We check the assumptions required in (b) of III.32. We take

$$P = X = (L^q, \text{ strong}),$$

$$f(u) = \text{ind}\,(U, u) + l[x(u)(T)] + \int_0^T L(t, x(u)(t), u(t))dt,$$

noticing that U is closed since K is and (18) holds. By (17) we have lower semicontinuity of $(u, x) \to \int_0^T L(t, x, u)dt$ with respect to the strong convergence of u in L^q and the uniform convergence of x on $[0, T]$. This yields lower semicontinuity of f by the closedness of U and (18). $\qquad\square$

Theorem 3. Given $z_1, z_2 \in R^m \times L^2 \times L^2, z_i = (y_i, u_i, x_i)$, $i = 1, 2$, define the inner product

(25) $$< z_1, z_2 >= \int_0^T (u_1' \, Q u_2 + x_1' P x_2)dt + y_1' E y_2.$$

The Hilbert space structure induced by (25) on $R^m \times L^2 \times L^2$ is equivalent to the usual one, by (23). Given $t_0 \in [0, T], v \in R^m, \; u \in L^2$, let

$$z(t_0, v, u)$$

denote the only solution in $[0, T]$ to $\dot{x} = g(t, x, u), \; x(t_0) = v$. Put
$\quad G = \{(x(T), u, x) \in R^m \times L^2 \times L^2 : u \in L^2, x = z(0, v, u) \text{ for some } v \in R^m, (v, u, x) \in K\}$.
$\quad$ If $(y_n, u_n, x_n) \in G$ with $y_n = x_n(T) \to y, u_n \to u$, and $x_n = z(0, v_n, u_n) \to x$, then $x_n(0) \to v = x(0)$, so by (21) $x = z(0, v, u)$ and $y = x(T)$, thus showing closedness of G. Furthermore, for any desired trajectory z^*, the pair $(\bar{v}, \bar{u})$ is an optimal control for the corresponding problem (1), (13), (20) iff there exists some $\bar{y} \in R^m$ such that, writing $\bar{x} = z(0, \bar{v}, \bar{u})$,

$$\bar{z} = (\bar{y}, \bar{u}, \bar{x}) \in G \text{ and } \|\bar{z} - z^*\| \le \|w - z^*\|, \; w \in G.$$

Then the conclusion follows by III.32.(b) (with $f = \text{ind}\, G, X = R^m \times L^2 \times L^2$). $\qquad\square$

Theorem 4. The necessary condition is proved by an obvious modification of theorem 1. Let us show sufficiency. Thus, we assume well-posedness (in either sense) for every desired trajectory. We consider the Hilbert space $H = R^m \times L^2 \times L^2$ equipped with the scalar product (25), which defines a Hilbert space structure equivalent to the natural one on H, by (23). Then (again) for every $z^* \in H$, $(\bar{v}, \bar{u})$ is an optimal control for the corrisponding problem (1), (13) iff $(\bar{x}(T), \bar{u}, \bar{x})$, where $\bar{x} = z(0, \bar{v}, \bar{u})$, minimizes the (squared) distance of z^* (in the sense of the norm induced by (25) on H) to

(26) $$G = \{(x(T), u, x) \in H : u \in L^2, x = z(0, v, u) \text{ for some } v \in R^m\},$$

(notation as in the preceding proof). In fact, given $(v, u) \in R^m \times L^2$, the corresponding performance (1) may be written as

$$(27) \quad \int_0^T [(u-u^*)'Q(u-u^*)+(x-x^*)'P(x-x^*)]dt+(x(T)-y^*)'E(x(T)-y^*) = \|z^*-w\|^2,$$

where $x = z(0, v, u)$ and $w = (x(T), u, x) \in G$.

Conversely, if $w = (x(T), u, x) \in G$ then $x = z(0, v, u)$ for some $v \in R^m$ and (27) holds.

5 Lemma G is convex.

Proof of lemma 5. Assume Hadamard well-posedness. Then G is a Chebychev set in H. Given any convergent sequence $z_n^* \to z_0^*$ in H, denote by $(\bar{v}_n, \bar{u}_n)$ the corresponding optimal control. Then by Hadamard well-posedness $\bar{v}_n \to \bar{v}_0$, $\bar{u}_n \to \bar{u}_0$ in L^2, yielding $z(0, \bar{v}_n, \bar{u}_n) \to z(0, \bar{v}_0, \bar{u}_0)$ in L^2, and $z(0, \bar{v}_n, \bar{u}_n)(T) \to z(0, \bar{v}_0, \bar{u}_0)(T)$ by (21). Then G is convex by propositition II.8. Assume Tykhonov well-posedness. Then G is again a Chebychev set. Given $z^* \in H$ with corresponding optimal control $(\bar{v}, \bar{u})$, let $(y_n, u_n, x_n) \in G$ be such that

$$\|(y_n, u_n, x_n) - z^*\| \to \inf\{\|z^* - z\| : z \in G\}.$$

Then $x_n = z(0, v_n, u_n)$ for some $v_n \in R^m$ and $y_n = x_n(T)$. By (27) and Tykhonov well-posedness

$$u_n \to \bar{u} \text{ in } L^2, \ v_n \to \bar{v},$$

yielding $x_n \to z(0, \bar{v}, \bar{u})$ uniformly in $[0, T]$ by (21), whence $y_n \to z(0, \bar{v}, \bar{u})(T)$. Convexity of G comes from Asplund [2, cor. 3, p. 238]. □

6 Lemma The following are equivalent facts:

$$z(0, \cdot, \cdot) \text{ is affine (as a mapping from } R^m \times L^2 \text{ to } L^2);$$

$$\text{there exists some } t_0 \in [0, T] \text{ such that } z(t_0, \cdot, \cdot) \text{ is affine};$$

$$z(t_0, \cdot, \cdot) \text{ is affine for every } t_0 \in [0, T].$$

Proof of lemma 6. A mapping s between real vector spaces is affine iff

$$s(bx + (1-b)y) = bs(x) + (1-b)s(y) \text{ for all } x, y \text{ and all } b \in R.$$

Now let $z(0, \cdot, \cdot)$ be affine. Given any $t_0 \in [0, T], v_i \in R^m, u_i \in L^2, i = 1, 2, b \in R$, consider

$$z_i = z(t_0, v_i, u_i), \ i = 1, 2; \ z_3 = z(t_0, bv_1 + (1-b)v_2), bu_1 + (1-b)u_2,) \ \bar{z}_i = z_i(0).$$

Let us show that

$$(28) \qquad\qquad\qquad z_3 = bz_1 + (1-b)z_2.$$

By the uniqueness in the large

$$bz_1 + (1-b)z_2 = bz(0, \bar{z}_1, u_1) + (1-b)z(0, \bar{z}_2, u_2)$$
$$= z(0, b\bar{z}_1 + (1-b)\bar{z}_2, bu_1 + (1-b)u_2),$$

moreover

$$z_3(t_0) = bv_1 + (1-b)v_2 = bz_1(t_0) + (1-b)z_2(t_0),$$

hence (28) follows (again by the uniqueness in the large). The proof is ended by exchanging the roles of 0 and t_0. □

End of proof of theorem 4. The graph of $z(T, \cdot, \cdot) : R^m \times L^2 \to L^2$ is the same as G given by (26). But G is convex by lemma 5. Then $z(T, \cdot, \cdot)$ is affine. So by lemma 6, $z(t_0, \cdot, \cdot)$ is affine for every $t_0 \in [0, T]$. Now fix any $t_0 \in [0, T] \setminus L$. Given v, $w \in R^m$, p, $q \in R^k$, $b \in R$, consider

$$x_1 = z(t_0, v, p), \quad x_2 = z(t_0, w, q), \quad x_3 = z(t_0, bv + (1-b)w, \; bp + (1-b)q).$$

By affinity of $z(t_0, \cdot, \cdot)$ for every $t \in [0, T]$ we get

$$b \int_{t_0}^{t} g(s, x_1(s), p)ds + (1-b) \int_{t_0}^{t} g(s, x_2(s), q)ds = \int_{t_0}^{t} g(s, x_3(s), bp + (1-b)q)ds.$$

Dividing by $t - t_0$ and letting $t \to t_0$, by (21) we obtain

$$bg(t_0, v, p) + (1-b)g(t_0, w, q) = g(t_0, bv + (1-b)w, bp + (1-b)q),$$

which proves affinity of $g(t, \cdot, \cdot)$. Therefore (19) holds and A, B, C there are continuous outside L by (22). $\square$

Section 3.   Hadamard Well-Posedness for Linear Convex Problems under Plant Perturbations.

In this section we characterize solution and value Hadamard well-posedness of the optimal control problem (1), (2), (3), or of a generalization of it, obtained by minimizing an integral convex criterion

$$(29) \qquad \qquad \int_0^T f[t, x(t), u(t)]dt$$

instead of (1), when bounded (regular) perturbations act on the plant coefficients A, B in (2). Furthermore, we are interested in the continuous behaviour of the adjoint states and of the solution to the corresponding Riccati equation.

The linear-quadratic optimal control problem (1), (2), (3), or the linear- convex problem (29), (2), (3) occupy a central position in optimal control theory and applications. Therefore its well-posedness analysis is of basic significance to the whole theory.

From the applied point of view, the choice of f in (29), or of P and Q in (1), is usually fixed by the designer, while the plant is subject to perturbations (from various origins). This motivates the setting of well-posedness we shall consider.

As often happens, an abstract approach is better suited to understand the whole matter, and to reveal connections with the theory already developed.

Abstract Approach

We consider two real Hilbert spaces X, Y, sequences of linear bounded operators

$$L_n : X \to Y, \; n = 1, 2, 3, \ldots$$

and an (unperturbed) one

$$L_0 : X \to Y.$$

Given $c \in (0, 1]$ we denote by C^* the set of all functions

$$F : X \times Y \to (-\infty, +\infty)$$

such that F is continuous together with its Gâteaux gradient with respect to the Y-valued variable, which we denote by $\nabla_2 F$; moreover

$$(30) \qquad F[\frac{1}{2}(u_1 + u_2), \frac{1}{2}(x_1 + x_2)] \leq \frac{1}{2} F(u_1, x_1) + \frac{1}{2} F(u_2, x_2) - (c/4)\|u_1 - u_2\|^2$$

for every $u_1, u_2 \in X$, $x_1, x_2 \in Y$.

We consider as well the subclass Q of C^* of all "quadratic" functions

$$F(u, x) = \|u - u^*\|^2 + \|x - y^*\|^2, \ u^* \in X, \ y^* \in Y.$$

Standing assumptions: C is a fixed set of functions F such that $Q \subset C \subset C^*$; the sequence L_n is bounded, i.e. there exists $b > 0$ such that

$$(31) \qquad\qquad \|L_n\| \leq b \text{ for every } n.$$

Given $F \in C$, put

$$(32) \qquad\qquad I_n(u) = F[u, L_n(u)], u \in X, \ n = 0, 1, 2, ...$$

Here $n = 0$ defines the unperturbed problem.

Since I_n is strongly convex, (X, I_n) is Tykhonov well-posed for any n, by corollary I.17. Therefore there exists an unique

$$\bar{u}_n = \arg\min (X, I_n).$$

For simple notation, no mention will be made of the dependence of $I_n, \bar{u}_n$ on F.

Since we are interested in the behaviour of the necessary optimality conditions under perturbations, we sketch below the relevant duality theory we shall need. All the basic results may be found in Barbu-Precupanu [1]. For any fixed n, we consider the primal problem, to minimize I_n given by (32) on X. Let

$$G_n(x, y) = F[x, L_n(x) - y], x \in X, \ y \in Y.$$

The dual problem (corresponding to the perturbation function G_n) is defined by maximizing on Y

$$\omega \to -G_n^*(0, \omega),$$

where G_n^* denotes the conjugate function.

This yields

$$(33) \qquad\qquad G_n^*(0, \omega) = F^*[L_n^*(\omega), -\omega], \ \omega \in Y^*,$$

where L_n^* denotes the Hilbert space adjoint operator to L_n. The marginal function

$$h_n(y) = \inf\{F[x, L_n(x) - y] : x \in X\}, \ y \in Y,$$

is everywhere finite since $x \to F[x, L_n(x) - y]$ is strongly convex, hence bounded from below (Polyak [1, th. 5]). By the continuity of F and Berge theorem (proposition IX.2), h_n is upper semicontinuous, moreover convex by Barbu-Precupanu [1, lemma 2.1 p. 198]. By th. 2.2 p. 201 there, the primal problem is stable, since h_n is everywhere continuous, hence subdifferentiable at 0. Therefore there exists at least one solution

$$\bar{\omega}_n \in \arg\max [Y^*, \, -G_n^*(0, \cdot)].$$

By Barbu-Precupanu [1, (2.13), p. 201] we get

(34) $$-\bar{\omega}_n = \nabla_2 \, F \, [\bar{u}_n, L_n(\bar{u}_n)].$$

This shows that $\bar{\omega}_n$ is the unique solution to the dual problem.

Now assume that $F \in Q$, and consider the graph of L_n,

$$\text{gph } L_n = \{(x, L_n(x)) : x \in X\},$$

which is a closed convex subset of $X \times Y$, the Hilbert product space equipped with the scalar product

$$< (x_1, y_1), \, (x_2, y_2) > \, = \, < x_1, x_2 > + < y_1, y_2 > .$$

Here we denote by $< \cdot, \cdot >$ either the scalar product or the duality pairing in X, Y or $X \times Y$.

Suppose that F corresponds to $u^* \in X$, $y^* \in Y$. Then $\bar{u}_n = \arg\min (X, I_n)$ iff $(\bar{u}_n, L_n(\bar{u}_n))$ is the best approximation to (u^*, y^*) from gph L_n within the Hilbert space $X \times Y$. This simple geometric interpretation allows us to apply theorem II.9. We see that value and solution Hadamard well-posedness are equivalent facts, moreover the weakest convergence of L_n giving raise to well-posedness (i.e. Mosco convergence of gph L_n) may be explicitly characterized.

Furthermore, convergence of the optimal solutions, for every $F \in Q$, turns out to be equivalent to that of the adjoint variables $\bar{\omega}_n$. Interestingly enough, Hadamard well-posedness for every convex problem in C is completely determined by well-posedness within the smaller class of the quadratic problems. In this way we get many equivalent characterization of Hadamard well-posedness, as follows.

7 Theorem *The following are pairwise equivalent conditions; here $\forall F$ means either for every F in Q, or for every F in C;*
(35) $\bar{u}_n \to \bar{u}_0 \ \forall \, F$;
(36) $\bar{u}_n \to \bar{u}_0$ and $L_n \bar{u}_n \to L_0 \bar{u}_0 \ \forall \, F$;
(37) $M - \lim \text{ gph } L_n = \text{ gph } L_0$ in $X \times Y$;
(38) $L_n(x) \to L_0(x), \, L_n^*(y) \to L_0^*(y)$ for every $x \in X$ and $y \in Y$;
(39) $M - \lim I_n = I_0 \ \forall \, F$;
(40) $\bar{\omega}_n \to \bar{\omega}_0 \ \forall \, F$;
(41) *weak seq. epi-* $\lim I_n = I_0$ *on* X *and* $I_n(x) \to I_0(x)$ *for every* $x \in X, \ \forall \, F$;
(42) *any asimptotically minimizing sequence for* (X, I_n) *converges strongly to* $\bar{u}_0 \ \forall \, F$;
(43) *val* $(X, I_n) \to$ *val* $(X, I_0) \ \forall \, F$.

Remark. We see from (35) ⇔ (40) that the convergence of the Lagrange multipliers is equivalent to that of the optimal solutions. In particular, theorem 7 states the equivalence between solution Hadamard well-posedness of the primal and dual problems in this setting.

Linear Convex Unconstrained Problems. Given $c > 0$ we denote by C, abusing notation, the set of all Carathéodory functions

$$f : [0,T] \times R^m \times R^k \to (-\infty, +\infty),$$

(i.e. $f(\cdot, x, u)$ is Lebesgue measurable for all x, u and $f(t, \cdot, \cdot)$ is continuous for all t) such that the following conditions are fulfilled:
$f(t, \cdot, \cdot)$ is convex for all t; the gradient $f_x(t, \cdot, \cdot)$ is continuous;
for every t, x, u

$$d(t) \leq f(t, x, u) \leq a_1 |x|^2 + b_1 |u|^2 + c_1(t),$$

$$|f_x(t, x, u)| \leq a_2 |x| + b_2 |u| + c_2(t);$$

for some constant a_i, b_i and some functions d, $c_1 \in L^1(0, T), c_2 \in L^2(0, T)$;

$$f[t, (1/2)(x_1 + x_2), (1/2)(u_1 + u_2)] \leq (1/2)f(t, x_1, u_1) + (1/2) f(t, x_2, u_2) - (c/4) |u_1 - u_2|^2.$$

We consider two fixed square symmetric matrices $P(t), Q(t)$ with $P \in L^\infty(0, T)$, $Q \in L^\infty(0, T)$ of dimension m, k respectively, such that for some $\alpha > 0$ and all vectors λ, μ of appropriate dimension,

(44) $$\lambda' P(t)\lambda \geq \alpha|\lambda|^2, \ \mu' Q(t)\mu \geq \alpha |\mu|^2 \text{ a.e. in } (0, T).$$

Given $f \in C$, we consider the following sequence of optimal control problems: to minimize (29) subject to

(45) $$\begin{cases} \dot{x} = A_n(t)x + B_n(t)u, & \text{a.e. in } (0, T), \\ x(0) = x^0; & n = 0, 1, 2..., \end{cases}$$

where $A_n \in L^1(0, T)$, $B_n \in L^2(0, T)$ are matrices of suitable dimensions, and $x^0 \in R^m$ is fixed. There exists exactly one optimal control $\bar{u}_n$ with optimal state $\bar{x}_n$ for every n. The corresponding *adjoint state* $q_n \in W^{1,1}(0, T)$ is defined by

(46) $$\begin{cases} \dot{q}_n + A_n(t)'q_n = f_x(t, \bar{x}_n(t), \bar{u}_n(t)), \text{a.e. in } (0, T), \\ q_n(T) = 0. \end{cases}$$

We shall denote by v_n *the optimal value function* of the $n-$ th problem, i.e. given $t \in [0, T]$ and $x \in R^m$, $v_n(t, x)$ is the optimal value of the problem, to minimize $\int_t^T f(s, y, u)ds$ subject to

$$\dot{y} = A_n(s)y + B_n(s)u, \text{ a.e. in } (t, T), \ y(t) = x.$$

The *optimal value* of the $n - th$ problem is then $v_n(0, x^0)$.

Remark. Under suitable assumptions about problem's data, it is known that v_n is the unique viscosity solution to the corresponding Hamilton-Jacobi-Bellman problem

$$v_t + \inf\{f(t, x, u) + v'_x(A_n(t)x + B_n(t)u) : u \in R^m\} = 0, 0 \leq t \leq T, \quad x \in R^m;$$

$$v(T, x) = 0, \quad x \in R^m;$$

see P.L. Lions [1]. See Fleming - Rishel [1, ch.IV] for the classical setting (and also Clarke [3, sec. 3.7]).

Standing assumptions: $0 < c \leq \alpha$; A_n is a bounded sequence in $L^1(0, T)$, B_n is a bounded sequence in $L^2(0, T)$.

The $n-$ th linear-quadratic (optimal regulator) problem with desired trajectory u^*, $y^* \in L^2(0, T)$ (of the appropriate dimension) is defined by minimizing

$$(47) \qquad \int_0^T [(u(t) - u^*(t))'Q(t)(u(t) - u^*(t)) + (x(t) - y^*(t))'P(t)(x(t) - y^*(t))]dt$$

subject to (45).

By (44), since $c \leq \alpha$ every cost functional (47) is of the form (29) with $f \in C$. The abstract theory developed above may be applied by considering the real Hilbert spaces X, Y obtained by $L^2(0, T)$ (made of R^m or R^k - valued functions), equipped with the scalar product

$$< u_1, u_2 >= \int_0^T u'_1 Q \ u_2 dt; \ < x_1, x_2 >= \int_0^T x'_1 P x_2 dt.$$

The *Riccati equation* with $m \times m$ matrix unknown $E \in W^{1,1}(0, T)$, associated with the optimal control problem (47), (45), is defined by

$$(48) \qquad \begin{cases} \dot{E} = P(t) - A_n(t)'E - EA_n(t) - EB_n(t)Q^{-1}(t)B_n(t)'E \text{ a.e. in } (0, T), \\ E(T) = 0. \end{cases}$$

The unique solution E_n to (48) plays a basic role in the synthesis of the optimal control by an affine feedback law, see Lee-Markus [1].

A straightforward application of theorem 7 gives

8 Theorem *Assume that $\int_0^t A_n \ ds \to \int_0^t A_0 \ ds$ uniformly in $[0, T]$. Then the following are equivalent facts:*
(49) $\bar{u}_n \to \bar{u}_0$ in $L^2(0, T)$ for every $f \in C$, equivalently for every $u^*, y^* \in L^2(0, T)$;
(50) $B_n \to B_0$ in $L^2(0, T - \varepsilon)$ for every $\varepsilon \in (0, T)$ and $B_n \rightharpoonup B_0$ in $L^2(0, T)$;
(51) *the optimal values converge for every $f \in C$, equivalently for every u^*, y^*;*
(52) $v_n(t, x) \to v_0(t, x)$ for every $t \in [0, T]$ and $x \in R^m$ for every $f \in C$, or equivalently for every $u^*, y^* \in L^2(0, T)$.
Moreover, under any of the above conditions (49)-(52), we have
(53) $E_n \to E_0$ uniformly in $[0, T - \varepsilon], 0 < \varepsilon < T$, for every u^*, y^* (hence pointwise in $[0, T]$).
Finally, if $A_n \to A_0$ in $L^1(0, T)$, then

(54) $q_n \to q_0$ in $W^{1,1}(0,T)$ for every $f \in C$, equivalently for every $u^*, y^* \in L^2(0,T)$, is equivalent to any of (49),...,(52).

Remarks. (i). From (52) we see that Hadamard well-posedness of the optimal control problems (in our linear-convex or quadratic setting) is equivalent to a form of well-posedness of the corresponding Hamilton-Jacobi-Bellman equations as far as plant perturbations are concerned. This is a further instance of the frequently observed phenomenon, according to which many fundamental aspects of optimal control problems may be revealed through an analysis based on a suitable dynamic programming approach.

(ii). Assumptions about the convergent behaviour of the uncontrolled plant, like uniform convergence of $\int_0^t A_n \, ds$, cannot be dropped in theorem 8. Indeed, (49) alone does force neither (51), nor the convergence of A_n (in any sense): take e.g. $B_n = 0 = B_0$. On the other hand, the following example shows that the convergence $A_n \to A_0$ in $L^1(0,T)$ is by no means necessary for (49) or (51).

9 Example. Let $B_n = B_0$, and for $n \geq 1$ put

$$A_n(t) = 2^n \sin(4^n \pi t) \text{ if } 2^{-n} \leq t \leq 2^{-n+1}, \ = 0 \text{ otherwise,}$$

$$A_0(t) = 0, \ 0 \leq t \leq 1.$$

Then (49) and (51) hold true (since $\int_0^t A_n \, ds \to 0$ uniformly), but $A_n \not\to 0$ in $L^1(0,1)$ since $\int_0^1 A_n \, y \, ds \not\to 0$ for $y = \sum_{n=1}^{\infty} 2^{-n} A_n$.

(iii) The explicit charaterization (50) shows an asymmetric behaviour of B_n near 0 and near T. This is due to the lack of state constraints at time T. By constraining $x(T)$ or by adding a final time term $[x(T) - x^*]'G[x(T) - x^*]$, $x^* \in R^m$, to the cost (47) (more generally, a term $g[x(T)]$ to (29)), we get, as a characterization, $B_n \to B_0$ in $L^2(0,T)$. The same is obtained in control constrained problems, as in the next subsection.

A Constrained Linear Regulator Problem. Given $u^*, y^* \in L^2(0,T)$ and a positive constant M, we consider the following optimal control problem: to minimize (47) subject to (45) and to the control constraint

$$\int_0^T |u(t)|^2 dt \leq M^2.$$

Standing assumptions: A_n is bounded in $L^1(0,T)$, B_n is bounded in $L^p(0,T)$ for some $p > 2$ and (44) is fulfilled.

We want to characterize strong convergence in $L^2(0,T)$ of the optimal controls and uniform convergence of the optimal states for every initial state x^0 and every desired trajectory. This is a more restrictive form of Hadamard well-posedness than before, since we require a stronger form of states convergence, and the initial state is no longer fixed. As a result we get a stronger form of coefficient convergence, as follows.

10 Theorem *The following are equivalent facts:*

(55) $\int_0^t A_n \ ds \to \int_0^t A_0 \ ds$ *uniformly in* $[0,T]$, *and* $B_n \to B_0$ *in* $L^2[0,T]$;

(56) *for every* $x^0 \in R^m$, *every* u^* *and* y^* *we have* $\bar{u}_n \to \bar{u}_0$ *in* $L^2(0,T)$ *and* $\bar{x}_n \to \bar{x}_0$ *uniformly in* $[0,T]$.

Example: *an ill-posed least squares problem.* We consider a least squares solution to an optimal output distance problem (of importance in many applications, including optimal control). Among all bounded controls which minimize the output deviation from a fixed desired output y, we seek the best one in the sense of least squares.

Given two Hilbert spaces X and Y, a linear bounded map $L : X \to Y$, a fixed $y \in Y$ and a positive constant M, we consider the (nonempty) set S of those $v \in X$ which minimize $u \to \|Lu - y\|$ subject to $\|u\| \leq M$. Then we minimize $\|v\|$ on S.

This problem is ill-posed as far as the dependence on L of the optimal solution is involved, with respect to uniform convergence of the map L, even in the finite-dimensional case. Indeed, consider $X = Y = R^2$, $M = 2$, $y = (1,1)$, $L_0(x_1, x_2) = (x_1, 0)$, $L_n(x_1, x_2) = (x_1, x_2/n)$.

Then by the Kuhn-Tucker theorem we get the optimal solutions $\bar{u}_0 = (1,0)$ (corresponding to L_0) and $\bar{u}_n$ with $|\bar{u}_n| = 2$. Therefore $\bar{u}_n \not\to \bar{u}_0$.

Well-Posedness in the Optimal Control of Dirichlet Problems. We consider a linear-quadratic problem for a (distributed) system governed by an elliptic Dirichlet problem, with control acting in the domain. The perturbations act on the coefficients of the elliptic (divergence form) operator which describes the plant.

We are given two constants $\alpha > 0$, $\omega > 0$, an open bounded set $\Omega \subset R^p$, and sequences

$$a_{ij}^n = a_{ji}^n \in L^\infty(\Omega), \ n = 0, 1, 2, ...; 1 \leq i, j \leq p,$$

such that the following uniform ellipticity assumption holds:

$$\alpha|t|^2 \leq \sum_{i,j=1}^p a_{ij}^n(x)t_i t_j \leq \omega|t|^2, \text{ a.e. } x \in \Omega \text{ and all } t \in R^p.$$

The optimal control problems we consider are defined as follows. Given n, $u^* \in L^2(\Omega), y^* \in H_0^1(\Omega)$, we minimize on $L^2(\Omega)$

$$\int_\Omega (u - u^*)^2 dx + \int_\Omega (z - y^*)^2 dx + \int_\Omega |\nabla z - \nabla y^*|^2 \ dx$$

subject to

(57)
$$-\sum_{i,j=1}^p \partial/\partial x_i(a_{ij}^n \partial z/\partial x_j) = u \text{ in } \Omega, z = 0 \text{ on } \partial\Omega.$$

In (57) the control u belongs to $L^2(\Omega)$, the unique weak solution z to $H_0^1(\Omega)$ (depending on n and u), and $n = 0$ defines the unperturbed problem. As before we denote by $\bar{u}_n$ the unique optimal control for the $n-$ th problem. From theorem 7, using known characterizations of the convergence of elliptic operators, we get

11 Theorem *The following are equivalent facts:*
(58) $\bar{u}_n \to \bar{u}_0$ *in* $L^2(\Omega)$ *for all* u^*, y^*;
  $a_{ij}^n \to a_{ij}^0$ *in* $L^2(\Omega)$, *and* $\sum_{i=1}^p \partial/\partial x_i \, a_{ij}^n \to \sum_{i=1}^p \partial/\partial x_i \, a_{ij}^0$ *in* $H_{loc}^{-1}(\Omega)$ *for all* j.

PROOFS.

Theorem 7. We denote by $(i)Q$ the statement (i) for every $F \in Q$, similarly $(i)C$, $35 \le i \le 43$. Of course, every C - statement implies the corresponding Q - statement.

(35) Q $\Rightarrow$ (38). Let $F \in Q$. By writing down $\nabla I_n(\bar{u}_n) = 0$ and remembering the isomorphic character of $I + L_n^* L_n$, where I denotes the identity operator on X, we get

(59) $$\bar{u}_n = (I + L_n^* L_n)^{-1} (u^* + L_n^* y^*).$$

By taking $y^* = 0$ in (59) we get

(60) $$L_n^* L_n \, x \to L_0^* L_0 \, x, \quad x \in X$$

since $I + L_n^* L_n$ are equibounded isomorphisms.
  Taking $u^* = 0$ in (59) and remembering (60) we get similarly

$$L_n^* y^* \to L_0^* y^*, \quad y^* \in Y.$$

Fix $x \in X$. Then for some subsequence $L_n x \rightharpoonup w$ by (31), whence

$$< L_n x, y > = < x, L_n^* y > \to < w, y > = < x, L_0^* y >$$

giving $L_n x \rightharpoonup L_0 x$. Then $L_n x \to L_0 x$ since by (60)

$$\|L_n x\|^2 = < x, L_n^* L_n x > \to \|L_0 x\|^2.$$

(38) $\Rightarrow$ (35) Q. Given $u \in X$, by (31)

$$L_n^* L_n \, u = L_n^*(L_n u - L_0 u) + L_n^* L_0 u \to L_0^* L_0 u,$$

giving pointwise convergence of the inverse operators, hence

$$\bar{u}_n \to \bar{u}_0$$

by (59).
  (36) Q $\Longleftrightarrow$ (35) Q. Since (35) $Q \Rightarrow$ (38),

$$L_n \bar{u}_n = L_n(\bar{u}_n - \bar{u}_0) + L_n \bar{u}_0 \to L_0 \bar{u}_0.$$

(36) Q $\Longleftrightarrow$ (37) $\Longleftrightarrow$ (43) Q by theorem II.9.

(35) Q $\Longleftrightarrow$ (39) Q. Assume (35) Q, and let $u_n \rightharpoonup u_0$ in X. Then by (38) Q, for every $y \in Y$

$$< L_n u_n, y > = < u_n, L_n^* y > \to < u_0, L_0^* y > = < L_0 u_0, y >$$

whence $L_n u_n \rightharpoonup L_0 u_0$. Then

$$\liminf I_n(u_n) \geq I_0(u_0).$$

For every $u_0 \in X$ we have $L_n\, u_0 \to L_0 u_0$, thus

$$I_n(u_0) \to I_0(u_0)$$

and (39) Q is proved. Conversely assume (39) Q. Then by (59)

$$< \nabla I_n(u),\ u - \bar{u}_n > = 2 < u - u^*,\ u - \bar{u}_n > \ +2 < L_n^*(L_n u - y^*),\ u - \bar{u}_n > $$
$$= 2 < (I + L_n\, L_n^*)\,(u - \bar{u}_n),\ u - \bar{u}_n > \ \geq 2\|u - \bar{u}_n\|^2,\ u \in X.$$

By I. (26) we get

$$I_n(u) \geq I_n(\bar{u}_n) + (1/2)\,\|u - \bar{u}_n\|^2.$$

By (59), $\bar{u}_n$ is bounded, hence by theorem IV.5

$$I_n(\bar{u}_n) \to I_0(\bar{u}_0).$$

Then

$$(61) \quad \limsup \|\bar{u}_n - \bar{u}_0\|^2 \leq 2 \limsup\, [I_n(\bar{u}_0) - I_n(\bar{u}_n)] = 2\limsup\, [I_n(\bar{u}_0) - I_0(\bar{u}_0)].$$

The sequence I_n is equi-locally Lipschitz since

$$I_n(x) - I_n(y) \leq C_1\|x - y\|^2 + C_2\|x - y\|(\|y\| \ + \ \|u^*\| \ + \ \|y^*\|),$$

C_1, C_2 suitable constants, any x and y.

By Mosco convergence, there exists $v_n \to \bar{u}_0$ such that $I_n(v_n) \to I_0(\bar{u}_0)$, implying $I_n(v_n) - I_n(\bar{u}_0) \to 0$. Thereby $I_n(\bar{u}_0) \to I_0(\bar{u}_0)$, and $\bar{u}_n \to \bar{u}_0$ by (61).

(35) $Q \iff$ (39) C. We know that (39) $C \Rightarrow$ (39) $Q \Rightarrow$ (35) Q. Conversely assume (35) Q. Given $u_n \to u$ in X, as before $L_n\, u_n \to L_0 u$, thus giving for every fixed $F \in C$

$$\liminf I_n(u_n) = \liminf F(u_n,\ L_n u_n) \geq I_0(u).$$

Moreover by (38) $L_n\, u \to L_0\, u,\ u \in X$, hence

$$F(u, L_n u) = I_n(u) \to I_0(u).$$

The above proof shows that

(35) $Q \Rightarrow$ (41) C; (35) $Q \iff$ (41) Q. Now we show that

(42) Q $\Longleftrightarrow$ (39) Q. Fix u^*, y^* and let $u_n \in X$ be such that

$$I_n(u_n) - \inf I_n(X) \to 0.$$

Assume (39) Q. Then for n sufficiently large

$$\|u_n - u^*\|^2 \leq 1 + \inf\ I_n(X) \leq 1 + \|u^*\|^2 + \|y^*\|^2.$$

So for some subsequence $u_n \rightharpoonup x_0$, whence by IV.(101)

$$I_0(x_0) \leq \liminf\ I_n(u_n) \leq \inf I_0(X).$$

Then by uniqueness $x_0 = \bar{u}_0$, and for the original sequence $u_n \rightharpoonup \bar{u}_0$. Since $I_n(u_n) \to I_0(\bar{u}_0)$ we have

$$\|u_n\|^2 - 2 < u_n, u^* > + \|L_n\ u_n\|^2 - 2 < L_n\ u_n, y^* >$$
$$\to \|\bar{u}_0\|^2 - 2 < \bar{u}_0, u^* > + \|L_0\ \bar{u}_0\|^2 - 2 < L_0\ \bar{u}_0, y^* >,$$

moreover

$$< u_n, u^* > \to < \bar{u}_0, u^* >, \quad < L_n\ u_n, y^* > \to < L_0\bar{u}_0,\ y^* >,$$

hence $\|u_n\|^2 + \|L_n\ u_n\|^2 \to \|\bar{u}_0\|^2 + \|L_0\ \bar{u}_0\|^2$, proving (42) Q. Conversely, (42) $Q \Rightarrow$ (35) $Q \Rightarrow$ (39) Q as already proved.

(39) C $\Rightarrow$ (35) C, (43) C and (42) C. Since every I_n is strongly convex, we have an unique

$$\bar{u}_n = \arg\min (X, I_n)$$

moreover I_n is strongly continuous, therefore by the strong convexity

$$I_n(\bar{u}_n) \leq I_n[\alpha u + (1-\alpha)\bar{u}_n] \leq \alpha I_n(u) + (1-\alpha)I_n(\bar{u}_n) - \alpha(1-\alpha)\ c\ \|u - \bar{u}_n\|^2$$

for every $u \in X$ and $\alpha \in (0,1)$, hence

$$I_n(u) \geq I_n(\bar{u}_n) + (1-\alpha)\ c\ \|u - \bar{u}_n\|^2.$$

Letting $\alpha \to 0$ we get

$$(62) \qquad I_n(u) \geq I_n(\bar{u}_n) + c\ \|u - \bar{u}_n\|^2, \ u \in X.$$

From Attouch [3, lemma 3.8 p. 275] we know that there exists a constant $a \geq 0$ such that

$$(63) \qquad I_n(x) + a\|x\| + a \geq 0, \ x \in X, \text{ every } n \text{ sufficiently large.}$$

Therefore, by (62)

$$F(0,0) = I_n(0) \geq I_n(\bar{u}_n) + c\|\bar{u}_n\|^2 \geq a\ \|\bar{u}_n\| + a + c\ \|\bar{u}_n\|^2,$$

which entails boundedness of $\bar{u}_n$. We show uniform coercivity of I_n. Let $y_n \in X$, $k \in R$ be such that $I_n(y_n) \leq k$ for every n. By (63), $I_n(\bar{u}_n)$ is bounded from below, hence by (62)

$$k \geq I_n(y_n) \geq I_n(\bar{u}_n) + c\ \|y_n - \bar{u}_n\|^2 \geq \text{ constant } + c\ \|y_n - \bar{u}_n\|^2,$$

so that y_n is bounded. Then we get the conclusion from theorem IV. 42.

(35) C ⇒ (36) C. Indeed, (35) $C \Rightarrow$ (35) $Q \Rightarrow$ (38), moreover $L_n \bar{u}_n = L_n(\bar{u}_n - \bar{u}_0) + L_n \, \bar{u}_0 \to L_0 \, \bar{u}_0.$

(35) C ⇒ (40) C. Given $F \in C$, by continuity of $\nabla_2 F$ and (34) we have $\bar{\omega}_n \to \bar{\omega}_0.$

(40) Q ⇒ (38). For any u^*, y^* consider

$$f(x) = \|x - u^*\|^2, \ g(y) = \|y - y^*\|^2.$$

Then by (33) and Barbu-Precupanu [1, (2.21) p. 203],

$$G_n^*(0, w) = f^*(L_n^* w) + g^*(-w) = (1/4)\|L_n^* w\|^2 + <L_n^* w, u^*> +(1/4)\|w\|^2 - <w, y^*>.$$

Since $\bar{\omega}_n = \arg\min(Y^*, G_n^*(0, \cdot))$, by expanding $\nabla G_n^*(0, \bar{\omega}_n) = 0$ we get

$$\bar{\omega}_n = (I + L_n L_n^*)^{-1}(2y^* - 2L_n u^*).$$

As in the proof of (35) $Q \Rightarrow$ (38), we get (38).

The proof is now complete. □

Theorem 8. By theorem 7, (49) $\Longleftrightarrow$ (51). The equivalence with (50) will be shown by theorem 7 again, proving that

(64) $L_n u \to L_0 u, \ u \in L^2(0, T)$ iff $B_n \rightharpoonup B_0$ in $L^2(0, T);$

(65) $L_n^* v \to L_0^* v, \ v \in L^2(0, T)$ iff $B_n \to B_0$ in every $L^2(0, T - \varepsilon),$

where of course

(66) $$(L_n u)(t) = \phi_n(t) \int_0^t \phi_n^{-1} \, B_n u ds.$$

Here ϕ_n is the fundamental matrix, principal at 0, for $\dot{x} = A_n(t)x$. To prove (64) we need the following

12 Lemma Let A_n be bounded in $L^1(0, T)$. Then the following are equivalent properties:

$$\int_0^t A_n ds \to \int_0^t A_0 \, ds \text{ uniformly in } [0, T];$$

$$\phi_n \to \phi_0 \text{ uniformly in } [0, T];$$

$$\phi_n^{-1} \to \phi_0^{-1} \text{ uniformly in } [0, T].$$

Taking lemma 12 for granted, and writing

$$(L_n u)(t) = [\phi_n(t) - \phi_0(t)] \int_0^t \phi_n^{-1} \, B_n \, u ds$$

$$+\phi_0(t) \int_0^t (\phi_n^{-1} - \phi_0^{-1}) B_n \, u ds + \phi_0(t) \int_0^t \phi_0^{-1} B_n \, u ds,$$

we see that weak convergence of B_n entails strong convergence of $L_n u$ for any u. Conversely, assume that $L_n u \to L_0 u$, $u \in L^2(0,T)$. By lemma 12 and equiboundedness of ϕ_n we get

$$\int_0^\cdot \phi_n^{-1} B_n\ uds \to \int_0^\cdot \phi_0^{-1} B_0\ uds\ \text{in}\ L^2(0,T),$$

and for some subsequence we have a.e. convergence in $(0,T)$. If $0 \le t' < t'' \le T$ we get

$$\left| \int_{t'}^{t''} \phi_n^{-1} B_n\ uds \right| \le \text{(constant)}\ \left(\int_{t'}^{t''} |u|^2 dt \right)^{1/2},$$

since B_n is bounded in $L^2(0,T)$. Then

$$\int_0^t \phi_n^{-1} B_n\ uds \to \int_0^t \phi_0^{-1} B_0\ uds,\ 0 \le t \le T,$$

that is $\phi_n^{-1} B_n \rightharpoonup \phi_0^{-1} B_0$ in $L^2(0,T)$. This implies $B_n \rightharpoonup B_0$ by lemma 12, thus proving (64). Let us now prove (65). From (66), a direct calculation gives

$$(L_n^* v)(t) = B_n(t)' \phi_n^{-1}(t)' \int_t^T \phi_n(s)' v(s) ds,\ \ v \in L^2(0,T).$$

Write

$$(L_n^* v)(t) = B_n(t)'[\phi_n^{-1}(t)' - \phi_0^{-1}(t)'] \int_t^T \phi_n' vds$$

$$+ B_n(t)' \phi_0^{-1}(t)' \int_t^T (\phi_n' - \phi_0') vds + B_n(t)' \phi_0^{-1}(t)' \int_t^T \phi_0' vds.$$

Assume that $L_n^* v \to L_0^* v$, $v \in L^2(0,T)$. Since the first and the second term in the right hand side converge to 0 in $L^2(0,T)$, we get

(67) $$B_n(\cdot)' \phi_0^{-1}(\cdot)' \int_\cdot^T \phi_0'\ vds \to B_0(\cdot)' \phi_0^{-1}(\cdot)' \int_\cdot^T \phi_0'\ vds.$$

Denote by $\int_\cdot^T L^2$ the subspace of $L^2(0,T)$ consisting of all functions $\int_\cdot^T wds, w \in L^2$, and by $\phi_0^{-1'} \int_\cdot^T L^2$ its image under the linear mapping $v \to \phi_0^{-1'} v$. Then (67) amounts to

$$B_n' z \to B_0' z\ \text{in}\ L^2(0,T)\ \text{for all}\ z \in \phi_0^{-1'} \int_\cdot^T L^2$$

(since $v \to \phi_0' v$ is an isomorphism). By integrating $(0,...,0,1,0,...,0)'$ (only one 1) we get

$$(T-t)B_n'\ \phi_0^{-1'} \to (T-t)B_0'\phi_0^{-1'}\ \text{in}\ L^2(0,T),$$

hence given $\varepsilon \in (0,T)$

(68) $$B_n \to B_0\ \text{in}\ L^2(0,T-\varepsilon).$$

Conversely, assume (68) and let $b_n = B_n' - B_0'$. Given $z \in \phi_0^{-1'} \int_\cdot^T L^2$ and $\varepsilon \in (0,T)$, we have

$$\int_0^T |b_n z|^2 dt = (\int_0^{T-\varepsilon} + \int_{T-\varepsilon}^T)|b_n z|^2 dt$$

$$\leq (\max |z|^2) \int_0^{T-\varepsilon} |b_n|^2 dt$$

$$+ |z(t_\varepsilon)|^2 \sup_n \int_0^T |b_n|^2 dt$$

for some point $t_\varepsilon \in [T - \varepsilon, T]$. Since z is continuous and $z(T) = 0$, the previous estimate shows that $b_n z \to 0$ in $L^2(0, T)$. Therefore (49), (50), (51) are equivalent conditions. Let us show equivalence with (52). Assume (50). Applying the previous results to the optimal control problem (29) (or (47)), (45) on the time interval $[t, T]$ with initial condition (t, x) instead of $(0, x^0)$ we see that for every $f \in C$ the corresponding optimal values converge. Therefore (50) $\Rightarrow$ (52). Of course (52) $\Rightarrow$ (51). Let us show that *(50) $\Rightarrow$ (53)*.

It is known that $E_n = -Y_n \, X_n^{-1}$, where

$$\frac{d}{dt}\begin{pmatrix} X_n \\ Y_n \end{pmatrix} = \begin{pmatrix} A_n & -B_n Q^{-1} B_n' \\ -P & -A_n' \end{pmatrix}\begin{pmatrix} X_n \\ Y_n \end{pmatrix}, \quad \begin{matrix} X_n(T) & = I, \\ Y_n(T) & = 0, \end{matrix}$$

see Anderson - Moore [1, p. 345]. But $B_n Q^{-1} B_n' \to B_0 Q^{-1} B_0'$ in every $L^2(0, T - \varepsilon)$ by (50), implying uniform convergence there of E_n by lemma 12 (and obviously pointwise convergence as stated, since ε is arbitrary). Now we show that
(49) $\Rightarrow$ (54), assuming strong convergence of A_n in $L^1(0, T)$. From (46)

$$q_n(t) = -\phi_n^{-1}(t)' \int_T^t \phi_n(s) f_x[s, \bar{x}_n(s), \bar{u}_n(s)] ds.$$

Now $\bar{x}_n \to \bar{x}_0$ in $L^2(0, T)$ from theorem 7. Moreover the Nemitskij operator

$$(x, u) \to f_x[\cdot, x(\cdot), u(\cdot)]$$

is continuous between L^2 spaces, see Krasnoselskii [1, th. 2.2, p.26], hence (remembering lemma 12) $q_n \to q_0$ uniformly in $[0, T]$. From (46) again, comparing $\dot{q}_n$ and $\dot{q}_0$ we get (54).

Conversely, assume (54) for every u^*, y^*, and strong convergence of A_n in $L^1(0, T)$. From (46) we see that

(69) $\qquad\qquad \bar{x}_n \to \bar{x}_0$ in $L^1(0, T)$ for every u^*, y^*.

By the optimality condition (see Fleming-Rishel [1, (11.30) p.43])

$$B_n' q_n = 2Q(\bar{u}_n - u^*),$$

thus we get boundedness of $\bar{u}_n$ in $L^2(0, T)$, and $\bar{x}_n$ is bounded in $L^2(0, T)$ too. For some subsequence $\bar{x}_n \rightharpoonup z$, say, in $L^2(0, T)$. Therefore $z = \bar{x}_0$ and the original sequence $\bar{x}_n \rightharpoonup \bar{x}_0$, giving $\bar{\omega}_n \rightharpoonup \bar{\omega}_0$, since by (34)

$$\bar{\omega}_n = -2P(\bar{x}_n - y^*).$$

Then (by theorem 7) we get (49) for every u^*, y^*. The proof of theorem 8 is then complete, except for the

Proof of lemma 12. Since a.e. in $(0, T)$

$$\frac{d}{dt}\phi_n^{-1} = -\phi_n^{-1}A_n,$$

it suffices to prove that, uniformly in $[0, T]$,

$$\int_0^t (A_n - A_0)ds \to 0 \text{ iff } \phi_n \to \phi_0.$$

To this end we write

(70) $$\phi_n(t) - \phi_0(t) = \int_0^t A_n(\phi_n - \phi_0)ds + \int_0^t (A_n - A_0)\phi_0 \ ds.$$

Assume uniform convergence of $\int_0^t A_n ds$. Integrating by parts we get

$$\int_0^t (A_n - A_0)\phi_0 ds = \int_0^t [(d/ds) \int_0^s (A_n - A_0)dr]\phi_0 ds$$

$$= \int_0^t (A_n - A_0)dr\phi_0(t) - \int_0^t \int_0^s (A_n - A_0)dr A_0\phi_0 ds.$$

An application of Gronwall's lemma in (70) yields uniform convergence of ϕ_n. Conversely, by (70), we see that

$$\int_0^t (A_n - A_0)\phi_0 ds \to 0 \text{ uniformly in } [0, T].$$

Then write $W_n = (A_n - A_0)\phi_0$, so that

$$\int_0^t (A_n - A_0)ds = \int_0^t [(d/ds] \int_0^s W_n dr]\phi_0^{-1}ds$$

$$= \int_0^t W_n ds \ \phi_0^{-1}(t) + \int_0^t (\int_0^s W_n dr)\phi_0^{-1}A_0 ds.$$

Since $\int_0^t W_n ds \to 0$ uniformly in $[0, T]$, this gives uniform convergence of $\int_0^t A_n ds$. □

Theorem 10. (55) $\Rightarrow$ *(56)*. By lemma 12, $\phi_n \to \phi_0$ uniformly in $[0, T]$. By theorems 7, 8 we get $M - \lim$ gph $L_n =$ gph L_0, where L_n is defined in (66). Set

$$S_n = \{(u, x) \in \text{ gph } L_n : \|u\| \le M\}.$$

We shall need the following

13 Lemma $M - \lim S_n = S_0$ in $L^2(0, T) \times L^2(0, T)$.

Assuming the conclusion of the lemma, we get by theorem II.9

$$\bar{u}_n \to \bar{u}_0, \ \bar{x}_n \to \bar{x}_0 \text{ both in } L^2(0, T)$$

for every u^*, y^*, x^0. Since

$$\bar{x}_n(t) = \phi_n(t)x^0 + \phi_n(t) \int_0^t \phi_n^{-1} B_n \; \bar{u}_n \; ds$$

we easily obtain uniform convergence of $\bar{x}_n$, taking in account lemma 12.

$(56) \Rightarrow (55)$. By theorem II.9 we see that

$$M - \lim S_n = S_0.$$

Moreover, boundedness of A_n implies uniform boundedness of ϕ_n. Now fix x^0 and take

$$u^* = 0, \; y^* = \phi_0 x^0.$$

Then $\bar{u}_n \to \bar{u}_0 = 0$ in $L^2(0,T)$, $\bar{x}_n \to y^* = \bar{x}_0$ uniformly. Therefore

$$|\phi_n(t)x^0 - \phi_0(t)x^0| \le \left| \phi_n(t) \int_0^t \phi_n^{-1}B_n \; \bar{u}_n \; ds \right| + |\bar{x}_n(t) - \phi_0(t)x^0|$$

$$\le (\text{const.}) \; \|\bar{u}_n\| + |\bar{x}_n(t) - y^*(t)| \to 0 \text{ uniformly in } [0,T] \text{ as } n \to +\infty.$$

Hence $\int_0^t A_n ds \to \int_0^t A_0 ds$ uniformly by lemma 12. Now we show convergence of B_n. Fix any subsequence of $\phi_n^{-1}B_n$. Since B_n and ϕ_n^{-1} are bounded, for some subsequence

$$\phi_n^{-1}B_n \to C \text{ in } L^2(0,T),$$

where $C \in L^2(0,T)$. Fix any $u \in L^2(0,T)$ such that $\|u\| \le M$.
    Then

$$L_n(u) \to \phi_0(\cdot) \int_0^{\cdot} C \; u \; ds \text{ in } L^2(0,T) \text{ and pointwise.}$$

Then for every t

$$\phi_0(t) \int_0^t C \; u \; ds = \phi_0(t) \int_0^t \phi_0^{-1} B_0 \; u \; ds$$

since $(u, L_n(u)) \in S_n$ and S_n converges in Mosco sense.
    Thus $C = \phi_0^{-1}B_0$, and the original sequence

(71) $$\phi_n^{-1}B_n \to \phi_0^{-1}B_0 \text{ in } L^2(0,T).$$

Fix i, j and denote by $f_n(t)$ the $(i,j)-$ th element of $\phi_n^{-1}(t)B_n(t)$, and by $v_n(t)$ the vector in R^k having all components equal to zero except the $j-$ th component, which is equal to $f_n(t)$. For some constant $c > 0$, $c \; v_n$ is an admissible control, i.e. has norm $\le M$. Therefore $(cv_n, \; L_n(cv_n)) \in S_n$. For some subsequence $L_n(cv_n) \to y_0$, say, in $L^2(0,T)$. Therefore by Mosco convergence $y_0 = L_0(cv_0)$, so that

(72) $$\phi_n(\cdot) \int_0^{\cdot} \phi_n^{-1}B_n \; v_n \; ds \to \phi_0(\cdot) \int_0^{\cdot} \phi_0^{-1} B_0 \; v_0 \; ds.$$

Put

$$g_n(t) = \int_0^t f_n^2 ds, \; 0 \le t \le T.$$

By the uniform convergence of ϕ_n, ϕ_n^{-1}, and (72)

$$g_n \to g_0 \text{ in } L^2(0,T).$$

The total variation of g_n in $[0,T]$ is $\le g_n(T) \le$ (const.) $\int_0^T |\phi_n^{-1} B_n|^2 ds \le$ constant, by (71). Then, by Helly's theorem, some subsequence of g_n converges everywhere in $[0,T]$ to some function of bounded variation, which necessarily agrees a.e. with g_0. In particular, there exists a sequence $t_k \uparrow T$, such that, as $n \to +\infty$,

$$(73) \qquad \int_0^{t_k} f_n^2 dt \to \int_0^{t_k} f_0^2 \, dt \text{ for every } k.$$

Write now

$$\int_0^T (f_n^2 - f_0^2)dt = (\int_0^T - \int_0^{t_k})f_n^2 dt + \int_0^{t_k}(f_n^2 - f_0^2)dt + (\int_0^{t_k} - \int_0^T)f_0^2 dt.$$

Boundedness of f_n in $L^p(0,T)$ and Hölder's inequality yield

$$\sup_n (\int_0^T - \int_0^{t_k})f_n^2 dt \to 0 \text{ as } k \to +\infty,$$

hence

$$(74) \qquad \int_0^T f_n^2 ds \to \int_0^T f_0^2 ds.$$

By arbitrariness of i,j, (71) and (74) (which entails convergence of the L^2-norms) it follows that $\phi_n^{-1} B_n \to \phi_0^{-1} B_0$ in $L^2(0,T)$, hence $B_n \to B_0$ in $L^2(0,T)$.

Proof of lemma 13. If n_j is a given subsequence of the positive integers and $z_j \in S_{n_j}$ with $z_j \to z$, then clearly $z \in S_0$. Conversely let $\|u\| \le M$. By Mosco convergence of gph L_n, there exists a sequence $v_n \to u$ such that $L_n v_n \to L_0 u$. If $\|u\| < M$, then $\|v_n\| < M$ for n sufficiently large, and

$$(v_n, L_n v_n) \to (u, L_0 u).$$

If $\|u\| = M$, then $M v_n/(\|v_n\| + 1/n) \to u$ with norms $\le M$, and

$$L_n(M v_n/(\|v_n\| + 1/n)) = M L_n v_n/(\|v_n\| + 1/n) \to Lu.$$

$\square$

Theorem 11. We use freely some standard facts about Sobolev spaces (see e.g. Adams [1]) and partial differential operators of elliptic type (see Gilbarg-Trudinger [1]). For any given n, the elliptic operator

$$A_n = -\sum_{ij=1}^p \frac{\partial}{\partial x_i}(a_{ij}^n \frac{\partial}{\partial x_j}) : H_0^1(\Omega) \to H^{-1}(\Omega)$$

is an isomorphism between $H_0^1(\Omega)$ and $H^{-1}(\Omega)$ with an uniformly bounded inverse A_n^{-1}, moreover A_n is selfadjoint. We apply theorem 7 with

$$L_n = \text{ restriction to } L^2(\Omega) \text{ of } A_n^{-1}.$$

Then $\|L_n\|$ are uniformly bounded since $L^2(\Omega)$ is continuously embedded in $H^{-1}(\Omega)$, and $L_n = L_n^*$. Therefore (58) amounts to

$$L_n u \to L_0 u \text{ in } H_0^1(\Omega), \text{ for every } u \in L^2(\Omega),$$

which in turn is equivalent to

$$L_n v \to L_0 v \text{ in } H_0^1(\Omega), v \in H^{-1}(\Omega)$$

by dense embedding of $L^2(\Omega)$ in $H^{-1}(\Omega)$ and boundedness of $\|A_n^{-1}\|$. But this latter convergence is equivalent to strong operator convergence of A_n to A_0 (Spagnolo [3, rem.10 p. 477]). Then the conclusion comes from Spagnolo [3, rem. 8 p. 476]. □

Section 4.   Well-Posedness of Constrained Linear-Quadratic Problems.

In this section we study linear-quadratic control problems with constraints. Our purpose is to estimate the changes of the optimal solution and the dual variables due to changes of the data. In such a way we obtain a form of well-posedness which is stronger than continuous dependence.

First we consider a problem with inequality state and control constraints:

$$(75) \qquad \text{minimize } I(x, u) = 0.5 \int_0^T [x'(t)Qx(t) + u'(t)Ru(t)]dt$$

on the set of those $x \in W^{1,\infty}(0, T)$, and $u \in L^\infty[0, T]$ such that

$$(76) \qquad \dot{x}(t) = Ax(t) + Bu(t), \quad \text{a. e. } t \in [0, T], \; x(0) = x^0,$$

$$Ku(t) + b \leq 0 \text{ and } Sx(t) + d \leq 0 \text{ for all } t \in [0, T],$$

where $x(t) \in R^n, u(t) \in R^k$, the initial state x^0 and the final time $T, 0 < T < +\infty$, are fixed, Q, R, A and B are constant matrices, $K \in R^{m \times k}, S \in R^{l \times n}$ and $b \in R^m, \; d \in R^l$.

We assume that only the plant matrices A and B are subject to perturbations.

Notation. Let $D_0 = (A_0, B_0)$ be fixed and let $D = (A, B)$. Everywhere in this section, any suitable neighbourhood of D_0 is denoted by N and c is a generic constant. We denote

$$< f, g >= \int_0^T f'g\,dt \text{ and } [f, g] = \int_0^T g'\,df,$$

and $V(f)$ is the total variation of f on $[0, T]$.

Our basic assumptions for the problem (75),(76) are:

(77) the matrix Q is positive semidefinite and R is positive definite;

(78) there exist a constant $\beta < 0$ and a continuous function $\tilde{u}$ on $[0, T]$ such that if $\tilde{x}$ solves (76) with $u = \tilde{u}$, $A = A_0$ and $B = B_0$ then

$$(K\tilde{u}(t) + b)_i \leq \beta \text{ and } (S\tilde{x}(t) + d)_j \leq \beta$$

for all $t \in [0, T]$, $i = 1, 2..., m$, $j = 1, 2, , ..., l$.

By (77), the quadratic functional is strictly convex and by (78) the feasible set is nonempty, hence there exists an unique solution (u_0, x_0) to (75), associated with $D_0 = (A_0, B_0)$. To characterize this solution we apply the duality theory.

Let us introduce the Lagrange functional

$$L(x; u; \Psi, \nu, \lambda) = I(x, u) + < \Psi, \dot{x} - Ax - Bu > \\ + < \lambda, Ku + b > + [\nu, Sx + d],$$

where Ψ and ν are of bounded variation on $[0, T]$ and continuous from the left, ν is increasing, $\nu(T) = 0, \lambda \in L^1(0, T)$ and $\lambda(t) \geq 0$ a.e. in $[0, T]$. According to the strong duality theorem from Hager-Mitter [1, th. 1] there exist dual multipliers Ψ_0, ν_0 and λ_0 such that

(79) $$L(x_0, u_0; \Psi_0, \nu_0, \lambda_0) \leq L(x, u; \Psi_0, \nu_0, \lambda_0)$$

for all $x \in W^{1,\infty}([0, T])$, $x(0) = x^0, u \in L^\infty(0, T)$, and the complementarity slackness condition holds

(80) $$< \lambda_0, Ku_0 + b >= 0 = [\nu_0, Sx_0 + d].$$

Let $q_0(t) = S'\nu_0(t) - \Psi_0(t)$ for $t \in (0, T), q_0(0) = q_0(0^+), q_0(T) = 0$. It is proved in Hager-Mitter [1, th. 4] that q_0 is absolutely continuous on $[0, T]$ and the following necessary optimality conditions hold:

(81) $$\dot{q}_0(t) = -A'[q_0(t) - S'\nu_0(t)] - Qx_0(t),$$

(82) $$Ru_0(t) + B_0'(q_0(t) - S'\nu_0(t)) + K'\lambda_0(t) = 0$$

for a. e. $t \in [0, T]$.

In our study we need the following regularity condition:

(83) there exists a constant $\gamma > 0$ such that

$$| \sum_{j \in r(t)} B' S_j' a_j + \sum_{i \in c(t)} K_i' b_i | \geq \gamma(|a| + |b|)$$

for all $t \in [0, T]$ and all vectors a and b, where $r(t)$ and $c(t)$ are the sets of active constraints, that is

$$r(t) = \{j \text{ from } 1, 2, ..., l : (Sx_0(t) + d)_j = 0\},$$

$$c(t) = \{i \text{ from } 1, 2, ..., m : (Ku_0(t) + b)_i = 0\},$$

and S_j and K_i are the $j-$ th and $i-$ th row of S and K, respectively.
It is proved in Hager [1] that, under (83), the optimal control u_0 can be regarded as
Lipschitz continuous on $[0, T]$, and (76) and the optimality conditions (81), (82) hold for
all $t \in [0, T]$.

Our first result is

14 Theorem *There exist a neighbourhood N of D_0 and a constant $c > 0$ such that for
each $D \in N$ there exist an optimal control u_D of (75), (76) and dual multipliers Ψ_D, ν_D
and λ_D that satisfy*

$$(84) \qquad \|u_D - u_0\|_{L^2} + \|\Psi_D - \Psi_0\|_{L^1} + \|\nu_D - \nu_0\|_{L^1} + \|\lambda_D - \lambda_0\|_{L^1} \leq c|D - D_0|.$$

Consider now a somewhat different problem without state constraints:

$$(85) \quad \text{minimize } I(x, u)$$
$$= \int_0^T \{0.5[x'(t)Q(t)x(t) + u'(t)R(t)u(t)] + x'(t)S(t)u(t) + r'(t)u(t) + s'(t)x(t)\}dt$$

subject to

$$(86) \qquad \dot{x}(t) = A(t)x(t) + B(t)u(t) + w(t) \text{ and } u(t) \in U, \text{ a.e. } t \in [0, T],$$
$$x(0) = x^0, x \in W^{1,\infty}([0, T])', u \in L^\infty(0, T),$$

where U is a nonempty closed and convex set in R^k. Now the plant matrices A and B
are fixed and the functions r, s, w, which are from L^∞, represent the perturbed data.
Assume that Q, R, S, A and B are in L^∞ and
(87) there exists $\alpha > 0$ such that

$$(x' \quad u') \begin{pmatrix} Q(t) & S(t) \\ S'(t) & R(t) \end{pmatrix} \begin{pmatrix} x \\ u \end{pmatrix} \geq \alpha|u|^2$$

for a.e. $t \in [0, T]$, all $x \in R^n$ and $u \in R^k$.
Then for each $p = (r, s, w)$ there exists an unique solution (x_p, u_p) to (85), (86). The
optimal control satisfies the maximum principle (see Fleming-Rishel [1, th. 11.6, p. 433]):

$$(88) \qquad [R(t)u_p(t) + S'(t)x_p(t) - B'(t)\Psi_p(t) + r(t)]'(u - u_p(t)) \geq 0$$

for all $u \in U$ and for a.e. $t \in [0, T]$, where Ψ_p is the adjoint variable, which solves

$$(89) \qquad \dot{\Psi}(t) = -A'(t)\Psi(t) + Q(t)x_p(t) + S(t)u_p(t) + s(t) \text{ for a.e. } t \in [0, T].$$

For the problem (85), under the condition (87), we will prove:

15 Theorem *The map $p \rightarrow (x_p, u_p, \Psi_p)$ is Lipschitz continuous from
L^∞ to $W^{1,\infty} \times L^\infty \times W^{1,\infty}$.*

In the next section, theorem 15 will be needed to get Lipschitz-type results for a
nonlinear optimal control problem.

PROOFS.

Theorem 14. Let $x[D, u]$ be the solution of (76) for given D and u. Since

(90) $$x[D, \tilde{u}] \to x[D_0, \tilde{u}] = \tilde{x} \text{ as } D \to D_0,$$

uniformly in $[0, T]$, there exists a neighbourhood N of D_0 such that

$$(Sx[D, \tilde{u}](t) + d)_i \leq \beta/2$$

for all $D \in N$, $t \in [0, T]$, and $i = 1, 2, ..., l$. Then (78) holds for D in N with $x[D, \tilde{u}]$, which implies that there exist an unique optimal solution (x_D, u_D) and dual multipliers Ψ_D, ν_D and λ_D that satisfy the analogous of (79) - (82). The estimate (84) will be obtained after a series of lemmas.

16 Lemma sup $\{(\|u_D\|_{L^\infty} + V(\nu_D) + \|\Psi_D\|_{L^\infty} + \|\lambda_D\|_{L^\infty}) : D \in N\} < +\infty.$

Proof. From (79) and (80) we have that for some $a > 0$,

$$0 \leq a\|u_D\|_{L^2}^2 \leq I(x_D, u_D) \leq I(x[D, \tilde{u}], \tilde{u}) + [\nu_D, \ Sx[D, \tilde{u}] + d] + \ < \lambda_D, K\tilde{u} + b >$$

$$\leq I(x[D, \tilde{u}], \tilde{u}) + (\beta/2)(\sum_{i=1}^{l} \int_0^T d\,\nu_D(t)_i + \sum_{j=1}^{m} \int_0^T \lambda_D(t)_j \ dt)$$

$$\leq I(x[D, \tilde{u}], \tilde{u}) \leq c < +\infty,$$

because of (90). Hence, for $D \in N$,

$$\|u_D\|_{L^2} + V(\nu_D) + \|\lambda_D\|_{L^1} \leq c < +\infty.$$

Gronwall's lemma, applied to (76) and (81), yields

$$\|x_D\|_{W^{1,\infty}} + \|q_D\|_{W^{1,\infty}} \leq c,$$

where $q_D = S'\nu_D - \Psi_D$ on $(0, T)$, $q_D(0) = q_D(0^+)$, $q_D(T) = 0$. Then

(91) $$\|\Psi_D\|_{L^\infty} \leq c.$$

From (80) and (82) with u_D, Ψ_D, and λ_D we have

$$(Ru_D(t) - B'\Psi_D(t))'(\tilde{u}(t) - u_D(t)) = -\lambda_D'(t)K'(\tilde{u}(t) - u_D(t)) = -\lambda_D'(t)(K\tilde{u} + b) \geq 0.$$

By (77)

$$(\tilde{u}(t) - u_D(t))' R(\tilde{u}(t) - u_D(t)) \geq a|\tilde{u}(t) - u_D(t)|^2$$

for some $a > 0$ and for all $t \in [0, T]$. Combining the last two inequalities we obtain for all $t \in [0, T]$

$$a|\tilde{u}(t) - u_D(t)| \leq |R\tilde{u}(t) - B'\Psi_D(t)|,$$

which, using (91), yields

(92) $$\|u_D\|_{L^\infty} \leq c$$

for $D \in N$. Define

$$H_D(u,t) = 0.5 \; u'Ru - \Psi_D'(t)Bu + \lambda_D'(t)(Ku + b).$$

Then (82) implies

$$H_D(u_D(t),t) \le H_D(\tilde{u}(t),t), t \in [0,T],$$

which, together with (91) and (92), yields

$$-c \le \lambda_D'(t)(K\tilde{u}(t) + b) \le \beta \sum_{j=1}^{m} \lambda_D(t)_j \le 0,$$

thus, for $D \in N, \|\lambda_D\|_{L^\infty} \le c.$ □

17 Lemma

$$\|u_D - u_0\|_{L^2} \le c|D - D_0|^{1/2}.$$

Proof. From the analogous inequality to (79), corresponding to the $D-$ problem, we have

(93) $I(x_D, u_D) \le I(x_0, u_0) + \; < \Psi_D, (A_0 - A)x_0 + (B_0 - B)u_0 > .$

On the other hand

$$I(x_D, u_D) \ge I(x_D, u_D) + \; < \Psi_0, \dot{x}_D - Ax_D - Bu_D > \\ + [\nu_0, Sx_D + d] + \; < \lambda_0, Ku_D + b > .$$

Expanding the expression in the right-hand side by Taylor's formula around (x_0, u_0) and using (77) we have

(94) $I(x_D, u_D) \ge I(x_0, u_0) + \; < \Psi_0, (A_0 - A)x_0 + (B_0 - B)u_0 > \\ + < Ru_0 - B'\Psi_0 + K'\lambda_0, u_D - u_0 > + < Qx_0 - A'\Psi_0, x_D - x_0 > \\ + < \Psi_0, \dot{x}_D - \dot{x}_0 > + [\nu_0, S(x_D - x_0)] + a\|u_D - u_0\|_{L^2}^2.$

Integrating by parts we obtain

(95) $< \Psi_0, \dot{x}_D - \dot{x}_0 > = \int_0^T (x_D - x_0)'\dot{q}_0 dt \\ - \int_0^{T^-} (x_D - x_0)'S'd\nu_0(t) + (x_D(T) - x_0(T))'S\nu_0(T^-),$

and

(96) $\int_0^{T^-} (S(x_D - x_0))'d\nu_0(t) = -(x_D(T) - x_0(T))'S'\nu_0(T^-) \\ + \int_0^T (x_D - x_0)'S'd\nu_0(t).$

Taking into account (81), (82), (95) and (96) in (94) one obtains

(97) $I(x_D, u_D) \ge I(x_0, u_0) + \; < \Psi_0, (A_0 - A)x_0 + (B_0 - B)u_0 >$

$$+ < (B_0 - B)'\Psi_0, u_D - u_0 > + < (A - A_0)'\Psi_0, x_D - x_0 > + \|u_D - u_0\|_{L^2}^2.$$

Then , from (93) and (97) we have

$$a\|u_D - u_0\|_{L^2}^2 \leq < \Psi_D - \Psi_0, (A_0 - A)x_0 + (B_0 - B)u_0 >$$
$$+ < (B - B_0)'\Psi_0, u_D - u_0 > + < (A_0 - A)'\Psi_0, x_D - x_0 > .$$

Taking into account that

$$\|x_D - x_0\|_{L^\infty} \leq c(\|u_D - u_0\|_{L^2} + |D - D_0|),$$

we obtain finally

(98) $$\|u_D - u_0\|_{L^2}^2 \leq c|D - D_0|(\|u_D - u_0\|_{L^2} + \|\Psi_D - \Psi_0\|_{L^1} + |D - D_0|).$$

The boundedness of u_D and Ψ_D, established in the previous lemma, completes the proof.

$\square$

18 Lemma There exist a constant $\bar{\delta} > 0$ and a natural number L such that the interval $[0, T]$ can be splitted into L subintervals $\Delta_k, k = 1, 2, ..., L$, to which sets of indices c^k and r^k correspond such that

$$c^k = c(\tau_k) \text{ and } r^k = r(\tau_k)$$

for some $\tau_k \in \Delta_k$, and moreover for every $t \in \Delta_k, k = 1, 2, ..., L$.

$$(Ku_0(t) + b)_i \leq -\bar{\delta} \text{ if } i \notin c^k,$$

$$(Sx_0(t) + d)_j \leq -\bar{\delta} \text{ if } j \notin r^k.$$

Proof. If all constraints are active at some $\bar{t} \in [0, T]$ then take $L = 1, \tau_k = \bar{t}$ and $\Delta_k = [0, T]$. Now assume that everywhere some of the constraints are nonactive. Take an arbitrary $t \in [0, T]$ and let

$$\delta_1(t) = \min\{-(Ku_0(t) + b)_j : j \notin c(t)\},$$

$$\delta_2(t) = \min\{-(Sx_0(t) + b)_i : i \notin r(t)\}$$

(the minimum over the empty set is $+\infty$). If there exists some $i \notin c(t)$, then by the continuity of u_0 there exists a neighbourhood $\Delta(t)$ of t such that

$$(Ku_0(s) + b)_j \leq -\frac{\delta_1(t)}{2} \text{ if } s \in \Delta(t) \text{ and } j \notin c(t).$$

Similarly,

$$(Sx_0(s) + d)_i \leq -\frac{\delta_2(t)}{2} \text{ if } s \in \Delta(t) \text{ and } i \notin r(t).$$

Put

$$\delta(t) = (1/2) \min\{\delta_1(t), \delta_2(t)\}.$$

Now from the covering $\{\Delta(t) : t \in [0, T]\}$ choose a finite subcovering $\Delta_k = \Delta(t_k), k = 1, 2, ..., L$, and let $\bar{\delta} = \min_k \delta(t_k)$.

$\square$

In the sequel we denote $M_k = $ cl $\Delta_k = [t_{k-1}, t_k], m^k = \{1, 2, ..., m\} \setminus c^k, n^k = \{1, 2, ..., l\} \setminus r^k, k = 1, 2, ..., L$.

19 Lemma For every $D \in N, t \in M_L$ and $i \in n^L$ one has $\nu_D(t)_j = 0$.

Proof. From the previous lemma

$$(Sx_0(t) + d)_i \leq -\bar{\delta}$$

for $t \in M_L, i \in n^L$. Since $x_D \to x_0$ uniformly in $[0, T]$ (which follows from lemma 17 and Gronwall's lemma) there exists a neighbourhood N of D_0 such that

$$(Sx_D(t) + d)_i < -\bar{\delta}/2$$

for $D \in N, t \in M_L$ and $i \in n^L$. Then the conclusion follows from the complementarity slackness (80).   □

20 Lemma $\max_k \int_{M_k} \sum_{j \in m^k} \lambda_D(t)_j dt \leq c|D - D_0|$.

Proof. We use (79) and lemma 16, obtaining

$$I(x_D, u_D) \leq I(x_0, u_0) + <\lambda_D, Ku_0 + b> +c|D - D_0|.$$

This inequality, combined with (97), yields

$$-c|D - D_0| \leq <\lambda_D, Ku_0 + b> \leq -\bar{\delta} \int_{M_k}' \sum_{j \in m_k} \lambda_D(t)_j dt \leq 0, k = 1, 2, ..., L.$$

     □

21 Lemma $\|\Psi_D - \Psi_0\|_{L^1} + \|\nu_D - \nu_0\|_{L^1} + \|\lambda_D - \lambda_0\|_{L^1} \leq c(\|u_D - u_0\|_{L^1} + |D - D_0|)$.

Proof. Denote $\Delta u = u_D - u_0, \Delta q = q_D - q_0$ etc. Since $\lambda_0(t)_j = 0$ for $t \in M_k$ and $j \in m^k$ by (80), we have from (82)

$$(99) \quad -\sum_{j \in c^k} K_j' \Delta\lambda(t)_j + \sum_{i \in r^k} B_0' S_j' \Delta\nu(t)_i = R\Delta u(t) + B_0' \Delta q(t) - \sum_{i \in n^k} B_0' S_j' \Delta\nu(t)_i$$

$$+(B - B_0)'(q_D(t) - S'\nu_D(t)) + \sum_{j \in m^k} K_j' \lambda_D(t)_j.$$

From lemmas 19 and 20, on the last interval M_L, applying (77), we obtain

$$(100) \quad \int_{M_L} \{\sum_{j \in c^k} |\Delta\lambda_j| + \sum_{i \in r^k} |\Delta\nu_i|\} dt$$

$$\leq c(\int_{M_L} [|\Delta q| + |\Delta u|] dt + |D - D_0|).$$

Since $(Sx_0(t) + d)_i \leq -\bar{\delta}$ for $i \in n^k, t \in M_k, k = 1, 2, ..., L$, there exists $\gamma_k > 0$ such that

$$(Sx_0(t) + d)_i \leq -\bar{\delta}/2 \text{ for } i \in n^k, t \in [t_{k-1}, t_k + \gamma_k].$$

Let

$$\bar{\gamma} = \min_k \{\min[\gamma_k, t_{k+1} - t_k]\}.$$

Since $x_D \to x_0$ uniformly in $[0, T]$ then

$$(Sx_D(t) + d)_i < -\bar{\delta}/4 \text{ for } i \in n^k, t \in [t_{k-1}, t_k + \bar{\gamma}].$$

and for $D \in N$. Then, by the complementarity slackness (80) we have

$$\Delta\nu(t)_i = \text{ const. on } [t_{k-1}, t_k + \bar{\gamma}], i \in n^k.$$

Hence

(101)
$$\int_{M_{k-1}} \sum\{|\Delta\nu(t)_i| : i \in n^{k-1}\}dt$$

$$= \frac{t_{k-1} - t_{k-2}}{\bar{\gamma}} \int_{t_{k-1}}^{t_{k-1}+\bar{\gamma}} \sum\{|\Delta\nu(t)_i| : i \in n^{k-1}\}dt$$

$$\leq c \int_{t_{k-1}}^{t_k} \sum\{|\Delta\nu(t)_i| : i = 1, ..., l\}dt.$$

Take $t \in M_{L-1}$. From (83) and (99)

$$\sum\{|\Delta\lambda(t)_j| : j \in c^{L-1}\} + \sum\{|\Delta\nu(t)_i| : i \in r^{L-1}\}$$
$$\leq c(|\Delta q(t)| + |\Delta u(t)| + \sum\{|\lambda_D(t)_j| : j \in m^{L-1}\}$$
$$+ \sum\{|\Delta\nu(t)|_i : i \in n^{L-1}\} + |D - D_0|).$$

Integrating both sides of this inequality on M_{L-1} and using lemma 20 and (101) we obtain

$$\int_{M_{L-1}} (\sum\{|\Delta\lambda_j| : j \in c^{L-1}\} + \sum\{|\Delta\nu_i| : i \in r^{L-1}\})dt$$

$$\leq c(\int_{M_{L-1}} [|\Delta q| + |\Delta u| + \sum\{|\lambda_{Dj}| : j \in m^{L-1}\}$$

$$+ \sum\{|\Delta\nu_i| : i \in n^{L-1}\}]dt + |D - D_0|)$$

$$\leq c(\int_{M_{L-1}} [|\Delta q| + |\Delta u|]dt + \int_{M_L} |\Delta\nu|dt + |D - D_0|).$$

Now we apply (100) and lemma 19 to the last inequality obtaining

$$\int_{M_{L-1}} (\sum\{|\Delta\lambda_j| : j \in c^{L-1}\} + \sum\{|\Delta\nu_i| : i \in r^{L-1}\})dt$$

$$\leq c(\int_{t_{L-1}}^T [|\Delta q| + |\Delta u|]dt + |D - D_0|).$$

Then, using (100) and (101) for $k = L$, and applying lemma 19, we have

$$\int_{M_{L-1}} (|\Delta\lambda| + |\Delta\nu|)dt \leq c(\int_{t_{L-1}}^T [|\Delta q| + |\Delta u|]dt + |D - D_0|).$$

Proceeding in the same manner, by induction we obtain for $k = 1, 2, ..., L$

$$(102) \qquad \int_{M_k} (|\Delta\lambda| + |\Delta\nu|) dt \le c\left(\int_{t_k}^{T} [|\Delta q| + |\Delta u|) dt + |D - D_0|\right).$$

If $t \in M_k$, by repeating the above argument we get

$$(103) \qquad \int_{t}^{t_k} |\Delta\nu| ds \le c\left(\int_{t}^{T} [|\Delta q| + |\Delta u|] ds + |D - D_0|\right).$$

Then, by (102) and (103), for any $t \in [0, T])$

$$(104) \qquad \int_{t}^{T} |\Delta\nu| ds \le c\left(\int_{t}^{T} [|\Delta q| + |\Delta u|] ds + |D - D_0|\right).$$

From (81), for every $t \in [0, T]$,

$$(105) \qquad |\Delta q(t)| \le c\left(\int_{t}^{T} [|\Delta q| + |\Delta x| + |\Delta\nu|] ds + |D - D_0|\right).$$

Applying (104) to (105) and using Gronwall's lemma we get

$$\|\Delta q\|_{L^\infty} \le c(\|\Delta u\|_{L^1} + |D - D_0|).$$

Now, by (102),

$$\|\Delta\lambda\|_{L^1} + \|\Delta\nu\|_{L^1} \le c(\|\Delta u\|_{L^1} + |D - D_0|),$$

hence $\|\Delta\Psi\|_{L^1} \le c(\|\Delta u\|_{L^1} + |D - D_0|).$ $\qquad\qquad \square$

End of the proof of theorem 14. Applying lemma 21 to (98) we obtain

$$\|\Delta u\|_{L^2}^2 \le c|D - D_0| \left(\|\Delta u\|_{L^2} + |D - D_0|\right).$$

Solving this inequality we get

$$\|\Delta u\|_{L^2} \le c|D - D_0|$$

which yields the desired estimate via lemma 21. $\qquad\qquad \square$

Theorem 15. Let $p_i = (r_i, s_i, w_i), i = 1, 2$, be given and (x_i, u_i) be the corresponding solution to (85), (86). Denote

$$H_i(u, t) = 0.5 u' R(t) u + x_i(t)' S(t) u - \Psi_i'(t) B(t) u + r_i'(t) u,$$

where Ψ_i is the adjoint variable associated with p_i. Then (87) implies

$$(\frac{\partial}{\partial u} H_1(u_1(t), t) - \frac{\partial}{\partial u} H_1(u_2(t), t))'(u_1(t) - u_2(t)) \ge \alpha |u_1(t) - u_2(t)|^2$$

for a.e. $t \in [0, T]$. By (88) we have

$$\frac{\partial}{\partial u} H_1(u_1(t), t)'(u_1(t) - u_2(t)) \le 0$$

and

$$\frac{\partial}{\partial u} H_2(u_2(t), t)'(u_1(t) - u_2(t)) \geq 0, \quad \text{a.e. } t \in [0, T].$$

By the last three inequalities we get

$$\alpha |u_1(t) - u_2(t)| \leq |\frac{\partial}{\partial u} H_2(u_2(t), t) - \frac{\partial}{\partial u} H_1(u_2(t), t)|,$$

that is

(106) $\alpha |u_1(t) - u_2(t)| \leq |S'(t)(x_2(t) - x_1(t)) + B'(t)(\Psi_1(t) - \Psi_2(t)) + r_2(t) - r_1(t)|.$

Applying Gronwall's lemma to (86) with $u = u_1 - u_2$ and $w = w_1 - w_2$ we obtain

(107) $\|x_1 - x_2\|_{L^\infty} \leq c(\|u_1 - u_2\|_{L^2} + \|w_1 - w_2\|_{L^\infty}).$

Repeating this procedure to (89) results in

(108) $\|\Psi_1 - \Psi_2\|_{L^\infty} \leq c(\|x_1 - x_2\|_{L^\infty} + \|u_1 - u_2\|_{L^2} + \|s_1 - s_2\|_{L^\infty}).$

Hence, by (106), (107)

(109) $\|u_1 - u_2\|_{L^\infty} \leq c(\|u_1 - u_2\|_{L^2} + \|p_1 - p_2\|).$

Denote by $I_i(x, u)$ the functional in (85) with $r = r_i$ and $s = s_i, i = 1, 2$. By the maximum principle in the Lagrangian form (Ioffe - Tihomirov [2, th. 1', p.135])

(110) $I_2(x_2, u_2) \leq I_2(x_1, u_1) + < \Psi_2, \dot{x}_1 - Ax_1 - Bu_1 - w_2 > .$

Furthermore, using (87), the convexity, and integrating by parts

(111) $I_2(x_2, u_2) = I_2(x_2, u_2) + < \Psi_1, \dot{x}_2 - Ax_2 - Bu_2 - w_2 >$
$\geq I_2(x_1, u_1) + < \Psi_1, \dot{x}_1 - Ax_1 - Bu_1 - w_2 >$
$+ < Ru_1 + S'x_1 - B'\Psi_1 + r_2, u_2 - u_1 > + \alpha \|u_1 - u_2\|_{L^2}^2$
$+ < Qx_1 + Su_1 - A'\Psi_1 + s_2 - \dot{\Psi}_1, x_2 - x_1 >$
$\geq I_2(x_1, u_1) + < \Psi_1, w_1 - w_2 > + < r_2 - r_1, u_2 - u_1 >$
$+ < s_2 - s_1, x_2 - x_1 > + \alpha \|u_1 - u_2\|_{L^2}^2.$

Combining (110) and (111)

$$\|u_1 - u_2\|_{L^2}^2 \leq c(\|r_1 - r_2\|_{L^\infty}\|u_1 - u_2\|_{L^2}$$
$$+ \|s_1 - s_2\|_{L^\infty}\|x_1 - x_2\|_{L^\infty} + \|w_1 - w_2\|_{L^\infty}\|\Psi_1 - \Psi_2\|_{L^\infty}).$$

Now (107) and (108), applied to the above inequality, yield

$$\|u_1 - u_2\|_{L^2}^2 \leq a\|u_1 - u_2\|_{L^2} + a^2,$$

where $a = c\|p_1 - p_2\|$. Solving this inequality and using (107) - (109) we complete the proof.   □

Section 5.   Lipschitz Behavior of the Optimal Control for Nonlinear Problems.

Let us consider a nonlinear optimal control problem with control constraints

$$(112) \qquad \text{minimize } I_p(x,u) = \int_0^T g_p(x(t), u(t))dt$$

subject to

$$(113) \qquad \dot{x}(t) = f_p(x(t), u(t)) \text{ and } u(t) \in U, \text{ a.e. } t \in [0,T],$$
$$x(0) = x_0, x \in W^{1,\infty}(0,T), u \in L^\infty(0,T),$$

where $x(t) \in R^n$, $u(t) \in R^k$, p is a parameter from the metric space P with a fixed element 0 therein and, for every $p \in P$, $f_p : R^n \times R^k \to R^n$, $g_p : R^n \times R^k \to R$, the final time $T, 0 < T < +\infty$, is fixed, U is a closed and convex subset of R^k.

We assume that there exists a (global) solution (x_0, u_0) to this problem corresponding to $p = 0$, and wish to obtain a nearby solution for each p near 0 such that the map "parameters $\to$ solution" is Lipschitz continuous. This is a strong form of solution Hadamard well-posedness.

To this end, *suppose* that there exist a closed set Δ in R^{n+k} and $\delta > 0$ such that $(x_0(t), u_0(t))$ lies in Δ for a.e. $t \in [0,T]$ and the distance from $(x_0(t), u_0(t))$ to the boundary of Δ is at least δ for a.e. $t \in [0,T]$. Let the first and second derivatives of $f_p(x,u)$ and $g_p(x,u)$ with respect to x and u exist on Δ and be uniformly continuous with respect to $(x,u) \in \Delta$ and p near 0. We assume also that $f_p(x,u)$ and the first derivatives of f and g with respect to x and u are Lipschitz continuous in p in a neighbourhood of $0 \in P$, uniformly in $(x,u) \in \Delta$. Finally, we suppose that for every u in a neighbourhood of u_0 and p in a neighbourhood of 0 there exists a solution x of (113) such that $(x(t), u(t)) \in \Delta$ for a.e. $t \in [0,T]$.

By Fleming-Rishel [1, p.54] the maximum principle holds at (x_0, u_0), that is

$$(114) \qquad H_{0u}(x_0(t), u_0(t), \Psi_0(t))'(v - u_0(t)) \geq 0, \text{ a.e. } t \in [0,T], \text{ for all } v \in U,$$

where the Hamiltonian is

$$H_0(x, u, \Psi) = g_0(x,u) - \Psi' f_0(x,u),$$

and Ψ_0 is the solution of the adjoint equation

$$(115) \qquad \dot{\Psi}(t) = H_{0x}(x(t), u(t), \Psi(t)), \text{ a.e. } t \in [0,T], \Psi(T) = 0,$$

associated with $x = x_0$ and $u = u_0$. Let

$$A = f_{0x}, B = f_{0u}, Q = H_{0xx}, S = H_{0xu}, R = H_{0uu},$$

where the derivatives are computed at (x_0, u_0, Ψ_0). Then we have

22 Theorem *Suppose that*

$$(116) \qquad \text{there exists } \alpha > 0 \text{ such that for all } z = (x, u) \in R^{n+k} \text{ and for a.e. } t \in [0,T]$$

$$z' \begin{pmatrix} Q(t) & S(t) \\ S'(t) & R(t) \end{pmatrix} z \geq \alpha |u|^2.$$

Then there exist neighbourhoods V of $0 \in P$ and W of (x_0, u_0) in $W^{1,\infty} \times L^\infty$ and a Lipschitz function

$$p \to (x_p, u_p)$$

from V to W such that, for every $p \in V$, (x_p, u_p) is the unique local solution of (112) in W.

To prove theorem 22 we observe that the relations (113), (114) and (115) can be written as a "generalized equation" of the form

(117)   find $z \in \Omega$ such that $T_p(z) \in F(z)$,

where Ω is a subset of a Banach space X, T maps $X \times P$ into a Banach space Y and F is a multifunction from X to Y. In our case $X = W^{1,\infty} \times L^\infty \times W^{1,\infty}, Y = L^\infty$,

$$\Omega = \{(x, u, \Psi) \in X : x(0) = x^0, \Psi(T) = 0, u(t) \in U \text{ a.e. in } [0,T]\},$$

(118) $T_p(x, u, \Psi) = \begin{pmatrix} H_{px}(x, u, \Psi) - \dot{\Psi} \\ f_p(x, u) - \dot{x}, \\ H_{pu}(x, u, \Psi) \end{pmatrix}, F(x, u, \psi) = \begin{pmatrix} 0 \\ 0 \\ N(u) \end{pmatrix},$

where $N(u) = \{\omega \in L^\infty : \omega'(t)(v - u(t)) \geq 0 \text{ for all } v \in U \text{ and a.e } t \in [0,T]\}$.

Then we apply the following more general result.

23 Theorem *Consider (117) and let $z_0 \in \Omega, y_0 = T_0(z_0) \in F(z_0)$. Suppose that the set Ω is closed and*

(119)   *for every p near $0, T_p(\cdot)$ is Fréchet differentiable around z_0*
              and the derivative $T'_p(z)$ is continuous in (p, z) at $(0, z_0)$;

(120)   *$T_p(z)$ is Lipschitz in p near 0 uniformly in z in a neighbourhood of z_0;*

(121)   *for every y near y_0 the problem*

(122)   *find $z \in \Omega$ such that $T'_0(z_0)(z - z_0) + y \in F(z)$*

has a unique solution $H(y)$ and the function $y \to H(y)$ is Lipschitz near y_0.

Then there exist neighbourhoods V of $0 \in P$ and W of z_0, and a Lipschitz function z from V to W such that, for every $p \in V, z(p)$ is the unique solution of (117) in W.

The condition (121) means exactly that the map

$$[-T'_0(z_0)(\cdot - z_0) + F(\cdot)]^{-1}$$

is single - valued and Lipschitz near y_0.

Assuming that the above theorem holds, let us complete the proof of theorem 22.

The assumptions for the functions f and g imply that the map T defined in (118) satisfies (119) and (120). The problem (122) has the following form: find $x \in W^{1,\infty}$, $u \in L^\infty$ and $\Psi \in W^{1,\infty}$ such that for a.e. $t \in [0,T]$

(123)
$$\begin{cases} \dot{\Psi} = -A'\Psi + Qx + Su + s + s_0, \\ \dot{x} = Ax + Bu + w + w_0, \\ (Ru + S'x - B'\Psi + r - r_0)'(v - u) \geq 0 \text{ for all } v \in U \end{cases}$$

and $x(0) = x^0, \Psi(T) = 0$, $(s, w, v) = y$. Then $y_0 = (0, 0, H_{0u}(x_0, u_0, \Psi_0))$, where

$$s_0 = \dot{\Psi}_0 + A'\Psi_0 - Qx_0 - Su_0, w_0 = \dot{x}_0 - Ax_0 - Bu_0,$$

$$r_0 = Ru_0 + S'x_0 - B'\Psi_0.$$

The conditions (123) are the necessary optimality conditions for the problem (85), (86) considered in the previous section. By (116), these conditions are also sufficient and the solution is unique. By theorem 15, (121) holds and then theorem 23 says that there is a Lipschitz function

$$p \to (x_p, u_p, \Psi_p)$$

from a neighbourhood of $0 \in P$ to a neighbourhood W of (x_0, u_0, Ψ_0) such that (x_p, u_p, Ψ_p) is the unique solution of the problem analogous to (113), (114) and (115) at p in W.

It remains to show that (x_p, u_p) is a local solution to the problem (112), (113). Because of (116) this can be proved in a standard way. For completeness, we sketch the proof.

Put $h = (x, u, \Psi), Q_p = H_{pxx}, S_p = H_{pxu}, R_p = H_{puu}$ and

$$M_p(h) = \begin{pmatrix} Q_p(h) & S_p(h) \\ S_p'(h) & R_p(h) \end{pmatrix}.$$

Choose p so close to 0 that

(124)
$$z' M_p(h) z \geq (\alpha/2)|u|^2$$

for all $z = (x, u) \in R^{n+k}$, for $t \in [0, T]$ and for all h in a neighbourhood of $h_p = (x_p, u_p, \Psi_p)$. This is possible by (116), and since $h_p \to h_0$ in L^∞ as $p \to 0$. Take any $u \in L^\infty$ and $s \geq 0$ sufficiently small such that $u_s = u_p + su$ is feasible and if x_s is a solution of (113) then $(x_s(t), u_s(t)) \in \Delta$ for a.e. $t \in [0, T]$. The assumptions for the function f_p readily imply that $x_s \to x_p$ in L^∞ as $s \to 0$.

An integration by parts gives

(125)
$$\int_0^T \Psi_p'(\dot{x}_s - \dot{x}_p) dt = \int_0^T \dot{\Psi}_p'(x_s - x_p) dt.$$

Using the Taylor expansion about (x_p, u_p) and the analogous of (114), (115) at p, and moreover applying (124) and (125) we have

$$I_p(x_s, u_s) = \int_0^T \{g_p(x_s, u_s) + \Psi_p'[\dot{x}_s - f_p(x_s, u_s)]\} dt$$

$$= I_p(x_p, u_p) + \int_0^T [\dot{\Psi}_p + H_{px}(h_p)]'(x_s - x_p) dt$$

$$+ \int_0^T H_{pu}(h_p)' su\, dt + \int_0^T \Delta z' M_p(\bar{h}) \Delta z\, dt$$

$$\geq I_p(x_p, u_p) + s^2 \|u\|_{L^2}^2 \geq I_p(x_p, u_p),$$

where $\Delta z = (x_s - x_p, su)$ and $\bar{h}$ is between h_p and (x_s, u_s, Ψ_p). This shows that for no $\bar{u}$ near u_p one has $I_p(\bar{x}, \bar{u}) < I_p(x_p, u_p)$ with $\bar{x}$ corresponding to $\bar{u}$ and p. Thus, (x_p, u_p) is a unique local solution to (112), (113).   □

Proof of theorem 23. We obtain theorem 23 as a corollary from an existence result for the generalized equation (117) with fixed parameter p.

Consider the problem (117) with fixed $p \in P$ and let $z_0 \in \Omega, y_0 \in Y, T(z) = T_p(z), L = T_0'(z_0)$. Define

$$D_\rho = \sup\{\frac{\|T(x) - T(z) - L(x - z)\|}{\|x - z\|} : x, z \in \Omega \cap B(z_0, \rho)\}, \delta = \|T(z_0) - y_0\|.$$

We need the following

24 Lemma Let γ, ρ, r and σ be real numbers that satisfy

(126) $$0 \leq \gamma D_\rho < 1, \rho \geq r > \gamma\delta/(1 - \gamma D_\rho), \sigma \geq rD_r + \delta.$$

Let H be a multifunction from $B(y_0, \sigma)$ to Ω such that:

(127) $$z_0 \in H(y_0);$$

(128) $H(y)$ is a closed and nonempty subset of solutions to

(129) $$L(z - z_0) + y \in F(z)$$

for each $y \in B(y_0, \sigma)$; and

(130) $$\text{haus}\,[H(y_1), H(y_2)] \leq \gamma\|y_1 - y_2\|$$

for every y_1 and $y_2 \in B(y_0, \sigma)$. Then (117) has a solution $z \in B(z_0, r)$.

If in addition there is only one solution of (129) for every $y \in B(y_0, \sigma)$ then z is the unique solution to (117) in $B(z_0, r)$, and the second condition in (126) can be weakened to $\rho \geq r \geq \gamma\delta/(1 - \gamma D_\rho)$.

The above lemma gives more than we need for the proof of theorem 23. However, we shall use this generality in section IX.2.

Proof of lemma 24. Define the multifunction

$$G(z) = H[T(z) - L(z - z_0)].$$

For any $z \in \Omega \cap B(z_0, r)$ we have

$$\|T(z) - L(z - z_0) - y_0\| \le \|T(z) - T(z_0) - L(z - z_0)\| +$$

$$+ \|T(z_0) - y_0\| \le \|z - z_0\| D_r + \delta \le \sigma,$$

hence G is defined on $\Omega \cap B(z_0, r)$. By the Lipschitz continuity of H

$$\text{haus } (G(z), G(x)) \le \gamma \|T(z) - T(x) - L(x - z)\| \le \gamma D_r \|x - z\|,$$

whenever x and $z \in \Omega \cap B(z_0, r)$. Hence G is Lipschitz there with Lipschitz constant $\lambda = \gamma D_r < 1$. Since $z_0 \in H(y_0)$ we have

$$\text{dist}(z_0, G(z_0)) \le \text{ haus } (H(y_0), H(T(z_0)) \le \gamma \|y_0 - T(z_0)\| = \gamma \delta.$$

Dividing this inequality by $(1 - \lambda)$ we see that

$$r > \text{ dist } (z_0, G(z_0))/(1 - \lambda).$$

Then we apply the contraction mapping principle for multifunctions, see Ioffe - Tihomirov [2, p.31], concluding that there exists $z \in G(z)$ with $z \in B(z_0, r)$. Observe that $z \in G(z)$ iff z is a solution of (117).

If the solution of (129) is unique for every $y \in B(y_0, \sigma)$ then G is single-valued on $\Omega \cap B(z_0, r)$ and we can apply the (single-valued) contraction mapping principle. Then there exists an unique $z \in \Omega \cap B(z_0, r)$ with $z = G(z)$, hence z is the unique solution to (117) on $B(z_0, r)$. □

End of the proof of theorem 23. Define

$$D_\rho(p) = \sup\{\frac{\|T_p(x) - T_p(z) - L(x - z)\|}{\|x - z\|} : x, \ z \in \Omega \cap B(z_0, \rho)\}.$$

Then

$$D_\rho(p) \le \sup\{\| \int_0^1 [T_p'(sx + (1 - s)z) - L](x - z)ds\| : x, z \in \Omega \cap B(z_0, \rho)\}$$

$$\le \sup\{\|T_p'(z) - T_0'(z_0)\| : \|z - z_0\| \le \rho\}.$$

Hence, by (119), $D_\rho(p) \to 0$ as $\rho \to 0$ and $p \to 0$. Put

$$M_\rho = \sup\{D_\rho(p) : p \in B(0, \rho)\}$$

and choose $\rho > 0$ and so small that $\gamma M_\rho < 1$, where γ is the Lipschitz constant of the map H from (130). Let

$$\delta(p) = \|T_p(z_0) - y_0\|.$$

Now choose s such that $0 < s \le \rho$ and if $p \in B(0, s)$ then

$$\rho \geq \frac{\gamma\delta(p)}{1 - \gamma M_\rho} \text{ and } \sigma \geq \frac{\gamma\delta(p)M_\rho}{1 - \gamma M_\rho} + \delta(p).$$

Take

$$r = \frac{\gamma\delta(p)}{1 - \gamma M_\rho}$$

and apply lemma 24 in the uniqueness case. Then for each $p \in B(0, s)$ there exists an unique solution z_p of (117) in $B(z_0, r)$.

In the proof of lemma 24 we showed that the map $G_p(z) = H[T_p(z) - L(z - z_0)]$ is a contraction on $\Omega \cap B(z_0, r)$ with constant $\lambda = \gamma M_\rho < 1$ for every fixed $p \in B(0, s)$ and moreover $z_p = G_p(z_p)$. Given p and $q \in B(0, s)$ we have

$$\|z_p - z_q\| = \|G_p(z_p) - G_q(z_q)\| \leq \|G_p(z_p) - G_q(z_p)\| + \|G_q(z_p) - G_q(z_q)\|$$
$$\leq \gamma\|T_p(z_p) - T_q(z_p)\| + \lambda\|z_p - z_q\|,$$

hence using (120)

$$\|z_p - z_q\| \leq \gamma\|T_p(z_p) - T_q(z_p)\|/(1 - \lambda) \leq \text{ (const.) dist } (p, q).$$

$\square$

Section 6.   Well-Posedness of Linear Time-Optimal Problems.

Hadamard Well-Posedness.

We consider value and solution Hadamard well-posedness of time-optimal problems for a linear differential inclusion

(131) $\dot{x}(s) \in A(s)x(s) + B(s)U$, a.e. $s \geq 0, x(0) = y$.

Here y and $x(s) \in R^m, U$ is a nonempty compact convex subset of R^k and $A, B \in L^1_{loc}([0, +\infty))$.

We denote by $R(t)$ the reachable set at the time $t \geq 0$ for (131), i.e. the set of all solutions to (131) in $[0, t]$ evaluated at time t. If $\dot{\emptyset} = A(t)\emptyset, \emptyset(0) = $ identity matrix, then

$$R(t) = \emptyset(t)y + \emptyset(t) \int_0^t \emptyset^{-1}(s)B(s)U \, ds,$$

where in the right-hand side we are using the Aumann integral of the multifunction $\emptyset^{-1}(\cdot)B(\cdot)U$ (see Artstein [1, p.875]). The corresponding *time-optimal function* T is defined by

$$T(x) = \inf\{t \geq 0 : x \in R(t)\}, x \in R^m.$$

Hence $T(x) < +\infty$ iff x is reachable.

We are interested in value Hadamard well-posedness under plant perturbations, i.e. the continuous behaviour of T as a function of A, B, and the behaviour of the optimal trajectories. As well known, the multifunction

$$R : [0, +\infty) \rightrightarrows R^m$$

is upper semicontinuous with nonempty compact convex values, and increasing, i.e.

$$R(t_1) \subset R(t_2) \text{ whenever } 0 \leq t_1 \leq t_2,$$

(see Hermes-La Salle [1, sec. 12]).

By abstracting precisely such properties, we will get a characterization of value Hadamard well-posedness with respect to plant perturbations.

Abstract Approach.

We are given a sequence of multifunctions

$$R_n : [0, +\infty) \rightrightarrows R^m, n = 0, 1, 2, ...,$$

with nonempty compact convex values. Each R_n is assumed increasing and upper semicontinuous. For each n we define

$$T_n(z) = \inf\{t \geq 0 : z \in R_n(t)\}, \text{ if } z \in \cup\{R_n(t) : t \geq 0\},$$

$$T_n(z) = +\infty, \text{ otherwise.}$$

The multifunction R_0 is called *proper* whenever $0 \leq t_1 < t_2$ implies $R_0(t_1) \subset \text{int } R_0(t_2)$.

An improper R_0 may give rise to ill-posed time-optimal problems, as shown by the following

Example. Let $A_n = 0, B_0(t) = 1$ if $0 \leq t < 1$ or $t \geq 2; B_0(t) = 0$ if $1 \leq t < 2; B_n(t) = B_0(t) - 1/n$ if $t \leq 1; B_n(t) = B_0(t)$ if $t > 1; y = 0; U = [-1, 1]$. The corresponding reachable sets are shown in the next figure.

It is easily seen that R_0 is not proper and haus $(R_n(t), R_0(t)) \to 0$ for every $t > 0$, while $T_0(1) = 1, T_n(1) \geq 2$ if $n \geq 1$.

We are interested in value Hadamard well-posedness of the time-optimal problem defined by R_0. The following property will be used:

(132) $t_j \to t_0$ in $[0, +\infty], n_j$ any subsequence of the positive integers imply

$$\cap_{j=1}^{\infty} R_{n_j}(t_j) \subset \liminf R_{n_j}(t_0).$$

25 Theorem *Let R_0 be a proper multifunction. Then the following are equivalent properties:*

(133) $K - \lim R_n(t) = R_0(t) \text{ for every } t \geq 0;$

(134) $x_n \to x_0 \text{ implies } T_n(x_n) \to T_0(x_0), \text{ and (132) holds.}$

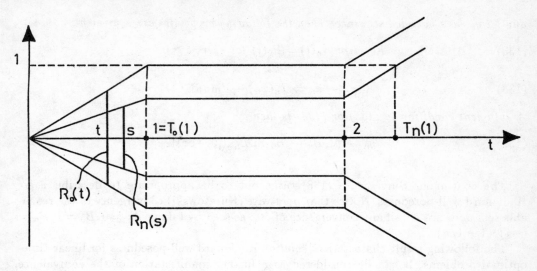

Linear Control Systems.

We consider a sequence of differential inclusions

(135) $\dot{x}(s) \in A_n(s)x(s) + B_n(s)U, x(0) = y_n,$

where $A_n, B_n \in L^1_{loc}([0, +\infty)), y_n$ is a given sequence in R^m, and $n = 0$ defines the original (unperturbed) problem. Again U is a compact (nonempty) convex subset of R^m.

Uniform convergence of $\int_0^t A_n ds$ to $\int_0^t A_0 ds$ together with $B_n \rightarrow B_0$ in $L^1_{loc}([0, +\infty])$ entails (132). If, moreover, B_n converges to B_0 strongly, then we get Kuratowski convergence (even Hausdorff) of the corresponding reachable sets $R_n(t)$ toward $R_0(t), t \geq 0$. A similar form of coefficient convergence is also relevant in the well-posedness analysis of linear regulator problems or, more generally, linear convex problems, as we saw in section 3.

The control system corresponding to $n = 0$ will be called *proper* iff the corresponding multifunction $R_0(\cdot)$ is, where $R_0(t)$ is the set of the reachable points at time t. So we get

26 Corollary *Let the original control system ($n = 0$) be proper. Assume that:*

(136) A_n *is bounded in* $L^1_{loc}([0, +\infty))$ *and*

$$\int_0^t A_n ds \rightarrow \int_0^t A_0 ds \text{ locally uniformly in } [0, +\infty);$$

(137) $B_n \rightarrow B_0 \text{ in } L^1_{loc}([0, +\infty)),$

and let y_n be a bounded sequence. Then the following properties are equivalent:

(133) $K - \lim R_n(t) = R_0(t)$ *for every $t \geq 0$;*

(138) $x_n \to x_0 \Rightarrow T_n(x_n) \to T_0(x_0).$

A sufficient condition to (133) or (138) is then

(139) $y_n \to y_0, B_n \to B_0$ *in* $L^1_{loc}([0, +\infty)).$

The continuous convergence (138) turns out to be appropriate to describe *value* Hadamard well-posedness. Notice that pointwise (Kuratowski) convergence of the reachable sets does not entail any convergence of B_n, as seen by taking $A_n = 0, B_0 = 1, B_n = (-1)^n(1 + 1/n)$.

The following result characterizes *solution* Hadamard well-posedness for linear time-optimal problems. It can be considered as a further manifestation of the equivalence between Hadamard and Tykhonov well-posedness outside the convex setting, at least under the assumptions of next corollary 29 (but notice that the time-optimal function is quasi-convex, since for (131)

(140) $\{x \in R^m : T(x) \leq t\} = R(t), t \geq 0,$

see Hajek [1, cor. 3 p. 341]).

Given the target point $\omega \in R^m$, denote by P_n the time-optimal problem corresponding to (135), i.e. to minimize $t \geq 0$ subject to $\omega \in R_n(t)$. It is well know that P_n has optimal trajectories if $T_n(\omega) < +\infty$ (see Hermes-La Salle [1. th. 13.1]).

27 Theorem *Assume properness of the original system ($n = 0$). Suppose that $T_0(\omega) < +\infty$ and A_n is bounded in $L^1_{loc}([0, +\infty))$. Then the following facts are equivalent:*

(141)   *the optimal trajectory for P_0 is unique;*

(142) *for every $y_n, A_n, B_n, y_n \to y_0, \int_0^t A_n ds \to \int_0^t A_0 ds$ locally uniformly in $[0, +\infty)$*

and $B_n \to B_0$ in $L^1_{loc}([0, +\infty))$ imply that every sequence x_n of optimal trajectories of P_n converges to some optimal trajectory of P_0, uniformly on the compact subsets of $[0, T_0(\omega))$.

PROOFS.

Theorem 25. We shall repeatedly use (140), which is proved as follows. Let $T(x) \leq t$. Then there exists a sequence $s_p \in (T(x), T(x) + 1/p)$ such that $x \in R(s_p)$ for every p. Since R has closed graph, it follows that

$$(143) \qquad\qquad x \in R[T(x)],$$

hence $x \in R(t)$. Conversely, if $x \in R(t)$, then of course $T(x) \leq t$.

(133) $\Rightarrow$ (134). Suppose $T_0(x_0) = +\infty$. Assume that, on the contrary, for some subsequence $T_n(x_n) \leq t < +\infty$, then by monotonicity of R_n we get $x_n \in R_n(t)$, giving $x_0 \in R_0(t)$, a contradiction. Suppose $T_0(x_0) < +\infty$. Arguing again by contradiction we have $T_n(x_n) \not\to T_0(x_0)$, thus there are two cases: for some subsequence either

$$(144) \qquad\qquad T_n(x_n) \leq t' < T_0(x_0),$$

or

$$(145) \qquad\qquad T_0(x_0) < t'' < T_n(x_n).$$

Assume (144). Then $x_n \in R_n(t')$, entailing $T_0(x_0) \leq t'$, a contradiction. As shown in the proof of (140), if $x \in \cup\{R_0(t) : t \geq 0\}$, then $x \in R_0[T_0(x)]$.

Assume (145). By properness, $x_0 \in$ int $R_0(t'')$. By (133) there exists $y_n \in R_n(t'')$ such that $y_n \to x_0$, while $x_n \notin R_n(t'')$. Some point $z_n \in \partial R_n(t'')$ belongs to the segment with ends x_n and y_n. Therefore $z_n \to x_0$. Moreover, there exists $v_n \notin R_n(t'')$ such that $|z_n - v_n| = 1$, and

$$(146) \qquad\qquad (v_n - z_n)'(x - z_n) \leq 0 \text{ for every } x \in R_n(t'').$$

Since z_n converges, v_n is bounded, thus for a subsequence $v_n \to v$, $|x_0 - v| = 1$. By (133), given $p \in R_0(t'')$ there exists $p_n \in R_n(t'')$ such that $p_n \to p$. By (146) with $x = p_n$

$$(147) \qquad\qquad (v - x_0)'(p - x_0) \leq 0, p \in R_0(t'').$$

If $v \in R_0(t'')$ we get $x_0 = v$, a contradiction. If $v \notin R_0(t'')$, then for some $a \in (0, 1)$ consider

$$p = av + (1 - a)x_0$$

and by (147) we get again $v = x_0$, a contradiction.

Summarizing, we showed that (133) implies $T_n \to T_0$ continuously. Now we show that *(133) $\Rightarrow$ (132).* If $x \in R_{n_j}(t_j)$ for every j, then $x \in R_0(t_0 + \epsilon)$ for every $\epsilon > 0$. By upper semicontinuity this gives $x \in R_0(t_0)$, hence $x \in \liminf R_{n_j}(t_0)$.

(134) $\Rightarrow$ (133). Given $t \geq 0$ and $x \in R_0(t)$, let us show that $x = \lim x_n$ with a suitable $x_n \in R_n(t)$. Set

$$n \in A \Leftrightarrow T_n(x) \leq T_0(x); n \in B \Leftrightarrow T_n(x) > T_0(x);$$

$$x_n = \begin{cases} x, & \text{if } n \in A, \\ \text{projection of } x \text{ on } R_n[T_0(x)], & \text{if } n \in B. \end{cases}$$

Suppose that B has infinitely many elements. Since $T_0(x) < +\infty$ by assumption, then $T_n(x) < +\infty$ for n sufficiently large. Thus $x \in R_n[T_n(x)]$ by (143). Then $x \in \liminf R_n[T_0(x)]$ by (132), yielding $x = \lim y_n$ for some $y_n \in R_n[T_0(x)]$. Then $x_n \to x$ since $|x - x_n| \leq |x - y_n|$. To end the proof, take $x_j \in R_{n_j}(t)$ with $x_j \to x$. Then $T_{n_j}(x_j) \leq t$, thus $T_0(x) \leq t$, whence $x \in R_0(t)$ by (140). $\qquad\square$

Corollary 26. We show that (136) and (137) imply (132). To this end let $x \in R_{n_j}(t_j)$ for every $j, t_j \to t_0, t_j \leq T$. Denote by $\emptyset_n$ the fundamental matrix of $\dot{x} = A_n(t)x$, principal at 0, and set

$$L_j = \emptyset_{n_j}^{-1} B_j, \quad F_j = \emptyset_{n_j}.$$

Then for some measurable u_j such that $u_j(t) \in U$ a.e.,

$$x = F_j(t_j)(y_{n_j} + \int_0^{t_j} L_j u_j\, dt).$$

By lemma 12, $F_j \to F_0$ uniformly on $[0, T]$, therefore F_j is uniformly bounded and equicontinuous there. By (137) we get

$$\int_{t_o}^{t_j} |B_{n_j}|\, ds \to 0 \text{ as } j \to +\infty,$$

so $z_j \to x$, where

$$z_j = F_j(t_0)(y_{n_j} + \int_0^{t_0} L_j u_j\, dt),$$

yielding (132). Finally, (136) and (139) imply (133) since for every fixed $t \geq 0$ we have

$$\emptyset_n(t)(y_n + \int_0^t \emptyset_n^{-1} B_n\ u\ ds) \to \emptyset_0(t)(y_0 + \int_0^t \emptyset_0^{-1} B_0\ u\ ds)$$

uniformly with respect to the admissible controls u (i.e. U-valued measurable functions on $[0, t]$). □

Theorem 27. (141) $\Rightarrow$ (142). Let u_n be a control function to which it corresponds $x_n : [0, T_n(\omega)] \to R^m$. Fix $s < T_0(\omega)$. Then the domain of x_n includes $[0, s]$ since $T_n(\omega) \to T_0(\omega)$ by corollary 26. By weak sequential compactness, some subsequence

$$u_n \rightharpoonup u \text{ in } L^\infty(0, s).$$

Moreover, x_n is equibounded and equicontinuous in $[0, s]$. This yields uniform convergence of x_n in $[0, s]$ by lemma 12. By a diagonal procedure we find a subsequence $x_n \to x_0$ uniformly on compact subsets of $[0, T_0(\omega))$, where x_0 is a trajectory of (135) with $n = 0$, which is defined on $[0, T_0(\omega))$. Let us show that x_0 is an optimal trajectory for P_0. To this end we need some more notation (in order to state Bellman's optimality principle in this special case).

Given $s \geq 0, y \in R^m$ we denote by $R_n(s, y)(t)$ the (reachable) set of the values at time $t \geq s$ of the trajectories of the differential inclusion in (135), which take on the value y at time s. Then set

$$T_n(s, y) = \left\{ \begin{array}{l} +\infty, \text{ if } \omega \notin \cup\{R_n(s, y)(t) : t \geq s\}, \\ \inf\{t \geq s : \omega \in R_n(s, y)(t)\} - s, \text{ otherwise.} \end{array} \right.$$

By Bellman's optimality principle and optimality of x_n

(148) $$T_n(0, y_n) = T_n[s, x_n(s)] + s, 0 \leq s < T_n(0, y_n).$$

Taking limits as $n \to +\infty$ in (148) and remembering corollary 26 we get

(149) $T_0(0, y_0) = T_0[s, x_0(s)] + s, 0 \le s < T_0(0, y_0)$.

Now x_0 can be defined by continuity at $T_0(0, y_0)$ (since $B_0 \in L^1[0, T_0(0, y_0)]$). Moreover, $T_0(\cdot, \cdot)$ is continuous (by a simple extension of theorem 25). By (149), as $s \to T_0(0, y_0)$ we get

$$x_0[T_0(0, y_0)] = \omega,$$

obtaining optimality of x_0. So (142) is proved.

   Conversely, assume (142). If x_1, x_2 are both optimal solutions to P_0, the sequence $x_1, x_2, x_1, x_2, ...$ is convergent, yielding (141).   □

Tykhonov Well-Posedness.

   We consider again the time-optimal problem of steering the target point ω starting from y at time 0 by the trajectories of the differential inclusion (131).
   We assume that

$$U \text{ is a polytope in } R^k,$$

i.e. the convex hull of a finite number of points, and (again) $A, B \in L^1_{loc}([0, +\infty))$.

   Let D denote the set of admissible controls, so that $u \in D$ iff there exists $t > 0$ such that u is measurable in $[0, t]$ and $u(s) \in {}_, U$ a.e. Given $u \in D$ we denote by $x(u)$ the corresponding trajectory. The sequence u_n is a minimizing one for our time-optimal problem iff $T(\omega) < +\infty, u_n \in D$ and there exists $t_n \ge 0$ such that

$$x(u_n)(t_n) = \omega; \ t_n \downarrow T(\omega) \text{ as } n \to +\infty.$$

   Let $D(0, T)$ be the subclass of D, formed by measurable functions on $[0, T], T > 0$. Given any minimizing sequence u_n, by weak-star sequential compactness we get for some subsequence

$$u_n \rightharpoonup v \text{ in } L^\infty(0, T), \text{ where } T = T(\omega).$$

Let $\emptyset$ denote the fundamental matrix of $\dot{x} = A(t)x$, principal at 0. Then for suitable constants c_1, c_2, c_3

$$|\omega - x(v)(T)| = |x(u_n)(t_n) - x(v)(T)|$$

$$\le c_1 |\emptyset(t_n) - \emptyset(T)| + c_2 \left| \int_0^T \emptyset^{-1} B(u_n - v) ds \right| + c_3 \int_T^{t_n} |B| ds,$$

yielding

$$x(v)(T) = \omega.$$

So v is an optimal control, showing well - posedness in the generalized sense (section I.6) with respect to the L^1 - topology of L^∞. Now suppose that an unique optimal control u^* exists. Of course the problem is Tykhonov well-posed with respect to the weak-star convergence in L^∞. The key property which allows us to gain Tykhonov well-posedness with respect to the *strong* topology of $L^1(0, T)$, say, is the bang-bang principle (see Sonneborn - Van Vleck [1]). Accordingly, we know that

(150) $u^*(t) \in \text{ extr } U, \text{ a.e. } t \in [0, T]$.

Then, for any minimizing sequence

$$u_n(t) \in U \text{ a.e. } , u_n \rightharpoonup u^* \text{ in } L^\infty(0,T).$$

We shall show that $u_n(t) \to u^*(t)$ a.e. because of (150), entailing strong convergence in $L^1(0,T)$.

This fact may be generalized (without reference to any control system), as follows. Given $T > 0$, set $M = D(0,T)$ and suppose we are given a proper functional

$$I : M \to (-\infty, +\infty].$$

28 Theorem *Assume that (M,I) is Tykhonov well-posed with respect to the L^1- convergence in $L^\infty(0,T)$. Suppose that*

(151) $$\arg \min (extr \; M, I) \neq \emptyset \text{ and } \inf I \, (extr \; M) = \inf I(M).$$

Then (M,I) is Tykhonov well-posed with respect to the strong convergence in every $L^p(0,T), 1 \le p < \infty$.

Coming back to the time-optimal control for the differential inclusion (151), we get the following

29 Corollary *If there exists an unique time-optimal control, the corresponding problem is Tykhonov well-posed as in theorem 28.*

PROOFS.

Theorem 28. Let u_n be a minimizing sequence for (M,I). Then, by (151), consider

$$u^* = \arg \; \min(M,I) = \arg \; \min (extr \; M, I).$$

Then by well-posedness
$$u_n \rightharpoonup u^* \text{ in } L^\infty(0,T).$$

Since $u^* \in$ extr M, by Aumann [1, prop 6.1] we get

$$u^*(t) \in extr \; U \text{ for a.e. } t.$$

Since every extremal point of U is an exposed one (being a vertex), for every $p \in$ extr U there exists $v = v(p) \in R^k$ such that $v(p) \neq 0$,

$$|v(p) - p| = 1, (v - p)'(x - p) \ge 0 \text{ for every } x \in U,$$

and the affine hyperplane through p, orthogonal to $v(p)$, meets U exactly at p. Define now

$$t \in L \text{ iff } u^*(t) \notin extr \; U,$$

$$g(t) = \begin{cases} u^*(t), & \text{if } t \in L, \\ v[u^*(t)] & \text{if } t \in [0,T] \setminus L. \end{cases}$$

g is then measurable since, for any open set $A \subset R^k$, $g^{-1}(A)$ agrees with $g^{-1}(A) \cap ([0,T] \setminus L)$ up to a set of measure 0, that is with

$$\cup \{u^{*-1}(p) : p \in \text{ extr } U, v(p) \in A\},$$

which is a finite union of measurable disjoints sets. Moreover

(152) $t \notin L, z \in U, (g(t) - u^*(t))'(z - u^*(t)) = 0 \Rightarrow z = u^*(t).$

Now set

$$f_n(t) = [g(t) - u^*(t)]'[u_n(t) - u^*(t)].$$

Then $f_n \geq 0$ a.e., therefore, by weak-star convergence of u_n, it follows that $f_n \to 0$ in $L^1(0,T)$ and in measure. Fix any subsequence of u_n. Then the corresponding subsequence of f_n has a further subsequence which converges to 0 a.e. For a.e. t, the corresponding subsequence of $u_n(t)$ has a further subsequence, which converges to $u(t)$, say, $u(t) \in U$. Since a.e. $f_n(t) \to [g(t) - u^*(t)]'[u(t) - u^*(t)] = 0$, then by (152) we get $u = u^*$ a.e. By the dominated convergence theorem we obtain $u_n \to u^*$ in $L^p(0,T), 1 \leq p < +\infty$ for the original sequence.   □

Section 7. Bounded Time Approximations for Problems on Unbounded Time Intervals.

We consider the following (infinite horizon) optimal control problem P_∞ : minimize

(153) $$I_\infty(u,x) = \int_0^{+\infty} f[t, x(t), u(t)]dt$$

subject to

(154) $$\begin{cases} \dot{x}(t) = & g[t, x(t)] + h[t, x(t)]u(t), \text{ a.e. in } (0, +\infty), \\ x(0) = & \bar{x}, \end{cases}$$

(155) $$(t, x(t)) \in E \text{ if } t \geq 0,$$

(156) $$u(t) \in V[t, x(t)] \text{ a.e. in } (0, +\infty),$$

where

$$V : [0, +\infty) \times R^p \rightrightarrows R^m$$

takes nonempty values. We assume that there exists at least one pair (u, x) fulfilling (154), (155), (156), such that $I_\infty(u, x) < +\infty$.

We are interested in conditions allowing to approximate value and optimal trajectories of the above problem (153),...,(156), by approximately solving a finite time horizon version of it. When this happens, we are in presence of a (Hadamard type) well-posedness property of the given problem, related to the asymptotic behavior, as $T \to +\infty$, of suboptimal trajectories obtained by controlling the system on (large but) bounded time intervals $[0, T]$. These bounded time approximations are often employed to computationally solve P_∞.

Two simplifying aspects in the above control problem are the lack of asymptotic state constraints (like e.g. $\lim_{t \to +\infty} x(t) \in S$, a preassigned set), and the affine dependence on the control variables in the state equations (154), allowing to obtain convergence of the controls.

Given $T > 0$ we shall consider the following bounded time approximation P_T of P_∞: to minimize

$$(157) \qquad I_T(u, x) = \int_0^T f(t, x, u)dt$$

subject to (154), (155), (156) restricted to $[0, T]$.

We shall write

$$(u, x) \in \varepsilon - \arg \min P_T$$

iff (u, x) is admissible for P_T, i.e. u is measurable and x is absolutely continuous on $[0, T], (u, x)$ fulfills there (154), (155), (156), and moreover

$$\int_0^T f(t, x, u)dt \leq v_T + \varepsilon,$$

where v_T is the optimal value of P_T.

We fix a sequence $T_n \uparrow +\infty$ and consider the corresponding problems

$$P_n = P_{T_n}.$$

Under natural conditions, it is possible to apply theorem IV. 5, thereby obtaining

30 Theorem *Assume the following conditions:*
(158) $f : [0, +\infty) \times R^p \times R^m \to [0, +\infty)$ *is a Carathéodory function; $f(t, x, \cdot)$ is convex, and $f(t, x, u) \geq q(|u|)$ for some continuous, increasing, convex $q : [0, +\infty) \to [0, +\infty)$ such that $q(z)/z \to +\infty$ as $z \to +\infty$;*
(159) *every component of g and every element of the $(p \times m)$ matrix h are Carathéodory functions on $[0, +\infty) \times R^p$; for every $T > 0, r > 0$ there exist $b \in L^1(0, T)$ and constants c_1, c_2 such that*

$$|g(t, x)| \leq b(t); |h(t, x)| \leq c_1, \text{ if } 0 \leq t \leq T, |x| \leq r;$$

$$|h(t, y) - h(t, z)| \leq c_2|y - z| \text{ if } 0 \leq t \leq T, |y| \leq r, |z| \leq r;$$

(160) E *is closed in $[0, +\infty) \times R^p$, and*

$$V(t, y) = \cap\{cl \text{ } co \text{ } V[t, B(y, \varepsilon)] : \varepsilon > 0\}$$

for every (t, y);
(161) *either the projection of E on R^p is bounded, or*

$$|g(t, x)| \leq m(t)(1 + |x|), |h(t, x)| \leq c(1 + |x|)$$

for all t, x, some constant c and $m \in L^1_{loc}(0, +\infty)$.

Then the following conclusions hold:

(162) $$v_T \uparrow v_\infty \text{ as } T \to +\infty :$$

(163)   *given $\varepsilon_n > 0, \varepsilon_n \to 0$ and $(u_n, x_n) \in \varepsilon_n - argmin\ P_n$, there exists*

$(u_\infty, x_\infty) \in argmin\ P_\infty$ such that for some subsequence and every $T > 0$ we have $u_n \to u_\infty$ in $L^1(0, T), x_n \to x_\infty$ uniformly in $[0, T], \dot{x}_n \to \dot{x}_\infty$ in $L^1(0, T)$.

The above conclusions may no longer hold when asymptotic state constraints are included and held fixed in the definition of P_T.

31 Counterexample. We want to minimize

$$\int_0^{+\infty} \dot{x}^2 dt \text{ subject to } x(0) = 0, \lim_{t \to +\infty} x(t) = 1.$$

Here $v_\infty = 0$, as readily seen by using $x_n(t) = t/n$ if $0 \le t \le n, = 1$ if $t \ge n$.
A reasonable (although rather rigid) definition of P_T in this case consists in minimizing

$$\int_0^T \dot{x}^2 dt \text{ subject to } x(0) = 0, x(T) = 1.$$

This gives the optimal value

$$v_T = 1/T \to v_\infty \text{ as } T \to +\infty$$

but the only minimizer $x_T(t) = t/T$ converges locally uniformly to the zero function, which is even not admissible for P_∞.

32 Example. There is a link between *bounded time approximations* and so-called *singular perturbations* studied in chapter VII. Consider the problem, to minimize

$$I(u) = \int_0^{+\infty} (e^{-t} - 2e^{-2t})u(t)dt$$

subject to $u(t) \in [-1, 1]$ a.e., u measurable. The optimal control is (a.e.)

$$u_\infty(t) = 1 \text{ for } 0 \le t \le \ln 2, = -1 \text{ for } \ln 2 < t < +\infty,$$

and the optimal value is $v_\infty = -0.5$. For $T > 0$ define a bounded time approximation:

(164)   minimize $I_T(u) = \int_0^T (e^{-t} - 2e^{-2t})u(t)dt$

subject to the same constraints. Then, if $T > \ln 2$, the optimal value is given by

$$v_T = e^{-T} - e^{-2T} - 1/2 \to v_\infty \text{ as } T \to +\infty,$$

hence we have value well-posedness.

Let $W(T)$ be the set of values of $I_T(u)$, u running through the admissible controls. Then (164) can be clearly rewritten as

(165) $\qquad\qquad\qquad$ minimize z subject to $z \in W(T)$.

Now consider the singularly perturbed system

(166) $\qquad\qquad\qquad\quad \begin{cases} \varepsilon\dot{y}_1 &= -y_1 + u, y_1(0) = 0, \\ \varepsilon\dot{y}_2 &= -2y_2 + u, y_2(0) = 0, \end{cases}$

where ε is a positive parameter and $u(t) \in [-1, 1]$ a.e., u measurable. Let $K(\varepsilon)$ be the reachable set at $t = 1$ of (166) with fixed ε, and let

$$R(\varepsilon) = \{y_1 - 2y_2 : (y_1, y_2) \in K(\varepsilon)\}.$$

Then the set

$$R(\varepsilon) = \{1/\varepsilon \int_0^1 (e^{-s/\varepsilon} - 2e^{-2s/\varepsilon})u(s)ds : u \text{ admissible for } (166)\}$$

concides with $W(T)$ provided $\varepsilon = 1/T$. Thus (165), hence (164), is equivalent to the problem

(167) $\qquad\qquad\qquad$ minimize $y_1(1) - 2y_2(1)$ subject to (166).

This is a Mayer's problem for the singularly perturbed system (166). The case $T = +\infty$ corresponds formally to $\varepsilon = 0$. Taking $\varepsilon = 0$ in (166), however, results in $y_1 = 2y_2 = u$, hence the optimal value of (167) with $\varepsilon = 0$ is $0 > v_\infty$. Thus, transforming the bounded time approximation into a singularly perturbed problem may destroy value well-posedness. This effect is deeply related to the singular behaviour of system (166) as $\varepsilon \to 0$, wich will be studied in chapter VII.

Proof of theorem 30. Denote by A_n the set of all admissible pairs for P_n, and by A_∞ that for P_∞. Then $(u, x) \in A_\infty$ iff (154), (155), (156) hold with u measurable and x locally absolutely continuous on $[0, +\infty)$. Let U denote the convergence space $L^1_{loc}(0, +\infty)$ equipped with the weak convergence on every bounded interval. Let X denote the convergence space $H^1_{loc}(0, +\infty)$ equipped with the uniform convergence on compact intervals and the weak convergence of derivatives in $L^1(0, T)$ for every $T > 0$. Write $I_n = I_{T_n}$. We shall show that

(168) $\qquad\qquad$ var $- \lim(I_n + \text{ind } A_n) = I_\infty + \text{ind } A_\infty$ within $U \times X$.

As readily seen, (168) is implied by the following three conditions:
(169) $u_n \to u$ in $U, x_n \to x$ in $X, (u_n, x_n) \in A_n$ for infinitely many $n \Rightarrow (u, x) \in A_\infty$;
(170) $u_n \to u$ in $U, x_n \to x$ in $X, \liminf I_n(u_n, x_n) < +\infty \Rightarrow \liminf I_n(u_n, x_n) \geq I_\infty(u, x)$;
(171) for every $(u, x) \in A_\infty$ such that $I_\infty(u, x) < +\infty$, there exists $(u_n, x_n) \in A_n$ (for every large n) such that $\limsup I_n(u_n, x_n) \leq I_\infty(u, x)$.

$\quad$ *To prove (170)*, consider a subsequence of the integrals (157) converging to their $\liminf$. Then for a further subsequence and some constant c

$$\int_0^{T_n} f(t, x_n, u_n)dt \le c.$$

Fix $T > 0$. Then for sufficiently large n

$$+\infty > c \ge \liminf \int_0^{T_n} f(t, x_n, u_n)dt \ge \lim \inf \int_0^T f(t, x_n, u_n)dt \ge \int_0^T f(t, x, u)dt$$

by lower semicontinuity of I_T (Ekeland-Temam [1, th. 2.1 p.226]). Since T is arbitrary we get

$$\int_0^{+\infty} f(t, x, u)dt \le \liminf \int_0^{T_n} f(t, x_n, u_n)dt,$$

which proves (170).

To prove (169), fix $T > 0$. Then suitable convex means ω_n of u_n converge to u strongly in $L^1(0, T)$ and a.e. in $(0, T)$, by Mazur's theorem (Dunford-Schwartz [1, V.3.14 p.422]). Given $\varepsilon > 0$, by the uniform convergence of x_n we find $x_n(t) \in B(x(t), \varepsilon)$ for all t and all sufficiently large n, hence by (156)

$$u_n(t) \in V[t, B(x(t), \varepsilon)], \text{ a.e. } t \in (0, T).$$

Thus

$$\omega_n(t) \in \text{ co } V[t, B(x(t), \varepsilon)]$$

hence $u(t) \in V[t, x(t)]$ a.e. by (160). Moreover $(t, x(t)) \in E, t \ge 0$. By (154)

$$x_n(t) = \bar{x} + \int_0^t g(s, x_n)ds + \int_0^t h(s, x_n)u_n(s)ds, 0 \le t \le T,$$

for all sufficiently large n. Then

$$\int_0^t g(s, x_n)ds \to \int_0^t g(s, x)ds$$

by (159). The same set of assumptions yields

$|\int_0^t [h(s, x_n) - h(s, x)]u_n ds| \le c_2 \int_0^T |u_n|ds \max\{|x_n(t) - x(t)| : 0 \le t \le T\} \to 0$
    as $n \to +\infty$,

since u_n is bounded in $L^1(0, T)$. Hence $(u, x) \in A_\infty$, thereby proving (169). (171) is obviously true with $(u_n, x_n) = (u, x)$ restricted to $[0, T_n]$. So (168) is proved. Moreover $T \to v_T$ is increasing. Indeed, if $T < S$ and $(u, x) \in A_S$, then the restriction of (u, x) to $[0, T]$ belongs to A_T. Provided $f(\cdot, x(\cdot), u(\cdot)) \in L^1(0, S)$ we have

$$\int_0^T f(t, x, u)dt \le \int_0^S f(t, x, u)dt$$

since $f \ge 0$. Hence $v_T \le v_S$. Now let (u_n, x_n) be as in (163). Then by theorem IV.5 we see that the sequence v_n of optimal values is bounded from above (by properness of $I_\infty + \text{ind } A_\infty$). Fix any $T > 0$, then for sufficiently large n and from (158)

$$\text{constant} \ge \int_0^{T_n} f(t, x_n, u_n)dt \ge \int_0^{T_n} q(|u_n|)dt \ge \int_0^T q(|u_n|)dt$$

showing weak sequential compactness of u_n in $L^1(0,T)$ by de la Vallée-Poussin's theorem (Ekeland-Temam [1, th. 1.3 p. 223]). If the projection of E on R^p is unbounded, the linear growth assumption (161) yields local uniform boundedness of x_n (since u_n is bounded in $L^1(0,T)$ for every $T > 0$). From (154) and (159) we then get local sequential compactness of x_n in X. An application of theorem IV. 5 ends the proof. $\qquad\qquad\square$

Section 8. Notes and Bibliographical Remarks.

To section 1. For theorem 1 see Zolezzi [10].

To section 2. Ill-posedness of standard optimal control problems was firstly pointed out in a general setting by Tykhonov [4] (see counterexample p. 180), where a regularization method was introduced. For strong convergence in L^1 of minimizing sequences see Berliocchi-Lasry [1, p. 173], and Fleming-Rishel [1, p. 182-183] where it is shown that the main sufficient conditions are affinity of $g(t,x,\cdot)$ and strict convexity of $L(t,x,\cdot)$.

Early results about Hadamard well-posedness belong to Kirillova [1], [2], who was among the first studying well-posedness in optimal control (more precisely, linear time optimal problems) and Cullum [1], [2] (who studied more general problems). See also Lee-Markus [1, th. 12 p.416]. Well-posedness criteria for various optimal control problems are contained in Gabasov-Kirillova [1, ch.7, sec. 10]. Generic well-posedness was obtained by Bidaut [1]. Theorems 3 and 4 are particular cases of Zolezzi [5].

A characterization of value Hadamard well-posedness for problems involving differential inclusions is obtained by Stassinopoulos-Vinter [1] through weak convergence of the right-hand sides. Results about Hadamard well-posedness for various optimal control problems are in Stassinopoulos [1], Tadumadze [1], Buttazzo-Dal Maso [1], Zolezzi [1], [3] and [5], Gičev [1], Korsak [1] and Chen Guang-Ya [1]. The last two papers obtain Hadamard well-posedness from Tykhonov's. For value Hadamard well-posedness under state constraints penalization, see Vinter-Pappas [1]. Well-posedness with respect to the initial state is studied by Abdyrakmanov-Kryazhimskii [1]. For well-posedness in the optimal control of variational inequalities see Giachetti [1] and Patrone [1]. See Papageorgiu [2] for related infinite-dimensional results.

For well-posedness under perturbations of deterministic control problems through an additive white noise term with small coefficient, see Fleming [2] and Fleming- Rishel [1, ch. VI, sec. 9]. See Visintin [1] for Tykhonov well-posedness in distributed optimal control problems.

To section 3. The Q-part (so to speak) of theorems 7 and 8 are due to Zolezzi [7], and the C-part to Bennati [2]. Example 9 was found by Buttazzo and Dal Maso. Particular cases of theorem 8 are in Nikolskii [1]. Theorem 10 is due to Pieri [1], and is one of the first characterizations of Hadamard well- posedness in optimal control. For the example thereafter see Lucchetti- Mignanego [2]. Theorem 11 is from Zolezzi [7].

Lemma 13 is a (very) particular case of the intersection lemma in Mosco [3] (which corrects the earlier statement of Mosco [2, lemma 1.4]).

For extensions within an abstract setting see Sonntag [3] and Zalinescu [1]. Extensions to semilinear problems are obtained by Bennati [1]. Well-posedness for minimum effort problems and more general cost functionals have been characterized by Lucchetti-Mignanego [1] (generalized by Cârjă [1] to the infinite-dimensional setting) and [2]. Linear regulator problems with a final- time constraint are treated by Pedemonte [2]. Well-posedness for the optimal control of the coefficients in boundary-value problems was obtained by Zolezzi [2]. Well-posedness in an infinite-dimensional (semi) linear setting is treated by Avgerinos-Papageorgiou [1] and Papageorgiou [4].

To section 4. Theorem 14 is based on Dontchev [1, ch. 2] where a more general convex problem is considered. For some extensions see Malanowski [2]. Theorem 15 is proved in Hager [2] under somewhat weaker assumptions. The differentiability properties of the optimal solutions to convex constrained problems are studied by Malanowski [1].

To section 5. Stronger forms of theorems 22 and 23 are proved in Dontchev-Hager [1]. Theorem 23 is actually a slightly extended version of theorem 2.1 in Robinson [3], see also Hager [2]. Lipschitz estimates for nonlinear problems with inequality control constraints are obtained by Alt [2] and Ito - Kunisch [1]. Lipschitz continuity properties for semilinear control problem are considered in Malanowski [1].

Theorem 22 can be extended to problems with perturbed constraining sets U_p, obtaining an estimate of the perturbed optimal solutions by $[\text{haus}(U_p, U_0)]^{1/2}$. If U_p is given by (in) equality constraints (satisfying a regularity condition) which are Lipschitz with respect to p, then the optimal solution becomes Lipschitz continuous as well.

To section 6. All the results in this section are due to Pieri [3], [4], and [2]. Theorem 27 extends a characterization of normal system due to Hajek [1]. Results for the infinite-dimensional setting can be found in Hoppe [1], Cârjă [1], [2] and [3]. Well-posedness under changes in the control region is examined by Bacciotti-Bianchini [1]. Continuity of the optimal - time function is equivalent to a form of local controllability, see Sussmann [2]. Changes in the nonlinear dynamics and convergence estimates are considered in Bardi-Sartori [1] (see also their references for Hölder continuity properties of the minimum-time function, and Vel'ov [1] on these topics). For the continuity of the minimum-time function with respect to parameters see Petrov [1].

To section 7. Theorem 30 is a particular case of Zolezzi [6]. For early asymptotic results see Bucy [1]; for extensions see Leizarowitz [1, sec. 7]. Related results may be found in Zolezzi [6] (nonlinear systems and asymptotic constraints), Mignanego [2] and [1] (distributed systems), Ewing [1], Flam- Wets [1] (discrete time), Gibson [1] and Pollock-Pritchard [1] (distributed systems). For linear regulator problems see Brunovsky-Komornik [1]. The convergence to finitely optimal solutions is investigated in Carlson [1].

Chapter VI.

RELAXATION AND VALUE HADAMARD WELL-POSEDNESS IN OPTIMAL CONTROL

Section 1.   Relaxation and Value Continuity in Perturbed Differential Inclusions.

In this chapter we consider nonlinear optimal control problems. We are interested in the continuous behavior of the optimal value function under perturbations.

An interesting answer to this problem may be obtained by relating the behavior of the optimal value of the conventional relaxed problem to the unperturbed one. In a sense which will be made explicit below, a necessary and sufficient condition to optimal value continuity under perturbations is the stable behavior of the value when changing the original control problem to its relaxed version.

Consider the following *unperturbed original* optimal control *problem* Q_0 : to minimize

$$f_0(x)$$

subject to

(1) $\begin{cases} x \in W^{1,1}(0,T); \; \dot{x}(t) \in D_0[t, x(t)] \text{ for a.e. } t \in (0,T); \\ x(0) = x_0; x(t) \in H_0 \text{ for all } t \in E. \end{cases}$

Here $D_0 : [0,T] \times R^m \rightrightarrows R^m$ is a given multifunction, and $T > 0, x_0 \in R^m$ are fixed; E is a fixed subset of $[0,T]$ and H_0 is a subset of R^m.

We shall consider classes of perturbations to Q_0 in the following form. Let P be a metric space, which plays the role of the parameter space, with a fixed element $0 \in P$. Given $p \in P$ the *optimal control problem $Q(p)$ with perturbation p* is defined as follows: to minimize

$$f(x,p)$$

subject to

(2) $\begin{cases} x \in W^{1,1}(0,T); \; \dot{x}(t) \in D(t, x(t), p) \text{ for a.e. } t \in (0,T); \\ x(0) = x_0; x(t) \in H(p) \text{ for all } t \in E. \end{cases}$

Here

$$D : [0,T] \times R^m \times P \rightrightarrows R^m, \quad H : P \rightrightarrows R^m$$

are given multifunctions.

The *relaxed problem Q^* of Q_0* is defined as follows: to minimize

$$f_0(y)$$

subject to

(3) $\quad \begin{cases} y \in W^{1,1}(0,T); \; \dot{y}(t) \in \text{ cl co } D_0[t, y(t)] \text{ for a.e. } t \in (0,T); \\ y(0) = x_0; y(t) \in \text{ cl } H_0 \text{ for all } t \in E. \end{cases}$

Given the perturbations $Q(p)$, define

$$v(p) = \text{ optimal value of } Q(p); \; v^* = \text{ optimal value of } Q^*.$$

The role of Q^*, as far as the behavior of $v(\cdot)$ at $p = 0$ is concerned, can be viewed as follows. We assume that

$$D_0(\cdot) = D(\cdot, 0), H_0 = H(0), f_0(\cdot) = f(\cdot, 0),$$

so that $Q_0 = Q(0)$ and $v(0)$ is the optimal value of Q_0. Of course

$$v^* \leq v(0).$$

Let $p_n \to 0$ in P be given. We are interested in getting

(4) $\qquad\qquad\qquad\qquad\qquad v(p_n) \to v(0).$

A function $x \in W^{1,1}(0,T)$ will be called $p-$ *admissible trajectory* whenever x fulfills (2). Consider a sequence x_n of p_n - admissible trajectories. Under standard conditions about problem's data we see that, for a subsequence, $x_n - y \in W^{1,1}(0,T)$ and y satisfies (3), i.e. any weak sequential cluster point of admissible trajectories is admissible for the relaxed problem Q^*. Then, under standard semicontinuity assumptions on f, we get

$$\liminf f(x_n, p_n) \geq f_0(y).$$

If we assume from the outset that some upper semicontinuity condition on the data is fulfilled (in order to guarantee IV.(9)), then putting

$$K_n = \text{ set of all } p_n - \text{ admissible trajectories},$$

$$K^* = \text{ set of all relaxed trajectories},$$

we get in $(W^{1,1}(0,T), \text{weak})$

$$\text{var-lim} [f(\cdot, p_n) + \text{ ind } (K_n, \cdot)] = f(\cdot, 0) + \text{ ind } (K^*, \cdot).$$

We assume sufficient compactness (in standard way) to apply theorem IV.5, thus obtaining

$$v(p_n) \to \text{ val } [K^*, f(\cdot, 0)].$$

Thanks to the arbitrariness of p_n we conclude that

$$v(p) \to v^* \text{ as } p \to 0.$$

Therefore obviously

$$v(p) \to v(0) \text{ iff } v^* = v(0).$$

This fact explains in a precise way why nonrelaxable problems (i.e. problems Q_0 such that $v^* < v(0)$) are ill-posed, a fact which is often stated in the optimal control literature as plausible or even intuitive (as in fact it is). For a given nonrelaxable problem Q_0 there exists some perturbation $Q(p)$ to it such that one has value Hadamard ill-posedness, i.e. for the corresponding optimal value function one has the discontinuous behavior

$$\lim_{p \to 0} v(p) < v(0).$$

Now we give a precise definition of the perturbations involved here. A perturbation $Q(p), p \in P$, to Q_0 is called *admissible* if the following conditions (5),..., (9) hold.

(5)   for every $p \in P$ there exist $p-$ admissible trajectories;

(6) $f : \Omega \times P \to (-\infty, +\infty)$ is lower semicontinuous at every point of $\Omega \times \{0\}$,
    where Ω is an open set of $C^0([0, T])$ containing all trajectories of Q^*;

(7) H is closed at $p = 0$;

(8) $D(t, x, 0)$ is compact nonempty for all (t, x); $D(\cdot, x, 0)$ is measurable for each x, $D(t, \cdot, 0)$ is upper semicontinuous in R^m for a.e. t; for every $\varepsilon > 0$ and every compact set $L \subset R^m$ there exists $\delta > 0$ such that for every p with dist $(p, 0) < \delta$ we can find $a_p \in L^1(0, T)$ so that for a.e. $t \in [0, T]$ and every $x \in H(P) \cap L$

$$D(t, x, p) \subset D(t, x, 0) + a_p(t)B \text{ and } \int_0^T a_p(t)dt \le \varepsilon,$$

where B is the closed unit ball of R^m. Furthermore, for every p there exist $A(\cdot, p), B(\cdot, p)$ uniformly bounded in $L^1(0, T)$ as $p \in P$, such that if $\omega \in D(t, x, p)$ then

$$|\omega| \le A(t, p)|x| + B(t, p), \text{ a.e. } t;$$

(9) $\limsup_{p \to 0} v(p) \le v^*.$

1 Theorem *Assume that relaxability holds for Q_0, i.e.*

(10) $v(0) = v^*.$

Then for every admissible perturbation we have

(11) $v(p) \to v(0) \text{ as } p \to 0.$

Conversely, suppose that for at least one admissible perturbation

(12) $\liminf v(p) \ge v(0).$

Then (10) holds.

In the next statements we write

$$X = (W^{1,1}[0,T], \text{ weak})$$

and use the self-explaining notations

$$\varepsilon - \arg \min Q(p), \ \arg \min Q^*.$$

2 Corollary *The relaxability condition (10) is equivalent to value Hadamard well - posedness (11), for at least one (and therefore for all) admissible perturbations to Q_0. Moreover for every $\varepsilon_n \to 0$ and $p_n \to 0$ we have*

$$\lim \sup [\varepsilon_n - \arg \min Q(p_n)] \subset \arg \min Q^*$$

within X.

The role of (9) is explained by

3 Proposition *For any admissible perturbation $Q(\cdot)$ to Q_0, among the three conditions (9), (10), (11) relative to $Q(\cdot)$, every two imply the third one.*

By proposition 3 we see that if Q_0 is relaxable (that is, (10) holds), then (9) is a necessary and sufficient condition to value Hadamard well- posedness. Conversely, assuming such a well-posedness, then (9) is a necessary and sufficient condition to relaxability. In this sense, relaxability and property (9) are the minimal conditions to get value Hadamard well-posedness.

When is condition (9) true?
If we assume further that

$$f(x, \cdot) \text{ is upper semicontinuous at } p = 0 \text{ for every } x \in \Omega,$$

then of course (9) obtains if

(13)   for every y fulfilling (3) and every $\varepsilon > 0$, there exists $\delta > 0$ such that for every p with $0 < \text{ dist } (0,p) < \delta$ we can find a p - admissible trajectory z such that $f(z,p) \leq f(y,p) + \varepsilon$.

Condition (13) amounts to the existence of viable solutions (see Aubin-Cellina [1,ch.IV]) to the perturbed differential inclusion

$$\dot{z}(t) \in D[t, z(t), p]$$

for small $p \neq 0$, with viability domain defined by

$$z(t) \in H(t,p) \text{ and } f(z,p) \leq f(y,p) + \varepsilon.$$

From a Berge - type theorem (proposition IX.2) we know that (9) obtains if f is upper semicontinuous in $\Omega \times \{0\}$ and $S : P \rightrightarrows C^0([0,T])$ given by

$$S(p) = \begin{cases} \text{set of p - admissible trajectories, } p \neq 0, \\ \text{set of admissible trajectories for } Q^*, p = 0 \end{cases}$$

is lower semicontinuous at $p = 0$.

More direct sufficient conditions to (9) are the following:
f is upper semicontinuous in $\Omega \times \{0\}$; there exist $y \in \arg \min Q^*$ and $a > 0$ such that

$$y(t) + aB \subset H(t,p) \text{ for every } t \in E \text{ and } p \neq 0;$$

the set of solutions to

(14) $$\dot{x}(t) \in D[t, x(t), p], \text{ a.e. } t \in [0,T]; x(0) = x^0$$

is lower semicontinuous at $p = 0$ as a multifunction from P to $C^0([0,T])$; the Filippov-Ważewski theorem holds for $p = 0$ (Clarke [3, cor. p.117]).

Proof. By upper semicontinuity of f and Filippov-Ważewski theorem, given $\varepsilon > 0$ there exists a solution to (14) with $p = 0$, such that

(15) $$f(x,0) \leq f(y,0) + \varepsilon, \quad \|x - y\| \leq a/2$$

(norm of $C^0([0,T])$). By lower semicontinuity, for every $p \neq 0$ there exists a solution z_p to (14) such that $z_p \to x$ in $C^0([0,T])$ as $p \to 0$. Then for small p

(16) $$\|x - z_p\| \leq a/2.$$

Therefore, for all $t \in E$
$$z_p(t) \in y(t) + aB \subset H(t,p).$$

So, for small p, the trajectory z_p is p - admissible. Therefore by (16) and (15)

$$\limsup v(p) \leq \limsup f(z_p, p) \leq f(y,0) + \varepsilon \leq v^* + \varepsilon.$$

□

4 Example. Minimize $-x(1)$ subject to

$$x(0) = 0, \quad \dot{x} = u - p, \quad |u| \leq 1, \quad x(1) = 0 \text{ or } |x(1) - 1| \leq p^\alpha.$$

Here the parameter $p \in [0,1)$ and $\alpha > 0$. Of course $v^* = -1 = v(0)$, i.e. the unperturbed problem (corresponding to $p = 0$) is relaxable. If $\alpha > 1$ there is no feasible control driving the state to fulfill $1 - x(1) \leq p^\alpha$. Therefore $v(p) = 0$ if $p > 0$, and the unperturbed problem is value Hadamard ill-posed. Of course condition (9) fails here. If $\alpha \leq 1$ then $v(p) = p - 1$, so Hadamard well-posedness obtains, and (9) is obviously fulfilled.

5 Example. The original problem is defined by minimizing

$$\int_0^1 (x^2 - u^2) dt$$

subject to

$$\dot{x} = u, x(0) = 0, |u| \leq 1, x(t) = 0 \text{ for all } t \in [0,1].$$

This problem is ill-posed, since for the perturbed constraint

$$|x(t)| \leq p, \quad p > 0, \ 0 \leq t \leq 1,$$

(same performance and dynamics) we get

$$v(p) \rightarrow -1 \text{ as } p \rightarrow 0,$$

while of course $v(0) = 0$. As a matter of fact, by taking $u_n(t) = \text{sgn}\,[\sin(\pi nt)]$ we see that

$$v^* \leq -1 < v(0) = 0,$$

i.e. the problem is not relaxable.

6 Example. Consider the problem, to minimize $f(x)$ subject to the state equation

$$x(0) = x_0, \dot{x} = g(t, x, u) + h(t, x, u, p), \text{ a.e. in } [0,T],$$

the control constraint $u \in U$, and the state constraint

$$x(t) \in H(p) \text{ for all } t \in [0,T], p \in P.$$

Assume the following conditions:
f is continuous with respect to uniform convergence; $g(\cdot), h(\cdot, p)$ are Carathéodory functions on $[0,T] \times R^m \times U$ for every p; $h(\cdot, 0) = 0$;

$$|g(t, x', u) - g(t, x'', u)| \leq L(t)|x' - x''| \text{ for every } t, x', x'', u$$
$$\text{and some } L \in L^1(0,T); g(\cdot, 0, 0) \in L^1(0,T);$$

$$|h(t, x, u, p)| \leq a_p(t) \text{ with } \int_0^T a_p\,dt \rightarrow 0 \text{ as } p \rightarrow 0;$$

for every measurable u and $p \in P$ the state equation defines uniquely $x \in W^{1,1}(0,T)$; for every $\varepsilon > 0$ there exists $\delta > 0$ such that

$$H(0) + \varepsilon B \subset H(p)$$

if $0 < \text{dist}(0,p) < \delta$.
Then condition (9) holds. To see this, take

$$D_0(t, x) = g(t, x, U).$$

Then Filippov - Ważewsky theorem applies (see Clarke [3, cor. p.117]: here the admissible trajectories for Q_0 are uniformly bounded). Let y be an admissible trajectory for the relaxed problem Q^*. Then for every $\alpha > 0$ there exist a feasible control u_α, and x_α such that

$$\dot{x}_\alpha(t) = g[t, x_\alpha(t), u_\alpha(t)] \text{ a.e. in } [0,T], x_\alpha(0) = x_0,$$

and

$$\max\{|x_\alpha(t) - y(t)| : 0 \le t \le T\} \le \alpha.$$

Now apply the control u_α to the perturbed state equation. For every p we then get $z_{\alpha p}$ such that

$$|z_{\alpha p}(t) - x_\alpha(t)| \le \int_0^t L(s)|z_{\alpha p}(s) - x_\alpha(s)|ds + \int_0^T a_p ds.$$

Therefore for all t and p, by Gronwall's inequality

$$|z_{\alpha p}(t) - y(t)| \le c_p + \alpha, \text{ where } c_p = (\int_0^T a_p ds)e^{\int_0^T L ds}.$$

Given $\varepsilon > 0$, by the continuity of f at y we get

$$f(z_{\alpha p}) \le f(y) + \varepsilon$$

for small p and α. Moreover $z_{\alpha p}$ is $p-$ admissible since

$$z_{\alpha p}(t) \in H(0) + (\alpha + c_p)B \subset H(p),$$

whence (9). If morever $H(\cdot)$ is closed at $p = 0$, we see that the perturbation is admissible. As a particular case, suppose that no state constraint exists, i.e. $H(p) = R^m$. Then by Filippov - Ważewski theorem and the continuity of f, the unperturbed problem is relaxable. Therefore by theorem 1, the above control problem is value Hadamard well-posed.

PROOFS.

Theorem 1. Given any admissible perturbation $Q(\cdot)$ consider any sequence $p_n \ne 0, p_n \to 0$ in P and put

$$f_n(\cdot) = f(\cdot, p_n), \ f_0(\cdot) = f(\cdot, 0);$$

$$x \in K_n \text{ iff } x \text{ is } p_n - \text{ admissible};$$

$$y \in K^* \text{ iff } y \text{ fulfills (3)}.$$

Notice that both $D(t, x, p)$ and $H(0)$ are closed sets. Finally put

$$I_n = f_n + \text{ ind } K_n, \ I^* = f_0 + \text{ ind } K^*, X = (W^{1,1}(0, T), \text{ weak }).$$

Steps of the proof:

(17) $$\text{var} - \lim I_n = I^* \text{ in } X;$$

(18) there exists in X a relatively compact sequence $x_n \in 1/n - \text{ arg min } (X, I_n)$.

Assume that (17), (18) have been proved. Then by theorem IV. 5

(19) $$v(p) \to v^*$$

since

$$v(p_n) = \text{ val } (K_n, I_n), \ v^* = \text{ val } (K^*, I^*).$$

Then by (19), (10)$\Rightarrow$(11). Conversely (12) implies that $v^* \geq v(0)$, but the opposite inequality holds. This will give the required conclusions. So we need to show (17) and (18).

Proof of (17). Assume that $x_n \to x$ in X. Then $x_n \to x$ uniformly on $[0, T]$. Moreover by (6)

$$\liminf f_n(x_n) \geq f_0(x).$$

If $x \in K^*$ then

$$\liminf I_n(x_n) = \liminf[f_n(x_n) + \text{ ind }(K_n, x_n)] \geq \liminf f_n(x_n) \geq f_0(x) = I^*(x).$$

If $x \notin K^*$ we want to show that $I_n(x_n) \to +\infty$. It suffices to show that

(20) $$x_n \notin K_n \text{ for all large } n.$$

We shall need the following

    7 Lemma Lim sup $K_n \subset K^*$ (in X).

Proof of lemma 7. Suppose that $y_n \in K_n$ for some subsequence, and $y_n \to y$ in X. Then

$$\dot{y}_n(t) \in D[t, y_n(t), p_n], \text{ a.e. } t \in (0, T); \; y_n(0) = x_0; \; y_n(t) \in H(p_n), t \in E.$$

Since y_n is uniformly bounded, by (8), given $\varepsilon > 0$, for all large n we find $c_n \in L^1(0, T)$ such that

(21) $$D(t, y_n(t), p_n) \subset D(t, y_n(t), 0) + c_n(t)B, \int_0^T c_n dt \leq \varepsilon.$$

Now consider the support function

$$s(t, x, q) = \text{ supp } [D(t, x, 0), q], q \in R^m.$$

From (21), for every $q \in R^m$

$$s[t, y_n(t), q] - q' \dot{y}_n(t) \geq -|q| c_n(t).$$

By the measurability of $D(\cdot, x, 0), s(\cdot, y_n(\cdot), q)$ is measurable as well. Therefore for every $t \in [0, T]$

(22) $$\int_0^t \{s[l, y_n(l), q] - q' \dot{y}_n(l)\} dl \geq -|q| \varepsilon.$$

By upper semicontinuity of $D(l, \cdot, 0), s(l, \cdot, q)$ is upper semicontinuous as well for a.e. $l \in (0, T)$. Therefore by (22)

$$-\varepsilon |q| \leq \limsup \int_0^t [s(l, y_n(l), q) - \dot{y}_n(l)' q] dl \leq$$

$$\leq \int_0^t [s(l, y(l), q) - q' \dot{y}(l)] dl,$$

since by (8) $D(l, y_n(l), 0)$ are uniformly bounded by some fixed integrable function. Thus for a.e. t and every $q \in R^m$

$$s(t, y(t), q) \geq q'\dot{y}(t)$$

whence

$$\dot{y}(t) \in \text{ co } D(t, y(t), 0).$$

But $y(t) \in H_0$ for $t \in E$ by (7), yielding $y \in K^*$. $\qquad \square$

End of proof of (17). Lemma 7 shows that $x \notin K^*$ and $x_n \to x$ imply (20). Therefore $I_n(x_n) = +\infty$ for all large n, yielding IV. (6). Now we prove IV. (9). Let $y \in K^*$ (the only case of interest). Given $\varepsilon > 0$, by (9) for every large n there exists a p_n - admissible trajectory x_n such that

$$I_n(x_n) = f_n(x_n) \leq f(y, 0) + \varepsilon = I^*(y) + \varepsilon$$

for n sufficiently large. Then IV. (9) holds, and (17) is completely proved.

Proof of (18). Let x_n fulfill

$$I_n(x_n) \leq v(p_n) + 1/n.$$

Since $v(p) < +\infty$ for all p, we get $x_n \in K_n$, whence

$$\dot{x}_n(t) \in D(t, x_n(t), p_n), \text{ a.e. } t, x_n(0) = x_0.$$

By (8) and Gronwall's lemma, x_n are uniformly bounded. Again by (8), we can find a sequence $a_n \to 0$ in $L^1(0, T)$ such that

$$\dot{x}_n(t) \in D(t, x_n(t), 0) + a_n(t)B,$$

where B is the unit ball in R^m. Hence by (8)

$$|\dot{x}_n(t)| \leq z(t) + a_n(t)$$

for all n, a.e. t, some $z \in L^1(0, T)$. Then (18) follows by Dunford-Pettis theorem (Dunford-Schwartz [1, IV.8.11]). $\qquad \square$

Corollary 2. Remember (17) and IV. (14). $\qquad \square$

Proposition 3. Remembering theorem 1, we need only to prove that if (9) fails, then (10) and (11) cannot hold simultaneously. Since (9) fails, there exist $\varepsilon > 0$ and a sequence $p_n \to 0$ such that

$$v(p_n) \geq v^* + \varepsilon.$$

If (10) and (11) were to hold together, then $v(p_n) \to v^*$, giving a contradiction. $\qquad \square$

Section 2.   An Abstract Approach.

We consider a convergence space X, two nonempty subsets E and F thereof, and two functions

$$I : E \to (-\infty, +\infty), J : F \to (-\infty, +\infty).$$

We shall say that (F, J) is a *weak variational extension* of (E, I) iff the following conditions hold:

(23)   there exists a continuous mapping $\theta : E \to F$ such that $cl\ \theta(E) \subset F$;

(24) $J[\theta(x)] \leq I(x)$ for every $x \in E$;

(25)   for every $y \in F$ there exists a sequence $x_n \in E$ such that lim sup $I(x_n) \leq J(y)$.

The link between relaxation stability and value Hadamard well-posedness can be presented in an abstract way as follows. We are given a variational pair (E, I) and a variational extension (F, J) thereof (i.e. the original optimization problem without constraints and its relaxed version). A further nonempty closed subset $H_0 \subset X$ is considered (i.e. the constraint comes in the picture). A suitable perturbation

$$H : P \rightrightarrows X$$

is introduced, where P is a metric space with a fixed element 0 (as in section 1), such that

$$H(0) = H_0$$

The question is whether the constrained problem is relaxable, i.e. does

$$\text{val}\ (F \cap H_0, \mathcal{J}) = \text{val}\ (E \cap H_0, I),$$

and how this property relates to value Hadamard well-posedness , i.e. does

$$\text{val}\ (E \cap H(p), I) \to \text{val}\ (E \cap H(0), I) \text{ as } p \to 0?$$

We say that (F, J) is a *strong variational extension* of (E, I) iff it is a weak one and morever x_n in (25) can be chosen such that

$$\theta(x_n) \to y.$$

8 Proposition *Let (F, J) be a weak variational extension of (E, I), with J lower semicontinuous. Then*

(26)   *val $(F, J) =$ val (E, I),*

and if x_n is any minimizing sequence to (E, I), then every cluster point of $\theta(x_n)$ belongs to arg min(F, J). If moreover (F, J) is a strong variational extension of (E, I), and

$$\{x \in E : I(x) \leq a\}$$

240

is sequentially compact for some $a > $ val (E, I), then arg min (F, J) agrees with the set of all cluster points of $\theta(x_n)$, x_n any minimizing sequence to (E, I).

Proposition 8 explains why this notion of (weak) variational extension is useful in this context. The next theorem generalizes (in part) the above results, showing in particular that suitable perturbations, acting on the constraints only, reveal ill-posedness of non relaxable problems. We shall work with the following assumption

(27) if $p_n \to 0$, $x_n \in E \cap H(p_n)$ and sup $I(x_n) < +\infty$, then $\theta(x_n)$ has at least one subsequence converging to a point in $H(0)$.

9 Theorem *Assume (27) and let (F, J) be a weak variational extension of (E, I) with J lower semicontinuous. Then every two of the following conditions imply the third one:*

(28) $\qquad\qquad$ val $(E \cap H(p), I) \to$ val $(E \cap H(0), I)$ as $p \to 0$;

(29) $\qquad\qquad$ val $(F \cap H(0), J) = $ val $(E \cap H(0), I)$;

(30) $\qquad$ *for every $y \in F \cap H(0)$ and $\varepsilon > 0$, for every sufficiently small $p \neq 0$*
$\qquad$ *there exists $x \in E \cap H(p)$ such that $I(x) \leq J(y) + \varepsilon$.*

PROOFS.

Proposition 8. Put

$$v = \text{ val } (E, I), \quad v^* = \text{ val } (F, J).$$

By (24) we see that $v^* \leq v$. Suppose now $v^* = -\infty$. Then for every $a > 0$ there exists $y \in F$ such that $J(y) \leq -a$. By (25) there exists a sequence $x_n \in E$ such that

$$v \leq \limsup I(x_n) \leq -a,$$

therefore $v = -\infty$. Suppose $v^* > -\infty$. Then given $a > 0$, pick $y \in F$ such that $J(y) \leq v^* + a$. Then letting x_n as in (25) we get $v \leq v^* + a$, hence $v \leq v^*$ and (26) is proved. Now let x_n be a minimizing sequence for (E, I) such that $\theta(x_n) \to y$ for some subsequence. Then $y \in$ cl $\theta(E)$ and therefore $y \in F$ by (23) . So by (24)

$$v^* = v \leq J(y) \leq \liminf J[\theta(x_n)] \leq \lim I(x_n) = v^*,$$

whence $J(y) = v^*$. Finally, let (F, J) be a strong extension. By coercivity, if x_n is a minimizing sequence for (E, I) then for some subsequence and by (23) we see that $\theta(x_n)$ has cluster points, and each of them belongs to arg min (F, J). Now let $y \in$ arg min (F, J). Take x_n as in (25) such that $\theta(x_n) \to y$. Then

$$v \leq \liminf I(x_n) \leq \limsup I(x_n) \leq J(y) = v^* = v$$

thus x_n is a minimizing sequence for (E, I). $\qquad\square$

Theorem 9. Assuming (30) we show that (28)$\Leftrightarrow$ (29). It suffices to show that, as $p \to 0$,

$$v(p) = \text{ val } (E \cap H(p), I) \to v^* = \text{ val } (F \cap H(0), J).$$

Fix $p_n \to 0$, $p_n \neq 0$, and put

$$v_n = v(p_n).$$

Suppose that $F \cap H(0) = \emptyset$. Then $v^* = +\infty$. If $\liminf v(p_n) < +\infty$, then for a subsequence and some $x_n \in E \cap H(p_n)$ we would have $I(x_n) \leq$ constant, therefore by (27) $\theta(x_n) \to z$, say, for some further subsequence, and

$$z \in [\text{ cl } \theta(E)] \cap H(0) \subset F \cap H(0),$$

a contradiction. So $v(p) \to v^*$ in this case. Now let $F \cap H(0) \neq \emptyset$. From (30) we know that $E \cap H(p) \neq \emptyset$ whenever $p \neq 0$ is sufficiently small. Pick $x_n \in E \cap H(p_n)$ such that

$$I(x_n) \leq v_n + 1/n \text{ if } v_n > -\infty, I(x_n) \leq -n \text{ if } v_n = -\infty.$$

By (30), $v_n \leq$ constant. So by (27), $\theta(x_n) \to y$ for some subsequence. Then remembering (23), (24), (27)

$$\liminf v_n \geq \liminf J[\theta(x_n)] \geq J(y) \geq v^*,$$

while by (30)

$$\limsup v_n \leq v^*,$$

hence $v_n \to v^*$. Therefore

$$v(p) \to v^*.$$

To end the proof, assume that (30) fails. Then we can find $y \in F \cap H(0), \varepsilon > 0$ and $p_n \to 0$ such that

$$I(x) > J(y) + \varepsilon$$

for every $x \in E \cap H(p_n)$. Therefore

$$v_n \geq v^* + \varepsilon$$

for all n sufficiently large. But then (28) and (29) cannot hold simultaneously. $\square$

Section 3.   Value Convergence of Penalty Function Methods.

We consider again problem Q_0 of section 1, to minimize $f(x)$ subject to

(31) $x \in W^{1,1}(0, T); \ \dot{x}(t) \in D[t, x(t)], \text{ a.e. } t \in [0, T], x(0) = x_0;$

(32) $x(t) \in H \text{ for all } t \in E.$

As before, E is a given nonempty subset of $[0, T]$, while H and $D(t, x)$ are subsets of R^m. Let

$$g : C^0([0, T]) \to [0, +\infty)$$

be a given continuous function with respect to the uniform topology, such that

$$g(x) = 0 \text{ iff } x(t) \in H \text{ for all } t \in E.$$

For example if H is closed one can take

$$g(x) = \sup \{ \text{dist } (x(t), H) : t \in E \}$$

for any fixed norm in R^m.

The penalty function method for reduction of the state constraints (31) we shall consider here, consists in replacing Q_0 by the parameterized *penalized problem*, to minimize

$$p(\varepsilon, x) = f(x) + \varepsilon g(x), \ \varepsilon > 0$$

subject to (31). Denote by

$$v(\varepsilon), \varepsilon > 0,$$

the optimal value of the penalized problem, and by v that of Q_0. A form of well-posedness of Q_0 is defined by the convergence

$$v(\varepsilon) \to v \text{ as } \varepsilon \to +\infty.$$

In the setting of theorem 1 we have state constraints

$$H(\varepsilon) = R^m \text{ if } \varepsilon > 0, H(+\infty) = H,$$

so that $H(\cdot)$ is not closed (except in the trivial case $H = R^m$). Thus theorem 1 cannot be applied. However an analogous result can be obtained by employing the *relaxed problem*, to minimize $f(y)$ subject to

(33) $$y(t) \in \text{ cl } H \text{ for all } t \in E;$$

(34) $$y \in W^{1,1}(0, T); \ \dot{y}(t) \in \text{ co } D[t, y(t)], \text{ a.e. } t \in [0, T], y(0) = x_0;$$

with optimal value v^*. Here $f : \Omega \to (-\infty, +\infty)$, where Ω is an open subset of $C^0([0, T])$ containing all trajectories of (34).

We shall use the following assumption:

(35) $D(t, x)$ is nonempty and compact for all $(t, x); D(\cdot, x)$ is measurable for all $x; D$ is locally integrably Lipschitz; there exist $a, b \in L^1(0, T)$ such that for every t and x

$$|\omega| \leq a(t)|x| + b(t) \text{ if } \omega \in D(t, x).$$

Relaxability is again equivalent to value convergence of the penalty method.

10 Theorem *Suppose that f is continuous, H is closed and (35) holds. Then*

$$v(\varepsilon) \to v \text{ as } \varepsilon \to +\infty \text{ iff } v = v^*.$$

Proof. It suffices to show that

(36) $$v(\varepsilon) \to v^* \text{ as } \varepsilon \to +\infty.$$

Case 1: $v^* = +\infty$.

Arguing by contradiction, there exists a sequence $\varepsilon_n \to +\infty$ such that

$$\limsup v(\varepsilon_n) < +\infty.$$

Then for each n there exists a solution x_n to (31) such that

$$(37) \qquad\qquad \limsup p(\varepsilon_n, x_n) < +\infty.$$

Then x_n is uniformly bounded by the sublinear growth of $D(t, \cdot)$. In turn this implies (by Dunford-Pettis theorem, see Dunford-Schwartz [1, IV.8.11]) weak sequential compactness of $\dot{x}_n$ in $L^1(0, T)$. Then for some subsequence $x_n \to x$, uniformly on $[0, T]$, for some $x \in W^{1,1}(0, T)$. As in the proof of lemma 7 we see that x solves (34) (notice that $D(t, \cdot)$ is upper semicontinuous for a.e. t, being locally integrably Lipschitz). Since $f(x_n)$ is bounded, we get by (37)

$$\limsup \varepsilon_n g(x_n) < +\infty,$$

hence $g(x) = 0$, i.e. $x(t) \in H$ for all $t \in E$. Then $v^* \le f(x) < +\infty$, a contradiction. Thus (36) is proved in this case.

Case 2: $v^* < +\infty$.

Fix any sequence $\varepsilon_n \uparrow +\infty$. Let y be any solution to (33), (34). By continuity of g at y (and $g(y) = 0$), for every n

$$\{z \in C^0([0, T]) : |g(z)| \le 1/n\varepsilon_n\}$$

contains a neighborhood of y. Then by Filippov-Ważewski theorem (Clarke [1], cor. p.117) there exists a sequence y_n of solutions to (31) such that

$$y_n \to y \text{ uniformly on } [0, T], \quad \varepsilon_n g(y_n) \to 0.$$

Thus $\limsup p(\varepsilon_n, y_n) \le f(y)$, hence

$$(38) \qquad\qquad \limsup v(\varepsilon) \le v^*.$$

Notice that $v(\varepsilon) > -\infty$ for every ε since for some fixed $A \in L^1(0, T)$

$$|\dot{x}(t)| \le A(t), \text{ a.e. } t \in (0, T),$$

for every solution x to (31). Now let x_n be a sequence of solutions to (31) such that

$$(39) \qquad\qquad p(\varepsilon_n, x_n) \le v(\varepsilon_n) + 1/n.$$

As in lemma 7, for some subsequence $x_n \to x$ uniformly on $[0, T]$, and x solves (34). Then by (38), (39)

$$\limsup \varepsilon_n g(x_n) < +\infty,$$

showing that x fulfills (33). Then by (39)

$$v^* \le f(x) = \lim f(x_n) \le \liminf p(\varepsilon_n, x_n) \le \liminf v(\varepsilon_n).$$

Together with (38) this implies (36). $\qquad\qquad\qquad\qquad\qquad\qquad\qquad\qquad \square$

Remark. A proof (similar to that of theorem 1) can be based on variational convergence. Alternatively, one can exploit the increasing behavior of $p(\cdot, x)$ and rely on the example after proposition IV.9. Under relaxability of Q_0 (or equivalently convergence of $v(\varepsilon)$) and the assumptions of theorem 10, any cluster point in $(W^{1,1}(0,T),$ weak) of asymptotically minimizing sequences for the penalized problems is an optimal solution to the relaxed problem.

Section 4. Relaxation and Value Convergence of Discrete Approximations.

We consider a slightly simplified version Q_0 of the problem in section 1, as follows: minimize $f(x)$ subject to

$$(40) \qquad x \in W^{1,1}(0,T),\ \dot{x}(t) \in D[x(t)] \text{ for a.e. } t \in (0,T), x(0) = x_0,$$

$$x(t) \in H \text{ for all } t \in [0,T].$$

We assume the following conditions (compare with (5), ..., (8)):

$f : \Omega \to (-\infty, +\infty)$ is continuous, where Ω is an open set of $C^0([0,T])$ containing all solutions to (40);

$D(x)$ is compact nonempty for all x; H is closed ; D is Lipschitz on the bounded sets of R^m; there exist constants a, b such that

$$|\omega| \le a|x| + b$$

whenever $\omega \in D(x),\ x \in R^m$.

Moreover, we suppose that the relaxed problem Q^*, in which $\dot{x} \in$ co $D(x)$, has admissible trajectories.

Now we approximate Q_0 by finite-dimensional problems. Let q be a fixed positive constant, $\{t_i\}$ be an uniform grid (for simplicity), with

$$t_{i+1} - t_i = T/N = h,\ N = 2, 3, 4, \dots.$$

Consider the following discrete approximation Q_N to Q_0 : to minimize $f(x)$ subject to

$$(41) \qquad \begin{cases} x \in C^0([0,T]) \text{ and piecewise linear across } \{t_i\}; \\ x(t_{i+1}) \in x(t_i) + hD[x(t_i)], i = 0, 1, ..., N-1, \\ x(0) = x_0; \end{cases}$$

$$x(t_i) \in H + qhB, i = 0, 1, ..., N,$$

where B is the closed unit ball in R^m.

Let v be the optimal value of Q_0 and v^N that of Q_N. We are interested in the property

$$v^N \to v \text{ as } N \to +\infty$$

which may be regarded as a form of well-posedness of the original problem Q_0. We will prove the following

11 Theorem *There exists a constant $q > 0$ such that the following are equivalent:*

$$v = v^* \ (i.e. \ Q_0 \ is \ relaxable);$$

$$v^N \to v \ as \ N \to +\infty.$$

In the proof we need the following two lemmas.

12 Lemma Every sequence x^N of solutions to (41) has a subsequence which is uniformly convergent as $n \to +\infty$ to some trajectory of the relaxed inclusion

$$\dot{x}(t) \in \ \text{co} \ D[x(t)], \ \text{a.e.} \ t \in (0, T).$$

Proof. We have

$$|x^N(t_{i+1})| \le (1 + ha)|x^N(t_i)| + hb.$$

Applying the discrete Gronwall lemma, we see that x^N is uniformly bounded. Then it is equi-Lipschitz. By a standard reasoning (similar to that in the proof of lemma 7) we get the conclusion. □

13 Lemma There exists a constant $C > 0$ such that for every solution x to (40) and for every $N = 2, 3, ...$ there is a solution x^N to (41) such that

$$\max \ \{|x^N(t) - x(t)| : 0 \le t \le T\} \le C/N.$$

Proof. Fix $p \in R^m$. Then for every $t \in [0, T - h]$ and suitable constants c_1, c_2 one has

$$[x(t + h) - x(t)]'p \le \int_t^{t+h} \text{supp} \ (D[x(s)], p) ds$$

$$\le \int_t^{t+h} \{ \text{supp} \ (D[x(s)], p) - \ \text{supp} \ (D[x(t)], p)\} ds + h \ \text{supp} \ (D[x(t)], p)$$

$$\le h \ \text{supp} \ (D[x(t)], p) + c_1|p| \int_t^{t+h} |x(s) - x(t)| ds$$

$$\le h \ \text{supp} \ (D[x(t)], p) + c_2|p|h^2.$$

We used the fact that since $D(\cdot)$ is Lipschitz then $\ \text{supp} \ (D(\cdot), p)$ is Lipschitz as well, and that x is Lipschitz on $[0, T]$. Then for every $p \in R^m$

$$\text{supp} \ (-x(t + h) + x(t) + hD[x(t)] + c_2h^2B, p) \ge 0,$$

which means that for every $t \in [0, T - h]$

(42) $$x(t + h) \in x(t) + hD[x(t)] + c_2h^2B.$$

Now let us construct the trajectory x^N. Take $x^N(0) = x_0$. Suppose that $x^N(t_i)$ is already obtained for some $i, 0 \le i \le N - 2$. Then we take $x^N(t_{i+1})$ as any nearest point to $x(t_{i+1})$ from the set

$$x^N(t_i) + hD[x^N(t_i)].$$

Clearly x^N satisfies (41). Then it is uniformly bounded (see the proof of lemma 12). Furthermore, we have from (42), using the Lipschitz continuity of $D(\cdot)$,

$$(43) \qquad x(t_{i+1}) \in x(t_i) - x^N(t_i) + x^N(t_i) + hD[x^N(t_i)] + c_3 h(|x(t_i) - x^N(t_i)| + h)B,$$

for a suitable constant c_3 independent of N. The following fact is obvious: if $u \in A + \rho B, A \subset R^m, \rho > 0$, and ω is any nearest point to u from A, then $|u - \omega| \le \rho$. Applying this conclusion to (43) with

$$u = x(t_{i+1}), A = x^N(t_i) + hD[x^N(t_i)], \omega = x^N(t_{i+1}),$$

$$\rho = (1 + c_3 h)|x(t_i) - x^N(t_i)| + c_3 h^2$$

we obtain for $i = 0, 1, ..., N - 1$

$$|x^N(t_{i+1}) - x(t_{i+1})| \le (1 + c_3 h)|x(t_i) - x^N(t_i)| + c_3 h^2.$$

Hence, by the discrete Gronwall lemma

$$\max\{|x^N(t_i) - x(t_i)| : 0 \le i \le N\} \le h \text{ (constant).}$$

It remains to use the Lipschitz continuity of x and x^N (uniformly in N) and this completes the proof. $\qquad \square$

Proof of theorem 11. Let x^* be any optimal trajectory of the relaxed problem Q^*, which exists because of the compactness of the set of the trajectories in $C^0([0, T])$. Fix $\varepsilon \in (0, 1)$. Applying the Filippov-Ważewski theorem for every N (Clarke [3 cor. p.117]) we choose a trajectory $\bar{x}_\varepsilon^N$ of (40) such that

$$\|\bar{x}_\varepsilon^N - x^*\| \le \varepsilon/N.$$

Now apply lemma 13 and choose a solution x_ε^N to (41) such that

$$\|x_\varepsilon^N - \bar{x}_\varepsilon^N\| \le c/N.$$

Then, taking $Tq = c + 1$ we obtain

$$x_\varepsilon^N(t_i) \in H + qhB, \ i = 0, 1, ..., N.$$

Hence (for such q), $v^N \le f(x_\varepsilon^N)$, and since $x_\varepsilon^N \to x^*$ uniformly as $N \to +\infty$ we get

$$(44) \qquad \limsup_{N \to +\infty} v^N \le v^*.$$

Notice that (44) implies that the problem Q_N has admissible trajectories. Choose a sequence x^N of such trajectories that satisfies

$$f(x^N) \le v^N + 1/N, N = 1, 2, ...$$

Applying lemma 12 we obtain

$$v^* \le \liminf f(x^N) = \liminf v^N.$$

Hence from (44)

$$v^N \to v^* \text{ as } N \to +\infty.$$

This completes the proof. $\qquad \square$

Section 5.   Notes and Bibliographical Remarks.

To section 1. The (conventional) relaxed problem was introduced in this context by Warga [1]. Ill-posedness of nonrelaxable problems is alluded to in Clarke [3, p.223 - 224]. For related results see Clarke [1]. The precise link between relaxability and value Hadamard well - posedness in optimal control problems was observed first by Dontchev-Morduhovič [1], where example 4 is presented. Results related to theorem 1, corollary 2 and proposition 3 are given in Dontchev [3] and Zolezzi [13]. Further results linking relaxability to well - posedness are given in Clarke [1] and Dontchev [2]. Related results in an infinite - dimensional setting are in Avgerinos-Papageorgiou [2] and Papageorgiou [1].

To section 2. The definition of variational extension is a modification of a concept introduced by Ioffe - Tikhomirov [1]. Proposition 8 may be found in Valadier [1]. Theorem 9 is related to the results of Zolezzi [13].

To section 3. Theorem 10 is due to Dontchev [3].

To section 4. Mordukhovich [1] firstly observed the equivalence between relaxability and value convergence of discrete approximations to optimal control problems. For extensions in the direction of nonsmooth analysis and approximate optimality conditions, see Mordukhovich [3]. Theorem 11 is due to Dontchev [3]. A similar result appeared independently in Mordukhovich [2]. Lemma 12 is generalized in Taubert [1] for a class of multistep methods. The method in the proof of lemma 13 is due to Pshenichny [1]. The (Hausdorff) distance (in an appropriate metric) between the sets of solutions to a differential inclusion and its discrete approximations is estimated in Dontchev-Farchi [1] by an averaged modulus of smoothness for multifunctions (see Sendov - Popov [1]).

Chapter VII.

SINGULAR PERTURBATIONS IN OPTIMAL CONTROL

Section 1.   Introduction and Example.

In this chapter we investigate the behaviour of the optimal value and controls for linear systems containing a scalar parameter in the derivatives of some of the states. Consider the following linear control system

(1)
$$\begin{cases} (1.1)\ \dot{x} & = A_1(t)x + A_2(t)y + B_1(t)u(t), \\ (1.2)\ \varepsilon\dot{y} & = A_3(t)x + A_4(t)y + B_2(t)u(t), \end{cases}$$

where $\varepsilon \geq 0, t \in [0,T], T$ with $0 < T < +\infty$ is fixed, $x \in R^n, y \in R^m, u \in R^k, A_i(t)$ and $B_j(t)$ are matrices of appropriate dimensions. Initial conditions are given by

(2)
$$\begin{cases} (2.1) & x(0) = x^0, \\ (2.2) & y(0) = y^0. \end{cases}$$

Throughout this chapter we assume that *the entries of A_i and B_j are $C^2([0,T])$* (although some of the results can be obtained under weaker conditions).

The perturbation represented by ε is called *singular* because for $\varepsilon = 0$ the differential equation (1.2) degenerates to a nondifferential one:

(3)
$$0 = A_3(t)x + A_4(t)y + B_2(t)u(t), \quad \text{a.e. } t \in (0,T).$$

The singularly perturbed systems can be regarded as modelling processes that run in different time scales. Indeed, if we change the time scale in (1.2) by setting in (1.2) $\tau = t/\varepsilon$ then $\varepsilon\dot{y} = \varepsilon dy/dt = dy/d\tau$ and the (small) parameter in the derivative disappears. In the world of y the time τ is running $1/\varepsilon$ times faster than the time t in the world of x. Therefore x is called "slow state" while y is the "fast state".

In this chapter we show that changing ε from a positive value to zero is a "singular" procedure, not only from a technical point of view. We see in the following example a standard singularly perturbed optimal control problem that is not value Hadamard well-posed.

1 Example. Consider the electric circuit in the figure $a)$ below (next page).

We suppose that the resistances ε are small. When neglected, one comes to the simplified circuit in the figure $b)$ below. The circuit is controlled by the imput voltage w. Using the voltages v_1 and v_2 as states and applying some standard physical laws we obtain the system

$$\varepsilon\dot{v}_1 = (1/C_1)\,(-v_1 + v_2),\ \varepsilon\dot{v}_2 = (1/C_2)\,(v_1 - 2v_2 + w).$$

In order to simplify the calculations we change the state space by taking

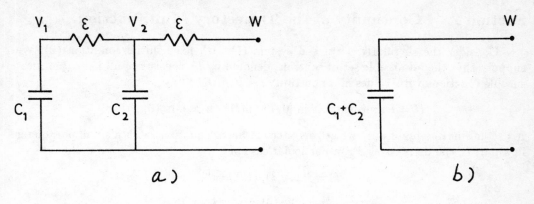

a)   b)

$$y_1 = (1/5)\,(v_1 + v_2),\, y_2 = (1/5)\,(-2v_1 + 3v_2),$$

and let $u = (2/5)w, C_1 = 1/3, C_2 = 1/2$. We get

$$\varepsilon \dot{y}_1 = -y_1 + u,\, \varepsilon \dot{y}_2 = -6y_2 + 3u.$$

Let $y_1(0) = y_2(0) = 0$ and the feasible controls be all measurable functions on $[0,1]$ with values in $[-1,1]$. Consider the optimal control problem, to minimize

$$I(u) = y_1(1)^2 + (y_2(1) + 0.5)^2.$$

For $\varepsilon = 0$ we have

$$y_1 = u,\, 2y_2 = u$$

and a formal substitution in I gives the following one-dimensional problem: to minimize

$$[u^2 + (0.5u + 0.5)^2]$$

subject to $u \in [-1,1]$. Its optimal value is $v(0) = 0.2$. Now take the control

$$u_\varepsilon(t) = \begin{cases} 1 & \text{for } 0 \le t \le 1 + \varepsilon \ln 0.5, \\ -1 & \text{for } 1 + \varepsilon \ln 0.5 < t \le 1. \end{cases}$$

Calculating the corresponding trajectory we obtain

$$I(u_\varepsilon) = 2^{-12} + 0(\varepsilon).$$

Thus, for the optimal value function $v(\varepsilon)$ we have, as $\varepsilon \to 0$,

$$\limsup v(\varepsilon) < v(0),$$

i.e. $v(\cdot)$ is not continuous at $\varepsilon = 0$.

We shall show in section 4 that the value function $v(\varepsilon)$ has a limit with $\varepsilon \to 0$ and this limit is the value of a certain minimum problem.

Section 2.   Continuity of the Trajectory Multifunction.

Consider the singularly perturbed system (1) with fixed initial condition (2). We suppose that the admissible set of controls, denoted by U, consists of all Lebesgue measurable functions with values in a nonempty set V in R^k :

(4) $$U = \{u \text{ measurable in } [0,T] : u(t) \in V \text{ a.e. in } (0,T)\}.$$

If V contains one point only we get a system of linear equations with a small parameter in a part of the derivatives, i.e. a particular case of

$$\dot{x} = f(x,y,t), x(0) = x^0,$$

$$\varepsilon \dot{y} = g(x,y,t), y(0) = y^0, t \in [0,T].$$

There is a fundamental theorem known as Tykhonov's theorem, giving a sufficient condition for convergence of solutions of this system as $\varepsilon \to 0$ to the solution of the degenerate system obtained for $\varepsilon = 0$.

Notation. $Re\ \sigma(A)$ is the maximal real part of the eigenvalues of the matrix A.

In the linear case Tykhonov's theorem claims the following.

2 Theorem *Let V consist of one point and $Re\ \sigma(A_4(t)) < 0$ for all $t \in [0,T]$. Then the solution $(x_\varepsilon, y_\varepsilon)$ of (1), (2) for $\varepsilon > 0$ satisfies*

$$x_\varepsilon \to x_0 \text{ uniformly in } [0,T];$$

$$y_\varepsilon \to y_0 \text{ uniformly in } [\delta,T]$$

for any $\delta > 0$, where (x_0,y_0) is the solution of (1.1), (2.1), (3).

We shall obtain theorem 2 from more general results presented in the sequel.

If V contains more than one point we have a control system which can be regarded as a differential inclusion. The behaviour of the (set of) trajectories becomes more sophisticated. As we shall see, generally speaking, Tykhonov's theorem cannot be extended (at least in the above setting) to differential inclusions.

Given $\varepsilon > 0$ and $u \in U$ denote by

$$z[u, \varepsilon] = (x[u,\varepsilon], y[u,\varepsilon])$$

the solution of (1), (2). Let $z[u,0]$ solve (1.1), (2.1), (3). We are interested in the continuity properties of the *trajectory mapping*

$$(u,\varepsilon) \to z[u,\varepsilon] = (x[u,\varepsilon]; y[u,\varepsilon])$$

for $(u,\varepsilon) \in L^1(0,T) \times [0,+\infty)$. In the considered linear case this map is obviously single valued.

Let $\varepsilon_k > 0, u_k \in L^1(0,T), k = 1, 2, ..., \varepsilon_k \to \varepsilon_0$ and $u_k \rightharpoonup u_0$ in $L^1(0,T)$. If $\varepsilon_0 > 0$ it is easy to show that $z[u_k, \varepsilon_k] \to z[u_0, \varepsilon_0]$ in $C^0([0,T])$. Now let $\varepsilon_0 = 0$ and let

$$z[u_k, \varepsilon_k] = (x[u_k, \varepsilon_k], y[u_k, \varepsilon_k]) \to (\bar{x}, \bar{y}) \text{ in } C^0([0, T]) \times (L^1(0, T), \text{ weak }).$$

Denoting $x_k = x[u_k, \varepsilon_k]$ and $y_k = y[u_k, \varepsilon_k]$ and passing to the limit with $k \to +\infty$ in the equality

$$\int_0^T |\int_0^t (A_3 x_k + A_4 y_k + B_2 u_k)\, ds| dt = \varepsilon_k \int_0^T |y_k - y^0| dt$$

we obtain that $\bar{x}, \bar{y}$ and u_0 satisfy (3). It is easy to see that $\bar{x}, \bar{y}$ and u_0 satisfy (1.1), (2.1), hence $(\bar{x}, \bar{y}) = z[u_0, 0]$. Thus we proved

3 Theorem *The trajectory mapping*

$$z = (x, y) : (L^1(0, T), \text{ weak }) \times [0, +\infty) \to C^0([0, T]) \times (L^1(0, T), \text{ weak })$$

is sequentially closed.

Consider now the multifunction "$\varepsilon \rightrightarrows$ set of trajectories of (1), (2)", i.e.

(5) $$\varepsilon \rightrightarrows Z(\varepsilon) = \{z[u, \varepsilon] = (x[u, \varepsilon], y[u, \varepsilon]) : u \in U\}.$$

If V is compact and convex then U is sequentially compact in $(L^1(0, T), \text{ weak })$, hence from theorem 3 we obtain

4 Corollary *Let V be convex and compact. Then the multifunction*

$$Z : [0, +\infty) \rightrightarrows C^0([0, T]) \times (L^1(0, T), \text{ weak })$$

is sequentially closed.

Notation. $Y_\varepsilon(t, s), 0 \le s \le t \le T, \varepsilon > 0$, is the transition matrix of $\varepsilon \dot{y} = A_4(t)y$, principal at $t = s$.

From now on to the end of this section we assume the following:
(6) the matrix $A_4(t)$ is invertible for every $t \in [0, T]$ and

$$\sup \{|Y_\varepsilon(t, s)| : \varepsilon > 0, 0 \le s \le t \le T\} < +\infty.$$

Remark. If A_4 is constant and invertible, then condition (6) is equivalent to Lyapunov stability of $\dot{y} = A_4 y$.

Under (6), by solving (3) with respect to y we obtain the following *reduced system*

(7) $$\left\{ \begin{array}{ll} (7.1) & \dot{x} = A_0(t)x + B_0(t)u(t), x(0) = x^0, \\ (7.2) & y(t) = -A_4^{-1}(t)[A_3(t)x(t) + B_2(t)u(t)], \text{ a.e. } t \in [0, T] \end{array} \right.$$

where

(8) $$A_0 = A_1 - DA_3, B_0 = B_1 - DB_2, D = A_2 A_4^{-1}.$$

In the propositions below, ε_k is an arbitrary sequence of positive numbers, $\varepsilon_k \to 0$, and u_k is any sequence of functions from $L^1(0,T)$ such that $u_k \to u_0$ in $L^1(0,T)$. We put

$$(x_k, y_k) = z[u_k, \varepsilon_k] \text{ and } (x_0, y_0) = z[u_0, 0]$$

i.e. the solution of (7) for $u = u_0$.

5 Proposition

(9) x_k is bounded in $C^0([0,T])$ and $\int_0^t x_k ds \to \int_0^t x_0 ds$ for every $t \in [0,T]$;

(10)   if $u_k = u_0$ for $k = 1, 2, ...$ then $x_k \to x_0$ in $C^0([0,T])$.

Now define the set

(11) $$M = \{t \in [0,T] : Re\ \sigma(A_4(t)) < 0\}.$$

6 Proposition *Assume that*

(12) $$meas\ M = T.$$

Then

(13) $$x_k(t) \to x_0(t) \text{ for every } t \in M;$$

(14) $$\int_0^t y_k ds \to \int_0^t y_0 ds \text{ for every } t \in M;$$

(15)   if $u_k = u_0, k = 1, 2, ...,$ where $u_0 \in L^\infty(0,T)$, then $y_k(t) \to y_0(t)$ for every $t \in M, t > 0$.

7 Proposition *Let $M = [0,T]$ and let $u_k \to u_0$ in $L^p(0,T), 1 \leq p < +\infty$. Then*

(16) $$x_k \to x_0 \text{ in } C^0([0,T]);$$

(17) $$y_k \to y_0 \text{ in } L^p(0,T);$$

(18)   if $u_k = u_0, k = 1, 2, ...,$ where $u_0 \in C^0([0,T])$, then for any $\delta > 0$, $y_k \to y_0$ in $C^0([\delta,T])$.

Remark. Now theorem 2 follows from (16) and (18).

For the trajectory multifunction defined in (5) we obtain

8 Theorem *Under (6) and (12) the multifunction*

$$Z : [0, +\infty) \rightrightarrows C^0([0, T]) \times (L^p(0, T), \ strong\ ), 1 \le p < +\infty,$$

is lower semicontinuous at $\varepsilon = 0$. If V is convex and compact and $M = [0, T]$ then

$$Z : [0, +\infty) \rightrightarrows C^0([0, T]) \times (L^1(0, T), \ weak)$$

is upper semicontinuous at $\varepsilon = 0$.

PROOFS.

Throughout the proofs we shall use many times the following
9 Basic Lemma. Let $-\infty < t_1 \le t_2 < +\infty$ and $A(t)$ be an $n \times n$ matrix with continuous entries on $[t_1, t_2]$ such that for some $\rho > 0$

$$\text{Re } \sigma(A(t)) \le -2\rho \text{ for all } t \in [t_1, t_2].$$

Let $\emptyset(t, \tau, \varepsilon)$ be the transition matrix of $\varepsilon \dot{y} = A(t)y$, principal at $t = s$. Then there exist $\sigma > 0$ and $\varepsilon_0 > 0$ such that

(19) $$|\emptyset(t, s, \varepsilon)| \le \sigma \exp\left(-\rho(t - s)/\varepsilon\right)$$

for all $t_1 \le s \le t \le t_2, 0 < \varepsilon < \varepsilon_0$.

Proof. We prove first that there exists a constant $c > 0$ such that

(20) $$|\exp(A(t)s)| \le c \exp(-\rho s)$$

for all $t_1 \le t \le t_2$ and $s \ge 0$. The columns ψ_i of $\exp(A(t)s)$ have the form $\psi_i(t, s) = P_i^t(s) \exp(\lambda_i^t s)$, where $P_i^t(s)$ are polynomials and λ_i^t are the eigenvalues of $A(t), i = 1, 2, ..., n$. Hence there exists $c_t > 0$ such that

$$|\exp(A(t)s| \le c_t \exp(-2\rho s) \text{ for all } s \ge 0.$$

Take some $t, \tau \in [t_1, t_2]$. Then, for any $s \ge 0$, from

$$\exp(A(\tau)s) = \exp(A(t)s) + \int_0^s \exp(A(t)(s - \theta))[A(\tau) - A(t)] \exp(A(\tau)\theta)d\theta$$

it follows that

$$|\exp(A(\tau)s)| \le c_t \exp(-2\rho s) + c_t \int_0^s \exp(-2\rho(s - \theta))|A(\tau) - A(t)|| \exp(A(\tau)\theta)|d\theta.$$

By Gronwall's lemma, for every $s \ge 0$,

$$|\exp(A(\tau)s)| \le c_t \exp[(-2\rho + c_t|A(\tau) - A(t)|)s].$$

Then, for every τ sufficiently close to t, one can write

$$| \exp (A(\tau)s)| \leq c_t \exp (-\rho s), s \geq 0.$$

By a compactness argument, choosing a finite covering of $[t_1, t_2]$, we obtain (20).

Now we are ready to prove (19). We have

$$\emptyset(t, s, \varepsilon) = \exp (A(t)(t - s)/\varepsilon) + (1/\varepsilon) \int_s^t \exp (A(t)(t - \tau)/\varepsilon)(A(\tau) - A(t))\emptyset(\tau, s, \varepsilon)d\tau$$

for $t_1 \leq s \leq t \leq t_2, \varepsilon > 0$. Denote $\omega(t, s, \varepsilon) = |\emptyset(t, s, \varepsilon)| \exp (\rho(t - s)/\varepsilon)$. Using (20) we obtain

$$\omega(t, s, \varepsilon) \leq c \exp (-\rho(t - s)/\varepsilon) + (c/\varepsilon) \int_s^t \exp (-\rho(t - \tau)/\varepsilon)|A(\tau) - A(t)|\omega(\tau, s, \varepsilon)d\tau.$$

Let $M(\varepsilon) = \sup \{\omega(t, s, \varepsilon) : t_1 \leq s \leq t \leq t_2\}$ and

$$\Delta^t(\varepsilon) = \max\{|A(\tau) - A(t)| : t - \sqrt{\varepsilon} \leq \tau \leq t\}.$$

Clearly, $\Delta^t(\varepsilon) \to 0$ as $\varepsilon \to 0$ uniformly in $t_1 \leq t \leq t_2$. If $\alpha = \max\{t_1, t - \sqrt{\varepsilon}\} \leq s \leq t$ then

$$\omega(t, s, \varepsilon) \leq c + \frac{c\Delta^t(\varepsilon)}{\epsilon} \int_\alpha^t \exp (-\rho\frac{t - \tau}{\varepsilon})\omega(\tau, s, \varepsilon)d\tau \leq c(1 + \frac{\Delta^t(\varepsilon)}{\rho}M(\varepsilon)).$$

Furthermore, if $s < t - \sqrt{\varepsilon}$, then

$$\omega(t, s, \varepsilon) \leq c + (2c/\varepsilon) \max_{t_1 \leq t \leq t_2} |A(t)|M(\varepsilon) \int_s^{t-\sqrt{\varepsilon}} \exp (-\rho\frac{t - \tau}{\varepsilon}d\tau)$$

$$+ (c/\varepsilon)\Delta^t(\varepsilon)M(\varepsilon) \int_{t-\sqrt{\varepsilon}}^t \exp(-\rho\frac{t - \tau}{\varepsilon})d\tau.$$

Combining the last two inequalities we obtain that for every $s, t, t_1 \leq s \leq t \leq t_2$, and $\varepsilon > 0$

$$\omega(t, s, \varepsilon) \leq c(1 + \delta(\varepsilon))M(\varepsilon),$$

where $\delta(\varepsilon) \to 0$ as $\varepsilon \to 0$, uniformly in $t_1 \leq s \leq t \leq t_2$. Taking the supremum on the left, yields the existence of $\varepsilon_0 > 0$ such that $\sup\{M(\varepsilon) : 0 < \varepsilon \leq \varepsilon_0\} < +\infty$. The proof is complete. $\quad\square$

Notation. Below, $0(\varepsilon)$ denotes any function of ε such that as $\varepsilon \to 0$, $\limsup |0(\varepsilon)|/\varepsilon < +\infty$. As before $Y_\varepsilon(t, s)$ is the transition matrix of $\varepsilon\dot{y} = A_4(t)y$, principal at $t = s$. If $\varepsilon_k \to 0$ we denote

$$Y_k(t, s) = Y_{\varepsilon_k}(t, s).$$

We start the *proof of proposition 5* with two lemmas.

10 Lemma Let u_k be bounded in $L^1(0, T)$, and let $\bar{x}_k$ and $\tilde{x}_k$ be defined by

$$\bar{x}_k(t) = x^0 + \int_0^t ((A_0(s) + D(t)Y_k(t, s)A_3(s))\bar{x}_k(s) + B_0(s)u_k(s))ds,$$

$$\frac{d}{dt}\tilde{x}_k = A_0(t)\tilde{x}_k + B_0(t)u_k(t), \tilde{x}(0) = x^0.$$

Then
$$\max\{|\bar{x}_k(t) - \tilde{x}_k(t)| : 0 \le t \le T\} = 0(\varepsilon_k).$$
Proof. By (6) and the boundedness of u_k in $L^1(0,T)$ we conclude that $\bar{x}_k$ and $\tilde{x}_k$ are bounded in $C^0([0,T])$ and $(d/dt)\,\tilde{x}_k$ is bounded in $L^1(0,T)$. Let $\Delta x_k = \bar{x}_k - \tilde{x}_k$. Then

$$\Delta x_k(t) = \int_0^t \{[A_0(s) + D(t)Y_k(t,s)A_3(s)]\Delta x_k(s) + D(t)Y_k(t,s)A_3(s)\tilde{x}_k(s)\}ds$$

$$= \int_0^t [A_0(s) + D(t)Y_k(t,s)A_3(s)]\Delta x_k(s)ds$$

$$- \varepsilon_k \int_0^t D(t)[\partial/\partial s Y_k(t,s)]A_4^{-1}(s)A_3(s)\tilde{x}_k(s)ds.$$

Integrating by parts and using Gronwall inequality we complete the proof. $\qquad\square$

11 Lemma Let u_k be bounded in $L^1(0,T)$ and let (x_k, y_k) be the solution of (1),(2) for $u = u_k$ and $\varepsilon = \varepsilon_k, \varepsilon_k > 0, \varepsilon_k \to 0$. Denote

$$\emptyset_k(t) = \int_0^t D(t)Y_k(t,s)B_2(s)u_k(s)ds.$$

Then there exists a constant $c_1 > 0$ such that for every $t \in [0,T]$

$$|x_k(t) - \tilde{x}_k(t)| \le c_1(|\emptyset_k(t)| + \int_0^t |\emptyset_k(s)|ds + \varepsilon_k).$$

Proof. We have

$$x_k(t) = x^0 + \int_0^t A_1(s)x_k(s)ds + \int_0^t A_2(s)Y_k(s,0)y^0 ds$$

$$+ (1/\varepsilon_k)\int_0^t \int_0^s A_2(s)Y_k(s,\tau)(A_3(\tau)x_k(\tau) + B_2(\tau)u_k(\tau))d\tau ds$$

$$+ \int_0^t B_1(s)u_k(s)ds = x^0 + \int_0^t A_1(s)x_k(s)ds + \int_0^t A_2(s)Y_k(s,0)y^0 ds$$

$$+ \int_0^t \int_\tau^t D(s)\frac{\partial}{\partial s}Y_k(s,\tau)ds[A_3(\tau)x_k(\tau) + B_2(\tau)u_k(\tau)]d\tau$$

$$+ \int_0^t B_1(s)u_k(s)ds = x^0 + \int_0^t (A_0(\tau) + D(t)Y_k(t,\tau)A_3(\tau))x_k(\tau)d\tau$$

$$+ \int_0^t [B_0(\tau) + D(t)Y_k(t,\tau)B_2(\tau)]u_k(\tau)d\tau + 0(\varepsilon_k),$$

where $0(\varepsilon_k)$ does not depend on $t \in [0,T]$. Hence

(21) $\quad x_k(t) - \bar{x}_k(t) = \int_0^t [A_0(\tau) + D(t)Y_k(t,\tau)A_3(\tau)][x_k(\tau) - \bar{x}_k(\tau)]d\tau + \emptyset_k(t) + 0(\varepsilon_k).$

Denote $\psi_k(t) = |x_k(t) - \bar{x}_k(t)| - |\emptyset_k(t)|$. Then there exists a constant $c_2 > 0$ such that

$$\psi_k(t) \le c_2(\int_0^t \psi_k(\tau)d\tau + \int_0^t |\emptyset_k(\tau)|d\tau + \varepsilon_k).$$

Using Gronwall's inequality and lemma 10 we get the desired estimate. $\qquad\square$

Proposition 5. Since $\emptyset_k$ is bounded in $C^0([0,T])$, by lemmas 10 and 11 we obtain that x_k is bounded in $C^0([0,T])$ as well. Denote $\Delta x_k = x_k - \bar{x}_k$. Then we have

$$\int_0^t \int_0^\tau D(\tau)Y_k(\tau,s)A_3(s)\Delta x_k(s)ds d\tau = \int_0^t \int_s^t D(\tau)Y_k(\tau,s)d\tau A_3(s)\Delta x_k(s)ds$$

$$= \varepsilon_k \int_0^t \int_s^t D(\tau)A_4^{-1}(\tau)\frac{\partial}{\partial\tau}Y_k(\tau,s)d\tau A_3(s)\Delta x_k(s)ds$$

$$= \varepsilon_k \int_0^t (D(t)A_4^{-1}(t)Y_k(t,s) - D(s)A_4^{-1}(s)$$

$$- \int_s^t [\frac{\partial}{\partial\tau}D(\tau)A_4^{-1}(\tau)]Y_k(\tau,s)d\tau)A_3(s)\Delta x_k(s)ds = 0(\varepsilon_k),$$

uniformly in $t \in [0,T]$. Using a similar argument

$$\int_0^t \emptyset_k(\tau)d\tau = \int_0^t \int_0^\tau D(\tau)Y_k(\tau,s)B_2(s)u_k(s)ds d\tau$$

$$= \varepsilon_k \int_0^t \int_s^t D(\tau)A_4^{-1}(\tau)\frac{\partial}{\partial\tau}Y_k(\tau,s)d\tau B_2(s)u_k(s)ds$$

$$= \varepsilon_k \int_0^t (D(t)A_4^{-1}(t)Y_k(t,s) - D(s)A_4^{-1}(s)$$

$$- \int_s^t \frac{\partial}{\partial\tau}(D(\tau)A_4^{-1}(\tau))Y_k(\tau,s)d\tau)B_2(s)u_k(s)ds = 0(\varepsilon_k).$$

Hence, from (21) and Gronwall's inequality

$$\int_0^t \Delta x_k(s)ds = 0(\varepsilon_k).$$

Then lemma 10 completes the proof of (9).

If $u_k = u_0$, from lemma 11 it is sufficient to prove that $\emptyset_k \to 0$ in $C^0([0,T])$. Take $\delta > 0$ and let $u_\delta \in C^1([0,T]), \|u_\delta - u_0\| < \delta$ (norm of $L^1(0,T)$). Then

$$|\int_0^t \emptyset_k d\tau| \le 0(\delta) + |\int_0^t D(t)Y_k(t,\tau)B_2(\tau)u_\delta(\tau)d\tau|$$

$$\le 0(\delta) + \varepsilon_k|D(t)A_4^{-1}(t)B_2(t)u_\delta(t) - D(t)Y_k(t,0)A_4^{-1}(0)u_\delta(0)$$

$$- \int_0^t D(t)Y_k(t,\tau)\frac{\partial}{\partial\tau}(A_4^{-1}(\tau)B_2(\tau)u_\delta(\tau))d\tau| \le 0(\delta) + \varepsilon_k c(\delta).$$

Since δ can be arbitrarily small, the proof is completed. $\qquad\square$

The *proof of proposition 6* will be given after

12 Lemma For every $t \in M \cap (0, T)$ there exist σ_t, ρ_t and $\delta_t > 0$ so that

$$(22) \qquad |Y_\varepsilon(\tau, s)| \leq \sigma_t \exp\left(-\rho_t \frac{\tau - \max\{s, t - \delta_t\}}{\varepsilon}\right)$$

for all $t - \delta_t \leq \tau \leq t + \delta_t$, $0 \leq s \leq \tau$, and $\varepsilon > 0$.

Proof. If $t \in M$ there exist ρ_t and $\delta_t > 0$ so that Re $\sigma(A_4(\tau)) < -2\rho_t$ for all $t - \delta_t \leq \tau \leq t + \delta_t$. Then, by (19), the inequality (22) holds with σ_t and $t - \delta_t \leq s \leq \tau \leq t + \delta_t$. If $s \in [0, t - \delta_t]$ then from (6)

$$|Y_\varepsilon(\tau, s)| \leq |Y_\varepsilon(\tau, t - \delta_t)||Y_\varepsilon(t - \delta_t, s)|$$
$$\leq K_0|Y_\varepsilon(\tau, t - \delta_t)| \leq K_0\sigma_t \exp\left(-\rho_t \frac{\tau - t + \delta_t}{\varepsilon_k}\right),$$

where K_0 is the supremum of $|Y_\varepsilon(t, s)|$ in (6). □

Proposition 6. **Proof of (13).** Taking into account lemma 11 it is sufficient to show that $\emptyset_k(t) \to 0$ for every $t \in M$. Let $t \in M, t > 0$. Fix $\alpha > 0, \alpha < \sigma_t$. For small ε_k we have

$$B_\alpha^k(t) = \{s \in [0, t] : |Y_k(t, s)| \geq \alpha\}$$
$$\subset \{s \in [0, t] : \exp\left(-\rho_t \frac{t - \max\{s, t - \delta_t\}}{\varepsilon_k}\right) \geq \frac{\alpha}{\sigma_t}\}$$
$$= \{s \in [0, t] : \max\{s, t - \delta_k\} \geq t + \frac{\varepsilon_k}{\rho_t} \ln \frac{\alpha}{\sigma_k}\}$$
$$= [t + \frac{\varepsilon_k}{\rho_t} \ln \frac{\alpha}{\sigma_t}, t].$$

Then, for every $t \in M, t > 0$ and $\alpha \in (0, \sigma_t)$

$$\text{meas } B_\alpha^k(t) \to 0 \text{ as } k \to +\infty.$$

By Dunford-Pettis theorem (Dunford-Schwartz [1, IV.8.11])

$$\int_{B_\alpha^k(t)} |u_k| ds \to 0 \text{ as } k \to +\infty.$$

We have

$$|\emptyset_k(t)| \leq c_3\left(\int_{B_\alpha^k(t)} |u_k| ds + \alpha\right) \to c_3\alpha$$

for every $t \in M$. Since α can be arbitrary small, $\emptyset_k(t) \to 0$ as $k \to +\infty$ for every $t \in M$.

Proof of (14). Denote $\Delta u_k = u_k - u_0, \Delta x_k = x_k - x_0, \Delta y_k = y_k - y_0$ and

$$p_k(t) = (-1/\varepsilon_k) \int_0^t Y_k(t, s) A_4(s) y_0(s) ds - y_0(t).$$

Then

$$\Delta y_k(t) = Y_k(t,0)y^0 + (1/\varepsilon_k)\int_0^t Y_k(t,\tau)(A_3(\tau)\Delta x_k(\tau) + B_2(\tau)\Delta u_k(\tau))d\tau + p_k(t).$$

Integrating by parts we have

$$\int_0^t p_k(s)ds = -\int_0^t [(1/\varepsilon_k)\int_\tau^t Y_k(s,\tau)ds A_4(\tau)y_0(\tau) + y_0(\tau)]d\tau$$

$$= -\int_0^t [\int_\tau^t A_4^{-1}(s)\partial/\partial s Y_k(s,\tau)ds A_4(\tau)y_0(\tau) + y_0(\tau)]d\tau$$

$$= -\int_0^t A_4^{-1}(t)Y_k(t,\tau)A_4^{-1}(\tau)y_0(\tau)d\tau$$

$$+ \int_0^t \int_\tau^t [d/ds A_4^{-1}(s)]Y_k(s,\tau)ds A_4(\tau)y_0(\tau)d\tau.$$

Using (22), if $t \in M$,

$$\int_0^t A_4^{-1}(t)Y_k(t,\tau)A_4^{-1}(\tau)y_0(\tau)d\tau = 0_t(\varepsilon_k).$$

For the second term we get

(23) $$\int_\tau^t [d/ds A_4^{-1}(s)]Y_k(s,\tau)ds = \varepsilon_k \int_\tau^t [d/ds A_4^{-1}(s)]A_4^{-1}(s)\partial/\partial s Y_k(s,\tau)ds$$

$$= \varepsilon_k[d/dt(A_4^{-1}(t))A_4^{-1}(t)Y_k(t,\tau) - d/d\tau(A_4^{-1}(\tau))A_4^{-1}(\tau)$$

$$- \int_\tau^t d/ds(d/ds(A_4^{-1}(s))A_4^{-1}(s))Y_k(s,\tau)ds] = 0(\varepsilon_k).$$

Hence

(24) $$\int_0^t p_k(s)ds = 0_t(\varepsilon_k).$$

Furthermore, again integrating by parts and using (9), (13) and (23)

(25) $$(1/\varepsilon_k)\int_0^t \int_0^s Y_k(s,\tau)A_3(\tau)\Delta x_k(\tau)d\tau$$

$$= \int_0^t \{A_4^{-1}(t)Y_k(t,\tau) - A_4^{-1}(\tau)$$

$$- \int_\tau^t [d/ds A_4^{-1}(s)]Y_k(s,\tau)ds\}A_3(\tau)\Delta x_k(\tau)d\tau = 0_t(\varepsilon_k).$$

Analogously,

(26) $$(1/\varepsilon_k)\int_0^t \int_0^s Y_k(s,\tau)B_2(\tau)\Delta u_k(\tau)d\tau$$

$$= \int_0^t A_4^{-1}(t)Y_k(t,\tau)B_2(\tau)\Delta u_k(\tau)d\tau - \int_0^t A_4^{-1}(\tau)B_2(\tau)\Delta u_k(\tau)d\tau$$

$$- \int_0^t \int_\tau^t [d/ds A_4^{-1}(s)]Y_k(s,\tau)ds B_2(\tau)\Delta u_k(\tau)d\tau.$$

Using the weak convergence of Δu_k and (23) we obtain convergence to zero of the second and the third term. Fix arbitrarily $\alpha > 0$. Then there exists $\lambda > 0$ such that for all $k = 1, 2, \ldots$

$$\int_{|\Delta u_k| \geq \lambda} |\Delta u_k| ds \leq \alpha.$$

Then, if $t \in M$, using (22)

$$|\int_0^t A_4^{-1}(t) Y_k(t, \tau) B_2(\tau) \Delta u_k(\tau) d\tau|$$

$$\leq c_4 \int_{|\Delta u_k| \geq \lambda} |\Delta u_k| dt + \lambda \int_0^t |A_4^{-1}(t) Y_k(t, \tau) B_2(\tau)| d\tau \to c_4 \alpha, \text{ as } k \to +\infty.$$

Hence

(27) $$\int_0^t A_4^{-1}(t) Y_k(t, s) B_2(s) \Delta u_k(s) ds \to 0 \text{ as } k \to +\infty.$$

By (24),..., (27) we obtain (14).

Proof of (15). We use the following

13 Lemma Let $g \in L^\infty(0, T)$ and let y_k solve

$$\varepsilon_k \dot{y} = A_4(t) y + g(t), y(0) = y^0.$$

Then

$$y_k(t) \to -A_4^{-1}(t) g(t) \text{ for a.e. } t \in [0, T].$$

Proof. Let $t \in M$ be a point of approximative continuity of g. Then there exists a measurable $E \subset [0, T]$ such that

$$\delta^{-1} \text{ meas } ([t - \delta, t] \cap E) = 1 - w(\delta),$$

where $w(\delta) \to 0$ with $\delta \to 0$ and g is continuous on E. Denote

$$h(\delta) = \sup\{|g(\tau) - g(t)|, \tau \in [t - \delta, t] \cap E\}.$$

Clearly, $\lim h(\delta) = 0$ as $\delta \to 0$. We have

$$y_k(t) = Y_k(t, 0) y^0 + (1/\varepsilon_k) \int_0^t Y_k(t, \tau) d\tau g(t) - (1/\varepsilon_k) \int_0^t Y_k(t, \tau)(g(t) - g(\tau)) d\tau.$$

For the first integral an integration by parts gives

(28) $$(1/\varepsilon_k) \int_0^t Y_k(t, \tau) d\tau \to -A_4^{-1}(t), \text{ as } k \to +\infty.$$

Choose $N > O$. Then denoting $Q_k = [t - N\varepsilon_k, t]$ we have

$$(1/\varepsilon_k) \int_0^t Y_k(t, \tau)[g(t) - g(\tau)] d\tau = (1/\varepsilon_k) \int_0^{t - N\varepsilon_k} \ldots + (1/\varepsilon_k) \int_{Q_k \cap E} \ldots + (1/\varepsilon_k) \int_{Q_k \cap E^c} \ldots,$$

where E^c is the complement of E. For the first integral we have from (22)

$$(29) \qquad (1/\varepsilon_k) \int_0^{t-N\varepsilon_k} Y_k(t,\tau)(g(t) - g(\tau))d\tau$$

$$\leq 2\|g\|(\sigma_t/\varepsilon_k) \int_0^{t-N\varepsilon_k} \exp\left(-\rho_t \frac{t - \max\{\tau, t - \delta_t\}}{\varepsilon_k}\right)d\tau$$

$$= 2\|g\|(\sigma_t/\varepsilon_k)\left(\int_0^{t-\delta_t} \exp\left(-\rho_t \delta_t/\varepsilon_k\right)d\tau + \int_{t-\delta_t}^{t-N\varepsilon_k} \exp\left(-\rho_t \frac{t-\tau}{\varepsilon_k}\right)d\tau\right)$$

$$\leq 2\|g\|\sigma_t(\frac{t-\delta_t}{\varepsilon_k} \exp\left(-\rho_t\delta_t/\varepsilon_k\right) + (1/\rho_t) \exp\left(-\rho_t N\right)),$$

all norms of $L^\infty(0,T)$. Take $\eta > 0$. One can choose *independently* ε_k and N so that the last expression becomes $< \eta$. Furthermore

$$(30) \qquad |(1/\varepsilon_k) \int_{Q_k \cap E} Y_k(t,\tau)[g(\tau) - g(t)]d\tau|$$

$$\leq (K_0/\varepsilon_k)h(N\varepsilon_k)\text{meas}\,([t - N\varepsilon_k, t] \cap E) \leq K_0 N h(N\varepsilon_k) < \eta$$

for ε_k small, where K_0 again is the supremum of $|Y_\varepsilon(t,s)|$ in (6). Finally

$$(31) \qquad |(1/\varepsilon_k) \int_{Q_k \cap E^c} Y_k(t,\tau)[g(\tau) - g(t)]d\tau|$$

$$\leq 2(K_0/\varepsilon_k)\|g\| \text{ meas }([t - \varepsilon_k N, t] \cap E^c) \leq 2K_0\|g\|Nw(N\varepsilon_k) < \eta$$

again for small ε_k. Combining (28)-(31) we complete the proof. $\qquad\square$

End of the proof of proposition 6. Let $u_k = u_0 \in L^\infty(0,T), k = 1, 2, \ldots$. Since $y_0 \in L^\infty(0,T)$, lemma 13 yields $p_k(t) \to 0$ as $k \to +\infty$ for a.e. $t \in [0,T]$. By (10) $\Delta x_k \to 0$ in $C^0([0,T])$. Using (22), for $t \in M$

$$|(1/\varepsilon_k) \int_0^t Y_k(t,s)A_3(s)\Delta x_k(s)ds|$$

$$\leq \max_{0 \leq s \leq t} |A_3(s)\Delta x_k(s)|(\sigma_t/\varepsilon_k) \int_0^t \exp\left(-\rho_t \frac{t - \max\{s, t - \delta_t\}}{\varepsilon_k}\right)ds$$

$$\leq c_t \max\{|\Delta x_k(s)| : 0 \leq s \leq T\} \to 0 \text{ as } k \to +\infty.$$

Furthermore, if $t \in M, t > 0$ then

$$Y_k(t,0) \to 0 \text{ as } k \to +\infty.$$

This completes the proof. $\qquad\square$

Proposition 7. Applying lemma 11 it is sufficient to show that $\emptyset_k \to 0$ in $C^0([0,T])$. Take any $\alpha > 0$. Then there exists $\lambda > 0$ such that

$$\int_{|u_k| \geq \lambda} |u_k| ds \leq \alpha, k = 1, 2, \ldots.$$

Using (19) we have

$$|\emptyset_k(t)| \leq c_5 [\lambda \int_0^t \exp{(-\rho \frac{t-s}{\varepsilon_k})} ds + \alpha] \to c_5 \alpha$$

as $k \to +\infty$, uniformly in $t \in [0,T]$. Thus, (16) is proved.

In order to show (17) we use (19) and the convolution inequality (Aubin [2, p. 131]):

$$\|\sigma/\varepsilon_k \int_0^{\cdot} \exp{(-\rho \frac{\cdot - \tau}{\varepsilon_k})} |u_k(\tau)| d\tau \|_p \leq \|(\sigma/\varepsilon_k) \exp{(-\rho \cdot /\varepsilon_k \|_1 \|u_k\|_p},$$

where $\| \cdot \|_p$ denotes the norm of $L^p(0,T)$. Hence y_k is bounded in $L^p(0,T)$ whenever u_k is bounded in $L^p(0,T)$. This and (14) yield (17).

We prove (18) by showing that if $f_k \to f_0$ in $C^0([0,T])$ then

$$(1/\varepsilon_k) \int_0^t Y_k(t,\tau) f_k(\tau) d\tau \to -A_4^{-1}(t) f_0(t)$$

uniformly in $[\delta, T,]$ for any $\delta > 0$. Using (19) we have

$$|(1/\varepsilon_k) \int_0^t Y_k(t,\tau) f_k(\tau) d\tau + A_4^{-1}(t) f_0(t)| \leq c_7 \max\{|f_k(t) - f_0(t)| : 0 \leq t \leq T\}$$

$$+ |\int_0^t [\partial/\partial\tau \, Y_k(t,\tau)] A_4^{-1}(\tau) f_0(\tau) d\tau - A_4^{-1}(t) f_0(t)|$$

for some $c_7 > 0$. Let $x > 0$ and $\bar{f} \in C^1([0,T])$ be such that $\max\{|f_0(t) - \bar{f}(t)| : 0 \leq t \leq T\} < x$. Then

$$|\int_0^t [\partial/\partial\tau Y_k(t,\tau)] A_4^{-1}(\tau) f_0(\tau) d\tau - A_4^{-1}(t) f_0(t)|$$

$$\leq x(\int_0^t |\partial/\partial\tau Y_k(t,\tau) A_4^{-1}(\tau)| d\tau + |A_4^{-1}(t)|) + |Y_k(t,0) A_4^{-1}(0)\bar{f}(0)| + C(x)\varepsilon_k.$$

Using (19) and passing to the limit with $\varepsilon_k \to 0$ and $x \to 0$ we complete the proof. $\quad\square$

Theorem 8. We start with

14 Lemma Let (6) and (12) hold. Then for any $u_0 \in U$ and $\delta > 0$ there exist $u_\delta \in U$ and $k(\delta) > 0$ such that

$$\text{meas } \{t \in [0,T] : u_\delta(t) \neq u_0(t)\} \leq \delta$$

and if $(x_\varepsilon^\delta, y_\varepsilon^\delta)$ solves the perturbed system (1), (2) for u_δ and $\varepsilon > 0$ then

$$\sup_{\varepsilon > 0} \max\{|y_\varepsilon^\delta(t)| : 0 \leq t \leq T\} \leq k(\delta).$$

Proof. Step 1. For every $\delta > 0$ there exists a finite number of closed intervals $\Delta_i^\delta, i = 1, 2, ..., N_\delta$ such that meas $\cup_i \Delta_i^\delta \geq T - \delta$ and $\Delta_i^\delta \subset M$. Let $\Delta_i^\delta = [t_i^-, t_i^+]$. Take any $v \in V$ and define

$$u_\delta(t) = u_0(t) \text{ if } t \in \Delta_i^\delta, = v \text{ otherwise.}$$

Apply u_δ to the perturbed system (1),(2) with $\varepsilon_k > 0, \varepsilon_k \to 0$. Let (x_k^δ, y_k^δ) be the corresponding solution.

Step 2. Put

$$g_k^\delta(t) = (1/\varepsilon_k) \int_0^t Y_k(t, \tau) B_2(\tau) u^\delta(\tau) d\tau.$$

We prove that

(32) $$\sup_k \max \{|g_k^\delta(t)| : 0 \leq t \leq T\} \leq K_1(\delta) < +\infty.$$

Notation. 0_k^δ denotes any sequence of real numbers such that $\sup_k |0_k^\delta| \leq K_2(\delta) < +\infty$.

Assume first that $t_1^- = 0$ and take some i. Then

$$g_k^\delta(t_i^-) = \sum_{j=1}^{i-1} [(1/\varepsilon_k) \int_{t_j^-}^{t_j^+} Y_k(t, \tau) B_2(\tau) u_0(\tau) d\tau + (1/\varepsilon_k) \int_{t_j^+}^{t_{j+1}^-} Y_k(t, \tau) B_2(\tau) v d\tau].$$

Since $\Delta_i^\delta \subset M$, using (22) we get

(33) $$|(1/\varepsilon_k) \int_{t_j^-}^{t_j^+} Y_k(t, \tau) B_2(\tau) u_0(\tau) d\tau|$$

$$\leq (K_0/\varepsilon_k) \int_{t_j^-}^{t_j^+} |Y_k(t_j^+, \tau) B_2(\tau) u_0(\tau)| d\tau$$

$$\leq \frac{\sigma_j k_1}{\varepsilon_k} \int_{t_j^-}^{t_j^+} \exp\left(-\rho_j \frac{t_j^+ - s}{\varepsilon_k}\right) ds = 0_k^\delta.$$

Next, an integration by parts gives us

(34) $$|(1/\varepsilon_k) \int_{t_j^+}^{t_{j+1}^-} Y_k(t, \tau) B_2(\tau) v d\tau|$$

$$= |\int_{t_j^+}^{t_{j+1}^-} [\partial/\partial\tau Y_k(t, \tau)] A_4^{-1}(\tau) B_2(\tau) v d\tau|$$

$$\leq |Y_k(t, \tau) A_4^{-1}(\tau) B_2(\tau) v|_{t_j^+}^{t_{j+1}^-}$$

$$- \int_{t_j^+}^{t_{j+1}^-} Y_k(t, \tau) d(A_4^{-1}(\tau) B_2(\tau) v)| = 0_k^\delta.$$

Combining (33) and (34) we see that

(35) $$\sup_{i} \max \{g_k^\delta(t_i^-), g_k^\delta(t_i^+)\} = 0_k^\delta.$$

One comes to the same conclusion assuming that $t_1^+ = 0$. Analogously to (33) and (34), using (35) we obtain

$$\sup_{i} \max[\max\{|g_k^\delta(t)| : t_i^- \leq t \leq t_i^+\}, \max\{|g_k^\delta(t)| : t_i^+ \leq t \leq t_{i+1}^-\}] = 0_k^\delta.$$

This proves (32).

Step 3. We show that

$$\max\{|\dot{x}_k^\delta(t)| : 0 \leq t \leq T\} = 0_k^\delta.$$

Denoting $\psi_k^\delta(t) = A_1(t)x_k^\delta(t) + B_1(t)u^\delta(t) + A_2(t)g_k^\delta(t) + A_2(t)Y_k(t,0)y^0$, we have

$$\dot{x}_k^\delta(t) = \psi_k^\delta(t) + (1/\varepsilon_k) \int_0^t A_2(t)Y_k(t,\tau)A_3(\tau)x_k^\delta(\tau)d\tau$$

$$= \psi_k^\delta(t) - \int_0^t A_2(t)[\partial/\partial\tau Y_k(t,\tau)]A_4^{-1}(\tau)A_3(\tau)x_k^\delta(\tau)d\tau$$

$$= \psi_k^\delta(t) - A_2(t)[A_4^{-1}(t)A_3(t)x_k^\delta(t) - Y_k(t,0)A_4^{-1}(0)A_3(0)x^0$$

$$- \int_0^t Y_k(t,\tau)[\partial/\partial\tau A_4^{-1}(\tau)A_3(\tau)]x_k^\delta(\tau)d\tau$$

$$- \int_0^t Y_k(t,\tau)A_4^{-1}(\tau)A_3(\tau)\dot{x}_k^\delta(\tau)d\tau]$$

$$= \phi_k^\delta(t) + \int_0^t A_2(t)Y_k(t,\tau)A_4^{-1}(\tau)A_3(\tau)\dot{x}_k^\delta(\tau)d\tau,$$

where $\max\{|\phi_k^\delta(t)| : 0 \leq t \leq T\} = 0_k^\delta$ (by (32) and the uniform boundedness of x_k^δ). Then Gronwall's inequality completes the proof of Step 3.

End of the proof of lemma 14. We have

$$|y_k^\delta(t)| = |Y_k(t,0)y^0 + (1/\varepsilon_k) \int_0^t Y_k(t,\tau)A_3(\tau)x_k^\delta(\tau)d\tau + g_k^\delta(t)|$$

$$\leq 0_k^\delta + |\int_0^t [\partial/\partial\tau Y_k(t,\tau)]A_4^{-1}(\tau)A_3(\tau)x_k^\delta(\tau)d\tau|$$

$$\leq 0_k^\delta + |\int_0^t Y_k(t,\tau)A_4^{-1}(\tau)A_3(\tau)\dot{x}_k^\delta(\tau)d\tau| = 0_k^\delta.$$

□

Proof of theorem 8. Take any $(x_0, y_0) \in Z(0)$ and let u_0 be a corresponding control. Take any positive sequences $\varepsilon_n \to 0$ and $\delta_m \to 0$. Applying lemma 14, there exists $u_m \in U$ such that

(36) $$u_m \to u_0 \text{ in } L^p(0,T) \text{ as } m \to +\infty$$

and if (x_n^m, y_n^m) solves the perturbed system for $u = u^m$ and $\varepsilon = \varepsilon_n$, then

$$\sup_{n} \max\{|y_n^m(t)| : 0 \le t \le T\} \le K_m < +\infty.$$

Moreover, by (10) and (15), for fixed m

(37a) $$x_n^m \to x^m \text{ in } C^0([0,T])$$

and

$$y_n^m(t) \to y^m(t), \text{ a.e. } t \in [0,T]$$

as $n \to +\infty$, where (x^m, y^m) is the solution of the reduced system (7) with $u = u^m$. Then, by Lebesgue dominated convergence theorem (Dunford-Schwartz [1, III.6.16])

(37b) $$y_n^m \to y^m \text{ in } L^p(0,T) \text{ as } n \to +\infty.$$

By (36), it is easy to see that

(38) $$\begin{cases} x^m \to x_0 & \text{in } C^0([0,T]), \text{ and} \\ y^m \to y_0 & \text{in } L^p(0,T) \text{ as } m \to +\infty. \end{cases}$$

Thus, combining (37a,b) and (38), and taking ε_n and δ_m sufficiently small, we conclude that $Z(\varepsilon_n)$ will meet any $C^0([0,T]) \times L^p(0,T)-$ neighbourhood of (x_0, y_0), thereby proving lower semicontinuity. The second part of the theorem follows from proposition 7 and the (sequential) compactness of U in $(L^1(0,T), \text{ weak})$, provided that V is convex and compact. □

Section 3. Continuity of the Reachable Set.

Throughout this section we consider the singularly perturbed system (1),(2) with admissible set of controls (4). We assume that the set V is convex and compact and that the conditions (6) and (12) from the previous section hold.

Denote by $S(t,\varepsilon)$ the *reachable set* at the time $t, 0 \le t \le T$, of the system (1),(2),(4) for $\varepsilon > 0$, i.e.

$$S(t,\varepsilon) = \{s \in R^{n+m} : s = z(t), z \in Z(\varepsilon)\}.$$

The projection of $S(t,\varepsilon)$ on R^n, i.e. the $x-$ part, will be denoted by $P(t,\varepsilon)$, i.e.

$$P(t,\varepsilon) = \{x \in R^n : (x,y) \in S(t,\varepsilon) \text{ for some } y \in R^m\}.$$

We write again the reduced system

(7.1) $$\dot{x} = A_0(t)x + B_0(t)u(t), x(0) = x^0,$$

(7.2) $$y = -A_4^{-1}(t)[A_3(t)x + B_2(t)u(t)],$$

where A_0 and B_0 are given in (8). Given $t \in (0,T]$ one can formally define a "reachable set" at t, corresponding to $\varepsilon = 0$, as

$$S(t,0) = \{(x,y) \in R^{n+m} : x \in P(t,0), y \in -A_4^{-1}(t)(A_3(t)x + B_2(t)V)\},$$

where $P(t,0)$ is the (well-defined) reachable set of (7.1) with admissible controls from (4).

Since V is nonempty and compact, for each $t \in [0,T]$ the sets $S(t,\varepsilon), \varepsilon > 0$, and $P(t,\varepsilon), \varepsilon \geq 0$, are nonempty, compact and convex. The following fact is a consequence of (10) and (16):

15 Corollary *For every $t \in M$ the multifunction $\varepsilon \rightrightarrows P(t,\varepsilon)$ is Hausdorff continuous at $\varepsilon = 0^+$.*

(For a definition of Hausdorff continuity, see Klein-Thompson [1, p. 73]).

For the set $S(t,\varepsilon)$, however, a statement like this one is not true in general, as we shall see now. Consider first the simplest case when (1) contains the (fast) variables y only, A_4 and B_2 are constant, Re $\sigma(A_4) < 0$ and $y(0) = 0$. Let V be the unit cube in R^k, i.e. we have

$$\varepsilon \dot{y} = A_4 y + B_2 u, y(0) = 0, |u_i| \leq 1, i = 1, 2, ..., k.$$

Then the set $S(t,\varepsilon)$ can be written as an Aumann's integral, see Aumann [1]:

$$S(t,\varepsilon) = (1/\varepsilon) \int_0^t \exp\,(A_4 s/\varepsilon) B_2 V ds.$$

Changing the variable under the integral we get

$$S(t,\varepsilon) = \int_0^{t/\varepsilon} \exp\,(A_4 \tau) B_2 V d\tau,$$

that is

(39) $$S(t,\varepsilon) = K(\theta), \theta = t/\varepsilon,$$

where $K(\theta)$ is the reachable set at time θ of the linear autonomous system

(40) $$\dot{y} = A_4 y + B_2 u, y(0) = 0, |u_i| \leq 1, i = 1, ..., k.$$

Since t is fixed, $\varepsilon \to 0$ means exactly $\theta \to +\infty$. The set

$$K = \cup_{\theta > 0} K(\theta)$$

consists of all points that can be attained by the state of (40) at some finite time, starting from the origin at $t = 0$ and using feasible controls. It is known, see e.g. Macki-Strauss [1, p. 31], that if the pair (A_4, B_2) is controllable then $0 \in$ int K. The map $\theta \rightrightarrows K(\theta)$ is monotone by inclusion, therefore it is not difficult to see that

$$K - \lim K(\theta) = \text{cl } K \text{ as } \theta \to +\infty.$$

Hence, by (39), for every fixed $t \in (0,T]$

$$K - \lim S(t,\varepsilon) = \text{cl } K \text{ as } \varepsilon \to 0.$$

However cl K may be different from $S(t,0) = -A_4^{-1} B_2 V$ as we see in the following

1 Example (continuation). Consider

$$\varepsilon \dot{y}_1 = -y_1 + u, y_1(0) = 0,$$

$$\varepsilon \dot{y}_2 = -6y_2 + 3u, y_2(0) = 0,$$

$$u(t) \in [-1, 1], T = 1.$$

Since the eigenvalues of the matrix A_4 are real, every point from the boundary of $S(t, \varepsilon)$ can be attained using a bang-bang control having no more than one switching (see Macki-Strauss [1, p. 77]). Fix $\varepsilon > 0$ and take any $t^* \in (0, 1)$. Define

$$u_\varepsilon(t) = \alpha \text{ if } 0 \le t \le t^*, = -\alpha \text{ if } t^* < t \le 1,$$

where $\alpha \in \{-1, 1\}$. Then the corresponding (y_1, y_2) satisfies

$$y_1 = \alpha(2 \exp{(-\frac{1-t^*}{\varepsilon})} - 1) + 0(\varepsilon),$$

$$y_2 = \alpha(\exp{(-6\frac{1-t^*}{\varepsilon})} - 1/2) + 0(\varepsilon).$$

Eliminating t^* and letting $\varepsilon \to 0$ we get

$$\text{cl } K = \{(y_1, y_2) : -1 \le y_1 \le 1, -1/2 + [(y_1 + 1)/2]^6 \le y_2 \le 1/2 - [(y_1 - 1)/2]^6\}.$$

On the other hand

$$S(1, 0) = \{(y_1, y_2) : -1 \le y_1 \le 1, y_2 = (1/2)y_1\}.$$

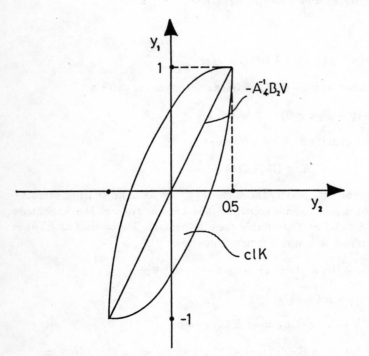

Consider now the general case, that is the system (1),(2) with admissible controls (4) under the conditions (6) and (12). Let $t \in M, t > 0$, where M is defined in (11). Introduce the set

$$W(t) = \int_0^{+\infty} \exp\,(A_4(t)s)B_2(t)V ds$$

$$= \{y \in R^m : y = \int_0^{+\infty} \exp\,(A_4(t)s)B_2(t)u(s)ds,$$

$$u \text{ locally integrable, } u(t) \in V \text{ for a.e. } t \in [0, +\infty)\}.$$

Since $t \in M$ we see on the basis of (19) that $W(t)$ is a compact subset of R^m. Let

(41) $$F(x,t) = -A_4^{-1}(t)A_3(t)x + W(t)$$

(a translate of $W(t)$) and define

(42) $$H(t) = \{(x,y) \in R^{m+n} : x \in P(t,0), y \in F(x,t)\}.$$

16 Theorem *For every $t \in M, t > 0$*

$$haus\,[S(t,\varepsilon), H(t)] \to 0 \text{ as } \varepsilon \to 0.$$

This result may be viewed as follows. For small $\varepsilon > 0$ the state y reacts faster than x to changes of the control. Changing rapidly the control on a small time interval one obtains big changes of y, while the variation of x remains small. Loosely speaking, on some fixed time interval $[0, \bar{t}]$ we need to take care only of x by reaching a neighbourhood of $\bar{t}$. For t near $\bar{t}$, by a fast change of the control we can do very much for y while x remains "frozen". When $\varepsilon \to 0$ the slow state x and the fast state y are controlled more and more independently. In the limit, on the interval $[0, \bar{t}]$, the system decomposes to
the reduced system (7.1) on $[0, \bar{t}]$ with reachable set $P(\bar{t}, 0)$;
an associate system of the form

(43) $$dy/ds = A_3(\bar{t})x(\bar{t}) + A_4(\bar{t})y + B_2(\bar{t})u(s),$$
$$y(0) = 0, u(s) \in V, \text{ a.e. } s \in [0, +\infty).$$

The set $F(x(\bar{t}), \bar{t})$ is the (Hausdorff) limit of the reachable set of (43) at the time s as $s \to +\infty$. Observe that the y -part of $S(\bar{t}, 0)$ consists of all points that are limits as $s \to +\infty$ of trajectories of (43) on $[0, +\infty)$. Then the discontinuity of the attainable mapping can be regarded as follows: the (Hausdorff) limit as $s \to +\infty$ of the set of all ends of trajectories of (43) on $[0, s]$ may be larger than the set of all limits of trajectories of (43) on $[0, +\infty)$.

Remark. The mathematical background of theorems 8 and 16 can be described in the following way. Consider first a sequence $f_n : [0, T] \to B$ of continuous functions with values in the closed unit ball B in R^m and suppose that

$$f_n \rightharpoonup f \text{ in } L^1(0, T) \text{ and } f_n(t) \to g(t), \text{ a.e. } t \in (0, T).$$

Then it is easy to show (e.g. applying Mazur's theorem) that $f = g$ a.e. in $(0, T)$. Now let F_n be a sequence of sets of continuous functions from $[0, T]$ to B and let

$$\text{seq. } K - \lim F_n = F \text{ in } (L^1(0, T), \text{ weak }).$$

Define for each $t \in [0, T]$

$$S_n(t) = \{f(t) : f \in F_n\}, n = 1, 2, \dots$$

and suppose that for every fixed $t \in [0, T]$.

$$K - \lim S_n(t) = S(t),$$

where $S(\cdot)$ is a closed-valued multifunction from $[0, T]$ to the subsets of B. By Lebesgue dominated convergence theorem for multifunctions (see Aumann [1, th. 5], and Artstein [1, th. 2.7]) one has

$$H - \lim \int_\Delta S_n \, dt = \int_\Delta S \, dt$$

for any measurable $\Delta \subset [0, T]$. This means that S_n converges weakly to S in the sense of Artstein [1]. Take any $f \in F$. Then there exists a sequence $f_n \in F_n, f_n \to f$ in $L^1(0, T)$. Clearly, f_n is a measurable selection of S_n. Then, by prop. 4.1.1 in Artstein [1], $f(t) \in S(t)$ a.e. in $(0, T)$.

It may be not true, however, that every measurable selection of S is a member of the set F. This is the case with the trajectory mapping $Z(\cdot)$, if V is a convex set. Take $\varepsilon_n \to 0, F_n = Z(\varepsilon_n), S_n(t) = S(t, \varepsilon_n)$ and $S(t) = H(t)$. By theorem 8, $K - \lim Z(\varepsilon_n) = Z(0)$ in $(L^1(0, T), \text{ weak })$, and by theorem 16, $H - \lim S(t, \varepsilon_n) = H(t)$, moreover $S(t, 0)$ is a subset of $H(t)$ but not always equal to it.

Now we obtain an estimate for the convergence rate of $S(t, \varepsilon)$ to $H(t)$.

17 Theorem *Let $M = [0, T]$. Then for any fixed $t_0 \in (0, T)$ and $\alpha \in (0, 1)$ there exist positive constants $L(t_0, \alpha)$ and $\varepsilon(t_0, \alpha)$ such that for every $\varepsilon \in (0, \varepsilon(t_0, \alpha))$ and $t \in [t_0, T]$*

$$\text{haus } [S(t, \varepsilon), H(t)] \leq L(t_0, \alpha)\varepsilon^\alpha.$$

The following example shows that α cannot be 1, in general.

18 Example. Consider

$$\dot{x} = u, \quad x(0) = 0; \quad \varepsilon\dot{y} = -y + u, \quad y(0) = 0, u(t) \in [-1, 1], T = 1.$$

We have $H(1) = [-1, 1] \times [-1, 1]$. Take $t_\varepsilon \in [0, 1]$ and

$$u_\varepsilon(t) = \beta \text{ if } t \in [0, t_\varepsilon], = -\beta \text{ if } t \in (t_\varepsilon, 1], \beta = 1 \text{ or } -1.$$

Then, setting $z_\varepsilon = \exp((t_\varepsilon - 1)/\varepsilon)$ one obtains

$$x_\varepsilon(1) = \beta(1 + 2\varepsilon \ln z_\varepsilon), \quad y_\varepsilon(1) = \beta(2z_\varepsilon - \exp(-1/\varepsilon) - 1).$$

We have, see the next figure, that

$$\text{haus } [H(1), S(1, \varepsilon)] = \text{ dist } [(1, -1), S(1, \varepsilon)]$$

$$= \min\{\Psi(z, \varepsilon) = |(2\varepsilon \ln z)^2 + (2z - \exp(-1/\varepsilon))^2|^{1/2} : \exp(-1/\varepsilon) \leq z \leq 1\}.$$

It is easy to see that for all sequences $\varepsilon_n \to 0$ and $z_n \in (0, 1], \Psi(z_n, \varepsilon_n)/\varepsilon_n$ is unbounded.

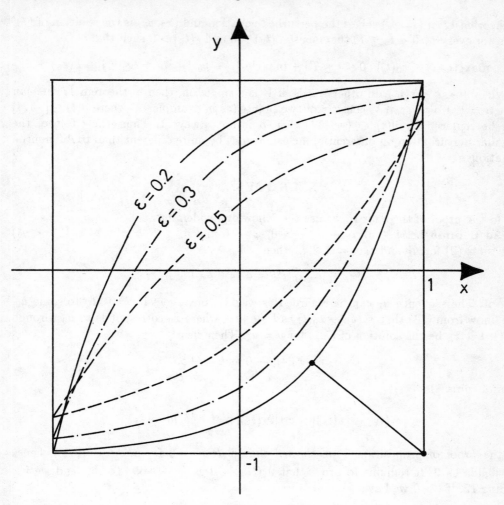

If A_4 and B_2 are constant in $[0, T]$ then the map $\varepsilon \rightrightarrows S(t, \varepsilon)$ turns out to be "upper Lipschitzian" at $\varepsilon = 0$. More precisely, the following corollary can be extracted from the proof of the theorem.

19 Corollary *Suppose that the matrices A_4 and B_2 are independent of t. Then given $t_0 \in (0, T]$ there exists $L > 0$ such that for every $\varepsilon \in (0, \varepsilon_0)$ and $t \in [t_0, T]$,*

$$S(t, \varepsilon) \subset H(t) + \varepsilon LB,$$

where ε_0 is as in (19).

Remark. Consider the singularly perturbed system (1),(2) and the corresponding reduced system (7), assuming that V is a singleton. It is easy to see that, on the condition of theorem 17, if $(x_\varepsilon, y_\varepsilon)$ solves the perturbed equation and (x_0, y_0) is the solution of (7), then for every $t_0, 0 < t_0 \leq T$, there exist $L(t_0) > 0$ and $\varepsilon(t_0) > 0$ such that

$$\max\{|x_\varepsilon(t) - x_0(t)| : 0 \leq t \leq T\} + \max\{|y_\varepsilon(t) - y_0(t)| : t_0 \leq t \leq T\} \leq L(\varepsilon_0)\varepsilon$$

provided $0 < \varepsilon < \varepsilon(t_0)$. In other words, if V is a singleton, then in theorem 17 one can take $\alpha = 1$. However, if V consists of two points (as in example 18, where $u(t) \in [-1,1]$ can be replaced by $u(t) \in \{-1,1\}$) then α is not always 1. Thus, the effect of the singular perturbation on differential inclusions may be quite different than to differential equations.

PROOFS.

In the proof of theorem 16 we use the following two lemmas.

20 Lemma Let $t \in M, t_k \to t, \varepsilon_k > 0, \varepsilon_k \to 0$ and $u_k \in U, k = 1, 2, \ldots$. If (x_k, y_k) solves (1),(2) for u_k and $\varepsilon_k, k = 1, 2, \ldots$, then

$$|x_k(t_k) - x_k(t)| \to 0 \text{ as } k \to +\infty.$$

Proof. The sequence u_k can be regarded as weakly convergent in $L^1(0, T)$ to some u_0. We know from (13) that $x_k(t) \to x_0(t)$ as $k \to +\infty$, where x_0 corresponds to u_0 through (7.1). Let $\tilde{x}_k$ be the solution of (7.1) for $u = u_k$. Then clearly

$$|\tilde{x}_k(t_k) - x_0(t)| \to 0 \text{ as } k \to +\infty.$$

From lemma 11

$$|x_k(t_k) - \tilde{x}_k(t_k)| \leq c_1(|\emptyset_k(t_k)| + \int_0^{t_k} |\emptyset_k| ds + \varepsilon_k).$$

In the proof of proposition 5 we showed that $|\emptyset_k(t)| \to 0$ for a.e. $t \in (0, T)$, hence $\int_0^{t_k} |\emptyset_k| ds \to 0$. It remains to prove that $\emptyset_k(t_k) \to 0$ as $k \to +\infty$. To this end we use lemma 12. If $t > 0$ we have

$$|\emptyset_k(t_k)| \leq c_t \left(\int_0^{t_k - \delta_t} \exp\left(-\rho_t \frac{t_k - t + \delta_t}{\varepsilon_k}\right) ds \right.$$
$$\left. + \int_{t_k - \delta_t}^{t_k} \exp\left(-\rho_t \frac{t_k - s}{\varepsilon_k}\right) ds \right) \leq c_t \exp\left(-\rho_t \frac{\delta_t}{2\varepsilon_k}\right) + 0(\varepsilon_k)$$

for $|t_k - t| < \delta_t/2$. This completes the proof. $\qquad \square$

21 Lemma Let $t^* \in M, \varepsilon_k \to 0$ and the functions y_k, y_k^* solve respectively

$$\varepsilon \dot{y}_k = A_4(t) + B_2(t)u_k(t) + g_k(t), y_k(0) = y^0,$$

$$\varepsilon \dot{y}_k^* = A_4(t^*)y_k + B_2(t^*)u_k(t) + g_k(t^*), y_k^*(0) = y^0,$$

where $u_k \in U$, the sequence g_k is bounded in $L^\infty(0, T)$, and if $t_k \to t^*, t_k < t^*$

$$\max\{|g_k(t) - g_k(t^*)| : t_k \leq t \leq t^*\} \to 0 \text{ as } k \to +\infty.$$

Then
$$|y_k(t^*) - y_k^*(t^*)| \to 0 \text{ as } k \to +\infty.$$

Proof. Clearly, from (19)
$$\sup_k \max\{|y_k^*(t)| : 0 \le t \le T\} < +\infty.$$

Denote $\Delta y_k = y_k - y_k^*, \Delta A_4 = A_4(t) - A_4(t^*), \Delta B_2 = B_2(t) - B_2(t^*), \Delta g_k = g_k(t) - g_k(t^*)$. Then
$$\varepsilon \Delta \dot{y}_k = A_4(t) \Delta y_k + \Delta A_4(t) y_k^*(t) + \Delta B_k(t) u_k(t) + \Delta g_k(t), \Delta y_k(0) = 0.$$

Let ρ^*, δ^* correspond to t^* as in lemma 12. We have
$$\Delta y_k(t^*) = (1/\varepsilon_k) \int_0^{t^*} Y_k(t^*, s)[\Delta A_4(t) y_k^*(s) + \Delta B_2(s) u_k(s) + \Delta g_k(s)] ds$$
$$= (1/\varepsilon_k)(\int_0^{t^*-\delta^*} \ldots + \int_{t^*-\delta^*}^{t^*-\sqrt{\varepsilon_k}} \ldots + \int_{t^*-\sqrt{\varepsilon_k}}^{t^*} \ldots).$$

It remains to estimate all three integrals on the basis of lemma 12, as follows:
$$|(1/\varepsilon_k) \int_0^{t^*-\delta^*} Y_k(t^*, s)(\ldots) ds| \le c^* \frac{t^* - \delta_k}{\varepsilon_k} \exp(-\rho^* \delta^*/\varepsilon_k) \to 0,$$
$$|(1/\varepsilon_k) \int_{t^*-\delta^*}^{t^*-\sqrt{\varepsilon_k}} Y_k(t^*, s)(\ldots) ds| \le (c^*/\rho^*) \exp(-\rho^*/\sqrt{\varepsilon_k}) \to 0,$$
$$|(1/\varepsilon_k) \int_{t^*-\sqrt{\varepsilon_k}}^{t^*} Y_k(t^*, s)(\ldots) ds|$$
$$\le (c^*/\varepsilon_k) \int_{t^*-\sqrt{\varepsilon_k}}^{t^*} \exp(-\rho^* \frac{t^* - s}{\varepsilon_k}) ds \, [\max\{|\Delta A_4(t) y_k^*(t)|$$
$$+ |\Delta B_2(t) u_k(t)| + |\Delta g_k(t)| : t^* - \sqrt{\varepsilon_k} \le t \le t^*\}] \to 0.$$
$$\square$$

Theorem 16. Take $(x^*, y^*) \in H(t)$ and let $x^* = x[u^*; t]$, i.e. u^* is a feasible control associated with x^* according to (7.1). By the definition of $H(t)$ there exists $v^*, v^*(s) \in V$ for a.e. $s \in [0, +\infty)$, such that
$$y^* = -A_4^{-1}(t) A_3(t) x^* + \int_0^{+\infty} \exp(A_4(t)s) B_2(t) v^*(s) ds.$$

Let $\varepsilon_k > 0, \varepsilon_k \to 0$, and
$$u_k(s) = \begin{cases} u^*(s), & s \in [0, t - \sqrt{\varepsilon_k}], \\ v^*[(t - s)/\varepsilon_k], & s \in (t - \sqrt{\varepsilon_k}, t]. \end{cases}$$

If (x_k, y_k) solves (1),(2) for $u = u_k$ and $\varepsilon = \varepsilon_k, k = 1, 2, \ldots$, then, by proposition 6 $x_k(t) \to x[u^*; t] = x^*$ and, by lemma 20, $x_k(t - \sqrt{\varepsilon_k}) \to x^*$. Taking into account lemma 21 it is sufficient to prove that $y_k^*(t) \to y^*$, where y_k^* solves

(44) $$\varepsilon_k dy/ds = A_4(t) y + B_2(t) u_k(s) + A_3(t) x_k(t), y(0) = y^0,$$

(t is fixed). By virtue of (19) we get

$$(45) \quad y_k^*(t) = \exp{(A_4(t)t/\varepsilon_k)}$$

$$+ (1/\varepsilon_k) \int_0^{t-\sqrt{\varepsilon_k}} \exp{(A_4(t)(t-s)/\varepsilon_k)}[A_3(t)x_k(t) + B_2(t)u_k(s)]ds$$

$$+ (1/\varepsilon_k) \int_{t-\sqrt{\varepsilon_k}}^{t} \exp{(A_4(t)(t-s)/\varepsilon_k)}[A_3(t)x_k(t) + B_2(t)v^*[(t-s)/\varepsilon_k)]ds$$

$$= y^* + 0_t(\varepsilon_k).$$

Finally

$$H(t) \subset S(t,\varepsilon_k) + 0_t(\varepsilon_k)B,$$

where B is the unit ball. Now let

$$\varepsilon_k > 0, \varepsilon_k \to 0, (x_k, y_k) \in S(t,\varepsilon_k), (x_k, y_k) \to (x^*, y^*).$$

We will show that $(x^*, y^*) \in H(t)$. Then, by a compactness argument, the proof will be completed.

Let (x_k, y_k) correspond to u_k and $\varepsilon_k, k = 1, 2, ...$ through (1), (2), (4). Without loss of generality let $u_k \rightharpoonup u_0$ in $L^1(0,T)$. Denote by $(x_k(\cdot), y_k(\cdot))$ the corresponding trajectory of (1), (2) with $\varepsilon = \varepsilon_k, k = 1, 2, ...$ From proposition 6, $x_k(t) \to x_0(t)$ where x_0 solves (7.1) for $u = u_0 \in U$. Thus, $x_0(t) = x^* \in P(t,0)$. Furthermore, by lemma 20, $x_k(t-\sqrt{\varepsilon_k}) \to x^*$ and lemma 21 applies. Therefore it is sufficient to prove that if y_k^* is determined by (44) then every cluster point of $y_k^*(t)$ is from $F(x^*, t)$. To this end, as in (45) we have

$$y_k^*(t) = -A_4^{-1}(t)A_3(t)x_k(t) + (1/\varepsilon_k) \int_0^t \exp{(A_4(t)(t-s)/\varepsilon_k)}B_2(t)u_k(s)ds + 0_t(\varepsilon_k)$$

$$= -A_4^{-1}(t)A_3(t)x^* + \int_0^{t/\varepsilon_k} \exp{(A_4(t)s)}B_2(t)u_k(t - \varepsilon_k s)ds + 0_t(\varepsilon_k)$$

$$\in F(x^*, t) + 0_t(\varepsilon_k)B.$$

Hence $y^* \in F(x^*, t)$. $\qquad\square$

For the proof of theorem 17 we need the following technical lemma, whose proof can be easily completed on the basis of lemmas 9,10 and 11.

22 Lemma There exists a constant $c > 0$ such that for any $\varepsilon, 0 < \varepsilon < \varepsilon_0$ and $u^1, u^2 \in U$ (possibly depending on ε), if $(x_\varepsilon^1, y_\varepsilon^1)$ solves (1), (2) with u^1, and x_ε^2 solves (7.1) with u^2, then

$$\max\{|x_\varepsilon^1(t) - x_\varepsilon^2(t)| : 0 \le t \le T\} \le c[\varepsilon + \max\{|\int_0^t A_2(s)A_4^{-1}(s)B_2(s)[u^1(s) - u^2(s)]ds|$$

$$+ |\int_0^t B_1(s)[u^1(s) - u^2(s)]ds| : 0 \le t \le T\}].$$

Proof of theorem 17. Let $t_0 \in (0,T]$ and $\alpha \in (0,1)$. Choose a constant $\bar\varepsilon > 0$ such that $\bar\varepsilon_0 \le \varepsilon_0$ and $\bar\varepsilon^\alpha < t_0$ (where ε_0 is as in (19)). Take some $t^0 \in [t_0, T]$ and let $(x^*, y^*) \in H(t^0)$. Then there exists $u_0 \in U$ such that if $x_0(\cdot)$ solves (7.1) with $u_0(\cdot)$ then

$x^* = x_0(t^0)$. Moreover, there exists a locally integrable function $v_0(\cdot)$, $v_0(t) \in V$ for a.e. $t \in [0, +\infty)$ such that

$$y^* = -A_4^{-1}(t^0)A_3(t^0)x^* + \int_0^{+\infty} \exp\,(A_4(t^0)s)B_2(t^0)v_0(s)ds.$$

Define the admissible control

$$u_\varepsilon(t) = u_0(t) \text{ if } t \in [0, t^0 - \varepsilon^\alpha] \cup (t^0, T], \; = v_0(\frac{t^0 - t}{\varepsilon}) \text{ if } t \in (t^0 - \varepsilon^\alpha, t^0].$$

Denote by $(x_\varepsilon, y_\varepsilon)$ the solution of (1) with u_ε. Then from lemma 22

(46) $$\max\{|x_\varepsilon(t) - x_0(t)| : 0 \leq t \leq t^0\} \leq c\varepsilon^\alpha.$$

Now let $\tilde{y}_\varepsilon$ solve

$$\varepsilon\dot{y} = A_4(t)y + B_2(t)u_\varepsilon(t), \quad y(0) = y^0,$$

and let $\bar{y}_\varepsilon$ be the solution of

$$\varepsilon\dot{y} = A_4(t^0)y + B_2(t^0)u_\varepsilon(t), \quad y(0) = y^0.$$

Denote $\Delta y_\varepsilon = \tilde{y}_\varepsilon - \bar{y}_\varepsilon$, $\Delta A_4(t) = A_4(t) - A_4(t^0)$, $\Delta B_2(t) = B_2(t) - B_2(t^0)$. Then some standard computations give us

(47)

$$|\Delta y_\varepsilon(t^0)| \leq (\sigma/\varepsilon) \int_0^{t^0 - \varepsilon^\alpha} \exp\,(-\rho\frac{t^0 - s}{\varepsilon})|\Delta A_4(t)\tilde{y}_\varepsilon(t) + \Delta B_2(t)u_\varepsilon(t)|dt$$

$$+ (M_1/\varepsilon)\int_{t^0 - \varepsilon^\alpha}^{t^0} \exp\,(-\rho\frac{t^0 - t}{\varepsilon})dt \max\{|\Delta A_4(t)| + |\Delta B_2(t)| : t^0 - \varepsilon^\alpha \leq t \leq t^0\}$$

$$\leq M_2(\exp\,(\frac{-\rho}{\varepsilon^{1-\alpha}}) + \varepsilon^\alpha),$$

where the constants M_1 and M_2 do not depend on t^0. We have

(48) $$|y_\varepsilon(t^0) - y^*| \leq |Y_\varepsilon(t^0, 0)y^0|$$

$$+ |(1/\varepsilon)\int_0^{t^0} Y_\varepsilon(t^0, t)A_3(t)x_\varepsilon(t)dt + A_4^{-1}(t^0)A_3(t^0)x^*|$$

$$+ |\Delta y_\varepsilon(t^0)| + |(1/\varepsilon)\int_0^{t^0} \exp\,(A_4(t^0)(\frac{t^0 - t}{\varepsilon}))B_2(t^0)u_\varepsilon(t)dt$$

$$- \int_0^{+\infty} \exp\,(A_4(t^0)t)B_2(t^0)v_0(t)dt|.$$

From (19)

(49) $$|Y_\varepsilon(t^0, 0)y^0| \leq M_3(t_0)\varepsilon,$$

and, integrating by parts,

(50) $\quad |(1/\varepsilon) \int_0^{t^0} Y_\varepsilon(t^0, t) A_3(t) x_\varepsilon(t) dt + A_4^{-1}(t^0) A_3(t^0) x^*| \leq M_4(\varepsilon + |x_\varepsilon(t^0) - x^*|).$

Furthermore,

$$|(1/\varepsilon) \int_0^{t^0} \exp\left(A_4(t^0)\left(\frac{t^0 - t}{\varepsilon}\right)\right) B_2(t^0) u_\varepsilon(t) dt - \int_0^{+\infty} \exp\left(A_4(t^0)t\right) B_2(t^0) v_0(t) dt|$$

$$\leq M_5 \exp\left(\frac{-\rho}{\varepsilon^{1-\alpha}}\right).$$

Finally, we obtain (46),

$$|y_\varepsilon(t^0) - y^*| \leq M_6(t_0)(\varepsilon^\alpha + \exp(-\rho\varepsilon^{\alpha-1})),$$

and $(x_\varepsilon(t^0), y_\varepsilon(t^0)) \in S(t^0, \varepsilon)$.

Now let $\varepsilon \in (0, \bar{\varepsilon}), (x_\varepsilon^*, y_\varepsilon^*) \in S(t^0, \varepsilon)$ and let u_ε be a corresponding control with corresponding trajectory $(x_\varepsilon, y_\varepsilon)$, i.e. $x_\varepsilon(t^0) = x_\varepsilon^*, y_\varepsilon(t^0) = y_\varepsilon^*$. Denote by x_ε^0 the solution of (7.1) with u_ε. By lemma 22

(51) $\quad\quad\quad\quad \max\{|x_\varepsilon(t) - x_\varepsilon^0(t)| : 0 \leq t \leq t^0\} \leq c\varepsilon,$

and thus $x_\varepsilon^0(t) \in P(t^0, 0)$ is in the $c\varepsilon$− neighbourhood of x_ε^*. Let $v \in V$ be arbitrarily chosen. Consider

$$v_\varepsilon(t) = u_\varepsilon(t^0 - \varepsilon t) \text{ if } t \in [0, t^0/\varepsilon), = v \text{ if } t \in [t^0/\varepsilon, +\infty),$$

and let

$$y_\varepsilon^0 = -A_4^{-1}(t^0) A_3(t^0) x_\varepsilon^0(t^0) + \int_0^{+\infty} \exp\left(A_4(t^0)t\right) B_2(t^0) v_\varepsilon(t) dt \in F(x_\varepsilon^0(t^0), t^0).$$

Let Δy_ε be as in (47). Then from (48)-(51)

(52) $\quad\quad\quad\quad\quad\quad |y_\varepsilon^* - y_\varepsilon^0| \leq M_7\varepsilon + |\Delta y_\varepsilon(t^0)|.$

Hence, from (47)

$$|y_\varepsilon^* - y_\varepsilon^0| \leq M_8(\varepsilon^\alpha + \exp\left(-\frac{\rho}{\varepsilon^{1-\alpha}}\right)),$$

which completes the proof. $\quad\quad\quad\quad\quad\quad\quad\quad\quad\quad\quad\quad\quad\quad\quad\quad\quad\quad\quad\quad\quad\quad$ □

Corollary 19. If $A_4(t) = A_4(t^0)$ and $B_2(t) = B_2(t^0)$ then, by (52), $|y_\varepsilon^* - y_\varepsilon^0| \leq M_7\varepsilon$. $\quad$ □

Section 4.   Hadamard Well and Ill-Posedness.

In this section we consider first a standard linear - quadratic problem subject to a singular perturbation: minimize

$$(53) \qquad I(x,y,u) = \int_0^T (x'Px + y'Ry + u'Qu)dt$$

subject to

$$(1.1) \qquad \dot{x} = A_1(t)x + A_2(t)y + B_1(t)u, x(0) = x^0,$$

$$(1.2) \qquad \varepsilon\dot{y} = A_3(t)x + A_4(t)y + B_2(t)u, y(0) = y^0,$$

$$\varepsilon \geq 0, T > 0, t \in [0,T],$$

$$(4) \qquad u \in U = \{u \in L^\infty(0,T) : u(t) \in V \text{ a.e. in } (0,T)\},$$

where the matrices P, Q and R are symmetric, P and Q are positive semidefinite, R is positive definite, V is compact and convex. For $\varepsilon = 0$ the differential equation (1.2) becomes a mixed equality constraint (3) and y may be no longer a continuous function. Denote by $v(\varepsilon)$ the optimal value of (53), $\varepsilon \geq 0$.

23 Proposition $v(0) \leq \lim\inf v(\varepsilon)$ as $\varepsilon \to 0$.

To prove value Hadamard well-posedness we need more conditions.

24 Theorem *If the assumptions (6) and (12) hold,*

$$v(\varepsilon) \to v(0) \text{ as } \varepsilon \to 0.$$

Next we focus our attention to Mayer's optimal control problem

$$(54) \qquad \text{minimize } \phi[x(T), y(T)] \text{ subject to } (1), (2) \text{ and } (4) \text{ for } \varepsilon \geq 0,$$

where $\phi : R^n \times R^m \to (-\infty, +\infty)$ is a given continuous function. Denote by $\omega(\varepsilon)$ the optimal value of (54). By theorem 16 we get

25 Theorem *Let (6) and (12) hold and $T \in M$. Let $\bar{\omega}$ be the optimal value of the following problem:*

$$(55) \qquad \text{minimize } \phi_0(x(T)) \text{ subject to } \dot{x} = A_0(t)x + B_0(t)u, x(0) = x^0, u \in U,$$

where

$$\phi_0(x) = \inf\{\phi(x,y) : y \in F(x,T)\}$$

and $F(x,T)$ is given in (41). Then

$$\omega(\varepsilon) \to \bar{\omega} \text{ as } \varepsilon \to 0.$$

Thus, the *true limit problem*, associated with (54), is the problem (55).

If we look at system (1) as a process with slow and fast states then we can interpret the limit problem as a decomposition of the original problem resulting from the separation of the time scales. For simplicity, let $\phi(x,y) = \phi_1(x) + \phi_2(y)$, and $A_3 = 0$. Then the true limit problem consists of the following two independent problems:

the *reduced order problem*, to minimize $\phi_1(x(T))$ subject to the constraints in (55);

an *associate problem*, to minimize $\phi_2(y)$ on the (Hausdorff) limit of the reachable set at t as $t \to +\infty$ of the time-invariant system

$$(56) \qquad dy/ds = A_4(T)y + B_2(T)u, y(0) = 0,$$
$$u(s) \in V, \text{ a.e. } s \in [0,t].$$

Ill-posedness comes from the associate problem and is due to the fact that the set of limits of trajectories of (56) as $s \to +\infty$ may be smaller than the limit of sets of points achieved at t as $t \to +\infty$.

There is a further singularity of the value function of Mayer's problem, related to Hoelder properties of the attainable mapping. If ϕ is locally Lipschitz then theorem 17 yields that for any $\alpha \in (0,1)$ there exist $l > 0$ and $\varepsilon_1 > 0$ such that

$$|v(\varepsilon) - \bar{\omega}| \le l\varepsilon^\alpha$$

whenever $0 < \varepsilon < \varepsilon_1$. From the next example (actually, a continuation of example 18) we see that the restriction $\alpha < 1$ is sharp. This means that the value $v(\varepsilon)$ may be "almost Lipschitz" but not right-differentiable at $\varepsilon = 0$.

26 Example. Minimize $|x(1) - 1| + |y(1) + 1|$ subject to

$$\dot{x} = u, \quad x(0) = 0, \quad t \in [0,1],$$
$$\varepsilon\dot{y} = -y + u, \quad y(0) = 0, \quad u(t) \in [-1,1].$$

The (true) limit problem is: minimize $|x(1) - 1|$ subject to $\dot{x} = u, x(0) = 0$ and $v(0) = \bar{\omega} = 0$. The value $v(\varepsilon)$ is proportional to the distance from the point $(1,-1)$ to the set $S(1,\varepsilon)$, see example 18. Then $(1/\varepsilon)v(\varepsilon) \to +\infty$ as $\varepsilon \to 0$.

Let the entries of A_4 and B_2 be constant, $m > k$ and $0 \in$ int V. We have already mentioned in section 3 that $W(t) = W$ has nonempty interior whenever the pair (A_4, B_2) is controllable, i.e. (Macki-Strauss [1, th. 3 p. 31])

$$(57) \qquad \text{rank } [B_2, A_4B_2, ..., A_4^{m-1}B_2] = m.$$

On the other hand int $(A_4^{-1}B_2V) = \emptyset$ since $m > k$. As far as controllability is a generic property (see Macki-Strauss [1, th. 6 p. 37]) one can determine a class of problems most of which are value Hadamard ill - posed.

27 Theorem *Consider the problem (54) under (6) and (12), assuming additionally that:*

the matrices A_4 and B_2 are constant;

$0 \in intV; m > k$ *(i.e. dim y > dim u);*
$$\phi(x, y) = \phi_1(x) + |y + A_4^{-1} A_3(T)x - p|,$$
where $\phi_1 : R^n \to (-\infty, +\infty)$ *is a continuous function and* $p \in R^m$. *Then for all matrices* A_4 *and* B_2 *except those in a set of Lebesgue measure zero (in the space of entries of* A_4 *and* B_2) *there exists* $\delta > 0$ *such that for all* p *in the open ball in* R^m *centered at the origin with radius* δ, *except those in a* k *- dimensional subspace (i.e. in a set of measure zero), the problem is value Hadamard ill - posed.*

The proof follows from the fact the that rank condition (57) is not fulfilled only for matrices from a set of measure zero. This condition implies $0 \in$ int W ($W = W(t) =$ const.). For every $p \in W$ we have

$$\phi_0(x) = \phi_1(x) \leq \phi_1(x) + \text{ dist } (p, -A_4^{-1}B_2V) \leq \phi(x, y)$$

for all $(x, y) \in S(t, 0)$. If $p \notin -A_4^{-1}B_2V$, then $\bar{\omega} < \omega(0)$. $\qquad \square$

PROOFS.

Proposition 23. Let $\varepsilon_k > 0, \varepsilon_k \to 0$ such that lim $\inf_{k \to +\infty} v(\varepsilon_k) < +\infty$. Then there exists a sequence $\bar{u}_k$ of optimal controls for $k = 1, 2, ...$ Without loss of generality, let $\bar{u}_k \rightharpoonup u_0$ in $L^1(0, T), u_0 \in U$. If $(\bar{x}_k, \bar{y}_k)$ is the corresponding trajectory of (1) then $\bar{y}_k$ is bounded in $L^2(0, T)$, since $v(\varepsilon_k) \geq \alpha \int_0^T |\bar{y}_k|^2 dt$ for some $\alpha > 0$. Choose a further subsequence $\bar{y}_k \rightharpoonup y_0$ in $L^1(0, T)$ as $k \to +\infty$. Then $x_k \to x_0$ in $C^0([0, T])$, where (x_0, y_0) satisfies (1.1), (3) with $u = u_0$ and $x(0) = x^0$, by theorem 3. Therefore $I(x_0, y_0, u_0) \geq v(0)$. It remains to use the lower semicontinuity of I in the appropriate topology. $\qquad \square$

Theorem 24. By repeating the argument in the proof of theorem 8 and taking into account the continuity properties of the cost functional one obtains upper semicontinuity of the value. Then proposition 23 completes the proof. $\qquad \square$

Section 5. Well- and Ill-Posedness in Semilinear Distributed Problems.

In this section we consider a quadratic optimal control problem for a singularly perturbed semilinear state equation. Under suitable monotonicity conditions we shall see that the formal limit problem agrees with the variational (true) limit problem. Under different assumptions ill-posedness will arise.

Let Ω be an open bounded subset of R^n. We consider a linear operator

$$A : D(A) \to L^2(\Omega)$$

which is selfadjoint on $L^2(\Omega)$, with domain $D(A)$ dense in $L^2(\Omega)$. Given $\varepsilon \geq 0, z^* \in L^2(\Omega)$ and some (fixed) function $g : R \to R$, consider the problem, to minimize

$$I(u, z) = \int_\Omega (z - z^*)^2 dx + \int_\Omega u^2 dx$$

on the set of all pairs (u, z) such that

(58)
$$\begin{cases} \varepsilon Az + g(z) = u \text{ a.e. in } \Omega, \\ z \in D(A), u \in L^2(\Omega). \end{cases}$$

Denote by $E(\varepsilon)$ the set of all pairs (u, z) satisfying (58), and by $v(\varepsilon)$ the optimal value:

$$v(\varepsilon) = \inf\{I(u, z) : (u, z) \in E(\varepsilon)\}, \varepsilon \geq 0.$$

We shall use the following conditions:

(59) g is continuous and $|g(z)| \leq M(1 + |z|)$ for some constant M and for every $z \in R$;

(60)
$$z \in D(A) \text{ implies } \int_\Omega (Az)g(z)dx \geq 0.$$

28 Theorem *Assume conditions (59) and (60). Then*

(61)
$$v(\varepsilon) \to v(0) \text{ as } \varepsilon \to 0.$$

If moreover g^2 is convex, then given $\varepsilon_k > 0, a_k > 0, \varepsilon_k \to 0, a_k \to 0$, if $(u_k, z_k) \in a_k - \arg\min(E(\varepsilon_k), I), k = 1, 2, ...,$ then

(62)
$$\begin{cases} z_k \to \bar{z}, u_k \to g(\bar{z}) \text{ in } L^2(\Omega), \\ I(u_k, z_k) \to v(0), \end{cases}$$

where for a.e. $x \in \Omega$

(63)
$$\bar{z}(x) = \arg\min[R, j(x, \cdot)], j(x, z) = [z - z^*(x)]^2 + g^2(z).$$

Proof. Step 1: $\limsup v(\varepsilon) \leq v(0)$ as $\varepsilon \to 0$.
Let $a > 0$ and $z \in D(A)$ be such that

$$I(g(z), z) \leq v(0) + a.$$

Then, as $\varepsilon \to 0$,

$$\limsup v(\varepsilon) \leq \limsup I(\varepsilon Az + g(z), z) = I(g(z), z) \leq v(0) + a.$$

This proves Step 1.
Step 2: $v(0) \leq \liminf v(\varepsilon)$ as $\varepsilon \to 0$.
Pick $z \in D(A)$. By (59), $\varepsilon Az + g(z) \in L^2(\Omega)$. By (60)

$$I(\varepsilon Az + g(z), z) = \int_\Omega \{(z - z^*)^2 + [\varepsilon Az + g(z)]^2\}dx \geq \int_\Omega (z - z^*)^2 dx + \int_\Omega g^2(z)dx,$$

hence

$$I(\varepsilon Az + g(z), z) \geq I(g(z), z) \geq v(0),$$

because $(g(z), z) \in E(0)$. Then $v(\varepsilon) \geq v(0)$ and Step 2 is proved.
Step 3: proof of (62).

Since $(u_k, z_k) \in a_k - \arg \min (E(\varepsilon_k), I)$, then

$$\int_\Omega (z_k - z^*)^2 dx \le v(\varepsilon_k) + a_k, k = 1, 2, ...$$

By (61)

(64)
$$z_k \rightharpoonup \omega \text{ in } L^2(\Omega)$$

for a subsequence. Then (60) and (62) yield existence of a sequence $l_k \to 0$ such that

(65)
$$\int_\Omega (z_k - z^*)^2 dx + \int_\Omega g(z_k)^2 dx \le I(u_k, z_k) \le v(0) + l_k$$
$$\le \int_\Omega (\omega - z^*)^2 dx + \int_\Omega g(\omega)^2 dx + l_k, k = 1, 2, ...$$

By (64) and the convexity of g^2

(66)
$$\lim \inf \int_\Omega g(z_k)^2 dx \ge \int_\Omega g(\omega)^2 dx.$$

Then from (65) and (66)

$$\lim \sup \int_\Omega (z_k - z^*)^2 dx \le \int_\Omega (\omega - z^*)^2 dx$$

implying

$$z_k \to \omega \text{ in } L^2(\Omega).$$

It remains to show that $\omega = \bar{z}$, where $\bar{z}$ is defined in (63).

The Nemitskij operator defined by g acts continuously on $L^2(\Omega)$, as follows from (59), (see Krasnoselskij [1, p. 30]), hence

(67)
$$g(z_k) \to g(\omega) \text{ in } L^2(\Omega).$$

By (65) and the definition of I we have

$$\int_\Omega u_k^2 dx \le I(u_k, z_k) \le \text{ constant },$$

so $\varepsilon_k A z_k$ is bounded in $L^2(\Omega)$ as well, because of (67). Moreover, for any $p \in D(A)$

$$\int_\Omega (\varepsilon_k A z_k) p dx = \varepsilon_k \int_\Omega z_k (Ap) dx \to 0,$$

hence $\varepsilon_k A z_k \rightharpoonup 0$ in $L^2(\Omega)$. Then, using (67), we have

$$u_k = \varepsilon_k A z_k + g(z_k) \rightharpoonup g(\omega) \text{ in } L^2(\Omega).$$

From (65) and the definition of I we see that

$$\lim \sup \int_\Omega u_k^2 dx \le \int_\Omega g(\omega)^2 dx,$$

thereby giving $u_k \to g(\omega)$ in $L^2(\Omega)$. Then (61) yields

(68) $$I(u_k, z_k) \to I(g(\omega), \omega) = v(0).$$

The function $\bar{z}$ is well defined in (63) by strict convexity of $j(x, \cdot)$, it is measurable and

$$j[x, \bar{z}(x)] \leq z^*(x)^2 + g(0)^2, \text{ a.e. } x \in \Omega.$$

Then $\bar{z} \in L^2(\Omega)$ and $g(\bar{z}) \in L^2(\Omega)$. Furthermore the pair $(g(\bar{z}), \bar{z})$ is the unique element $(u, y) \in L^2(\Omega) \times L^2(\Omega)$ giving $I(u, y) = v(0)$. So by (68) $\omega = \bar{z}$, proving (62). □

Theorem 28 applies when $A = -\Delta$ (the Laplace operator) with Dirichlet or Neumann boundary conditions, and g is an increasing function such that $g(0) = 0$.

In the following example (60) fails, and we introduce pointwise control constraints.

29 An ill-posedness example. We consider

$$\Omega = (0, T), T = 2^{5/6} \int_0^1 dt/\sqrt{t - t^4}.$$

Put $c = 3/\sqrt[3]{2}$ and minimize

(69) $$\int_0^T (z + c)^6 dx + \int_0^T u^2 dx$$

subject to

(70) $$\begin{cases} z'' + n^2 z^3 = n^2 u \text{ a.e. in } (0, T), n = 1, 2, 3, ..., z(0) = z(T) = 0, \\ u \in L^2(0, T), 1 \leq u(x) \leq 2 \text{ a.e in } (0, T). \end{cases}$$

Denote by v_n the optimal value of (69), (70). Here $Az = z''$ with Dirichlet boundary conditions, $g(z) = z^3, \varepsilon = 1/n^2$. Put

$$v_0 = \inf\{\int_0^T (z + c)^6 dx + \int_0^T u^2 dx : z^3 = u, 1 \leq u(x) \leq 2 \text{ a.e.}\}.$$

Of course the infimum is obtained by minimizing $z \to (z+c)^6 + z^6$ subject to $1 \leq z \leq \sqrt[3]{2}$, hence

(71) $$v_0 = T(1 + c)^6 + T.$$

We show that

(72) $$\limsup v_n < v_0.$$

Proof of (72). Consider the unique solution w to the Cauchy problem

(73) $$w'' + w^3 = 1, w(0) = w'(0) = 0.$$

A first integral for (73) is

$$w'(x)^2 + (1/2)w(x)^4 = 2w(x)$$

which shows a priori uniform boundedness of w and w'. It follows that (73) has, as claimed, an unique solution w defined on the whole real line and fulfilling

$$(74) \qquad \qquad 0 \leq w(x) \leq \sqrt[3]{4}.$$

By Arnold [1, p. 88-89], w is periodic of period

$$T = 2 \int_0^{\sqrt[3]{4}} \frac{dx}{\sqrt{2x - x^4/2}}.$$

Consider $z_n(x) = w(nx), n = 1, 2, 3, \ldots$. Then

$$(1/n^2)z_n'' + z_n^3 = 1, z_n(0) = z_n'(0) = 0.$$

Therefore

$$v_n \leq \int_0^T (w(nx) + c)^6 dx + T$$

$$= (1/n) \int_0^{nT} (w(x) + c)^6 dx + T = \int_0^T (w(x) + c)^6 dx + T.$$

Remembering (71), it suffices to prove that

$$(75) \qquad \qquad \int_0^T (w(x) + c)^6 dx < T(1 + c)^6$$

To prove (75) consider

$$h(u) = (\sqrt[3]{u} + c)^6.$$

Integrating (73) and using periodicity of w we get

$$\int_0^T w(x)^3 dx = T,$$

hence (75) can be rewritten as

$$(76) \qquad \qquad (1/T) \int_0^T h[w(x)^3]dx < h((1/T) \int_0^T w(x)^3 dx).$$

But (76) follows by the (strict) Jensen's inequality, since $0 \leq w(x)^3 \leq 4$ by (74) and h is strictly concave on $[0, 4]$. $\qquad \qquad \square$

Section 6.   Notes and Bibliographical Remarks.

To section 1. The reader can find a number of results concerning mainly the analytic theory of singular perturbations in control in the surveys of Kokotovic- O'Malley-Sannuti [1], O'Malley [1], Vasil'eva-Dmitriev [1], Pervozwanski- Gaitsgory [1], Kolmanovsky [1], Chernousko-Kolmanovsky [1], Saksena-O'Reilly- Kokotovic [1], Kokotovic [1], and in the recent books of Bensoussan [1] and Kokotovic-Khalil-O'Reilly [1]. Example 1 is based on a similar example in Kokotovic-Khalil-O'Reilly [1, p. 6].

To section 2. For statements, proofs and refinements of Tykhonov's theorem, see Tykhonov [1], Vasil'eva [1], Hoppensteadt [1].Results related to propositions 5,6 and 7 are announced in Dontchev-Veliov [5] and Veliov-Dontchev [1]. Lemma 9 is due to Flatto-Levinson [1].

To section 3. Theorem 16 is an extension of Dontchev-Veliov [1], where it is supposed that $M = [0, T]$. Theorem 17 is from Dontchev-Slavov [1].

To section 4. Dmitriev [1] was the first to observe that singularly perturbed optimal control problems of Mayer's type may be value Hadamard ill-posed. This effect was justified later in Dontchev-Veliov [1] who gave the form of the true limit problem as it is presented in theorem 25. Theorem 24 is from Dontchev-Veliov [5], where it is given in a more general form. Other results concerning solution Hadamard well-posedness are given in Dontchev [1, ch. 3.]

To section 5. Theorem 28 and example 29 are due to Haraux-Murat [1], where further results may be found. For generalizations, see Buttazzo-Dal Maso [2].

Extensions. Singularly perturbed linear control systems containing stable and unstable fast subsystems are studied in Dontchev-Veliov [4] and Gičev [2]. A special case when exp $(A_4 t)$ is periodic is considered in Dontchev-Veliov [3]. The behaviour of time-optimal controls is investigated in Dontchev-Gičev [1]. In Dontchev-Veliov [2] singular perturbation techniques are applied to study asymptotic properties of linear differential inclusions.

Chapter VIII.

WELL-POSEDNESS IN THE CALCULUS OF VARIATIONS

Very few well-posedness results (in Tykhonov or Hadamard sense) are available in the classical calculus of variations. The main research efforts were addressed to get existence, necessary conditions and regularity of the minimizers. Moreover, Hadamard well-posedness for e.g. the Lagrange problem in the calculus of variations is related to somewhat sophisticated topologies (or convergences) on the integrands (coming from epi-convergence theory) which were developed only recently.

We shall see, however, that Tykhonov and Hadamard well-posedness are deeply related to the classical issues of existence, necessary conditions and regularity of minimizers.

Section 1.   Integral Functionals Without Derivatives.

Tykhonov well-posedness. We are given a Lebesgue measurable set $T \subset R^s$, a multi-function

$$K : T \rightrightarrows R^m$$

and a function

$$f : T \times R^m \to (-\infty, +\infty].$$

We denote by X the metric space of all measurable selections of K which belong to $L^1(T)$ (if any), equipped with the $L^1(T)$ metric. We wish to characterize Tykhonov well-posedness of (X, I), where

$$(1) \qquad\qquad I(x) = \int_T f[t, x(t)]dt.$$

We shall freely employ some standard terminology and definitions which may be found in Rockafellar [6].

Standing assumptions: f is a normal integrand, K is a measurable multifunction with nonempty closed values.

The integral functional (1) is defined according to the convention of Rockafellar [6].

We shall show that $x_0 \in \arg\min(X, I)$ iff x_0 is a measurable selection of K and $x_0(t) \in \arg\min[K(t), f(t, \cdot)]$ for a.e. $t \in T$, provided that I is proper on X. Under some more assumptions, $x_0 \in X$. Therefore the (infinite-dimensional) optimization problem (X, I) is, in an obvious sense, equivalent to the simpler finite- dimensional problem $[K(t), f(t, \cdot)]$ with parameter $t \in T$. This suggests in a natural way that Tykhonov well-posedness of (X, I) and of $[K(t), f(t, \cdot)]$, $t \in T$, should be related too.

Assume that $x_0(t) \in \arg\min[K(t), f(t, \cdot)]$, $t \in T$ for some measurable x_0. Remembering theorem I.11 we consider $\varepsilon - \arg\min[K(t), f(t, \cdot)]$ denoted as

(2) $$P(t, \varepsilon) = \{x \in K(t) : f(t, x) \leq f(t, x_0(t)) + \varepsilon\}, \ t \in T; \ \varepsilon > 0.$$

Suppose (for simplicity) that

(3) $$\sup\{|x| : x \in K(t)\} \leq a(t) \text{ for some } a \in L^1(T),$$

and let $I(x_0)$ be finite. Fix $\varepsilon > 0$ and $y \in \varepsilon - \arg\min(X, I)$. Consider

$$A = \{t \in T : f(t, y(t)) > f(t, x_0(t)) + \sqrt{\varepsilon}\}.$$

Then $\int_A f(t, y) dt \geq \sqrt{\varepsilon} \text{ meas } A + \int_A f(t, x_0) dt$, but

$$\int_A f(t, y) dt \leq \int_A f(t, x_0) dt + \varepsilon,$$

thus meas $A \leq \sqrt{\varepsilon}$ and $y(t) \in P(t, \sqrt{\varepsilon})$ whenever $t \notin A$. Then given $y, z \in \varepsilon - \arg\min(X, I)$ there exists $E \subset T$ such that meas $E \leq 2\sqrt{\varepsilon}$ and

$$\int_T |y - z| dt \leq \int_T \text{ diam } P(t, \sqrt{\varepsilon}) dt + \int_E |y - z| dt$$

(since diam $P(\cdot, \sqrt{\varepsilon}) \in L^1(T)$, as easily seen). So we get by (3)

(4) $$\text{diam } [\varepsilon - \arg\min(X, I)] \leq \int_T \text{ diam } P(t, \sqrt{\varepsilon}) dt + 2 \int_E a \, dt.$$

Assume now Tykhonov well-posedness of $(K(t), f(t, \cdot))$ for a.e. $t \in T$. Then by theorem I.11 we get

$$\int_T \text{ diam } P(t, \sqrt{\varepsilon}) dt \to 0 \text{ as } \varepsilon \to 0.$$

Hence from (4) we obtain well-posedness of (X, I) by theorem I.11 again, provided I is lower semicontinuous and bounded from below on X. The converse holds true, in fact we have

1 Theorem *Assume that T is of finite measure,*

(5) *for every $q \in L^1(T)$, every measurable selection of*

$$t \rightrightarrows \{x \in K(t) : f(t, x) \leq q(t)\}$$

belongs to $L^1(T)$, and

(6) $$\inf f(\cdot, K(\cdot)) \in L^1(T).$$

Then (X, I) is Tykhonov well-posed iff $[K(t), f(t, \cdot)]$ is for a.e. $t \in T$.

Hadamard well-posedness. Now we come to Hadamard well-posedness of minimum problems involving I defined by (1). Setting

$$g(t, x) = f(t, x) \text{ if } x \in K(t), = +\infty \text{ otherwise },$$

then g is a normal integrand (provided f is and it is proper). Thus we get equivalence of the two problems (X, I) and $(L^1(T), J)$ where

$$J(x) = \int_T g[t, x(t)]dt$$

(provided J is proper). This reduction to unconstrained optimization problems allows us to fix the following (simpler) setting.

Given f and a suitable function space Y with $f \in Y$, we are interested in finding convergences (or topologies) on Y giving solution and value Hadamard well-posedness of (Y, I).

We shall consider only convex problems. In this case we know from section IV.6 that epi-convergence plays a fundamental role as far as Hadamard well-posedness is involved, while (at least in quadratic cases) Mosco convergence has a basic role, as we saw in section IV.9.

We shall use the following conditions.

(7) T is a measure space equipped with a countably finite positive measure dt;
$1 < p < \infty, 1/p + 1/q = 1$;

(8) $f_n : T \times R^m \to (-\infty, +\infty]$ is a sequence of normal integrands such that there exist an a.e. finite function $a \in L^p(T), w \in L^1(T)$ and a sequence $a_n \in L^p(T)$ such that

$$|a_n(t)| \le a(t), f_n[t, a_n(t)] \le w(t), \text{ a.e. } t \in T;$$

(9) there exist $b \in L^q(T), z \in L^1(T), b_n \in L^q(T)$ such that

$$|b_n(t)| \le b(t), f_n^*[t, b_n(t)] \le z(t), \text{ a.e. } t \in T.$$

Here $f_n^*(t, u) = \sup\{y'u - f_n(t, y) : y \in R^m\}$. We set

$$I_n(u) = \int_T f_n[t, u(t)]dt, I(u) = \int_T f[t, u(t)]dt.$$

2 Theorem *Let f_n be a sequence of normal convex integrands, and $f : T \times R^m \to (-\infty, +\infty]$. Assume conditions (7), (8), (9). If*

$$M - \lim f_n(t, \cdot) = f(t, \cdot) \text{ in } R^m \text{ for every } t \in T$$

then

$$M - \lim I_n = I \text{ in } L^p(T).$$

For narrower classes of convex integrands, we obtain sharper results, in the form of direct characterizations of epi-convergence, as follows. We shall need the following condition:

(10) T is a bounded measurable set in R^s and $f, f_n : T \times R^m \to R$ are normal convex integrands.

3 Theorem *Assume condition (10). If there exist $p > 1$ and $c > 0$ such that $0 \leq f_n(t,x) \leq c|x|^p$ for every x, n and a.e. $t \in T$, then*

(11) *strong epi lim $I_n = I$ in $L^p(T)$ iff $f_n(\cdot, u) \to f(\cdot, u)$ in $L^\infty(T)$ for every $u \in R^m$.*

If there exist $a > 0$ and $p > 1$ such that $f_n(t,x) \geq a|x|^p$ for every x, n and a.e. $t \in T$, $f_n(t,0) = f(t,0) = 0$, then

(12) *weak epi-lim $I_n = I$ in $L^p(T)$ iff $f_n^*(\cdot, u) \to f^*(\cdot, u)$ in $L^\infty(T)$ for every $u \in R^m$.*

From theorem 3 we get

4 Corollary *Assume condition (10) and suppose that*

$$c_1|x|^p \leq f_n(t,x) \leq c_2|x|^p, c_1 > 0, p > 1 \text{ and } f(t,0) = 0$$

for every n, x and a.e. $t \in T$. Then

$M - lim\ I_n = I$ in $L^p(T)$ iff $f_n(\cdot, u) \to f(\cdot, u)$ and $f_n^*(\cdot, x) \to f^*(\cdot, x)$ in $L^\infty(T)$ for every u and x.

5 Example. Let $A(t), A_n(t)$ be a sequence of square symmetric measurable matrices on T, which fulfills (10), with eigenvalues between c_1 and $c_2, c_1 > 0$, and consider

$$J_n(u) = \int_T u(t)' A_n(t) u(t) dt, u \in L^2(T).$$

Then by corollary 4, $M - \lim J_n = J$ with

$$J(u) = \int_T^\cdot u(t)' A(t) u(t) dt$$

iff $A_n \to A$ and $A_n^{-1} \to A^{-1}$ in $L^\infty(T)$, since setting $f_n(t,x) = x' A_n(t) x$, then $f_n^*(t,x) = (1/4) x' A_n(t)^{-1} x$.

PROOFS.

Theorem 1. Assume well-posedness of (X, I) and let $x_0 = \arg\min(X, I)$. Then I is proper on X. We need the following

6 Lemma For a.e. $t \in T$ we have $x_0(t) \in \arg\min[K(t), f(t, \cdot)]$.

Taking the lemma for granted we see that $f[\cdot, x_0(\cdot)] \in L^1(T)$ since $f(t, x_0(t)) = \inf f[t, K(t)]$ a.e. By theorem I.11 it suffices to show that diam $P(t, \varepsilon) \to 0$ as $\varepsilon \to 0$ for a.e. $t \in T$. We begin by proving that

(13) $$P(t, \varepsilon) \text{ is compact for all } \varepsilon \text{ and a.e. } t \in T.$$

Arguing by contradiction, there exists a subset $E \subset T$ of positive measure, such that $P(t, \varepsilon)$ (which is closed) is unbounded for all $t \in E$ and some $\varepsilon > 0$. Let b be measurable on E such that

$$0 < b(t) < +\infty \text{ if } t \in E \text{ and } \int_E b dt = +\infty.$$

By the measurable selection theorem (see Rockafellar [6]) we can find some measurable function ω on E such that

$$\omega(t) \in P(t, \varepsilon), |\omega(t)| \geq b(t), t \in E.$$

Now define $u = \omega$ on $E, u = x_0$ on $T \setminus E$. Then u is a measurable selection of $P(\cdot, \varepsilon)$. By (5)

$$+\infty > \int_T |u| dt \geq \int_{T \setminus E} |x_0| dt + \int_E b dt = +\infty,$$

a contradiction. This yields compactness of all sets $P(t, \varepsilon)$, therefore

$$\text{diam } P(t, \varepsilon) = |y(t) - z(t)|, t \in T$$

for some measurable selections y, z of $P(\cdot, \varepsilon)$. Both y and z are in X by (5). Write

$$M(\varepsilon) = \varepsilon - \arg\min(X, I).$$

So by (2) $y, z \in M(\varepsilon \text{ meas } T)$, yielding

$$\int_T |y - z| dt = \int_T \text{ diam } P(t, \varepsilon) dt \leq \text{ diam } M(\varepsilon \text{ meas } T).$$

By theorem I.11 we know that diam $M(\varepsilon \text{ meas } T) \to 0$ as $\varepsilon \to 0$, therefore diam $P(\cdot, \varepsilon) \to 0$ in $L^1(T)$ as $\varepsilon \to 0$. Then diam $P(t, \varepsilon) \to 0$ a.e. in T since it is an increasing function of ε. This implies well-posedness of $[K(t), f(t, \cdot)]$ for a.e. $t \in T$ again by theorem I.11 (since by lemma 6, $f[t, x_0(t)] = \inf f[t, K(t)]$).

Proof of lemma 6 . Set
$$v(t) = \inf f[t, K(t)], t \in T.$$

Arguing by contradiction, assume that

$$f(t, x_0(t)) > v(t), t \in E, \text{ some } E \text{ with meas } E > 0.$$

Then consider

$$E_1 = \{t \in E : -\infty < v(t) < +\infty, \text{ and } \inf f(t, K(t)) \text{ is not attained }\};$$

$$E_2 = \{t \in E : -\infty < v(t) < +\infty, \text{ and } \inf f(t, K(t)) \text{ is attained }\}.$$

We shall use in the rest of the proof some (standard) facts about measurability and measurable selections of multifunctions, which may be found in Rockafellar [6]. Setting

$$A = \{t \in E : -\infty < v(t) < +\infty\},$$

$$B = \{t \in E : f(t, x) \leq v(t) \text{ for some } x \in K(t)\}$$

we see that $E_2 = A \cap B, A$ is (Lebesgue) measurable, while B is the projection on E of

$$\text{gph } K \cap \{(t, x) \in E \times R^m : f(t, x) \leq v(t)\}$$

which is measurable with respect to the sigma-algebra generated by the products of Lebesgue sets in E and Borel sets in R^m, since f is a normal integrand. Thus B is (Lebesgue) measurable, therefore E_2 too. Also E_1 is, since it can be represented as $\{t \in E : -\infty < v(t) < +\infty, f(t, x) > v(t) \text{ for every } x \in K(t)\}$. Using the measurable selection theorem, we fix $\varepsilon > 0$ and find some measurable function ω on E such that

$$\omega(t) \in K(t), t \in E,$$

$$f(t, \omega(t)) = v(t), t \in E_2; f(t, \omega(t)) \leq v(t) + \varepsilon, t \in E_1,$$

$$f(t, \omega(t)) \leq -1/\varepsilon \text{ on the set of measure 0 where } v(t) = -\infty.$$

Then consider the measurable function

$$u = \omega \text{ in } E, u = x_0 \text{ in } T \setminus E.$$

We get $I(x_0) \leq \int_T f(t, u)dt$, since this inequality is obvious if $f(\cdot, u(\cdot)) \notin L^1(T)$ (remembering that $f[t, u(t)] \geq v(t)$ a.e.), and comes from (5) if $f[\cdot, u(\cdot)] \in L^1(T)$. Then

$$\int_E f(t, x_0)dt \leq \int_{E_1} f(t, \omega)dt + \int_{E_2} f(t, \omega)dt \leq \varepsilon \text{ meas } E_1 + \left(\int_{E_1} + \int_{E_2}\right)v \, dt.$$

By Lusin's theorem, there exist a closed set $F \subset E$ with positive measure, and some $k > 0$ such that

$$f(t, x_0(t)) \geq k + v(t), t \in F.$$

Then for ε sufficiently small

$$\int_E f(t, x_0)dt \leq \varepsilon \text{ meas } E_1 + \int_E f(t, x_0)dt - k \text{ meas } F < \int_E f(t, x_0)dt,$$

a contradiction. □

End of proof of theorem 1 . Assume now $[K(t), f(t, \cdot)]$ well-posed for a.e. $t \in T$. By th. 2K and cor. 1C of Rockafellar [6], there exists an a.e. unique measurable function x_0 such that

(14) $$x_0(t) \in K(t), f[t, x_0(t)] \leq f(t, u) \text{ if } u \in K(t), \text{ a.e. } t \in T.$$

Then $x_0 \in X$ by (5) and (6). Of course $x_0 \in \arg\min(X, I)$. Let u_n be any fixed minimizing sequence for (X, I). By (14) we get

(15) $$f[\cdot, u_n(\cdot)] \to f[\cdot, x_0(\cdot)] \text{ in } L^1(T).$$

Therefore by well-posedness of $[K(t), f(t, \cdot)]$ a.e., by (14) and (15) some subsequence fulfills

(16) $$u_n(t) \to x_0(t), f[t, u_n(t)] \to f[t, x_0(t)] \text{ a.e. in } T.$$

Arguing by contradiction, assume that $u_n \not\to x_0$ in $L^1(T)$. Put $x_n = u_n - x_0$. Then by Vitali's theorem (Dunford-Schwartz [1, III. 6. 15]) there exists $\varepsilon > 0$ such that for every $a > 0$ we can find n and a measurable subset E of T such that

$$\text{meas } E < a, \int_E |x_n| dt \geq 2\varepsilon.$$

Fix E_{11}, n_1 such that $\int_{E_{11}} |x_{n_1}| dt \geq 2\varepsilon$. For some $a_1 > 0$, meas $E < a_1$ implies $\int_E |x_{n_1}| dt < \varepsilon/2$. Then for some measurable subset E_{21} and some $n_2 > n_1$ we have

$$\text{meas } E_{21} < a_1, \int_{E_{21}} |x_{n_2}| dt \geq 2\varepsilon.$$

Putting $E_{12} = E_{11} \setminus E_{21}$ we obtain

$$\int_{E_{12}} |x_{n_1}| dt = (\int_{E_{11}} - \int_{E_{11} \cap E_{21}}) |x_{n_1}| dt \geq 2\varepsilon - \varepsilon/2.$$

For some $a_2 > 0$, meas $E < a_2$ implies

$$\int_E |x_{n_1}| dt < \varepsilon/4 \text{ and } \int_E |x_{n_2}| dt < \varepsilon/4.$$

Then for some measurable subset E_{31} and some $n_3 > n_2$ we have

$$\text{meas } E_{31} < a_2, \int_{E_{31}} |x_{n_3}| dt \geq 2\varepsilon.$$

Putting $E_{13} = E_{12} \setminus E_{31}, E_{22} = E_{21} \setminus E_{31}$ we obtain

$$\int_{E_{13}} |x_{n_1}| dt = (\int_{E_{12}} - \int_{E_{12} \cap E_{31}}) |x_{n_1}| dt \geq 2\varepsilon - \varepsilon/2 - \varepsilon/4;$$

$$\int_{E_{22}} |x_{n_2}| dt \geq 2\varepsilon - \varepsilon/4.$$

Continuing in this way, we obtain a subsequence of x_n, and countably many sequences of measurable subsets of T,

$$E_{1k}, E_{2k}, ..., E_{nk}, ...,$$

such that for every n and k

$$E_{nk+1} \subset E_{nk} \quad , \int_{E_{nk}} |x_n| dt \geq \varepsilon.$$

Put $E_n = \cap_k E_{nk}$. Then E_n are disjoint (since $E_{nk} \cap E_{ph} = \emptyset$ if $p < n < h$) and

$$\int_{E_n} |x_n| dt = \int_{E_n} |u_n - x_0| dt \geq \varepsilon.$$

Moreover, for every n there exists $b_n > 0$ such that

$$\int_F |u_n - x_0| dt < \varepsilon/2 \text{ whenever meas } F < b_n.$$

By Egorov's theorem, given n we can find $F_n \subset E_n$ such that meas $E_n \leq$ meas $F_n + b_n, f[\cdot, u_k(\cdot)] \to f[\cdot, x_0(\cdot)]$ uniformly on F_n. Then for some further subsequence

$$(17) \qquad \int_{F_n} |u_n - x_0| dt \geq \varepsilon/2,$$

moreover

$$(18) \qquad f(t, u_n(t)) \leq f(t, x_0(t)) + 1 \text{ if } t \in F_n.$$

Now define

$$\bar{u}(t) = x_0(t) \text{ if } t \notin \cup_{n=1}^{\infty} F_n; \bar{u}(t) = u_n(t) \text{ if } t \in F_n.$$

By (18) and (5), (6) we get $\bar{u} \in L^1(T)$. On the other hand, by (17)

$$\int_T |\bar{u} - x_0| dt = \sum_{n=1}^{\infty} \int_{F_n} |u_n - x_0| dt = +\infty,$$

a contradiction. □

Theorem 2. We shall use freely some results from Rockafellar [6] about measurability of multifunctions, normal integrands and integral functionals. More explicitly put

$$I(f_n, u) = I_n(u), I(f, u) = I(u).$$

Step 1. f is a normal convex integrand.

The convexity of $f(t, \cdot)$ comes from the convexity of each $f_n(t, \cdot)$, and the lower semicontinuity from remark IV. 21. For a.e. t, $a_n(t)$ has some convergent subsequence in R^m to y say, therefore from Mosco convergence of f_n,

$$(19) \qquad +\infty > w(t) \geq \liminf f_n[t, a_n(t)] \geq f(t, y), |y| \leq a(t),$$

thus $f(t, \cdot)$ is not identically $+\infty$. Consider now

$$E_n(t) = \text{ epi } f_n(t, \cdot), E(t) = \text{ epi } f(t, \cdot), t \in T.$$

By Attouch [3, th. 1.39 p.98] we get $K - \lim E_n(t) = E(t), t \in T$, therefore $E(t)$ is closed (Kuratowski [1 p.245]). By theorem II.9

$$\text{dist } [x, E_n(t)] \to \text{ dist } [x, E(t)], t \in T \text{ and every } x \in R^m.$$

This entails measurability of dist $[x, E(\cdot)]$, thus showing measurability of $E(\cdot)$. Therefore f is a normal integrand.

Step 2. Every I_n and I are proper on $L^p(T)$, moreover

$$I^*(f, \cdot) = I(f^*, \cdot) \text{ on } L^q(T).$$

We show firstly properness of $I(f, \cdot)$. Fix $u \in L^p(T)$. Then for a.e. $t \in T$ there exist $x_n \to u(t)$ in R^m such that

$$f_n(t, x_n) \to f[t, u(t)].$$

Let $v_n(t)$ be the unique minimum point on R^m of

$$x \to g_n(t, x) + (1/p)|x - u(t)|^p,$$

where

(20) $$g_n(t, x) = \max\{0, f_n(t, x) - f[t, u(t)]\}.$$

Then

$$g_n[t, v_n(t)] + (1/p)|v_n(t) - u(t)|^p \le g_n(t, x_n) + (1/p)|x_n - u(t)|^p$$

showing that

(21) $$v_n(t) \to u(t), f_n[t, v_n(t)] \to f[t, u(t)], \text{ a.e. } t \in T.$$

Moreover

$$f_n[t, v_n(t)] \ge v_n(t)'b_n(t) - f^*[t, b_n(t)] \ge -|v_n(t)|b(t) - z(t),$$

giving as $n \to +\infty$

(22) $$f[t, u(t)] \ge h(t) \text{ for some } h \in L^1(T) \text{ and a.e. } t \in T.$$

So $I(f, \cdot)$ is well defined and $I(f, \cdot) > -\infty$ everywhere. We show that $I(f, \cdot)$ is somewhere finite. To this aim we consider

$$F(t) = \{x \in R^m : f(t, x) \le w(t), |x| \le a(t)\}.$$

By (19) we see that $F(t) \ne \emptyset$ for a.e. t. Then the multifunction $F : T \rightrightarrows R^m$ has measurable selections $c \in L^p(T)$, therefore $I(c) < +\infty$. The properness of $I(f, \cdot)$ then implies $I^*(f, \cdot) = I(f^*, \cdot)$. A similar reasoning gives the conclusion for I_n.

Step 3. Fix $u \in L^p(T)$ such that $I(f, u) < +\infty$. Then $f[\cdot, u(\cdot)] \in L^1(T)$. Consider v_n as defined in step 2. Then by (8) and (20)

$$(1/p)|u(t) - v_n(t)|^p + g_n[t, v_n(t)] \le (1/p)|u(t) - a_n(t)|^p + g_n[t, a_n(t)]$$
$$\le (\text{const.}) (|u(t)|^p + |a(t)|^p) + \max\{0, w(t) - f[t, u(t)]\} = k(t),$$

where $k \in L^1(T)$. Then by (21) and the dominated convergence theorem

$$v_n \to u \text{ in } L^p(T); g_n[t, v_n(t)] \to 0 \text{ in } L^1(T).$$

Therefore $\int_T g_n(t, v_n)dt \to 0$, yielding

$$\max\{0, I(f_n, v_n) - I(f, u)\} \to 0.$$

This shows that

(23) $$\limsup I(f_n, v_n) \le I(f, u).$$

Now let $y_n \rightharpoonup y$ in $L^q(T)$. Fix $u \in L^p(T)$ and consider $v_n \to u$ there fulfilling (23). We have

$$I^*(f_n, y_n) \ge <y_n, v_n> - I(f_n, v_n)$$

and letting $n \to +\infty$

$$\liminf I^*(f_n, y_n) \ge <y, u> - I(f, u).$$

Taking the supremum with respect to u we get

(24) $$\liminf I^*(f_n, y_n) \geq I^*(f, y).$$

Similarly to (23), we see that for every $v \in L^q(T)$ there exists $u_n \to v$ in $L^q(T)$ such that $\limsup I^*(f_n, u_n) \leq I^*(f, v)$, since $I^*(f_n, \cdot) = I(f_n^*, \cdot)$. Then, similarly to (24), we get

$$x_n \rightharpoonup x \text{ in } L^p(T) \Rightarrow \liminf I_n(f_n, x_n) \geq I(f, x)$$

since $I^{**}(f_n, \cdot) = I(f_n^{**}, \cdot) = I(f_n, \cdot)$ and the same is true for f, as we see from step 2. $\square$

Theorem 3. **Proof of the first half.** Assume that $f_n(\cdot, u) \to f(\cdot, u)$ for every u. Since from the assumptions

$$0 \leq I_n(u) \leq c\|u\|^p, u \in L^p(T),$$

the sequence I_n is locally equi-Lipschitz continuous on $L^p(T)$ (this is proved as in Rockafellar [3, th. 10.6]). By example IV.10 it suffices to prove that $I_n \to I$ pointwise on $L^p(T)$. By local Lipschitz continuity, it suffices to show that

(25) $$I_n(u) \to I(u) \ \forall u \in C^0(cl \ T)$$

(see Dunford-Schwartz [1, IV.8.19]). Fix $u \in C^0(cl \ T)$. By equiboundedness and convexity, the sequence

$$\text{ess sup } \{f_n(t, \cdot) : t \in T\}$$

is (again) locally equi-Lipschitz, as well as ess sup $\{f(t, \cdot) : t \in T\}$ since convergence of $f_n(\cdot, u)$ yields

$$0 \leq f(t, x) \leq c|x|^p.$$

Since $u(cl \ T)$ is compact in R^m, there exists a constant $L > 0$ such that

$$|f_n(t, y) - f_n(t, z)| \leq L|y - z|, |f(t, y) - f(t, z)| \leq L|y - z|$$

for a.e. t, every n and all y, z in $u(T)$. Given $\varepsilon > 0$ let $V_1, ..., V_r$ be a measurable partition of $u(T)$ such that diam $V_k \leq \varepsilon$ for all k, and fix $y_k \in V_k$. Put

$$t \in T_k \text{ iff } t \in T \text{ and } u(t) \in V_k.$$

Then $T_1, ..., T_r$ is a measurable partition of T, and

$$|u(t) - y_k| \leq \varepsilon \text{ if } t \in T_k, k = 1, ..., r.$$

Moreover $|\int_{T_k}[f_n(t, y_k) - f(t, y_k)]dt| < \varepsilon/r$ for $k = 1, ..., r$ and all sufficiently large n. Thus

$$|\int_T \{f_n[t, u(t)] - f[t, u(t)]\}dt| \leq \sum_{k=1}^r |\int_{T_k} [f_n(t, u) - f_n(t, y_k)]dt|$$

$$+ \sum_{k=1}^r |\int_{T_k} [f_n(t, y_k) - f(t, y_k)]dt| + \sum_{k=1}^r |\int_{T_k} \{f(t, y_k) - f[t, u(t)]\}dt|$$

$$\leq 2L \sum_{k=1}^r \int_{T_k} |y_k - u(t)|dt + \varepsilon \leq 2L\varepsilon \text{ meas } T + \varepsilon$$

for all sufficiently large n. This proves (25).

Conversely, assume that

$$\text{strong epi-lim } I_n = I \text{ in } L^p(T).$$

For every $x \in R^m$, $f_n(\cdot, x)$ is a bounded sequence in $L^\infty(T)$. Let Q denote the set of points in R^m with rational coordinates. By a diagonal procedure we can find a subsequence such that

$$f_n(\cdot, x) \to g(\cdot, x) \text{ in } L^\infty \text{ for every } x \in Q,$$

where $g : T \times Q \to R$. For every measurable $E \subset T$ we get by assumption

$$0 \le (1/\text{meas } E) \int_E f_n(t, x) dt \le c|x|^p.$$

Then $0 \le g(t, x) \le c|x|^p$, a.e. $t \in T$ and every x. An analogous reasoning shows that $g(t, \cdot)$ is locally Lipschitz continuous (since $f_n(t, \cdot)$ are locally equi-Lipschitz) on Q. Therefore $g(t, \cdot)$ can be extended by continuity on the whole R^m, moreover it is convex since $f_n(t, \cdot)$ is. Fix $z \in L^1(T), x \in R^m$, and let $x_k \in Q$ be such that $x_k \to x$. Then

$$\left| \int_T [f_n(t, x) - g(t, x)]z(t)dt \right| \le \left| \int_T [f_n(t, x) - f_n(t, x_k) + g(t, x_k) - g(t, x)]z(t)dt \right|$$

$$+ \left| \int_T [f_n(t, x_k) - g(t, x_k)]z(t)dt \right| \le (\text{ const. })|x - x_k| + \left| \int_T [f_n(t, x_k) - g(t, x_k)]z(t)dt \right|,$$

yielding

$$f_n(\cdot, x) \to g(\cdot, x) \text{ in } L^\infty(T) \text{ for every } x \in R^m.$$

By the first part of the proof and uniqueness of epi-limit (proposition IV.9) it follows that

$$(26) \qquad \int_T f(t, u)dt = \int_T g(t, u)dt \text{ for every } u \in L^p(T).$$

Fix $x \in R^m$, let $E \subset T$ be measurable and consider

$$u(t) = x \text{ if } t \in E, = 0 \text{ if } t \notin E.$$

Then by (26) and continuity of $g(t, \cdot)$, $f(t, \cdot)$ we get

$$f(t, x) = g(t, x) \text{ for a.e. } t \in T, \text{ every } x \in R^m.$$

Hence the original sequence $f_n(\cdot, x) \to f(\cdot, x)$.

Proof of the second half. We know from theorem IV.34 that weak epi-lim $I_n = I$ on $L^p(T)$ iff strong epi-lim $I_n^* = I^*$ on $L^q(T), 1/p + 1/q = 1$. Since (by Rockafellar [6, th. 3C])

$$I_n^*(u) = \int_T f_n^*[t, u(t)]dt, u \in L^q(T),$$

the conclusion follows from (11).   □

Section 2.   Lagrange Problems.

We consider Tykhonov well-posedness of the simplest (Lagrange) problem in the calculus of variations: to minimize

$$(27) \qquad \int_0^T f[x(t), \dot{x}(t)]dt$$

on the set of all absolutely continuous functions $x : [0, T] \to R^m$ such that

$$(28) \qquad x(0) = \bar{x}, x(T) = y,$$

where $\bar{x}, y \in R^m$, and $\bar{x}$ is fixed throughout.

The basic approach is a dynamic programming one. We assume the following about $f = f(x, u)$:

$$(29) \qquad f \in C^3(R^{2m}) \text{ and } f_{uu}(x, u) \text{ is everywhere positive definite.}$$

(30) there exists a continuous $g : [0, +\infty) \to R$ such that $g(t)/t \to +\infty$ as $t \to +\infty$ and $f(x, u) \ge g(|u|)$ for every x and u.

We define the value function

$$V : [0, +\infty) \times R^m \to R$$

as

$$V(T, y) = \inf\{\int_0^T f[x(t), \dot{x}(t)]dt : x \in W^{1,1}(0, T), x(0) = \bar{x}, x(T) = y\}.$$

By (29), (30) it is well known that V is finite-valued and $V(T, y)$ is attained for every T, y (see Dacorogna [1]). We denote by

$$Q(T, y)$$

the corresponding problem (27), (28). Every optimal solution $x \in \text{arg min } Q(T, y)$, for any T, y, belongs to $C^2(0, T)$ and fulfills the Euler-Lagrange equation

$$(d/dt)f_u(x(t), \dot{x}(t)) = f_x(x(t), \dot{x}(t)), 0 \le t \le T,$$

that by (29) may be written equivalently in normal form

$$\ddot{x}(t) = E[x(t), \dot{x}(t)], 0 \le t \le T, \text{ where } E = f_{uu}^{-1}(f_x - f_{ux}\dot{x}).$$

See Fleming-Rishel [1, ch. I].

The regularity properties of V at (T, y) determine the Tykhonov well-posedness of $Q(T, y)$: in a sense, there is equivalence between uniqueness of the minimizer for $Q(T, y)$ and differentiability of V at (T, y). Under (29), (30) uniqueness of the minimizer for $Q(T, y)$ entails Tykhonov well-posedness (in a restricted sense), a further example of an often observed phenomenon. As in other aspects of the calculus of variations, the value function provides, through its smoothness, a complete picture of the well-posedness of the optimization problem, not only in Tykhonov's sense, but also for Hadamard well-posedness with respect to the end point (T, y). These facts are related to theorem III.35 (here the natural parameter is $p = (T, y)$).

7 Theorem *Assume that (29), (30) hold. Let $T > 0$. Then V is differentiable at (T, y) iff $Q(T, y)$ is Tykhonov well-posed in the following sense: every minimizing sequence which is bounded in $W^{1,\infty}(0, T)$ is strongly convergent in $W^{1,2}(0, T)$.*

8 Theorem *V is locally Lipschitz continuous in $(0, +\infty) \times R^m$ if (29), (30) hold.*

From theorem 8 (and Rademacher's theorem) we see that $Q(T, y)$ is Tykhonov well-posed (as in theorem 7) for a dense set of points (T, y). In fact the following stronger genericity result holds.

9 Theorem *$Q(T, y)$ is Tykhonov well-posed as in theorem 7 for a dense G_δ set of points (T, y) in R^m, provided that (29) and (30) hold.*

Tykhonov well-posedness for multiple integrals will now be studied. Let Ω be a bounded open connected set in R^m. Consider a Carathéodory function

$$f = f(x, p) : \Omega \times R^m \to R,$$

and (for a given q) a fixed nonempty closed convex set $K \subset W_0^{1,q}(\Omega)$. We write

$$I(u) = \int_\Omega f(x, \nabla u) dx.$$

10 Theorem *Assume that*

(31) $f(x, \cdot)$ *is strictly convex for a.e. $x \in \Omega$;*

(32)   *for a.e. x and every p, $C|p|^q + a(x) \le f(x, p)$ where $q > 1, C > 0, a \in L^1(\Omega)$.*

Then (K, I) is Tykhonov well-posed in $(W_0^{1,q}(\Omega)$, strong).

PROOFS.

Theorem 7. We shall use the following

11 Lemma Given any compact $K \subset (0, +\infty) \times R^m$, there exist constants A, B such that

$$|x(t)| \le A, |\dot{x}(t)| \le B, 0 \le t \le T,$$

for every $x \in$ arg min $Q(T, y)$ if $(T, y) \in K$.

Proof of lemma 11. Fix K, let $(T, y) \in K$, and consider any $x \in$ arg min $Q(T, y)$. We have

(33)
$$\int_0^T g(|\dot{x}|)dt \le \int_0^T f(x, \dot{x})dt \le \int_0^T f[\bar{x} + (\frac{y - \bar{x}}{T})t, \frac{y - \bar{x}}{T}]dt$$
$$\le H \text{ (a constant independent of } T, y, x).$$

For some $a > 0, g(z) \ge z$ if $z \ge a$. Put

$$t \in L \text{ iff } |\dot{x}(t)| \ge a.$$

Then by (30) and (33)

$$|x(t)| \le |\bar{x}| + (\int_L + \int_{[0,T] \setminus L})|\dot{x}(t)|dt \le |\bar{x}| + \int_L g(|\dot{x}|)dt + aT \le A,$$

where A is a fixed constant (independent of T, y, x). Now we prove the equi-Lipschitz behaviour. Given $q > 0$ put

$$t \in E \text{ iff } |\dot{x}(t)| \le q.$$

Of course E depends on q. We can assume without loss of generality that g is increasing. Then by (33)

$$H \ge (\int_E + \int_{[0,T] \setminus E})g(|\dot{x}|)dt \ge g(0) \text{ meas } E + (T - \text{ meas } E)g(q).$$

Hence

(34)
$$\text{meas } E \ge \frac{Tg(q) - H}{g(q) - g(0)} > T/2$$

provided $g(q) \ge (2H/T_0) - g(0)$, where $0 < T_0 \le T$ for all $(T, y) \in K$. Therefore we can fix q (independent of T, y, x) such that (34) holds. Now for each $M > q$ we put

$$t \in F \text{ iff } q < |\dot{x}(t)| < M, t \in G \text{ iff } |\dot{x}(t)| \ge M.$$

Let us consider $z \in AC([0, T])$ such that $z(0) = 0$ and

$$\dot{z}(t) = 1 \text{ if } t \in F, \dot{z}(t) = |\dot{x}(t)| \text{ if } t \in G.$$

Finally we put

$$\dot{z}(t) = r \text{ if } t \in E$$

where r is a real constant (depending of T) such that

$$z(T) = T.$$

Therefore

$$\int_0^T \dot{z}\,dt = T = r \text{ meas } E + T - \text{ meas } E - \text{ meas } G + \int_G |\dot{x}|\,dt,$$

hence

(35)
$$(1 - r) \text{ meas } E = \int_G (|\dot{x}| - 1)\,dt.$$

By (34)

$$\text{meas } G \leq \frac{H - Tg(0)}{g(M) - g(0)},$$

so that

$$\text{meas } G \to 0 \text{ as } M \to +\infty, \text{ uniformly with respect to } T.$$

Let $M > 1$ be so large that

$$\int_G (|\dot{x}| - 1)\,dt \leq T/4.$$

Then by (34) and (35), $0 \leq 1 - r \leq 1/2$, hence

(36)
$$1 \geq r \geq 1/2.$$

Now consider

$$u(t) = x[z^{-1}(t)], 0 \leq t \leq T,$$

which is admissible for $Q(T, y)$ since x is smooth and $z(0) = 0, z(T) = T$. Then

$$\int_0^T f(x, \dot{x})\,dt \leq \int_0^T f(u, \dot{u})\,dt_e =$$

$$(\text{changing variable by } s = z^{-1}(t)) = \int_0^T f[x(s), \dot{x}(s)/\dot{z}(s)]\dot{z}(s)\,ds,$$

which may be rewritten as

$$\int_E f(x, \dot{x})\,dt + \int_F f(x, \dot{x})\,dt + \int_G f(x, \dot{x})\,dt$$

$$\leq \int_E rf(x, \dot{x}/r)\,ds + \int_F f(x, \dot{x})\,ds + \int_G f(x, \dot{x}/|\dot{x}|)|\dot{x}|\,ds.$$

Hence

(37)
$$\int_E [rf(x, \dot{x}/r) - f(x, \dot{x})]\,dt + \int_G [P|\dot{x}| - g(|\dot{x}|)]\,dt \geq 0,$$

where the constant P, independent of M, is given by

$$P = \max\{|f(x, w)| : |x| \leq A, |w| \leq 1\}.$$

To estimate the first term in (37), write

$$rf(x, \dot{x}/r) - f(x, \dot{x}) = (r - 1)f(x, \dot{x}/r) + f(x, \dot{x}/r) - f(x, \dot{x}).$$

For almost every $t \in E$ we have $|\dot{x}/r| \leq 2q$ by (36). Let S, N (independent of M) be given by

$$S = \max\{|f(x, \omega)| : |x| \leq A, |\omega| \leq 2q\},$$

$$N = \max\{|f_u(x, \omega)| : |x| \leq A, |\omega| \leq 2q\}.$$

Then we get for a.e. $t \in E$

$$|rf(x, \dot{x}/r) - f(x, \dot{x})| \leq (1 - r)(S + 2Nq).$$

Then by (37)

$$(1 - r) \text{ meas } E(S + 2Nq) + \int_G [P|\dot{x}| - g(|\dot{x}|)]dt \geq 0.$$

Remembering (35) we obtain, for all M sufficiently large,

(38)
$$\int_G [g(|\dot{x}|) - |\dot{x}|(P + S + 2Nq)]dt \leq 0.$$

By (30) there exists $B > 0$ such that

$$g(h) \geq h(1 + P + S + 2Nq) \text{ whenever } h \geq B.$$

Hence meas $G = 0$ for every $M \geq B$ by (38). Since $\dot{x}$ is continuous, this implies for every $t \in [0, T]$

$$|\dot{x}(t)| \leq B$$

for every T, y, x. $\square$

Proof of theorem 7. Suppose that V is differentiable at (T, y). Let K be a compact subset of $(0, +\infty) \times R^m$ with $(T, y) \in K$. Then there exist $a \in R, b \in R^m$ such that

$$V(t, z) = V(T, y) + a(t - T) + b'(z - y) + o(|t - T| + |z - y|).$$

Then given $\varepsilon > 0$, if z is sufficiently near to y

(39)
$$V(T, z) \geq V(T, y) + b'(z - y) - \varepsilon|y - z|.$$

Let $x \in \arg\min Q\ (T, y)$ and set

$$u(t) = x(t) + (t/T)\ (z - y).$$

Then $u(0) = \bar{x}, u(T) = z$, therefore by (39)

(40)
$$\int_0^T f[x(t) + (t/T)\ (z - y), \dot{x}(t) + (1/T)\ (z - y)]dt \geq V(T, z)$$

$$\geq \int_0^T f(x, \dot{x})dt + b'(z - y) - \varepsilon|z - y|.$$

By lemma 11, using the mean value theorem for f, there exists a constant c (depending only on K and ε) such that

(41)
$$\int_0^T \{f[x(t) + (t/T)\,(z-y), \dot{x}(t) + (1/T)\,(z-y)] - f[x(t), \dot{x}(t)]\}dt$$

$$\leq \int_0^T [f_x(x(t), \dot{x}(t))'(t/T)\,(z-y) + f_u(x(t), \dot{x}(t))'(z-y)/T]dt + \varepsilon|y-z|$$

if $|z-y| \leq c$.

By the Euler - Lagrange equation

$$\int_0^T [f_x(x, \dot{x})'(t/T)\,(z-y) + f_u(x, \dot{x})'(z-y)\,/T]dt$$

$$= \int_0^T d/dt[f_u(x, \dot{x})'\,(t/T)\,(z-y)]dt = f_u[y, \dot{x}(T)]'(z-y),$$

hence by (40) and (41)

$$(b - f_u[y, \dot{x}(T)])'(z-y) \leq 2\varepsilon|z-y|$$

if z is sufficiently near to y. It follows that

$$|b - f_u[y, \dot{x}(T)]| \leq 2\varepsilon,$$

hence for every $x_1, x_2 \in \arg\,\min Q(T, y)$

$$|f_u[y, \dot{x}_1(T)] - f_u[y, \dot{x}_2(T)]| \leq 4\varepsilon.$$

Then by (29) and lemma 11
$$|\dot{x}_1(T) - \dot{x}_2(T)| \leq M\varepsilon$$

for some constant M (depending only on K). Arbitrariness of ε gives

$$\dot{x}_1(T) = \dot{x}_2(T) = p, \text{ say.}$$

Thus we see that both x_1, x_2 solve the Cauchy problem

$$\begin{cases} \ddot{x}(t) & = E(x(t), \dot{x}(t)), 0 \leq t \leq T, \\ x(T) & = y, \dot{x}(T) = p, \end{cases}$$

which has a unique solution by (29). This shows uniqueness of the minimizer of $Q(T, y)$. To complete the proof of Tykhonov well-posedness, let x_n be a minimizing sequence for $Q(T, y)$ such that

$$\text{ess sup } |\dot{x}_n(t)| \leq L < +\infty, n = 1, 2, 3, \cdots.$$

Since $\sup \int_0^T g[|\dot{x}_n(t)|]dt < +\infty$, we get weak sequential compactness of $\dot{x}_n$ in $L^1(0, T)$ (see Ekeland-Temam [1, th.1.3 p.223]). Then by weak sequential lower semicontinuity of I (Cesari [2, th. 2.18.i p.104]) we get

$$x_n \to x \text{ uniformly in } [0, T] \text{ and weakly in } W^{1,1}(0, T),$$

where $x = \arg \min Q(T, y)$. Let us show strong convergence of the derivatives in $L^2(0, T)$. We have

$$(42) \qquad \int_0^T [f(x_n, \dot{x}_n) - f(x, \dot{x})]dt = \int_0^T [f(x_n, \dot{x}_n) - f(x, \dot{x}_n) + f(x, \dot{x}_n) - f(x, \dot{x})]dt.$$

By equiboundedness of $\dot{x}_n(t)$

$$(43) \qquad |f(x_n, \dot{x}_n) - f(x, \dot{x}_n)| \le (\text{const.}) \, |x_n - x|,$$

and by (29)

$$(44) \qquad f(x, \dot{x}_n) \ge f(x, \dot{x}) + f_u(x, \dot{x})'(\dot{x}_n - \dot{x}) + a|\dot{x}_n - \dot{x}|^2, a > 0.$$

Integrating by parts and using the Euler - Lagrange equation we get

$$(45) \qquad \int_0^T f_u(x, \dot{x})'(\dot{x}_n - \dot{x})dt = - \int_0^T f_x(x, \dot{x})'(x_n - x)dt.$$

Then by (42), (43), (44), (45)

$$\int_0^T f(x_n, \dot{x}_n)dt \ge \int_0^T f(x, \dot{x})dt - (\text{const.}) \max_t |x_n(t) - x(t)| + a \int_0^T |\dot{x}_n - \dot{x}|^2 dt.$$

This ends the proof of the first half of theorem 7.

Conversely assume Tykhonov well-posedness of $Q(T, y)$. We know that V is well - defined near (T, y). Put

$$x = \arg \min Q(T, y).$$

By the maximum principle (Fleming-Rishel [1, II.8]), for every $t \in [0, T]$, $u \to \bar{p}(t)'u + f(x(t), u)$ takes on the minimum value on R^m only at $u = \dot{x}(t)$, where

$$\bar{p}(t) = \int_t^0 f_x(x, \dot{x})ds.$$

Fix any $p, x \in R^m$ and define

$$\Theta(u) = p'u + f(x, u), \quad q(p, x) = \arg \min (R^m, \Theta).$$

By (29) q is (everywhere) well-defined. Of course

$$(46) \qquad p + f_u(x, q) = 0.$$

By the implicit function theorem we see that q is continuously differentiable near $(y, \bar{p}(T))$. Given $S > 0, p$ and $\omega \in R^m$ consider the solution z of

$$(47) \qquad \left\{ \begin{array}{l} (d/dt)f_u(z, \dot{z}) = f_x(z, \dot{z}), 0 \le t \le S, \\ z(S) = \omega, \dot{z}(S) = q(p, \omega) \end{array} \right.$$

and set

$$u(t) = z(t) + [\bar{x} - z(0)]\frac{S-t}{S},$$

$$L(S, \omega, p) = \int_0^S f(u, \dot{u})dt.$$

Solutions to smooth Cauchy's problems in normal form, like (47) depend on parameters in a continuously differentiable way: see e.g. Hartman [1, V.3]. Then L is C^1 near $(T, y, \bar{p}(T))$. Since (47) has uniqueness by (29), and x satisfies (47) with $S = T, \omega = y, p = \bar{p}(T)$, then

(48) $V(T, y) = L[T, y, \bar{p}(T)].$

Now let $h \in R, k \in R^m$ with $|h| + |k|$ sufficiently small. Take any $x_{hk} \in$ arg min $Q(T + h, y + k)$. Then

(49) $L(T + h, y + k, p_{hk}) = V(T + h, y + k),$

where

$$p_{hk} = \int_{T+h}^0 f_x(x_{hk}, \dot{x}_{hk})dt.$$

Moreover

$$L(T, y, p_{hk}) \geq V(T, y); V(T + h, y + k) \leq L[T + h, y + k, \bar{p}(T)],$$

hence by (48), (49)

$$L(T + h, y + k, p_{hk}) - L(T, y, p_{hk}) \leq V(T + h, y + k) - V(T, y)$$
$$\leq L[T + h, y + k, \bar{p}(T)] - L[T, y, \bar{p}(T)].$$

From the mean value theorem for $L = L(t, x, p)$,

(50) $hL_t^* + k'L_x^* \leq V(T + h, y + k) - V(T, y) \leq h\bar{L}_t + k'\bar{L}_x.$

Here $L_t^*, \bar{L}_t$ etc. denote partial derivatives of L evaluated at suitable points (t^*, y^*, p_{hk}), $(\bar{t}, \bar{y}, \bar{p}(T))$ etc. To show differentiability òf V at (T, y) it suffices by (50) to prove that

(51) $p_{hk} \to \bar{p}(T)$ as $h \to 0, k \to 0$

since L is continuously differentiable. To show (51), let $h_n \to 0, k_n \to 0$ and consider the corresponding sequence $x_n = x_{h_n k_n}$. Fix $S \geq T + h_n$ for all n, and extend x_n by constancy and continuity on $[0, S]$. For some subsequence $x_n \rightharpoonup z$, say, in $W^{1,1}(0, S)$ and uniformly in $[0, S]$ due to (30). By the lower semicontinuity of the integral functional (27) and lemma 11,

(52) $\int_0^T f(z, \dot{z})dt \leq$ lim inf $\int_0^{T+h_n} f(x_n, \dot{x}_n)dt$

         $=$ lim inf $V(T + h_n, y + k_n) = V(T, y).$

The last equality is a consequence of theorem 8 (we shall prove next). Taking theorem 8 for granted, (52) implies $z = x =$ arg min $Q(T, y)$. By lemma 11 and the Euler-Lagrange equation, we see that $\ddot{x}_n$ are uniformly bounded. Then setting $p_n = p_{h_n k_n}$, we obtain by Arzelá - Ascoli' s theorem

$$q(p_n, y + k_n) = \dot{x}_n(T + h_n) \to \dot{x}(T) = q(\bar{p}(T), y).$$

This yields $p_n \to \bar{p}(T)$ by (46). Hence (51) is obtained, ending the proof.   □

Theorem 8. Let K be a fixed compact subset of $(0, +\infty) \times R^m$. Pick $(T_1, y_1), (T_2, y_2) \in K$ and consider $x_i \in$ arg min $Q(T_i, y_i), i = 1, 2$. Then put

$$z(t) = x_2(tT_2/T_1) + (t/T_1)(y_1 - y_2).$$

Since $z(0) = \bar{x}, z(T_1) = y_1$,

$$V(y_1, T_1) \leq \int_0^{T_1} f(z, \dot{z})dt.$$

Minimizers have equibounded second derivatives by lemma 11 and the Euler- Lagrange equation. Then there exists some constant M (depending on K only) such that

$$\int_0^{T_1} f(z, \dot{z})dt \leq \int_0^{T_2} f(x_2, \dot{x}_2)dt + M(|y_1 - y_2| + |T_1 - T_2|).$$

We used here the local Lipschitz continuity of f. By exchanging the roles of x_1, x_2 in the definition of z, we conclude the proof. $\square$

Theorem 9. Fix any compact ball K in $(0, +\infty) \times R^m$. Let

$$F_n = \{(T, y) \in K : |\dot{x}_1(T) - \dot{x}_2(T)| < 1/n \text{ for every}$$
$$x_1, x_2 \in \text{ arg min } Q(T, y)\}; n = 1, 2, 3, \ldots.$$

Assume that

(53)   for every n, F_n is open dense in K.

Then $G = \cap_{n=1}^{\infty} F_n$ will be a dense G_δ subset of K. Pick $(T, y) \in G$, then if $x_1, x_2 \in$ arg min $Q(T, y)$ we see that $\dot{x}_1(T) = \dot{x}_2(T)$. As in the proof of theorem 7, if follows that $Q(T, y)$ is Tykhonov well- posed. Thus it suffices to prove (53).

Density of F_n. By theorem 8, V is a.e. differentiable (Rademacher's theorem). By theorem 7 it follows that $Q(T, y)$ is Tykhonov well-posed for a dense set of points (T, y), thereby showing density of F_n in K.

Openness of F_n. Arguing by contradiction, assume that there exists $(T, y) \in F_n$ such that we can find sequences $T_k \to T$, $y_k \to y$, x_k^1 and $x_k^2 \in$ arg min $Q(T_k, y_k)$ fulfilling

(54) $$|\dot{x}_k^1(T_k) - \dot{x}_k^2(T_k)| \geq 1/n.$$

By lemma 11, for some subsequence

$$\dot{x}_k^i(T_k) \to p_i, \text{ say }, i = 1, 2.$$

Now let ω_1, ω_2 be the solutions to the Euler-Lagrange equation such that

$$\omega_i(T) = y, \quad \dot{\omega}_i(T) = p_i,$$

with maximal existence domain $(T - a_i, T + b_i), i = 1, 2$. We shall prove that

(55)
$$T - a_i < 0;$$

(56) the restriction of ω_i to $[0, T]$ belongs to arg min $Q(T, y)$.

Then by (54)
$$|p_1 - p_2| = \lim |\dot{x}_k^1(T_k) - \dot{x}_k^2(T_k)| \geq 1/n,$$

and by (56) $(T, y) \notin F_n$, a contradiction. So we need only to prove (55) and (56). Put
$$z_k^i(t) = x_k^i(t + T_k - T), T - T_k \leq t \leq T, i = 1, 2.$$

Then z_k^i is a solution to the Euler-Lagrange equation. Moreover
$$z_k^i(T) = x_k^i(T_k) \to y, \dot{z}_k^i(T) = \dot{x}_k^i(T_k) \to p_i.$$

Then, given $\varepsilon > 0$, for k sufficiently large we can extend z_k^i to a solution of the Euler-Lagrange equation on $[T - T_k, T] \cup [T - a_i + \varepsilon, T + b_i - \varepsilon]$, moreover on the latter set
$$z_k^i(t) \to \omega_i(t), \; \dot{z}_k^i(t) \to \dot{\omega}_i(t)$$

(see Hartman [1, ch.2, sec.3]). By lemma 11, the sequence x_k^i is equi-Lipschitz continuous, and $\dot{x}_k^i$ too (from the Euler-Lagrange equation and lemma 11 again). Then for every k sufficiently large, x_k^i can be extended to a solution of the Euler-Lagrange equation on $[0, T_k] \cup [T - a_i + \varepsilon, T + b_i - \varepsilon]$, and on this last interval
$$x_k^i \to \omega_i, \dot{x}_k^i \to \dot{\omega}_i \text{ both uniformly,}$$

moreover, for some subsequence, x_k^i and $\dot{x}_k^i$ converge uniformly on $[0, T]$. By maximality of the domain we get (55). Moreover $\omega_i(0) = \lim x_k^i(0) = \bar{x}$, while $\omega_i(T) = y$ by the definition of ω_i. Hence
$$V(T_k, y_k) = \int_0^{T_k} f(x_k^i, \dot{x}_k^i)dt \to \int_0^T f(\omega_i, \dot{\omega}_i)dt = V(T, y)$$

by uniform convergence and continuity of V (theorem 8). This proves the optimality of ω_i, hence (56). □

Theorem 10. Let u_n be any minimizing sequence for (K, I). By coercivity (assumption (32)), u_n is bounded in $W_0^{1,q}(\Omega)$, hence for some subsequence $u_n \rightharpoonup u \in K$. A standard semicontinuity thorem (see Dacorogna [1, th.3.4 p.74]) implies that $u \in$ arg min (K, I). Thus for the original sequence
$$\nabla u_n \rightharpoonup \nabla u \text{ in } L^q(\Omega),$$

(57)
$$\int_\Omega f(x, \nabla u_n)dx \to \int_\Omega f(x, \nabla u)dx.$$

Having fixed $x \in \Omega$ we consider
$$g(p) = (1/2)f(x, p) + (1/2)f[x, \nabla u(x)] - f(x, \frac{\nabla u(x) + p}{2}).$$

We claim that

(58) $\qquad (R^m, g)$ is Tykhonow well-posed.

Write for short

$$q = \nabla u(x), f(p) = f(x, p).$$

Of course

$$g(p) \geq 0 = g(q), p \in R^m.$$

Given $t > s > 0$ and $y \in R^m$ we have

$$\lambda(q + (s/2)y) + (1 - \lambda)(q + ty) = q + sy \text{ with } \lambda = 2(s - t)/(s - 2t),$$

$$\alpha(q + (s/2)y) + (1 - \alpha)(q + ty) = q + (t/2)y \text{ with } \alpha = t/(2t - s).$$

Hence by convexity

$$f(q + (t/2)y) \leq \frac{t}{2t - s} f(q + (s/2)y) + \frac{t - s}{2t - s} f(q + ty),$$

$$(1/2)f(q + sy) \leq \frac{t - s}{2t - s} f(q + (s/2)y) + \frac{s}{4t - 2s} f(q + ty).$$

Adding, we get

$$(1/2)f(q + ty) + (1/2)f(q) - f(q + (t/2)y)$$
$$\geq (1/2)f(q + sy) + (1/2)f(q) - f(q + (s/2)y)$$

which means that g increases along any ray issued from q. Hence by strict convexity

$$\inf \{g(p) : |p - q| \geq \varepsilon\} > 0$$

for every $\varepsilon > 0$, proving (58). Since

$$I(u) \leq I(\frac{u_n + u}{2}) \leq (1/2)I(u_n) + (1/2)I(u)$$

we obtain from (57)

$$I(\frac{u_n + u}{2}) \to I(u),$$

hence

(59) $\qquad \displaystyle\int_\Omega g(\nabla u_n)dx \to 0$

yielding for some subsequence

$$g[\nabla u_n(x)] \to 0 \text{ a.e. in } \Omega.$$

Then by (58)

(60) $\qquad \nabla u_n(x) \to \nabla u(x) \text{ a.e.}$

By (32), for some $s \in L^1(\Omega)$

$$|\nabla u_n - \nabla u|^q \le f(x, \nabla u_n) + s(x) = f_n(x), \text{ say.}$$

Fatou's lemma yields

$$\liminf \int_\Omega (f_n(x) - |\nabla u_n - \nabla u|^q)dx = \int_\Omega f(x, \nabla u)dx$$

$$+ \int_\Omega s\, dx - \limsup \int_\Omega |\nabla u_n - \nabla u|^q dx \ge \int_\Omega f(x, \nabla u)dx + \int_\Omega s\, dx,$$

whence

$$\nabla u_n \to \nabla u \text{ in } L^q(\Omega).$$

This entails strong convergence of u_n to u.   □

Section 3.   Integral Functionals with Dirichlet Boundary Conditions.

We consider Hadamard well-posedness of minimum problems for multiple integrals of the following form: to minimize

(61) $$\int_\Omega f[x, \nabla u(x)]dx$$

subject to

(62) $$u \in W^{1,p}(\Omega), u - \phi \in W_0^{1,p}(\Omega).$$

We are interested in the continuous dependence of the optimal value and arg min on the integrand f, for a fixed domain Ω and any boundary datum $\phi \in W^{1,p}(\Omega)$.

As we know from chapter IV, such a well-posedness is related in a natural way to epi-convergence for the integral functionals

$$I : L^p(\Omega) \to (-\infty, +\infty]$$

defined by

(63) $$I(u) = \begin{cases} \int_\Omega f[x, \nabla u(x)]dx, & \text{if } u - \phi \in W_0^{1,p}(\Omega), \\ +\infty, & \text{otherwise.} \end{cases}$$

We shall see that epi-convergence of (63) is implied by epi-convergence of the integral functionals (free of boundary conditions)

(64) $$F(u) = \begin{cases} \int_\Omega f[x, \nabla u(x)]dx, & \text{if } u \in W^{1,p}(\Omega), \\ +\infty & \text{otherwise} \end{cases}$$

under suitable assumptions about the integrands f.

More precisely, we are given a bounded open subset Ω of R^m, a real number $p > 1$ and constants $0 < \alpha \le \omega$. We consider the set M of all functions

$$f : R^m \times R^m \to [0, +\infty)$$

such that $f(\cdot, y)$ is Lebesgue measurable for all y, $f(x, \cdot)$ is convex for all x, and

(65) $$\alpha|y|^p \le f(x, y) \le \omega(1 + |y|^p)$$

for every x and y.

We shall denote by $f_n, n = 0, 1, 2, 3, ...$, any fixed sequence in M, and by I_n, F_n the corresponding integral functionals (63), (64) for any fixed $\phi \in W^{1,p}(\Omega)$.

The next theorem shows that epi-convergence of the free functionals F_n implies the same convergence when the Dirichlet boundary data are taken in account.

12 Theorem *For any $\phi \in W^{1,p}(\Omega)$, epi-lim $F_n = F_0$ implies epi-lim $I_n = I_0$, both in $(L^p(\Omega),$ strong $)$.*

The choice of $(L^p(\Omega),$ strong $)$ as the basic functional space where epi-convergence takes place is due to the following reason. According to the next proposition, this mode of convergence is equivalent to $(W^{1,p}(\Omega),$ weak $)$ epi-convergence, which is the right one to be employed due to the equicoercivity (65).

13 Proposition *For any $\phi \in W^{1,p}(\Omega)$,*

$$epi - lim \, I_n = I_0 \text{ in } (L^p(\Omega), \text{ strong }) \text{ iff } epi - lim \, I_n = I_0 \text{ in } (W^{1,p}(\Omega), \text{ weak }).$$

From corollary 4 we get a sufficient condition for Mosco convergence, as follows. Abusing notation, the restriction of F_n to $W^{1,p}(\Omega)$ will be denoted again by F_n.

14 Proposition *Let $f_n = f_n(x, y)$ be measurable in x, convex in y and such that*

$$\alpha|y|^p \le f_n(x, y) \le \omega|y|^p \text{ for every } x \text{ and } y \, , f(x, 0) = 0.$$

Then $f_n(\cdot, y) \rightharpoonup f_0(\cdot, y)$ and $f_n^*(\cdot, z) \rightharpoonup f_0^*(\cdot, z)$ in $L^\infty(\Omega)$ for every y and z, imply

$$M - lim \, F_n = F_0 \text{ in } (W^{1,p}(\Omega), \text{ weak }).$$

Now let $f_n \in M$. Since epi - convergence of I_n to I_0 in $(W^{1,p}(\Omega),$ weak $)$ is intimately linked to Hadamard well-posedness, it is natural to try to relate the values of the limit integrand f_0 to those of f_n. In the one-dimensional case we have the following explicit characterization.

15 Theorem *Let $m = 1$ and Ω be an open bounded interval. Then epi-lim $F_n = F_0$ in $(W^{1,p}(D),$ weak $)$ for every open interval $D \subset \Omega$ iff $f_n^*(\cdot, z) \rightharpoonup f_0^*(\cdot, z)$ in $L^\infty(\Omega)$ for every real z.*

16 Example. Let $m = 1, p = 2$ and

$$f_n(x, y) = a_n(x)y^2,$$

where

$$0 < \alpha \le a_n(x) \le \omega, n = 1, 2, ..., \text{ and a. e. } x \in \Omega.$$

Then

(66)
$$F_n(u) = \int_\Omega a_n(x) u'^2 dx, u \in W^{1,2}(\Omega).$$

From theorem 15 we see that

$$\text{epi - lim } F_n = F_0 \text{ in } (H^1(\Omega), \text{ weak }) \text{ iff } 1/a_n \to 1/a_0 \text{ in } L^\infty(\Omega).$$

This completes example IV. 3.

When $m > 1$, no general explicit condition is known directly relating f_0 to f_n under $(W^{1,p}(\Omega), \text{ weak })$ epi-convergence of the corresponding integral functionals. The difficulty is illustrated by the following striking result (Marino-Spagnolo [1]). Let

$$f_0(x, y) = \sum_{i,j=1}^m a_{ij}^0(x) y_i y_j$$

where $a_{ij}^0 = a_{ji}^0 \in L^\infty(\Omega)$ and $\alpha|y|^2 \le f_0(x, y) \le \omega|y|^2$, a.e. $x \in \Omega$ and every $y \in R^m$. Consider

$$F_0(u) = \sum_{i,j=1}^m \int_\Omega a_{ij}^0(x) \frac{\partial u}{\partial x_i} \frac{\partial u}{\partial x_j} dx, u \in H^1(\Omega).$$

Then there exist a constant $c > 0$ and a sequence of real-valued functions $a_n \in L^\infty(\Omega)$ such that $\alpha/c \le a_n(x) \le c\omega$, a.e. $x \in \Omega$, and

$$\text{epi - lim } F_n = F_0 \text{ in } (H_0^1(\Omega), \text{ weak }),$$

where

$$F_n(u) = \sum_{j=1}^m \int_\Omega a_n(x)(\frac{\partial u}{\partial x_i})^2 \, dx.$$

However, it is possible to get an indirect characterization of f_0 from the values of F_n which is of varational nature, as follows. Given $f \in M$, consider the set Q of all open subsets of Ω, and let

$$G : L_{loc}^p(R^m) \times Q \to [0, +\infty]$$

be defined by

(67)
$$G(u, A) = \begin{cases} \int_A f(x, \nabla u) \, dx, & \text{if } u|_A \in W^{1,p}(A), \\ +\infty, & \text{otherwise.} \end{cases}$$

Let $f_n \in M$ be given, with corresponding integral functionals G_n defined by (67) with f_n instead of f. For any $y \in R^m$ denote by $L(y)$ the linear function $x \to y'x, x \in R^m$. If for every $A \in Q$

$$\text{epi - lim } G_n(\cdot, A) = G_0(\cdot, A) \text{ in } (L^p(A), \text{ strong })$$

then from theorems 12 and IV. 5 we know that

$$(68) \qquad \text{val } [L(y) + W_0^{1,p}(A), G_n(\cdot, A)] \to \text{val } [L(y) + W_0^{1,p}(A), G_0(\cdot, A)]$$

for every y and A. .

Conversely, we shall see that (68) implies epi-convergence. This gives a characterization of epi-convergence of integral functionals and well- posedness, which is of interest for the following reason. The knowledge of

$$\text{val } (y, A, G_0) = \text{val } [L(y) + W_0^{1,p}(A), G_0(\cdot, A)]$$

for every y and $A \in Q$ completely identifies f_0. Let us verify these facts on the one-dimensional quadratic problems of example 16.

17 Example. Given a_n and f_n as in example 16, we get immediately from the Euler-Lagrange equation, for every subinterval $A \subset \Omega$,

$$(69) \qquad \text{val } (y, A, G_n) = \frac{y^2 (\text{ meas } A)^2}{\int_A dx/a_n(x)}.$$

Thus we get $f_0(x, y) = y^2 a_0(x)$ for a.e. x and every y, by firstly letting $n \to +\infty$ in (69), and then shrinking A to $\{x\}$ (in a suitable way) in the expression

$$\frac{1}{\text{meas } A} \lim_{n \to +\infty} \text{ val } (y, A, G_n).$$

In the general case the procedure is exactly the same. We denote by $B(x, r)$ the open ball of center x and radius r in R^m.

18 Theorem *If epi-lim $G_n(\cdot, A) = G_0(\cdot, A)$ in $(L^p(A)$, strong) for every $A \in Q$, then (68) holds. Conversely, if for every $y \in R^m$ and every ball $B \subset \Omega$ the sequence val (y, B, G_n) is convergent, then there exists $f_0 \in M$ such that*

$$epi \text{ - } lim \ G_n(\cdot, A) = G_0(\cdot, A) \ in \ (L^p(A), \ strong \ )$$

for every $A \in Q$. Moreover for a.e. $x \in \Omega$ and every $y \in R^m$

$$(70) \qquad f_0(x, y) = \lim_{r \to 0} \frac{1}{\text{meas } B(x, r)} \ val \ (y, B(x, r), G_0).$$

Putting together theorems 12, 18, IV. 5 and proposition 13 we obtain a characterization of Hadamard well-posedness.

19 Corollary *Let $f_n \in M$ be given. Then*

$$epi - lim \ F_n = F_0 \ in \ (L^p(A), \ strong \ ), A \in Q,$$

implies for every $\phi \in W^{1,p}(A)$

$$val \ (\phi + W_0^{1,p}(A), I_n) \to val \ (\phi + W_0^{1,p}(A), I_0),$$

$$lim \ sup \ arg \ min \ (\phi + W_0^{1,p}(A), I_n) \subset \ arg \ min \ (\phi + W_0^{1,p}(A), I_0)$$

both in $(L^p(A),$ strong $)$ and in $(W^{1,p}(A),$ weak $)$. Conversely, if for every $y \in R^m$ and every ball B

$$val\ (y, B, G_n) \to\ val\ (y, B, G_0)$$

then epi-lim $F_n = F_0$ in $(L^p(A),$ strong $), A \in Q$.

Application: Hadamard well-posedness of minimum problems for quadratic integrals. Let

$$(71) \qquad F(u) = \begin{cases} \sum_{ij=1}^{m} \int_{\Omega} a_{ij}(x)\ \partial u / \partial x_i\ \partial u / \partial x_j\,dx, & u \in H^1(\Omega), \\ +\infty, & u \in L^2(\Omega) \setminus H^1(\Omega), \end{cases}$$

where the $m \times m$ matrix (a_{ji}) fulfills

$$(72) \qquad \begin{cases} a_{ji} = a_{ij} \in L^\infty(\Omega) \text{ for every } i, j, \text{ and} \\ \alpha|z|^2 \leq \sum_{i,j=1}^{m} a_{ij}(x)z_i z_j \leq \omega|z|^2 \end{cases}$$

for a.e. $x \in \Omega$ and every $z \in R^m$.

Denote by E the set of all matrices (a_{ij}) fulfilling (72). We equip E with the following convergence:

$(a_{ij}^n) \to (a_{ij}^0)$ iff for the corresponding functionals F_n given by (71) we have

$$epi\text{ - }lim\ F_n = F_0 \text{ in } (L^2(\Omega),\ \text{strong }).$$

Using theorem IV. 47 and corollary 19 we get

20 Corollary *For any $\phi \in H^1(\Omega)$ the mappings*

$$(a_{ij}) \to\ val\ (\phi + H_0^1(\Omega), F),$$

$$(a_{ij}) \to arg\ min\ (\phi + H_0^1(\Omega), F)$$

are continuous on E, the latter as a mapping to $(L^2(\Omega),$ strong $)$ or $(H^1(\Omega),$ weak $)$. Moreover, the convergence on E is the weakest one giving continuity of either mappings

$$(a_{ij}) \to\ val\ (H_0^1(\Omega), F(u) + \int_\Omega uy\ dx),$$

$$(a_{ij}) \to arg\ min\ (H_0^1(\Omega), F(u) + \int_\Omega uy\ dx)$$

for every $y \in L^2(\Omega)$, and

$$(a_{ij}) \to\ val\ (\phi + H_0^1(A), G(\cdot, A)) \text{ for every } A \in Q \text{ and every } \phi \in H^1(A).$$

Finally, E is a compact space and $(a_{ij}^n) \to (a_{ij}^0)$ in E implies

$$epi\text{ - }lim\ G_n(\cdot, A) = G_0(\cdot, A) \text{ in } (L^2(A),\ \text{strong })$$

for every $A \in Q$, where

$$G_n(u, A) = \begin{cases} \sum_{ij=1}^m \int_A a_{ij}^n(x)\partial u/\partial x_i \, \partial u/\partial x_j \, dx, & \text{if } u \in H^1(A), \\ +\infty, & \text{if } u \in L^2(A) \setminus H^1(A). \end{cases}$$

The compactness of E comes from the compactness of epi-convergence, and has some bearing upon well-posedness, as follows. Given any sequence $(a_{ij}^n) \in E$, put

$$u_n = \arg \min (\phi + H_0^1(\Omega), F_n), \phi \in H^1(\Omega).$$

Then, by uniform ellipticity (72), u_n has a weak cluster point u in $H_0^1(\Omega)$. The compactness of E then guarantees that every such u is again the minimizer of $(\phi + H_0^1(\Omega), F)$ for a suitable F of the form (71) and some matrix $(a_{ij}) \in E$.

Roughly speaking, the appropriate convergence for well-posedness in E allows to preserve in the limit (up to subsequences) the minimizing nature of those cluster points within the same class of quadratic functionals. Of course this property is stronger than that asserted by lemma 23 below. It implies closure under epi-convergence of the set of quadratic functionals (71), a smaller set than that defined by (64) as $f \in M$.

Counterexample. Either solution or value Hadamard well-posedness alone for every boundary datum ϕ does not force epi-convergence. Indeed, for $m = 1$ let $a \in L^\infty(0,1)$ be such that

$$0 < \alpha \leq a(x) \leq 2a(\bar{x}) \leq \omega, \text{ a.e. } x,$$

and consider the sequence $a, 2a, a, 2a, \dots$. Of course the corresponding integral functionals (66) do not epi-converge, by example 16, while for any A, B, the minimizer of (66) with boundary conditions $u(0) = A, u(1) = B$ is independent of n and given by

$$u(x) = A + \left(\frac{B-A}{\int_0^1 dt/a}\right) \int_0^x dt/a.$$

A counter-example about value well-posedness is obtained by taking $\Omega = (0,2)$,

$$a_n(x) = \begin{cases} 1, & \text{if } 0 \leq x < 1, \\ 1/3 & \text{if } 1 \leq x \leq 2, \end{cases} \quad a_0(x) = 1/2.$$

The optimal value for minimizing (66) with boundary conditions $u(0) = A, u(2) = B$ is then given by

$$(B-A)^2 / \int_0^2 dx/a_n = (B-A)^2/4$$

independent of $n = 0, 1, 2, \dots$.

PROOFS.

Theorem 12. Following the definition, we firstly show that

(73) $$u_n \to u \text{ in } L^p(\Omega) \Rightarrow \lim \inf I_n(u_n) \geq I_0(u),$$

assuming that the lim inf is finite. For some subsequence $I_n(u_n) < +\infty$, yielding $u_n - \phi \in W_0^{1,p}(\Omega)$. This by (65) implies boundedness of u_n in $W^{1,p}(\Omega)$, therefore by (73) $u_n \to u$, in $W^{1,p}(\Omega)$, hence $u - \phi \in W_0^{1,p}(\Omega)$. Thus

$$\lim \inf \ I_n(u_n) = \lim \inf F_n(u_n) \geq F_0(u) = I_0(u).$$

Now we show that

(74) for every $z \in L^p(\Omega)$ with $I_0(z) < +\infty$ there exists
$z_n \to z$ in $L^p(\Omega)$ such that lim sup $I_n(z_n) \leq I_0(z)$.

By epi-convergence of F_n, given z there exists $u_n \to z$ in $L^p(\Omega)$ such that

(75) $$F_n(u_n) \to F_0(z).$$

By (63) we know that $u \in W^{1,p}(\Omega)$ and $z - \phi \in W_0^{1,p}(\Omega)$, hence $z \in W^{1,p}(\Omega)$. But (unfortunately) there is no reason why $u_n - \phi \in W_0^{1,p}(\Omega)$. So we need to suitably modify u_n near $\partial\Omega$. To this purpose we need the following

21 Lemma For every $\varepsilon > 0$, every compact $K \subset \Omega$ and every open A such that

(76) $$K \subset A \subset\subset \Omega$$

there exist a constant $T > 0$, an integer $q > 0$ and functions $a_i \in C_0^\infty(\Omega)$ such that for every $i = 1, ..., q$, $0 \leq a_i \leq 1$, $a_i(x) = 1$ in a neighborhood of cl A, and for every $f \in M$, every $u, v \in W^{1,p}(\Omega)$ we have

(77) $$\min_{1 \leq i \leq q} \int_\Omega f[x, \nabla(a_i u + (1 - a_i)v)]dx$$

$$\leq (1 + \varepsilon)[\int_\Omega f(x, \nabla u)dx + \int_{\Omega \backslash K} f(x, \nabla v)dx + \varepsilon + T \int_\Omega |u - v|^p dx.$$

Now we fix $\varepsilon > 0$ and a compact $K \subset \Omega$ such that

(78) $$\omega \int_{\Omega \backslash K} (1 + |\nabla z|^p)dx \leq \varepsilon.$$

Taking lemma 21 for granted, we fix A as in (76) and put $u = u_n, v = z$ in the lemma, thereby obtaining by (77) a sequence $a_n \in C_0^\infty(\Omega)$ such that

(79) $$\int_\Omega f_n[x, \nabla(a_n u_n + (1 - a_n)z)]dx \leq (1 + \varepsilon)[\int_\Omega f_n(x, \nabla u_n)dx$$

$$+ \int_{\Omega \backslash K} f_n(x, \nabla z)dx] + \varepsilon + T \int_\Omega |u_n - z|^p dx.$$

Put $z_n = a_n u_n + (1 - a_n)z$. Then $z_n - \phi \in W_0^{1,p}(\Omega)$, thus by (79) and (65)

$$I_n(z_n) = F_n(z_n) = \int_\Omega f_n(x, \nabla z_n)dx \leq (1 + \varepsilon)[\int_\Omega f_n(x, \nabla u_n)dx$$

$$+ \omega \int_{\Omega \backslash K}(1 + |\nabla z|^p)dx] + \varepsilon + T \int_\Omega |u_n - z|^p dx.$$

Then by (78)

$$\limsup\ I_n(z_n) \leq (1 + \varepsilon)(\ \limsup\ F_n(u_n) + \varepsilon) + \varepsilon.$$

The conclusion (74) is then obtained by letting $\varepsilon \to 0$ and remembering (75), since $F_0(z) = I_0(z)$. The proof of the theorem will be ended by the
proof of lemma 21. Choose q such that

$$\max \{\frac{2^p \omega}{\alpha q}, \frac{\omega}{q}\ \text{meas}\ \Omega\} \leq \varepsilon.$$

Now let $B_1, ..., B_{q+1}$ be open subsets of Ω such that

$$A \subset\subset B_1 \subset\subset B_2 \subset\subset ... \subset\subset B_{q+1} \subset\subset \Omega.$$

As well known, for every $i = 1, ..., q$ we can find a function $a_i \in C^\infty(\Omega) \cap C_0^\infty(B_{i+1})$ such that $a_i = 1$ near B_i, and $0 \leq a_i \leq 1$ everywhere. Fix any u, v, f, and write for every i

(80)
$$\int_\Omega f[x, \nabla(a_i u + (1 - a_i)v]dx = (\int_{B_i} + \int_{\Omega \backslash B_{i+1}} + \int_{B_{i+1} \backslash B_i})f(...)dx$$

$$= \int_{B_i} f(x, \nabla u)dx + \int_{\Omega \backslash B_{i+1}} f(x, \nabla v)dx + s_i.$$

Then by (65), writing $E_i = B_{i+1} \backslash B_i$, we get

(81)
$$s_i \leq \omega \int_{E_i} |\nabla [a_i u + (1 - a_i)v]|^p dx + \omega\ \text{meas}\ E_i.$$

By convexity of $|\cdot|^p$ on R^m we get a.e. in Ω

$$|\nabla [a_i u + (1 - a_i)v]|^p = 2^p |\frac{a_i \nabla u + (1 - a_i) \nabla v}{2} + \frac{(u - v) \nabla a_i}{2}|^p$$

$$\leq 2^p(a_i|\nabla u|^p + (1 - a_i)|\nabla v|^p + |u - v|^p|\nabla a_i|^p).$$

Therefore by (65) the integral term in (81) is

$$\leq 2^p \int_{E_i}(|\nabla u|^p + |\nabla v|^p + |u - v|^p|\nabla a_i|^p)dx$$

$$\leq (2^p/\alpha) \int_{E_i} [f(x, \nabla u) + f(x, \nabla v)]dx$$

$$+ 2^p\ \max \{|\nabla a_i(x)|^p : x \in \Omega, i = 1, ..., q\} \int_{E_i} |u - v|^p dx.$$

Observe that

$$\cup_{i=1}^{q}(B_{i+1} \setminus B_i) \subset \Omega \setminus K,$$

and put

$$T = \frac{2^p \omega}{q} \max |\nabla a_i|^p.$$

Then by (81) and the previous estimate, remembering the choice of q, we obtain

$$\min_{1 \leq i \leq q} s_i \leq (1/q) \sum_{1=i}^{q} s_i \leq \varepsilon (\int_{\Omega} f(x, \nabla u) dx$$

$$+ \int_{\Omega \setminus K} f(x, \nabla v) dx) + \varepsilon + T \int_{\Omega} |u - v|^p dx.$$

Coming back to (80) we get (77).   □

Proposition 13. We shall denote by I_n (as in the statement) both the functionals (63) and their restrictions to $W^{1,p}(\Omega)$. Assume epi-convergence in $L^p(\Omega)$. Let $u_n \rightharpoonup u$ in $W^{1,p}(\Omega)$ be with lim inf $I_n(u_n) < +\infty$. Then $u_n - \phi \in W_0^{1,p}(\Omega)$ for a subsequence such that $I_n(u_n)$ converges to lim inf $I_n(u_n)$. Since the embedding of $W^{1,p}(\Omega)$ in $L^p(\Omega)$ is compact (Brezis [2, p. 169]), $u_n \rightharpoonup u$ in $W^{1,p}(\Omega)$ implies lim inf $I_n(u_n) \geq I_0(u)$. Moreover, given $u \in W^{1,p}(\Omega)$ with $I_0(u) < +\infty$, there exists $u_n \to u$ in $L^p(\Omega)$ such that

$$\limsup I_n(u_n) \leq I_0(u).$$

Then ∇u_n is bounded in $L^p(\Omega)$ by (65), and $u_n - \phi \in W_0^{1,p}(\Omega)$. By Poincaré inequality (Brezis [2, p. 174]), u_n is bounded in $W^{1,p}(\Omega)$. Then by reflexivity of $W^{1,p}(\Omega)$ we conclude that $u_n \rightharpoonup u$ there. So epi-lim $I_n = I_0$ in $(W^{1,p}(\Omega)$, weak).

Conversely, assume epi-convergence in $(W^{1,p}(\Omega)$, weak). Let $u_n \to u$ in $L^p(\Omega)$ be with lim inf $I_n(u_n) < +\infty$. As before , $u_n \rightharpoonup u$ in $W^{1,p}(\Omega)$ for a suitable subsequence, implying lim inf $I_n(u_n) \geq I_0(u)$. Finally if $u \in L^p(\Omega)$ and $I_0(u) < +\infty$, then $u \in W^{1,p}(\Omega)$, there exists $u_n \rightharpoonup u$ in $W^{1,p}(\Omega)$ and $I_n(u_n) \to I_0(u)$. Then $u_n \to u$ in $L^p(\Omega)$ by compact embedding.   □

Proposition 14. Put

$$G_n(\omega) = \int_{\Omega} f_n[x, \omega(x)] dx$$

for every $(R^m -$ valued$)$ $\omega \in L^p(\Omega)$. Corollary 4 tells us that

(82) $M - \lim G_n = G_0$ in $L^p(\Omega).$

(Every f_n is a normal convex integrand by Rockafellar [6, cor. 2E]). Let now $u_n \rightharpoonup u$ in $W^{1,p}(\Omega)$. Since $\nabla u_n \rightharpoonup \nabla u$ in $L^p(\Omega)$, from (82) we obtain

$$\liminf F_n(u_n) \geq F_0(u_0).$$

By theorem 10.6 of Rockafellar [2] (which holds in Banach spaces) we see that G_n is locally equi-Lipschitz continuous on $L^p(\Omega)$. Therefore $G_n \to G_0$ pointwise by (82), and the same holds for F_n.   □

Theorem 15. **Proof of the "if" part.**

Assume convergence of $f_n^*(\cdot, z)$ for every z. Fix any subinterval $D = (a, b) \subset \Omega$ and write

$$u' = \nabla u, \ L^p = L^p(a, b), \ W^{1,p} = W^{1,p}(a, b).$$

Now consider for any $k \in R$

$$X = \{v \in W^{1,p} : v(b) = k\}.$$

We shall use the following preliminary result.

22 Lemma Given any k and $f \in M$, the conjugate of the restriction H of F (defined in (64)) to X is given by

$$H^*(u) = \int_a^b [k \ u(x) + f^*(x, - \int_a^x u \ dt)] \ dx,$$

$u \in L^q$, where $1/p + 1/q = 1$.

Taking this lemma for granted, we see that for any $u \in L^q$,

$$g_n \rightharpoonup g_0 \text{ in } L^\infty$$

where

$$g_n(x) = f_n^*(x, - \int_a^x u \ dt),$$

by the same reasoning as in the proof of the first half of theorem 3. (Proceeding as in that proof (same notations), for any $\varphi \in L^1$ and every $\varepsilon > 0$ we get

$$| \int_D (g_n - g_0)\varphi \ dx| \leq 2L\varepsilon \int_D |\varphi| \ dx + \varepsilon$$

for every sufficiently large n.) Therefore $H_n^*(u) \to H_0^*(u)$, $u \in L^q$, by lemma 22. This yields epi-convergence of the restrictions H_n to H_0 in (L^p, weak) by theorem IV. 26, for every k. This in turn is equivalent to epi-convergence in $(W^{1,p}, \text{weak})$ or (L^p, strong) (as in proposition 13). To verify this, it suffices to remember equicoercivity (65), and to notice that $u_n \rightharpoonup u$ in L^p together with boundedness of u_n' in L^p imply $u_n \rightharpoonup u$ in $W^{1,p}$. Therefore for every $u \in W^{1,p}$ there exists $u_n \rightharpoonup u$ in $W^{1,p}$ such that

$$\limsup F_n(u_n) \leq F_0(u)$$

(take X with $k = u(b)$). Moreover if $u_n \rightharpoonup u$ in $W^{1,p}$, then $w_n \in X$ with $k = u(b)$, where

$$w_n(x) = u_n(x) - u_n(b) + u(b)$$

and $w_n \rightharpoonup u$ in $W^{1,p}$ since $u_n \to u$ uniformly. Moreover $F_n(w_n) = F_n(u_n)$. Hence

$$\liminf F_n(u_n) \geq F_0(u),$$

yielding epi-$\lim F_n = F_0$ in $(W^{1,p}, \text{weak})$.

The proof of the first half of the theorem will be ended by the **proof of lemma 22.** For every $u \in L^q$

$$H^*(u) = \sup\{\int_a^b [uy - f(x, y')]dx : y \in X\}$$

$$= \sup\{\int_a^b [u(y + k) - f(x, y')dx] : y \in Z\}$$

where $y \in Z$ iff $y \in W^{1,p}$ and $y(b) = 0$. Hence

$$H^*(u) = \int_a^b ku\ dx + \sup\{\int_a^b [uy - f(x, y')dx] : y \in Z\}.$$

It suffices to show that this sup equals

$$\int_a^b f^*(x, -\int_a^x udt)dx.$$

Assume in addition that $f(x, \cdot)$ is continuously differentiable for every x. Consider

$$v(x) = -\int_a^x udt.$$

Then integrating by parts we get

$$\int_a^b uy\ dx = \int_a^b vy'\ dx,\ y \in Z.$$

Therefore

(83)
$$\sup\{\int_a^b [uy - f(x, y')]dx : y \in Z\}$$

$$= \sup\{\int_a^b [vy' - f(x, y')]dx : y \in Z\},$$

which is attained (by (65)) at y such that the Euler-Lagrange equation holds, i.e. $y \in Z$ and in the sense of distributions

$$(d/dx)[v - f_y(x, y')] = 0,\quad f_y[a, y'(a)] = 0.$$

(see Cesari [2, sec. 2.2]). Thus

(84)
$$v(x) = f_y[x, y'(x)] \text{ for a. e. } x.$$

On the other hand, for any fixed x

$$f^*(x, z) = \sup\{zw - f(x, w) : w \in R\}$$

is attained (by (65)) at w such that

Then by (83) and (84)

$$\sup\{\int_a^b [vy' - f(x,y)]dx : y \in Z\} = \int_a^b f^*[x, v(x)]dx,$$

yielding the conclusion in the particular case of a smooth $f(x, \cdot)$.

The general case can be obtained as follows. Let $\alpha \in C_0^1(-1, 1)$ be even and ≥ 0, such that $\int_{-1}^1 \alpha \, dx = 1$. Put

$$\alpha_j(t) = j\alpha(jt) \; ; f_j(x,z) = \int_{-1}^1 \alpha_j(t)f(x, z - t)dt, \; j = 1, 2, 3, \dots.$$

Then $f_j(x, \cdot)$ is convex and continuously differentiable, moreover by Jensen's inequality

(85) $$f_j(x, z) \geq f(x, z - \int_{-1}^1 t\alpha_j(t)dt) = f(x, z)$$

since $t\alpha_j(t)$ is odd. Since $f_j(x, \cdot) \to f(x, \cdot)$ uniformly on bounded sets, by theorem IV. 26 (with $X = R$)

(86) $$f_j^*(x, z) \to f^*(x, z) \text{ for every } x \in (a, b) \text{ and } z \in R.$$

The above argument for smooth $f(x, \cdot)$ yields

(87) $$H_j^*(u) + kv(b) = \int_a^b f_j^*(x, - \int_a^x u \, dt)dx, \; u \in L^q$$

for the conjugate H_j^* of the restriction of F_j to Z. Taking conjugates in (85) we get for every j

$$f_j^*(x, z) \leq c_1|z|^q$$

for a suitable constant c_1. From (65)

$$f_j(x, w) \leq \omega + \int_{-1}^1 \alpha_j(t)|w - t|^p dt,$$

hence for every j

$$f_j^*(x, z) \geq -\omega + \sup\{zw - 2^p \int_{-1}^1 \alpha_j(t)(|t|^p + |w|^p)dt\} \geq c_2 + c_3|z|^q$$

for suitable constants c_2, c_3. Then remembering (86) we see that the dominated convergence theorem applies to (87), yielding

$$\int_a^b f^*(x, - \int_a^x u \, dt)dx \leq H^*(u) + kv(b)$$

since $H_j^* \leq H^*$ by (85). From (83) we get the opposite inequality

$$H^*(u) + kv(b) \leq \int_a^b f^*[x, v(x)]dx$$

ending the proof of lemma 22.

Proof of the "only if" part. Fix any subsequence of f_n. The sequence $f_n^*(\cdot, z)$ is bounded in $L^\infty(\Omega)$ for every $z \in R$. Let T be the set of all rational numbers. Then by a diagonal procedure we can find

$$g : \Omega \times T \to R$$

such that for some further subsequence and every $z \in T$

(88) $$f_n^*(\cdot, z) \to g(\cdot, z) \text{ in } L^\infty(\Omega).$$

By convexity of $f_n^*(x, \cdot)$ we get convexity on T of $g(x, \cdot)$. From (65)

$$g_1(z) \le f_n^*(x, z) \le g_2(z)$$

where g_1 is the conjugate to $\alpha|\cdot|^p$, g_2 that to $\omega(1+ |\cdot|^p)$. Hence for every measurable $E \subset \Omega$ and every $z \in T$

$$g_1(z) \text{ meas } E \le \int_E f_n^*(x, z)dx \le g_2(z) \text{ meas } E$$

yielding

(89) $$g_1 \le g \le g_2.$$

By convexity, g can be extended by continuity on $\Omega \times R^m$ in such a way that (89) holds there. We want to show that

(90) $$f_n^*(\cdot, z) \to g(\cdot, z) \text{ in } L^\infty(\Omega), z \in R.$$

Given z, let $z_k \to z$ be with $z_k \in T$. By theorem 10.6 of Rockafellar [2]

$$|f_n^*(x, z) - f_n^*(x, z_k)| \le L|z - z_k|,$$
$$|g(x, z) - g(x, z_k)| \le L|z - z_k|$$

for some constant L, every k and n. Then for every $u \in L^1(\Omega)$

$$|\int_\Omega [f_n^*(x, z) - g(x, z)]u(x)dx|$$

$$\le 2L \int_\Omega |u|dx \, |z - z_k| + |\int_\Omega [f_n^*(x, z_k) - g(x, z_k)]u(x)dx|.$$

Letting $k \to +\infty$ we get (90) by (88). Now consider

$$h(x, y) = \sup\{zy - g(x, z) : z \in R\}.$$

Then by (89) we get $h \in M$. By (90) and the "if" part, we see that

$$\int_D h(x, y')dx = \int_D f_0(x, y')dx, \; y \in W^{1,p}(D)$$

for every open interval $D \subset \Omega$. Taking $y(x) = z_k x$, z_k any dense sequence in R, we get

$$h(x, z_k) = f_0(x, z_k) \text{ for a.e. } x \in \Omega,$$

hence by continuity $h(x, z) = f_0(x, z)$ for every z and a.e. x. By arbitrariness of the subsequence we get (90) with $g = f_0^*$ for the original sequence. □

Theorem 18. The first statement has been already proved. Assume that val (y, B, G_n) converges as $n \to +\infty$ for every y and every ball $B \subset \Omega$. We shall rely on the following

23 Compactness lemma. The space of all functionals G defined by (67) is compact if endowed with the following convergence:

$G_n \to G_0$ iff epi - $\lim G_n(\cdot, A) = G_0(\cdot, A)$ in $(L^p(A)$, strong) for every $A \in Q$.

For a proof of this lemma see Dal Maso-Modica [1, cor. 1.22 p.368].

By the compactness lemma 23, it suffices to show that if G', G'' are cluster points of G_n with respect to epi-convergence, then $G' = G''$. Indeed, by (68) and the assumed convergence of the values,

$$\text{val } (y, B, G') = \text{val } (y, B, G'')$$

for every y and B. Let G' correspond to f', G'' to f'', through (67). Assuming (70) we get

$$f'(x, y) = f''(x, y) \text{ for a. e. } x \text{ and every } y.$$

So the proof of the theorem will be ended by the

proof of (70). Step 1: (70) holds if there exists a constant S such that

$$f_0(x, y) = \alpha |y|^p \text{ if } x \in \Omega \text{ and } |y| \geq S.$$

Indeed, let us fix $x_0 \in \Omega$. Then for every y and $A \in Q$

$$\min \{ \int_A f_0(x_0, \nabla u) dx : u - L(y) \in W_0^{1,p}(A) \} \leq f_0(x_0, y) \text{ meas } A.$$

On the other hand, if $u - L(y) \in W_0^{1,p}(A)$, by Jensen's inequality

$$(1/\text{meas } A) \int_A f_0(x_0, \nabla u) dx \geq f_0(x_0, (1/\text{meas } A) \int_A \nabla u \, dx) = f_0(x_0, y),$$

since $\int_A \nabla u \, dx = \int_A \nabla(u - L) dx + \int_A y \, dx$. Hence

$$\min \{ \int_A f_0(x_0, \nabla u) dx : u - L(y) \in W_0^{1,p}(A) \} = f_0(x_0, y) \text{ meas } A.$$

Then writing

$$v(y, A) = \text{ val } (y, A, G_0)$$

we get

$$|f_0(x_0, y) - \frac{v(y, A)}{\text{meas } A}|$$

$$= \frac{1}{\text{meas } A} | \min \{ \int_A f_0(x_0, \nabla u) dx : u - L(y) \in W_0^{1,p}(A) \}$$

$$- \min \{ \int_A f_0(x, \nabla u) dx : u - L(y) \in W_0^{1,p}(A) \} |$$

$$\leq \frac{1}{\text{meas } A} \sup \{ | \int_A [f_0(x_0, \nabla u) - f_0(x, \nabla u)] dx | : u \in W^{1,p}(A) \}$$

$$\leq \frac{1}{\text{meas } A} \int_A \sup \{ |f_0(x_0, q) - f_0(x, q)| : q \in R^m \} dx.$$

Put

$$s(x, q) = |f_0(x_0, q) - f_0(x, q)|.$$

Then

(91) $$s(x, q) = 0 \text{ if } |q| \geq S.$$

Step 1 will be proved by showing that for a.e. $x_0 \in \Omega$

$$\frac{1}{\text{meas } B(x_0, r)} \int_{B(x_0, r)} \sup\{s(x, q) : q \in R^m\} dx \to 0 \text{ as } r \to 0.$$

Since $f_0(x, \cdot)$ is locally Lipschitz continuous, uniformly with respect to $x \in \Omega$ (Rockafellar [2, th. 10.6]) there exists a constant $k > 0$ such that

(92) $$|s(x, q') - s(x, q'')| \leq k|q' - q''|$$

for every x, q', q''. Fix a countable dense set D in R^m and write

$$m(r) = \text{meas } B(x, r).$$

By the Lebesgue differentiation theorem (Dunford-Schwartz [1, III.12.8]), for a.e. $x_0 \in \Omega$ and every $q \in D$

(93) $$\frac{1}{m(r)} \int_{B(x_0, r)} s(x, q) dx \to 0 \text{ as } r \to 0.$$

Given $\varepsilon > 0$, there exists a finite number of points $q_1, ..., q_l \in D$ such that for every $q \in R^m$ with $|q| \leq S$ we have

$$\inf \{|q - q_i| : 1 \leq i \leq l\} < \varepsilon.$$

Then by (91) and (92)

$$\sup \{s(x, q) : q \in R^m\} \leq k\varepsilon + \sum_{i=1}^{l} s(x, q_i)$$

and by (93), for a.e. $x_0 \in \Omega$

$$\limsup_{r \to 0} \frac{1}{m(r)} \int_{B(x_0, r)} \sup\{s(x, q) : q \in R^m\} dx \leq k\varepsilon.$$

Letting $\varepsilon \to 0$ we get the conclusion in step 1.

Step 2. There exists a sequence $g_h \in M$ such that $f_0 = \sup_h g_h$ and $g_h(x, y) = \alpha|y|^p$ for a.e. x and $|y|$ sufficiently large.

To this aim, for every $h = 1, 2, 3, ...,$ define

$$\omega_h(x, y) = \inf \{f_0(x, z) + h\,|z - y| : z \in R^m\}.$$

The above infimum with respect to $z \in R^m$ will be left unchanged by taking instead $z \in D$, any countable dense set in R^m, since $f_0(x, \cdot)$ is continuous. This shows measurability of each $\omega_h(\cdot, y)$. Moreover by (65)

$$\omega_h(x, y) \leq \omega + h|y|.$$

Then defining

$$g_h(x, y) = \max \{\alpha|y|^p, \omega_h(x, y)\}$$

we see that, for every h, $g_h \in M$ and there exists a positive constant S_h such that

$$g_h(x, y) = \alpha|y|^p \text{ if } |y| \geq S_h \text{ and a. e. } x.$$

Now $\omega_h \leq \omega_{h+1} \leq f_0$. We claim that

(94)
$$\sup_h \omega_h = f_0,$$

hence $\sup_h g_h = f_0$.

The proof of (94) runs as follows. Let us fix x, y. By (65) the infimum defining $\omega_h(x, y)$ is attained at some point z_h. Since

$$f_0(x, y) \geq \omega_h(x, y) \geq f_0(x, z_h) \geq \alpha|z_h|^p$$

we see that z_h is bounded. By convexity and (65), $f_0(x, \cdot)$ is locally Lipschitz, therefore for some constants c_1, c_2

$$h|y - z_h| = \omega_h(x, y) - f_0(x, z_h) \leq f_0(x, y) - f_0(x, z_h) \leq c_1|y - z_h| \leq c_2,$$

hence $z_h \to y$. Therefore $f_0(x, z_h) \to f_0(x, y)$ and by the previous inequality, $h|y - z_h| \to 0$. Hence $\omega_h(x, y) \to f_0(x, y)$. This completes step 2.

By step 1, (70) holds for every g_h. Denote by G_h the corresponding integral functionals defined by (67). As $f_0 \geq g_h$, we have for a.e. x

$$g_h(x, y) = \lim_{r \to 0} \frac{1}{m(r)} \text{val } (y, B(x, r), G_h) \leq \liminf_{r \to 0} \frac{1}{m(r)} \text{val } (y, B(x, r), G_0).$$

Letting $h \to +\infty$ we get

(95)
$$f_0(x, y) \leq \liminf_{r \to 0} \frac{1}{m(r)} \text{val } (y, B(x, r), G_0).$$

We need the converse inequality. Obviously

(96)
$$\frac{1}{m(r)} \int_{B(x,r)} f_0(t, y)dt = \frac{1}{m(r)} G_0[L(y)] \geq \frac{1}{m(r)} \text{val } (y, B(x, r), G_0).$$

Fix any countable dense set D in R^m. Then there exists a measurable $N \subset \Omega$ with meas $N = 0$, such that, as $r \to 0$

(97) $$\frac{1}{m(r)} \int_{B(x,r)} f_0(t,y)dt \to f_0(x,y) \text{ if } x \notin N \text{ and } y \in D.$$

But (97) holds for every $y \in R^m$ since $f(x, \cdot)$ is locally Lipschitz continuous, uniformly with respect to $x \in \Omega$. Then by (96) and (97)

$$\limsup_{r \to 0} \frac{1}{m(r)} \text{ val } (y, B(x,r), G_0) \leq f_0(x,y)$$

for a.e. x and every $y \in R^m$.   □

Corollary 20. Only the last assertion needs proof. Let (a_{ij}^n) be any sequence in E. By the compactness lemma 23, there exists $f_0 \in M$ such that for a subsequence and every $A \in Q$

$$\text{epi - } \lim G_n(\cdot, A) = G_0(\cdot, A) \text{ in } (L^2(A), \text{ strong})$$

where

$$G_n(u, A) = \sum_{i,j=1}^m \int_A a_{ij}^n(x) \frac{\partial u}{\partial x_i} \frac{\partial u}{\partial x_j} dx,$$

$$G_0(u, A) = \int_A f_0(x, \nabla u)dx, \quad u \in H^1(A).$$

Then by theorem 18, for that subsequence

(98) $$\text{val } (y, A, G_n) \to \text{ val } (y, A, G_0).$$

We show that for fixed $n \geq 1$ and A,

$$y \to q_n(y) = \frac{1}{\text{meas } A} \text{ val } (y, A, G_n)$$

is an equi-positive definite quadratic form, by proving that for all $y, z \in R^m$ and $t \in R$

(99) $$\alpha |y|^2 \leq q_n(y) \leq \omega |y|^2,$$

(100) $$q_n(ty) = t^2 q_n(y),$$

(101) $$q_n(y + z) + q_n(y - z) = 2q_n(y) + 2q_n(z).$$

As well known, (99), (100), (101) characterize quadratic forms with eigenvalues between α and ω.

To prove (99) we take $u(x) = y'x$ and obtain val $(y, A, G_n) \leq G_n(u, A)$, hence

$$q_n(y) \leq \frac{1}{\text{meas } A} \sum_{i,j=1}^m \int_A a_{ij}^n(x) y_i \, y_j \, dx \leq \omega |y|^2.$$

On the other hand, take $v(x) = u(x) + y'x$ with $u \in H_0^1(A)$. Then $\int_A \partial u/\partial x_i \, dx = 0$, so that

$$\sum_{i,j=1}^{m} \int_A a_{ij}^n(x) \frac{\partial v}{\partial x_i} \frac{\partial v}{\partial x_j} \, dx \geq \alpha \int_A |\nabla v|^2 \, dx$$

$$= \alpha \left(\int_A |\nabla u|^2 \, dx + |y|^2 \text{ meas } A \right),$$

hence

$$\text{val } (y, A, G_n) \geq \alpha |y|^2 \text{ meas } A$$

and (99) is proved. To prove (100), let $u, \, v \in H^1(A)$ be such that

$$\text{val } (y, A, G_n) = G_n(u, A), \quad u - L(y) \in H_0^1(A),$$

$$\text{val } (ty, A, G_n) = G_n(v, A), \quad v - L(ty) \in H_0^1(A).$$

Then from the Euler - Lagrange equations and the uniqueness of the minimizers we get $v = tu$, hence (100). To prove (101), let $u_k, k = 1, ...4$, be the solution to

$$-\sum_{i,j=1}^{m} \frac{\partial}{\partial x_i} \left(a_{ij}^n \frac{\partial u_k}{\partial x_j} \right) = 0 \text{ in } A$$

such that on ∂A

$$u_1(x) = (y + z)'x, \ u_2(x) = (y - z)'x, \ u_3(x) = y'x, \ u_4(x) = z'x.$$

By uniqueness of the solution to each of the above Dirichlet problems

$$u_1 = u_3 + u_4, \ u_2 = u_3 - u_4,$$

hence (101) is proved. By (98), we see that (in the limit) (99), (100), (101) hold for q_0. In particular, for every fixed $x \in \Omega$ and $r > 0$ sufficiently small, the same formulas hold for

$$y \to \frac{1}{\text{meas } B(x,r)} \text{ val } (y, B(x,r), G_0).$$

By letting $r \to 0$, from (70) we see that, for a.e. $x \in \Omega$, $f_0(x, \cdot)$ still verifies (99), (100), (101). This means that

$$f_0(x, y) = \sum_{i,j=1}^{m} a_{ij}(x) y_i y_j, \text{ a.e. } x \in \Omega \text{ and every } y \in R^m,$$

for some (a_{ij}) satisfying (72). Therefore we have shown that, for some subsequence,

$$(a_{ij}^n) \to (a_{ij}) \text{ in } E,$$

moreover for every $A \in Q$

(102)   epi-lim $G_n(\cdot, A) = G(\cdot, A)$ in $(L^2(A), \text{ strong})$, G corresponding to (a_{ij}).

Now suppose that

$$\text{epi - lim } G_n(\cdot, \Omega) = G_0(\cdot, \Omega).$$

For some subsequence, some matrix $(b_{ij}) \in E$ and every $A \in Q$ we have (102), where now G corresponds to (b_{ij}). Then

$$G_0(\cdot, \Omega) = G(\cdot, \Omega) \text{ on } H_0^1(\Omega).$$

It follows that the corresponding Euler-Lagrange operators

$$L = -\sum_{i,j=1}^{m} \frac{\partial}{\partial x_i}\left(a_{ij}^0 \frac{\partial}{\partial x_j}\right), \ K = -\sum_{i,j=1}^{m} \frac{\partial}{\partial x_i}\left(b_{ij} \frac{\partial}{\partial x_j}\right) : H_0^1(\Omega) \to H^{-1}(\Omega)$$

must agree, hence

(103) $$\int_\Omega \sum (a_{ij}^0 - b_{ij})\varphi_i \varphi_j \, dx = 0, \ \varphi \in C_0^1(\Omega)$$

where $\sum$ means $\sum_{i,j=1}^m$ and $\varphi_i = \partial\varphi/\partial x_i$. Fix $z \in C_0^1(\Omega)$, $y \in R^m$ and put

$$a = y'x, \ u(x) = z(x)\cos a, \ v(x) = z(x)\sin a.$$

We get

$$\sum b_{ij} u_i u_j = \sum b_{ij}(z_i z_j \ \cos^2 a + z^2 y_i y_j \ \sin^2 a - 2y_i z z_j \ \cos \ a \ \sin \ a);$$
$$\sum b_{ij} v_i v_j = \sum b_{ij}(z_i z_j \ \sin^2 a + z^2 \ y_i y_j \ \cos^2 a + 2y_i z z_j \ \cos \ a \ \sin \ a).$$

Adding, we obtain

$$\sum b_{ij}(u_i u_j + v_i v_j) = \sum b_{ij}(z_i z_j + z^2 y_i y_j)$$

and the analogous equality with a_{ij}^0. Then by (103)

$$\int_\Omega \sum b_{ij} \ y_i y_j \ z^2 \ dx = \int_\Omega \sum a_{ij}^0 \ y_i y_j \ z^2 \ dx.$$

By density of the subspace spanned by $\{z^2 : z \in C_0^1(\Omega)\}$ it follows that for all $y \in R^m$,

$$\sum (a_{ij}^0 - b_{ij})y_i y_j = 0 \text{ a.e. in } \Omega.$$

Hence $a_{ij}^0 = b_{ij}$ for every i, j. □

Section 4.   Problems with Unilateral Constraints.

We are interested in Hadamard well-posedness of problems involving the minimization of integral functionals (61) with $f \in M$, subject to the constraint

(104) $$u \geq g$$

for a given function $g : \Omega \to [-\infty, +\infty]$, under perturbations of both f and g.

Strong solution Hadamard well-posedness for quadratic functionals is obtained from theorem IV.48 by imposing M - convergence of both the integral functionals and the convex sets. A reasonable approach to get value and weak solution Hadamard well-posedness, based on theorem IV.5, should require weak epi-convergence of the integral functionals and M - convergence of the convex sets obtained in (104) by perturbing g. For general convex sets (not necessarily of the unilateral type (104)) this may fail, as shown by

24 Counterexample (to value well-posedness).

Consider l^2 equipped by its standard Hilbert space structure, and put

$$J_n(u) = \|u - e_n\|^2, e_1 = (1, 0, 0, ...),$$
$$e_2 = (0, 1, 0, ...), ...; K_n = \{x \in l^2 : <x, e_j> = 0 \text{ for every } j \geq n\}, n \geq 1.$$

Then it is readily verified that $M - \lim K_n = l^2$. Moreover the conjugate is given by

$$J_n^*(u) = \frac{1}{4}\|u\|^2 + <u, e_n>, \ u \in l^2,$$

so that

$$\text{weak epi-lim } J_n = J_0, \text{ where } J_0(u) = \|u\|^2,$$

by theorem IV.26. Since $J_n(u) = 1 + \|u\|^2 - 2 <u, e_n>$ we obtain

$$\text{val } (K_n, J_n) = 1 \text{ if } n \geq 1, \ \text{val } (l^2, J_0) = 0.$$

Due to the special kind of the constraint (104) (where the meaning of the inequality should be carefully defined for nonsmooth functions g) the above approach does in fact provide a sufficient condition to well-posedness, analogous to theorem IV.48, as follows.

We consider p, Ω and M defined as in section 3, and sequences $f_n \in M, g_n \in W_0^{1,p}(\Omega)$. Put

$$K_n = \{u \in W_0^{1,p}(\Omega) : u(x) \geq g_n(x) \text{ a.e. in } \Omega\},$$
$$J_n(u) = \int_\Omega f_n(x, \nabla u)dx, \ u \in W_0^{1,p}(\Omega).$$

25 Theorem *Suppose that $\partial\Omega$ is of class C^1 and*

$$weak \ epi - \lim J_n = J_0 \ in \ W_0^{1,p}(\Omega),$$

(105) $$M - \lim K_n = K_0 \ in \ W_0^{1,p}(\Omega).$$

Then

(106) $$weak \ epi - \lim (J_n + ind \ K_n) = J_0 + ind \ K_0 \ in \ W_0^{1,p}(\Omega),$$

and therefore in $W_0^{1,p}(\Omega)$

$$weak \ seq. \ \lim \sup arg \min (K_n, J_n) \subset arg \min (K_0, J_0),$$
$$val \ (K_n, J_n) \to val \ (K_0, J_0).$$

A simple sufficient condition to (105) is that

$$g_n \to g_0 \text{ in } W_0^{1,p}(\Omega)$$

since then, as readily verified,

$$K_n = P + g_n \xrightarrow{M} P + g_0 = K_0$$

where P denotes the closed convex set of a.e. nonnegative elements of $W_0^{1,p}(\Omega)$.

To explain the role of (105) as a well-posedness condition, consider the particular case $p = 2$ and a fixed functional

$$(107) \qquad J_n(u) = J_0(u) = \sum_{i,j=1}^{m} \int_{\Omega} a_{ij}(x) \frac{\partial (u-y)}{\partial x_i} \frac{\partial (u-y)}{\partial x_j} dx - \int_{\Omega} yu \, dx$$

for any $y \in H_0^1(\Omega)$ and some matrix (a_{ij}) satisfying (72). Then

$$u \to \sum_{i,j=1}^{m} \int_{\Omega} a_{ij}(x) \frac{\partial u}{\partial x_i} \frac{\partial u}{\partial x_j} dx$$

is the square of a norm $\| \cdot \|$ on $H_0^1(\Omega)$, which is equivalent to the usual one by Poincaré's inequality. Therefore

$$J_0(u) = \|u - y\|^2$$

and from theorem II.9 we see that (105) is necessary and sufficient for either solution or value Hadamard well-posedness for every $y \in H_0^1(\Omega)$.

PROOFS.

Theorem 25. From the assumptions we see that $u_n \rightharpoonup u$ in $W_0^{1,p}(\Omega)$ implies

$$\liminf[J_n(u_n) + \text{ind}\,(K_n, u_n)] \geq J_0(u) + \text{ind}\,(K_0, u).$$

Therefore to prove (106) it suffices to show that

(108)   for every $v \in K_0$ there exists $v_n \in K_n$ such that $v_n \rightharpoonup v$ and
$\limsup J_n(v_n) \leq J_0(v)$.

Consider

$$H = \{v \in W_0^{1,p}(\Omega) : v(x) > g_0(x) \text{ a. e. in } \Omega\}.$$

Of course $H \subset K_0$. Take $h \in C^1(\text{ cl } \Omega)$ such that $h = 0$ on $\partial\Omega$ and $h > 0$ in Ω. Then $v + \varepsilon h \in H$ provided that $v \in K_0$ and $\varepsilon > 0$, and $v + \varepsilon h \to v$ as $\varepsilon \to 0$. Therefore cl $H = K_0$ within $W_0^{1,p}(\Omega)$.

Suppose that (108) has been proved for every $v \in H$. Fix $v \in K_0$ and let $v_i \to v$ be such that $v_i \in H$. For every i consider $v_{ni} \in K_n$ such that

$$v_{ni} \rightharpoonup v_i \text{ and } J_n(v_{ni}) \to J_0(v_i) \text{ as } n \to +\infty.$$

Then in the product space $L^p(\Omega) \times R$ we have

$$x_{ni} = (v_{ni}, J_n(v_{ni})) \to_{n \to +\infty} y_i = (v_i, J_0(v_i)) \to_{i+\infty} z = (v, J_0(v)).$$

Let d be any metric which induces on $L^p(\Omega) \times R$ the usual strong convergence. For every $k = 1, 2, 3, ...$, there exist indices N_k, M_k such that

$$d(y_i, z) \le 2^{-k} \text{ if } i \ge N_k,$$
$$d(x_{n N_k}, y_{N_k}) \le 2^{-k} \text{ if } n \ge M_k;$$
$$N_{k+1} > N_k, \ M_{k+1} > M_k.$$

Therefore, given n sufficiently large, there exists an unique k such that $M_k \le n < M_{k+1}$. Put $i(n) = N_k$,

$$a_n = x_{n\,i(n)}.$$

Then $i(\cdot)$ is (not necessarily strictly) increasing and

$$d(a_n, z) = d(x_{n N_k}, z) \le d(z, y_{N_k}) + d(y_{N_k}, x_{n N_k}) \le 2^{-k} + 2^{-k} \text{ if } n \ge M_k,$$

proving that $a_n \to z$. Thus we have found a sequence $u_n \in K_n$ such that

$$u_n \rightharpoonup v, \ J_n(u_n) \to J_0(v).$$

Therefore we need to prove (108) only when $v \in H$. By assumption, there exist $z_n \in K_n$, $u_n \in W_0^{1,p}(\Omega)$ such that

(109)
$$z_n \to g_0, \ u_n \rightharpoonup v, \ J_n(u_n) \to J_0(v).$$

Consider

$$v_n = \max\{u_n, z_n\}.$$

Of course $v_n \in K_n$. Let us show that $v_n \overset{\varepsilon}{\rightharpoonup} \max\{v, g_0\} = v$ and

(110)
$$\limsup J_n(v_n) \le J_0(v)$$

thereby proving (108). To this aim consider

$$A_n = \{x \in \Omega : \ u_n(x) < z_n(x)\}.$$

If meas $A_n \nrightarrow 0$, there would exist $\varepsilon > 0$ such that for some subsequence

(111)
$$\text{meas } A_n \ge \varepsilon.$$

Let $B_k = \cup_{n=k}^{\infty} A_n$ and $C = \cap_{k=1}^{\infty} B_k$. Then B_k decreases to C, and meas $C = 0$ since $u_n - z_n \to v - g_0$ in $L^p(\Omega)$ and $v > g_0$ a.e. in Ω. Hence meas $B_k \to 0$, contradicting (111). Therefore

(112)
$$\text{meas } A_n \to 0.$$

Using theorem A. 1 from Kinderlehrer-Stampacchia [1] and (112), we readily obtain

$$v_{nx} \to v_x \text{ in } L^p(\Omega),$$

hence $v_n \rightharpoonup v$ in $W_0^{1,p}(\Omega)$. Using again theorem A.1 from Kinderlehrer-Stampacchia [1] and remembering (65) we can write

$$J_n(v_n) = \int_{\Omega \backslash A_n} f_n(x, \nabla u_n) dx + \int_{A_n} f_n(x, \nabla z_n) dx$$

$$\leq J_n(u_n) + \omega(\text{meas } A_n + \int_{A_n} |\nabla z_n|^p \, dx).$$

But $\int_{A_n} |\nabla z_n|^p dx \to 0$ since $|\nabla z_n|^p \to |\nabla g_0|^p$ in $L^1(\Omega)$ and (112). Thus, remembering (109), we get (110). Therefore (106) is proved. The remaining conclusions follow immediately from theorem IV.5.   □

Section 5.   Well-Posedness of the Rietz Method and Ill - Posedness in Least Squares.

Let X be a fixed real Banach space. We consider a subset $E \subset X$, a proper functional

$$I : X \to (-\infty, +\infty]$$

and a sequence E_n of subsets of X. The classical Rietz method tries to approximate

$$\text{arg min } (E, I) \text{ and val } (E, I)$$

by

$$\text{arg min } (E_n, I) \text{ and val } (E_n, I)$$

assuming E_n finite-dimensional and properly related to (E, I) in order to have simpler approximating problems than the original one.

Put

(113) $$f_n = I + \text{ ind } E_n, \quad f_0 = I + \text{ ind } E.$$

Under suitable (semi) continuity properties of I and convergence of E_n, we get Mosco convergence of f_n towards f_0. This yields convergence properties of val (E_n, I) to val (E, I) and $\varepsilon_n - \text{arg min } (E_n, I)$ to arg min (E, I) as $\varepsilon_n \to 0$. If, moreover, we have Tykhonov well-posedness, then better convergence properties may be obtained, as follows.

26 Theorem *Suppose that*

(114) $$M - \lim E_n = E;$$

(115)   *I is weakly sequentially lower semicontinuous and strongly continuous;*

(116)   *$\{u \in E : I(u) \leq t\}$ is weakly sequentially compact for every real t.*

Then for any $\varepsilon_n > 0$ such that $\varepsilon_n \to 0$,

(117) $$M - \lim f_n = f_0;$$

$$weak\ seq.\ lim\ sup\ \mathcal{E}_n - arg\ min\ (E_n, I) \subset arg\ min\ (E, I);$$

$$val\ (E_n, I) \to val\ (E, I).$$

If moreover (K, I) is Tykhonov well-posed for all $K \in Conv\ (X)$, $E \in Conv\ (X)$ and I is convex and uniformly continuous on bounded sets, then

(118) $$x_n \in \mathcal{E}_n - arg\ min\ (E_n, I) \Rightarrow x_n \to arg\ min\ (E, I).$$

The Lavrentiev phenomenon; an example of ill - posedness (in the Rietz method).

Given $T > 0$, A and B in R^n, let

$$X = W^{1,1}(0, T), E = \{x \in W^{1,1}(0, T) : x(0) = A, x(T) = B\},$$

$$I(x) = \int_0^T f(t, x, \dot{x}) dt,$$

where $f : [0, T] \times R^n \times R^n \to (-\infty, +\infty)$ is given.

A conventional version of the finite element method tries to approximate val (E, I) by constructing minimizing sequences with Lipschitz continuous elements (or belonging to $W^{1,p}(0, T)$ with $p > 1$), satisfying the same boundary conditions. Thus we try to approximate (E, I) by the (constant) sequence (F, I), where (e.g)

$$F = \{x \in W^{1,\infty}(0, T) : x(0) = A, x(T) = B\}.$$

A necessary condition for such a good behaviour is that

$$val\ (E, I) \geq val\ (F, I)$$

since $F \subset E$. The next example shows that we may well have

(119) $$val\ (F, I) > val\ (E, I)$$

in spite of the known fact that F is dense in E with respect to uniform convergence. This is a value Hadamard ill-posedness example, since for the constant sequence $E_n = F$ we have

$$M - \lim E_n = E \text{ in } C^0(0, T) \text{ (by density)},$$

$$\lim val\ (E_n, I) > val\ (E, I).$$

27 Example (the Lavrentiev phenomenon). Consider

$$I(x) = \int_0^1 (x^3 - t)^2 \dot{x}(t)^6 dt, A = 0, B = 1.$$

Of course

$$val\ (E, I) = I(y) = 0$$

where $y(t) = t^{1/3}$. We shall show that there exists a constant $k > 0$ such that

(120) $$I(x) \geq k \text{ for every } x \in F,$$

hence (119) will follow. To this aim, fix any $\alpha \in (0, 1]$ and consider

$$D = \{(t, x) \in R^2 : 0 \leq t \leq 1, 0 \leq x \leq \alpha t^{1/3}\}.$$

Fix any $x \in F$ and put

$$t_1 = \sup \{t \in [0, 1] : x(t) = 0\}.$$

If $t_1 = 0$ let L be a Lipschitz constant for x. Then we have

$$x(t) = \int_0^t \dot{x} \, ds \leq Lt < \alpha t^{1/3}$$

if $0 \leq t \leq \delta$ for some small δ. Put

$$t_2 = \inf \{t \geq \delta : x(t) = \alpha t^{1/3}\}.$$

Then

$$t_2 > t_1, 0 \leq x(t) \leq \alpha t^{1/3} \text{ if } t_1 \leq t \leq t_2, x(t_2) = \alpha t_2^{1/3}.$$

If $t_1 > 0$, consider

$$t_2 = \inf \{t \geq t_1 : x(t) = \alpha t^{1/3}\}.$$

Again $t_1 < t_2$ since $x(t_1) = 0$ and $x(t_2) = \alpha t_2^{1/3} > 0$. In any case

$$(t, x(t)) \in D \text{ if } t_1 \leq t \leq t_2,$$

whence

$$(x^3 - t)^2 \geq t^2(1 - \alpha^3)^2$$

yielding

$$I(x) \geq \int_{t_1}^{t_2} t^2 \dot{x}^6 dt (1 - \alpha^3)^2.$$

The problem of minimizing

$$\int_{t_1}^{t_2} t^2 \dot{x}^6 dt \text{ subject to } x \in W^{1,1}(t_1, t_2), x(t_1) = 0, x(t_2) = \alpha t_2^{1/3},$$

has an unique global minimizer given by the solution to the Euler-Lagrange equation

$$(d/dt)(t^2 \dot{x}^5) = 0, x(t_1) = 0, x(t_2) = \alpha t_2^{1/3}.$$

The minimizer can be obtained explicitly, yielding

$$I(x) \geq (3/5)^5 (1 - \alpha^3)^2 \alpha^6 \frac{t_2^2}{(t_2^{3/5} - t_1^{3/5})^5}.$$

Taking the maximum value of $(1 - \alpha^3)^2 \alpha^6$ when $0 \leq \alpha \leq 1$ and noticing that $t_2^{2/5} \geq t_2^{3/5} - t_1^{3/5}$ we get (120).

In this example, strong continuity of I fails (compare with theorem 26).

An important related problem is the following. We are given an infinite dimensional real separable Hilbert space X, and a linear bounded operator

$$A : X \to X.$$

Given $u \in A(X)$ we are interested in solving the operator equation

(121) $$Ax = u$$

by a least squares method of the following type. We consider an increasing sequence of finite-dimensional linear subspaces E_n of X, with $\cup_{n=1}^{\infty} E_n$ dense in X, and minimize on E_n

(122) $$y \to I(y, u) = (1/2)\|Ay - u\|^2.$$

Suppose that A is injective. Then an unique

$$x_n = \text{arg min } [E_n, I(\cdot, u)]$$

exists, and we try to obtain approximately the solution x to (121) by using (some computed approximation to) x_n. For a fixed u, this is a particular case of the Rietz method. In practice, u is not exactly available (e.g. by noisy data, computational imprecision or numerical (finitary) representation). So we introduce some perturbations u_n of u.

We are interested in the strong convergence of this (widely used) method, i.e. whether (for I given by (122) and x the unique solution to (121))

(123) $$u_n \to u \Rightarrow \text{arg min } (E_n, I(\cdot, u_n)) \to x.$$

For fixed n, the least squares solution

(124) $$x_n = \text{arg min } [E_n, I(\cdot, u_n)]$$

is clearly given by

(125) $$x_n = A_n^{-1}(p[u_n, A(E_n)]),$$

where A_n denotes the restriction of A to E_n and $p[u_n, A(E_n)]$ is the best approximation to u_n from the (finite-dimensional) linear subspace $A(E_n)$. If A is an isomorphism, then $M-\lim A(E_n) = X$ since $M-\lim E_n = X$, as readily verified. Then Lipschitz continuity of $p[\cdot, A(E_n)]$, theorem II. 9 and continuity of A^{-1} yield $x_n \to x$. If $A(X)$ is closed, then $[X, I(\cdot, u)]$ is strongly Tykhonov well-posed (by theorem I. 12 and Dunford-Schwartz [1, VI. 6.1]). Let $z_n \in E_n$ be such that $z_n \to x$. Then

$$\|Ax_n - u\| \le \|Az_n - u\| \le \|Az_n - u_n\| + \|u_n - u\|$$

yielding $I(x_n, u) \to 0$. Hence $x_n \to x$ by well-posedness.

Suppose now that A is one-to-one and compact with (infinite-dimensional) dense range, to which u belongs, and fix any sequence E_n as before. Then the above arguments yielding $x_n \to x$ break down. Thus A^{-1} cannot be continuous (since $A(X)$ is not closed), hence some minimizing sequence for $[X, I(\cdot, u)]$ will exist which does not converge to $A^{-1}(u)$. (If in particular A is monotone, we know that (121) is Hadamard ill-posed from theorem II. 33). Therefore, from ill-posedness of the least squares problem associated with the (classicaly Hadamard) ill-posed equation (121) we can expect unstability of the least squares solutions under perturbations of the datum u.

28 Theorem *Let $A : X \to X$ be an injective linear compact map with infinite dimensional dense range. Suppose A is positive selfadjoint and let $u \in A(X)$, and $Ax = u$. Then there exists $u_n \to u$ such that $\|x_n - x\| \to +\infty$, where x_n is given by (124).*

More is true: convergence of x_n toward x may fail even for exact data, i.e. $u_n = u$, and suitable choice of the approximating sequence E_n (as it could be guessed from corollary II. 15): see Seidman [1, th 3. 1].

PROOFS.

Theorem 26. By proposition IV. 41 and (115) we get

$$M - \lim\, f_n = M - \lim\, (I + \text{ind } E_n)$$
$$= I + M - \lim\, \text{ind } E_n = I + \text{ind } E = f_0.$$

Remembering (116), the remaining assertions in (117) follow from theorem IV. 5. Now let x_n be as in (118). Then

$$x_n \in E_n, \quad I(x_n) \le \varepsilon_n + \text{val } (E_n, I).$$

The conclusion is obvious if $\inf I(E) = \inf I(X)$. If $\inf I(E) > \inf I(X)$, as in the proof of theorem II. 13 we obtain a closed hyperplane H and a sequence $u_n \in H$ such that

$$u_n - x_n \to 0, I(u_n) \to \inf\, I(H) = I(x),$$

where $x = \arg\,\min\,(E, I)$. Hence (118) follows by well-posedness of (H, I). $\qquad\square$

Theorem 28. Let $A_n : E_n \to A(E_n)$ be the restriction of A to E_n. Clearly A_n is invertible. For any n there exists $z_n \in A(E_n)$ such that

$$\|z_n\| = 1, \ \|A_n^{-1} z_n\| = \|A_n^{-1}\|.$$

Then setting

$$v_n = z_n \|A_n^{-1}\|^{-1/2}, \ s_n = A_n^{-1} v_n,$$

we get $s_n \in E_n, v_n \in A(E_n)$, and

(126) $$\|s_n\| = \|A_n^{-1}\|^{1/2}, \|v_n\| = \|A_n^{-1}\|^{-1/2}.$$

Let

$$y_n = A_n^{-1}(p[u, A(E_n)]) = \arg\,\min\,[E_n, I(\cdot, u)].$$

Both inequalities

$$\|y_n + s_n\| < \|s_n\|, \ \|y_n - s_n\| < \|s_n\|$$

cannot hold, otherwise squaring and adding we would get a contradiction. Therefore at least one of $y_n \pm s_n$, say ω_n, satisfies

(127) $$\|\omega_n\| \geq \|s_n\|.$$

Now consider

$$u_n = u + A(\omega_n - y_n).$$

Then the corresponding least squares solution arg min $[E_n, I(\cdot, u_n)]$ is given by (125) as

$$x_n = A_n^{-1}\{p[u, A(E_n)] + A_n(\omega_n - y_n)\} = \omega_n.$$

We shall use

29 Lemma. $\|A_n^{-1}\| \to +\infty$ as $n \to +\infty$.

Hence by (126), (127) and lemma 29

$$\|x_n\| \geq \|s_n\| \to +\infty.$$

Then $\|x_n - x\| \to +\infty$, while

$$\|u_n - u\| = \|A(\omega_n - y_n)\| = \|As_n\| = \|v_n\| \to 0$$

by (126) and lemma 29.

Proof of lemma 29. Let $\alpha_n > 0$ denote the sequence of the eigenvalues of A with corresponding normalized eigenfunctions y_n. Then

$$Ay_n = \alpha_n y_n, \ \|y_n\| = 1; \ \alpha_n \to 0$$

(Brezis [2, th.VI.8]). Fix any subsequence of E_n. By density of $\cup_{n=1}^{\infty} E_n$, there exists a further subsequence of E_n, and for every n some $z_n \in E_n$ such that $\|y_n - z_n\| \leq 1/n$, whence

$$\|Az_n\| \leq \|A(z_n - y_n)\| + \|Ay_n\| \leq \frac{\|A\|}{n} + \alpha_n.$$

Then $\omega_n = z_n/\|z_n\| \in E_n$, and

$$\|A\omega_n\| \leq \|Az_n\|/(1 - (1/n)) \to 0$$

yielding (Anselone [1, p.5])

$$\|A_n^{-1}\|^{-1} = \inf \{\|Ax\| : x \in E_n, \|x\| = 1\} \leq \|A\omega_n\|,$$

hence $\|A_n^{-1}\| \to +\infty$. By the arbitrariness of the subsequence, this holds for the original sequence. $\square$

Section 6.   Notes and Bibliographical Remarks.

To section 1. Theorem 1 is due to Zolezzi [5]. For a related result in L^p spaces using strict convexity see Visintin [2]. For an extension of theorem 1 to Banach spaces see Papageorgiou [3]. Theorem 2 is due to Joly-De Thelin [1] and has been extended to the infinite-dimensional setting by Salvadori [1]. Theorem 3 and corollary 4 are particular cases of Marcellini-Sbordone [1].

To section 2. The basic connection between uniqueness of solutions to $Q(T, y)$ and differentiability of V at (T, y), leading to theorem 7, is due to Fleming [1] and Kuznetzov-Šiškin [1]. Extensions to optimal control problems are in Fleming-Rishel [1, ch.VI,sec.9]. For theorems 8 and 9 see Henry [1], whose proof we followed. Lemma 11 is an unpubblished result of Ekeland. For sharper results about smoothness of V see Fleming [1], Henry [1], and Sussmann [1, th. 4] for piecewise analyticity (provided f is analytic).

Lower semicontinuity of the optimal value function in a general setting was obtained by Rockafellar [3]. Convergence of minimizing sequences is studied in Cesari [1]. Generalized well-posedness of some integral functionals over unbounded domains is obtained by P. L. Lions [2]; the convergence of minimizing sequences in studied in Esteban-Lions [1]. A version of theorem 10 (whose proof was suggested by L. Boccardo) may be found in Visintin [2], and has been extended by Ball-Marsden [1, sec. 4] to vector-valued unknown. The trick which obtains strong convergence from weak one (implied here by strict convexity) was used by Pieri [4] and is developed by Visintin [1] (see also Olech [2, lemma 1]). Coupling theorem 10 with II. 13 one obtains Hadamard well-posedness for multiple integrals with moving constraints, see Zolezzi [15].

To section 3. Theorem 12 (in a more general form) may be found in Dal Maso [3, sec. 20]. Proposition 14 and theorem 15 are due to Marcellini-Sbordone [1] (related results are in Fusco [1]). Theorem 18 may be found in Dal Maso- Modica [2]. This characterization of epi-convergence generalizes the basic results of De Giorgi-Spagnolo [1] for quadratic integrals, relating epi-convergence to convergence of the energies. The compactness of the space of quadratic integrals and the local character of epi-convergence (corollary 20) were firstly established by Spagnolo [2] as a G-compactness property.

Hadamard well-posedness under weaker coercivity assumptions is obtained in Chiadó-Piat [1]. Links between epi-convergence of I_n and convergence of the subdifferentials of f_n (remember theorem IV. 34) are obtained in Defranceschi [1]. For well-posedness of $\int_\Omega |\nabla u|^2 dx + \int_\Omega u^2 d\mu$ with respect to changes in the measure μ, see Dal Maso-Mosco [1].

For further results, see Marcellini [3], Acerbi-Buttazzo [1], Dal Maso- Defranceschi [1]. Among the numerous results in homogenization, see e.g. Marcellini-Sbordone [2], Marcellini [3] and Mortola-Profeti [1].

Ulam's problem, related to stability of minimizers under additive perturbations of the integrand, has been investigated by Bobylev [1] in the one-dimensional case.

To section 4. A characterization of (105) (after properly defining the meaning of the inequality $u \geq g$ for general functions g) is obtained by Dal Maso [2] in terms of convergence of the capacities of suitable level sets. Some direct sufficient conditions are also obtained there. Counterexample 24 is due to Marcellini [2]. As pointed out by Boccardo [2], the convergence of arg min can happen without epi-convergence of the corresponding integral functionals.

Theorem 25 is a particular of Attouch [1]: see also Boccardo-Murat [1]. For the "diagonal" construction in the proof of theorem 25 (yielding $x_{ni(n)} \to z$) see Attouch [2]. For further results, see Carbone-Colombini [1], Dal Maso [1], Marcellini- Sbordone [2], Carbone [1].

To section 5. Results related to theorem 26 can be found in Sonntag [3]. An error estimate based on the modulus of Tykhonov well-posedness of (E, I) is obtained in Tjuhtin [1]. The example exhibiting the Lavrentiev phenomenon is due to Maniá [1]: see also Cesari [2], Loewen [1], Ball-Mizel [1]. The Lavrentiev phenomenon can be regularized (as a byproduct of variational convergence techniques) by decoupling $\int_0^T f(t, x, \dot{x})dt$ to $\int_0^T f(t, x, y)dt$, minimizing this latter functional subject to $\int_0^T |\dot{x} - y|dt \leq \varepsilon$ and then letting $\varepsilon \to 0$: see Ball-Knowles [1] and Zolezzi [15]. Theorem 28 is due to Seidman [1].

Chapter IX.

HADAMARD WELL-POSEDNESS IN MATHEMATICAL PROGRAMMING

Section 1.   Continuity of the Value and the Arg min Map.

Given two topological spaces X, Y, an extended real - valued function

$$f : X \times Y \to [-\infty, +\infty]$$

and a multifunction

$$A : Y \rightrightarrows X$$

we shall denote the value of the problem $[A(y), f(\cdot, y)]$, $y \in Y$ (sometimes called *marginal function* of the problem) by

$$\text{val } (y) = \text{ val } [A(y), f(\cdot, y)] = \inf\{f(x, y) : x \in A(y)\},$$

and the set of solutions by

$$\text{arg min } (y) = \{x \in A(y) : f(x, y) = \text{ val } (y)\}.$$

Here Y is the parameter space and X is the space of the decision variable. We are interested in the continuity properties of

$$\text{val } : Y \to [-\infty, +\infty], \text{ arg min } : Y \rightrightarrows X.$$

If $A(y) = \emptyset$ then val $(y) = +\infty$.
Throughout, $y_0 \in Y$ is any fixed point.

1 Proposition *Suppose that*
(1) *f is lower semicontinuous at each point of $A(y_0) \times \{y_0\}$;*
(2) *$A(y_0)$ is compact;*
(3) *the multifunction A is upper semicontinuous at y_0.*
  Then $y \to$ val (y) is lower semicontinuous at y_0.

2 Proposition *Suppose that*
(4) *f is upper semicontinuous at each point of $A(y_0) \times \{y_0\}$;*
(5) *A is lower semicontinuous at y_0.*
  Then $y \to$ val (y) is upper semicontinuous at y_0.

Combining propositions 1 and 2 we obtain continuity of the value. In some cases (e.g. applications to optimal control) it is more useful to employ different topologies on X.

The Hadamard well-posedness of minimum problems, as far as the topological setting is involved, has one of its basic results in the following

3 Theorem *Let t_1, t_2 be two topologies for X (not necessarily comparable). Suppose that*
(i) *f and A satisfy conditions (1), (2), (3) of proposition 1 for (X, t_1);*
(ii) *f and A satisfy conditions (4), (5) of proposition 2 for (X, t_2).*
   Then
(6) *$y \to$ val (y) is continuous at y_0;*
(7) *$y \rightrightarrows$ arg min (y) is upper semicontinuous at y_0 with respect to the topology t_1 (in which A is upper semicontinuous).*

In proposition 1 we have one condition more than in proposition 2, namely compactness of $A(y_0)$. The following example shows that this requirement cannot be dropped (in general).

4 Example. Let $X = Y = (-\infty, +\infty)$, $f(x, y) = xy$, $A(y) = X$ for all y. Then

$$\text{val } (y) = -\infty \text{ whenever } y \neq 0,$$

$$\text{val } (y) = 0 \text{ for } y = 0.$$

However, if f does not depend on the parameter y, we can achieve symmetry between the lower and upper semicontinuity parts of the value Hadamard well-posedness theorem.

5 Theorem *Let A be continuous at y_0 and*

$$f : X \to [-\infty, +\infty]$$

be continuous at every point of $A(y_0)$. Then $y \to$ val (y) is continuous at y_0.

The assumptions of theorem 5 are not sufficient for upper semicontinuity of the arg min map, as shown in the following
6 Example. $X = (-\infty, +\infty)$, $Y = [-1, 1]$, $y_0 = 0$, $A(0) = [0, +\infty)$; $A(y) = [y, y^{-2}]$ if $y \neq 0$. The multifunction A is continuous at y_0. The function

$$f(x) = x^2 \text{ if } x \leq 1, \ = \frac{1}{x} \text{ if } x \geq 1,$$

attains its minimum on $A(y)$ in the set $\{y, y^{-2}\}$ for $y > 0$, while arg min $(y) = \{0\}$ if $y \leq 0$. Thus, arg min is not upper semicontinuous at $y = 0$. Observe that $A(0) \setminus A(y)$ is not bounded for $y \neq 0$.

According to the equivalence between IV. (15) and IV. (27) (see p. 129), value well-posedness obtains iff every limit of asymptotically minimizing sequences for the perturbed problems solves the original problem, and moreover every solution to it can be approximated by means of suitable asymptotically minimizing sequences, thereby linking value and solutions Hadamard well-posedness. More precisely, we have

7 Theorem *Let X be a metric space, and Y be first countable. Suppose that A is continuous at y_0, $A(y_0)$ is closed and f is continuous at $A(y_0) \times \{y_0\}$, with a nonempty arg min (y_0). Then val $(\cdot)$ is continuous at y_0 iff for every $y_n \to y_0$ we have*

$$\text{arg min } (y_0) = \cap \{ \text{ lim inf}[\varepsilon - \text{ arg min } (y_n)] : \varepsilon > 0 \}.$$

8 Theorem *Let X be a complete metric space, and Y a first countable Hausdorff topological space. Let W_n be a countable base of neighborhoods of y_0. Suppose that*

$$f : X \to [-\infty, +\infty]$$

is continuous on $A(y_0)$, A is continuous at y_0, $A(y_0)$ is closed, and
(8) the Kuratowski number $\alpha(B_n) \to 0$ as $n \to +\infty$, where

$$B_n = \cup \{ A(y_0) \setminus A(y) : y \in W_n \}.$$

Then $y \rightrightarrows$ arg min (y) is upper semicontinuous at y_0.

Theorems 3, 5 and 8 provide sufficient conditions to Hadamard well - posedness. The following partial converses to theorems 5 and 8 furnish necessary well - posedness conditions.

9 Theorem *Let X be a normal space and $A(y_0)$ be closed.*
(9) Suppose that for every bounded continuous function $f : X \to (-\infty, +\infty)$ the value val $(A(\cdot), f)$ is continuous at y_0. Then A is continuous at y_0.
(10) Suppose that for every bounded continuous function $f : X \to (-\infty, +\infty)$, arg min $(A(\cdot), f)$ is upper semicontinuous at y_0. Then A is upper semicontinuous at y_0.
(11) Suppose that $A(y) \neq \emptyset$ for every $y \in Y$, and for every bounded continuous function $f : X \to (-\infty, +\infty)$, arg min $(A(\cdot), f)$ is upper semicontinuous at y_0. Then A is continuous at y_0.

Now we get another solution well-posedness result where the compactness condition (2) is replaced by well-posedness in the generalized sense (section I. 6).

10 Theorem *Let X be a complete metric space and Y be a first countable topological space. Let $f : X \to (-\infty, +\infty)$ be continuous on X, let A and val $(A(\cdot), f)$ be both upper semicontinuous at y_0. Moreover, let the problem $(A(y_0), f)$ be well-posed in the generalized sense. Then $y \rightrightarrows$ arg min (y) is upper semicontinuous at y_0.*

If f depends on the parameter y, the conclusion of the above theorem is no longer true.
11 Example. Let $X = (-\infty, +\infty), Y = [0, +\infty), f(x, y) = \min \{x^2, (x - 1/y)^2 - y\}$ for $y > 0$, $f(x, 0) = x^2$, $A(y) = [0, +\infty)$ for every y. We have: val $(\cdot)$ continuous at $y_0 = 0$ and

$$\text{arg min } (y) = \frac{1}{y} \text{ if } y > 0, = 0 \text{ if } y = 0,$$

therefore not upper semicontinuous at y_0.

12 Theorem *Let X be a Banach space, Y be a first countable topological space. Let*

$$f : X \times Y \to (-\infty, +\infty)$$

be continuous on $X \times Y$, moreover $f(x, \cdot)$ be continuous on Y, uniformly for x on bounded sets. Let $f(\cdot, y)$ be convex for every $y \in Y$, $A(\cdot)$ and val $(\cdot)$ be upper semicontinuous at y_0, with $A(y_0)$ convex. Suppose that the problem $[A(y_0), f(\cdot, y_0)]$ is well - posed in the generalized sense with respect to the strong convergence. Then $y \rightrightarrows arg \ min \ (y)$ is upper semicontinuous at y_0.

When both X and Y are metric spaces (X with metric d) and $A(y) = X$ everywhere, equi - well - posedness of $[X, f(\cdot, y)]$ is a sufficient condition for continuity of the optimal solution. Following IV. (42), we call $[X, f(\cdot, y)]$ *equi - well - posed* iff for every $y \in Y$

$$\text{val } (y) > -\infty, \ \text{arg min } (y) \ \text{is a singleton,}$$

and given $\varepsilon > 0$ there exists $\delta > 0$ such that for every $y \in Y$

(12) $$f(x, y) - \text{val } (y) < \delta \Rightarrow d[x, \text{arg min } (y)] < \varepsilon.$$

13 Proposition $y \to \ arg \ min \ (y)$ *is continuous on Y provided $[X, f(\cdot, y)]$ is equi-well-posed and*

(13) $$f(x, \cdot) \ \text{is upper semicontinuous on } Y \ \text{for every } x \in X;$$

(14) $$y \to \ val \ (y) \ \text{is lower semicontinuous on } Y.$$

Proposition 13 is independent on (but related to) theorem IV. 18.

PROOFS.

Proposition 1. The cases val $(y_0) = -\infty$ and $A(y_0) = \emptyset$ are obvious, the latter since $A(y) = \emptyset$ near y_0 by upper semicontinuity. Let $A(y_0) \neq \emptyset$ and pick $\varepsilon > 0$. From the lower semicontinuity of f, for every $x \in A(y_0)$ there are neighborhoods $W(x)$ of y_0 and $N(x)$ of x such that

(15) $$f(x, y_0) \leq f(z, y) + \varepsilon$$

whenever $z \in N(x)$ and $y \in W(x)$. Since $A(y_0)$ is compact, from the open covering $\cup\{N(x) : x \in A(y_0)\}$ one can extract a finite subcovering $\cup_{i=1}^{p} N(x_i)$ of $A(y_0)$. By (3) there is a neighborhood B of y_0 such that

(16) $$A(y) \subset \cup_{i=1}^{p} N(x_i)$$

whenever $y \in B$. Take

$$C = B \cap \cap \{W(x_i) : i = 1, ..., p\}.$$

For every $y \in C$, either $A(y) = \emptyset$ and then trivially

$$\text{val }(y_0) \leq \text{ val }(y) + \varepsilon,$$

or $A(y) \neq \emptyset$. Then, for every $x \in A(y)$, from (16) we get $x \in N(x_j)$ for some j between 1 and p. Clearly $y \in W(x_j)$. Then by (15)

$$f(x_j, y_0) \leq f(x, y) + \varepsilon.$$

This entails

$$\text{val }(y_0) \leq f(x, y) + \varepsilon \text{ if } x \in A(y),$$

hence

$$\text{val }(y_0) \leq \text{ val }(y) + \varepsilon$$

for $y \in C$. The proof is now complete. $\square$

Proposition 2. If $A(y_0) = \emptyset$ there is nothing to prove. Now let $A(y_0) \neq \emptyset$.
   Case 1: val $(y_0) > -\infty$.
Take any $\varepsilon > 0$. Choose $x_0 \in A(y_0)$ such that

$$(17) \qquad\qquad f(x_0, y_0) \leq \text{ val }(y_0) + \varepsilon.$$

From the upper semicontinuity of f there exist neighborhoods B of x_0 and C of y_0 such that

$$(18) \qquad\qquad f(x, y) \leq f(x_0, y_0) + \varepsilon,$$

whenever $x \in B$ and $y \in C$. Since A is lower semicontinuous at y_0, there is a neighborhood D of y_0 such that

$$(19) \qquad\qquad A(y) \cap B \neq \emptyset$$

for every $y \in D$. Consider now

$$(20) \qquad\qquad N = C \cap D.$$

Let $y \in N$. By (19) there exists $x \in A(y)$ fulfilling (18). Then by (17) and (18)

$$\text{val }(y) \leq f(x, y) \leq f(x_0, y_0) + \varepsilon \leq \text{ val }(y_0) + 2\varepsilon,$$

showing upper semincontinuity at y_0.
   Case 2: val $(y_0) = -\infty$.
Then for every real m there exists $x_m \in A(y_0)$ such that

$$f(x_m, y_0) < m.$$

Now it is sufficient to repeat steps (18), (19), (20) from the proof of the case 1 in order to obtain the following. If $y \in N$ there exists $x \in A(y)$ such that

$$\text{val }(y) \leq f(x, y) \leq f(x_m, y_0) + \varepsilon \leq m + \varepsilon.$$

$$T(y) = \{x \in X : f(x, y) \le v(y)\}.$$

From now on we work within the space (X, t_1). We will prove that for any open set $V \supset \arg \min (y_0)$ there is a neighborhood N of y_0 such that

(21) $\arg \min (y) \subset V$

whenever $y \in N$. If $\arg \min (y_0) = X$ then (21) is obvious. Assume that this is not the case. If $A(y_0) \subset V$ then (21) follows by upper semicontinuity of A at y_0. So, without loss of generality, we can assume that

$$K = \{x \in X : x \in A(y_0), x \notin V\}$$

is nonempty and compact. Take $x \in K$. Since $K \cap T(y_0) = \emptyset$, then $f(x, y_0) > \text{val} (y_0)$. From lower semicontinuity of f on $A(y_0) \times \{y_0\}$ and continuity of $\text{val} (\cdot)$ at y_0, there exist neighborhoods $N(x)$ of x and $W(x)$ of y_0 such that

$$f(z, y) > \text{val} (y)$$

whenever $z \in N(x)$, $y \in W(x)$. In other words

$$N(x) \cap T(y) = \emptyset$$

when $y \in W(x)$. Of course $\cup \{N(x) : x \in K\}$ is an open covering of K. Compactness of K yields the existence of a finite set $x_1, ..., x_p$ of points in K such that

$$K \subset B = \cup \{N(x_i) : i = 1, ..., p\}$$

and

(22) $T(y) \cap B = \emptyset$

whenever $y \in \cap \{W(x_i) : i = 1, ..., p\}$. We know that $A(y_0) = K \cup [V \cap A(y_0)]$, hence $A(y_0) \subset B \cup V$. From upper semicontinuity of A at y_0 there exists a neighborhood C of y_0 such that

(23) $A(y) \subset B \cup V$ if $y \in C$.

Define

$$M = C \cap \cap \{W(x_i) : i = 1, ..., p\}.$$

Then for $y \in M$, by (22) and (23)

$$\arg \min (y) = A(y) \cap T(y) \subset (B \cup V) \cap T(y) = [B \cap T(y)] \cup [V \cap T(y)] \subset V.$$

Thus (21) is proved.   □

Theorem 5. Thanks to proposition 2, we need to prove only lower semicontinuity of val $(\cdot)$ at y_0. If $A(y_0) = \emptyset$ then val $(y_0) = $ val $(y) = +\infty$ near y_0 since $A(y) = \emptyset$ there, and we are done.

Case 1: val $(y_0) < +\infty$.

Let $x \in A(y_0)$ and $\varepsilon > 0$ be such that $1/\varepsilon > $ val (y_0). From continuity of f, one can find a neighborhood $N(x)$ of x such that

$$\inf\{f(z) : z \in N(x)\} \geq \begin{cases} 1/\varepsilon & \text{if } f(x) = +\infty, \\ -\infty & \text{if } f(x) = -\infty, \\ f(x) - \varepsilon, & \text{otherwise.} \end{cases}$$

Then, since val $(y_0) < +\infty$

(24) $$\inf\{f(z) : z \in N(x)\} \geq \text{ val }(y_0) - \varepsilon.$$

Now let

$$M = \cup\{N(x) : x \in A(y_0)\}.$$

Since M is an open neighborhood of $A(y_0)$, there exists an open set S such that $y_0 \in S$ and

$$A(y) \subset M \text{ whenever } y \in S.$$

If $y \in S$ one has

(25) $$\inf\{f(z) : z \in M\} \leq \text{ val }(y).$$

On the other hand, from (24),

$$\inf\{f(z) : z \in M\} \geq \text{ val }(y_0) - \varepsilon$$

since

$$\inf\{f(z) : z \in M\} = \inf\{\inf f[N(x)] : x \in A(y_0)\}.$$

Thus

$$\text{val }(y) \geq \text{val }(y_0) - \varepsilon$$

and the proof is complete for this case.

Case 2: val $(y_0) = +\infty$.

Then for every $\varepsilon > 0$ and $x \in A(y_0)$ one can find a neighborhood N of x such that

$$\inf\{f(z) : z \in N\} > \frac{1}{\varepsilon}.$$

Then by (25) we have

$$\text{val }(y) \geq \inf\{f(z) : z \in M\} \geq \frac{1}{\varepsilon} \text{ whenever } y \in S,$$

showing that

$$\text{val }(y) \to +\infty \text{ as } y \to y_0.$$

$\square$

Theorem 7. Remembering that IV. (15) and IV. (27) are equivalent (p. 129), we need only to prove that seq. epi-lim $I_n = I$, where

$$I_n(\cdot) = f(\cdot, y_n) + \text{ ind } [A(y_n), \cdot],$$

$$I(\cdot) = f(\cdot, y_0) + \text{ ind } [A(y_0), \cdot],$$

and $y_n \to y_0$. Given $u_n \to u$ in X, let $u \in A(y_0)$. Then

$$\liminf I_n(u_n) \geq \liminf f(u_n, y_n) = I(u).$$

Let $u \notin A(y_0)$. There exist in X disjoint neighborhoods of u and $A(y_0)$ (Kelley [1, th. 10 p. 120]). Hence, by upper semicontinuity, $u_n \notin A(y_n)$ for every large n, yielding

$$+\infty = \liminf I_n(u_n) \geq I(u).$$

Now fix any $u \in A(y_0)$ (the only case of interest). By lower semicontinuity, there exist $x_n \in A(y_n)$ such that $x_n \to u$, hence

$$\limsup I_n(x_n) = \limsup f(x_n, y_n) = f(u, y_0) = I(u).$$

$\square$

Theorem 8. Assume that arg min is not upper semicontinuous at y_0. Then there exist a closed set F with $F \cap \text{arg min}(y_0) = \emptyset$, and sequences $y_n \to y_0$, $x_n \in \text{arg min}(y_n)$, such that $x_n \in F$, $n = 1, 2, \dots$.

Case 1: $x_n \in A(y_n) \setminus A(y_0)$ for all sufficiently large n.
If x_n does not possess any convergent subsequence, then for every open V the set

$$Z = V \setminus \cup\{x_n : n = 1, 2, \dots\}$$

is open. Taking V such that $A(y_0) \subset V$ we get $A(y_n) \not\subset Z$, hence A is not upper semicontinuous at y_0, a contradiction. Thus x_n must contain a convergent subsequence, say $x_n \to x_0$. Now we use the following known fact : if A is upper semicontinuous at y_0, $A(y_0)$ is closed and $y_n \to y_0$, then $x_n \in A(y_n)$, $x_n \to x_0$ imply $x_0 \in A(y_0)$ (see Klein-Thompson [1, proof of th. 7.1.15 p. 78]). Hence $x_0 \in A(y_0)$ and

$$\lim \text{val}(y_n) = \lim f(x_n) = f(x_0).$$

From continuity of val (theorem 5) we get val$(y_0) = f(x_0)$, which means that $x_0 \in$ arg min(y_0), a contradiction since $x_0 \in F$.

Case 2: for some subsequence, $x_n \in A(y_n) \cap A(y_0)$.
We consider two subcases a), b).

a) There exists some integer p such that

$$x_n \in A(y_p) \cap A(y_0) \text{ for all } n \geq p.$$

Then val$(y_n) \geq$ val$(y_p) \geq$ val(y_0). If val$(y_p) =$ val(y_0) then $x_p \in$ arg min(y_0), a contradiction since $x_p \in F$. Hence

$$\limsup \text{val}(y_n) > \text{val}(y_0),$$

contradicting theorem 5.

b) For every n there exists $k_n \geq n$ such that

$$z_n = x_{k_n} \notin A(y_n) \cap A(y_0).$$

Then $z_n \in A(y_0) \setminus A(y_n)$, hence for some further subsequence

$$z_n \in \cup \{A(y_0) \setminus A(y) : y \in W_n\}.$$

From (8), z_n must have a convergent subsequence

$$u_n \to x_0.$$

Then $x_0 \in F \cap A(y_0)$, moreover for the corresponding subsequence of y_n we get

$$\lim \text{val } (y_n) = \lim f(u_n) = f(x_0)$$

and $\lim \text{val } (y_n) = \text{val } (y_0)$ by theorem 5. Hence $x_0 \in \text{arg min } (y_0)$, a contradiction. $\square$

Theorem 9. **Proof of (9).** Assume that A is not upper semicontinuous at y_0. Then there exists a closed set F with $A(y_0) \cap F = \emptyset$ such that in every neighborhood of y_0 one can find some y such that $A(y) \cap F \neq \emptyset$. If $A(y_0) = \emptyset$, take $f(x) = 0$ on X, then val $(A(\cdot), f)$ is not continuous. If $A(y_0) \neq \emptyset$ then we can apply Urysohn's lemma (Kelley [1, p. 115]) to $A(y_0)$ and F. We get a continuous function $f : X \to [0,1]$ such that

$$f(x) = 0 \text{ if } x \in F, \ f(x) = 1 \text{ if } x \in A(y_0).$$

Then val $(y_0) = 1$ while val $(y) = 0$ for some y arbitrarily near to y_0, a contradiction.

Now suppose that A is not lower semicontinuous at y_0. Then there exists an open set V such that $V \cap A(y_0) \neq \emptyset$ and for every open U with $y_0 \in U$ there exists $y \in U$ for which

$$A(y) \subset W = X \setminus V.$$

Take $\bar{x} \in A(y_0) \cap V$ and apply Urysohn's lemma to $\bar{x}$ and W. In this way we get a continuous function $f : X \to [0,1]$ such that

$$f(\bar{x}) = 0, \ f(x) = 1 \text{ on } W.$$

Hence val $(y_0) = 0$, val $(y) = 1$ near y_0, a contradiction.

Proof of (10). It is completely analogous to the first part of that of (9). Arguing by contradiction, if $A(y_0) = \emptyset$ we take $f(x) = 0$ on X, and then arg min $(y_0) = \emptyset$ while arg min $(y) \neq \emptyset$ near y_0. If $A(y_0) \neq \emptyset$ we consider F as in the proof of (9). Then applying Urysohn's lemma we conclude that (for some continuous f)

$$\text{arg min } (A(y_0), f) = A(y_0), \ \text{arg min } (A(y), f) \cap F \neq \emptyset$$

for y near y_0, a contradiction.

344

Proof of (11). Suppose that A is not lower semicontinuous. Then there exist $x_0 \in A(y_0)$ and some open V with $x_0 \in V$ such that for every open U containing y_0 there exists $y \in U$ verifying

$$A(y) \subset W = X \setminus V.$$

By Urysohn's lemma, there exists a continuous function $f : X \to [0,1]$ such that

$$f(x_0) = 0, \ f(u) = 1 \text{ for every } u \in W,$$

thereby contradicting upper semicontinuity of $\arg \min (A(\cdot), f)$, since $\arg \min (y_0) \cap W = \emptyset$, while $\arg \min (y) \subset W$ arbitrarily near to y_0. □

Remark. The condition $A(y) \neq \emptyset$ on Y is essential in the case (11). Take $A(y_0) = \{x_0\}$ and $A(y) = \emptyset$ for $y \neq y_0$. Then for any f, $\arg \min$ is upper semicontinuous but $A(\cdot)$ is not lower semicontinuous at y_0.

Theorem 10. Assume the contrary. As in the proof of theorem 7, there exist a closed set F such that

$$(26) \qquad\qquad \arg \min (y_0) \cap F = \emptyset$$

and sequences $y_n \to y_0$, $x_n \in \arg \min (y_n)$, $x_n \in F$.

Case 1. If $x_n \in A(y_n) \setminus A(y_0)$ for all sufficiently large n, then x_n should have a convergent subsequence, otherwise upper semicontinuity is contradicted, as shown in the proof of theorem 7. So let $x_n \to x_0$. Then $x_0 \in \text{cl } A(y_0)$ by Klein-Thompson [1, prop. 7.3.5 and th. 7.1.15]. So there exists a sequence $z_n \in A(y_0)$ with $z_n \to x_0$. From the continuity of f one has $f(x_n) - f(z_n) \to 0$. But $f(x_n) = \text{val } (y_n)$, and by assumption

$$\limsup \text{val } (y_n) \leq \text{val } (y_0).$$

Hence $\limsup f(z_n) \leq \text{val } (y_0)$, that is $\lim f(z_n) = \text{val } (y_0)$. Thus z_n is a minimizing sequence to $(A(y_0), f)$. By generalized well-posedness, z_n has a cluster point in $\arg \min (y_0)$. On the other hand, $z_n \to x_0 \in F$, contradicting (26).

Case 2. If $x_n \in A(y_0)$ for some subsequence, then, since $f(x_n) \geq \text{val } (y_0)$ and

$$\limsup f(x_n) = \limsup \text{val } (y_n) \leq \text{val } (y_0),$$

one has that x_n is a minimizing sequence for $(A(y_0), f)$. This yields (for some subsequence)

$$x_n \to z \in \arg \min(y_0)$$

contradicting (26). □

Theorem 12. Assume the contrary. Then we can find a closed set F and sequences

$$y_n \to y_0, \ x_n \in \arg \min (y_n), \ x_n \in F \text{ with } \arg \min (y_0) \cap F = \emptyset.$$

Case 1. If $x_0 \in A(y_n) \setminus A(y_0)$ for all sufficiently large n, then x_0 has a convergent subsequence (otherwise upper semicontinuity would be contradicted, as shown in the proof of theorem 7), say $x_n \to x_0$. As in the proof of theorem 10, there exists a sequence $z_n \in A(y_0)$ such that $z_n \to x_0$. We have

$$f(x_n, y_n) - f(z_n, y_0) \to 0, \quad f(x_n, y_n) = \text{ val } (y_n),$$

hence

$$\limsup f(z_n, y_0) \leq \text{ val } (y_0),$$

showing that z_n is a minimizing sequence for $[A(y_0), \; f(\cdot, y_0)]$. By generalized well - posedness, $x_0 \in \arg \min (y_0)$. But $x_0 \in F$, a contradiction.

Case 2. Let $x_n \in A(y_0)$ for some subsequence. If x_n is bounded, then from

$$f(x_n, y_n) - f(x_n, y_0) \to 0, \quad f(x_n, y_n) = \text{ val } (y_n), \; \limsup \text{ val } (y_n) \leq \text{ val } (y_0),$$

we get that x_n is a minimizing sequence of $[A(y_0), \; f(\cdot, y_0)]$, thereby contradicting generalized well-posedness. Now let

$$\|x_n\| \to +\infty$$

for some subsequence.

Clearly, the generalized well-posedness implies the boundedness of $\arg \min (y_0)$. Take $x_0 \in \arg \min(y_0)$. Then denote

$$r = \sup\{ \; \|x - x_0\| \; : x \in \arg \; \min \; (y_0)\},$$

and let

$$z_n = a_n x_n + (1 - a_n)x_0, \quad a_n = \frac{2r}{\|x_n - x_0\|}.$$

From the convexity

$$f(z_n, y_n) \leq a_n f(x_n, y_n) + (1 - a_n) \; f(x_0, y_n),$$

hence

(27) $$\limsup f(z_n, y_n) \leq \limsup a_n \text{ val } (y_n) + \limsup(1 - a_n) \; f(x_0, y_n) \leq \text{ val } (y_0).$$

Clearly z_n is bounded, so $f(z_n, y_n) - f(z_n, y_0) \to 0$. Moreover $z_n \in A(y_0)$ and by (27)

$$\limsup f(z_n, y_0) \leq \text{ val } (y_0),$$

hence z_n is a minimizing sequence for $[A(y_0), f(\cdot, y_0)]$. On the other hand

$$\|z_n - x_0\| = 2r,$$

hence z_n does not possess strong cluster points in $\arg \; \min \; (y_0)$, contradicting generalized well-posedness.   □

346

Proposition 13. We fix any $y_0 \in Y$ and show continuity at y_0. Put

$$x(y) = \arg\ \min\ (y),\ x_0 = \arg\ \min\ (y_0).$$

Given $\varepsilon > 0$, by (12) it suffices to prove that

(28) $$f(x_0, y) -\ \text{val}\ (y) < \delta \text{ if } y \text{ is near } y_0.$$

Write

$$f(x_0, y) -\ \text{val}\ (y) = f(x_0, y) - f(x_0, y_0) +\ \text{val}\ (y_0) -\ \text{val}\ (y).$$

Then if y is near to y_0, by (13)

$$f(x_0, y) - f(x_0, y_0) < \delta/2$$

and by (14).

$$\text{val}\ (y_0) -\ \text{val}\ (y) < \delta/2$$

thus proving (28). □

Section 2.   Stability of Multifunctions Defined by Inequalities and Value Well - Posedness.

Let

$$g_j : R^n -\cdot [-\infty, +\infty], j = 1, 2, ..., m,$$

be given extended real - valued continuous functions on R^n. Consider the multifunction

$$A : R^m \rightrightarrows R^n$$

defined by

$$A(y) = \{x \in R^n : g_j(x) \leq y_j,\ j = 1, ..., m\},\ y \in R^m.$$

In this section we are concerned with continuity properties of A, which play a key role in the Hadamard well-posedness of the mathematical programming problems with inequality constraints (as we saw in section 1) with respect to the right-hand term y of the inequalities defining the feasible set $A(y)$. We characterize semicontinuity of A as follows.

Let y_0 be a given point of R^m. Denote

$$I(y) = \{x \in R^n : g_j(x) < y_j, j = 1, ..., m\},\ y \in R^m.$$

14 Proposition *The following are equivalent:*

(29) $$A(y_0) = cl\ I(y_0);$$

(30) $$A \text{ is lower semicontinuous at } y_0.$$

15 Proposition *The following are equivalent:*

(31)   *there exists $y_1 > y_0$ such that the set $A(y_1) \setminus A(y_0)$ is bounded;*

(32)   *A is upper semicontinuous at y_0.*

Combining theorems 5, 8, 9 and propositions 14, 15 we obtain the following corollaries.

16 Theorem *The following are equivalent:*
(i) *(29) and (31) hold;*
(ii) *for every continuous function $f : R^n \to [-\infty, +\infty]$, $y \to \ val\ (A(y), f)$*
is continuous at y_0.

17 Theorem *Let (29) hold. Suppose that there exists a neighbourhood N of y_0 such that the set $A(y_0) \setminus A(y)$ is bounded whenever $y \in N$. Then the following are equivalent:*
(i) *(32) holds;*
(ii) *for every continuous function $f : R^n \to [-\infty, +\infty]$ the multifunction*
$y \rightrightarrows \ arg\ min\ (A(y), f)$ is upper semicontinuous at y_0.

Let us consider now the multifunction

$$E : R^m \times R^q \rightrightarrows R^n,$$

defined by a system of inequalities and equalities, as follows:

$$E(u, t) = \{x \in R^n : g(x) \le u \text{ and } h(x) = t\},$$

where $g : R^n \to R^m$, $h : R^n \to R^q$ are given vector-valued functions and $(u, t) = p$ represents the parameter. Given $x_0 \in E(0)$ we will estimate the distance from x_0 to the set $E(p)$ for p near 0.

We *assume* that the functions g and h are continuously differentiable at x_0.
Denote the set of active inequality constraints at x_0 by J_A, i.e.

$$J_A = \{j \in \{1, 2, ..., m\} : g_j(x_0) = 0\}.$$

For $r \in R^m$ let

$$|r|_\infty = \ \max\ \{|r_i| \ : i = 1, 2, ..., \ m\}$$

and let r_- be the negative part of r, that is the vector whose $i - th$ component is $\min\{0, r_i\}$, $i = 1, 2, ..., m$.

We will prove the following characterization theorem.

18 Theorem *Let $x_0 \in E(0)$. The following are equivalent:*
(33) *there exists $\bar{z} \in R^n$ such that $\nabla g_j(x_0)'\bar{z} < 0$ for all $j \in J_A$ and $\nabla h_i(x_0)'\bar{z} = 0$ for $i = 1, 2, ..., q$; the vectors $\nabla h_i(x_0), i = 1, 2, ..., q$, are linearly independent;*
(34) *there exist positive constants c and α such that for every $p = (u, t) \in R^{m+q}$ with $max\ \{|u_-|_\infty, |t|_\infty\} < \alpha$ there exists $x_1 \in E(p)$ such that*

$$|x_1 - x_0| \leq c \ max \ \{|[u - g(x_0)]_-|_\infty, \ |t|_\infty\}.$$

Condition (33) is known as the Mangasarian-Fromovitz constraint qualification. The condition (34) can be viewed as a quantitive stability property of the feasible set in nonlinear programming, a property closely related to well - posedness. In fact, consider the problem to minimize a locally Lipschitz function over $E(p)$. Suppose that x_0 is a global optimal solution to the problem with $p = 0$, and x_0 satisfies the Mangasarian - Fromovitz condition (33). Then from (34) we obtain

$$val \ (p) \geq \ val(0) + \ const. \ |p|$$

for all p near 0. This is a sufficient condition for calmness (which in turn implies the existence of Lagrange multipliers), see Clarke [3, sec. 6.4].

PROOFS.

Proposition 14. Assume (29). If $I(y_0) = \emptyset$, then $A(y_0) = \emptyset$ and there is nothing to prove. Assume $I(y_0) \neq \emptyset$ and suppose that A is not lower semicontinuous at y_0. Then there is an open set V with $V \cap A(y_0) \neq \emptyset$ such that for any neighborhood U of y_0 there is $y \in U$ such that

(35) $$A(y) \cap V = \emptyset.$$

Since $V \cap cl \ I(y_0) \neq \emptyset$ by (29), then $V \cap I(y_0) \neq \emptyset$. Take $\bar{x} \in V \cap I(y_0)$. Then

$$g_j(\bar{x}) < y_{0j}, \ j = 1, 2, ..., m.$$

Therefore for every y sufficiently close to y_0 one has clearly $\bar{x} \in A(y)$. This gives $V \cap A(y) \neq \emptyset$ for every y near y_0, contradicting (35). Thus A is lower semicontinuous at y_0.

Now suppose that A is lower semicontinuous at y_0. If $A(y_0) = \emptyset$ then $I(y_0) = \emptyset$, and (29) is trivially fulfilled. Let $A(y_0) \neq \emptyset$. Given any open set V with $V \cap A(y_0) \neq \emptyset$, by lower semicontinuity one can find $\varepsilon > 0$ so small that if $u = (1, 1, ..., 1) \in R^m$, then

$$V \cap A(y_0 - \varepsilon u) \neq \emptyset.$$

Since $A(y_0 - \varepsilon u) \subset I(y_0)$, then $V \cap I(y_0) \neq \emptyset$. Thus

$$V \cap A(y_0) \neq \emptyset \Rightarrow V \cap I(y_0) \neq \emptyset,$$

moreover $I(y_0) \subset A(y_0)$ which is closed. This implies (29). $\qquad \square$

Remark. In the proof above one can substitute an arbitrary topological space to R^n.

Proposition 15. Suppose that there exists a sequence $y_k \rightarrow y_0$ such that $A(y_k) \backslash A(y_0)$ is unbounded for every k. Then for every k there exists $x_k \in A(y_k) \setminus A(y_0)$ such that $|x_k| > k$. Clearly $\cup \{x_k : k = 1, 2, ...\}$ is a closed set. Take an open $V \supset A(y_0)$. Then

$$W = V \setminus \cup \{x_k : k = 1, 2, ...\}$$

is also open and

$$A(y_0) \subset W \text{ but } A(y_k) \not\subset W,$$

contradicting (32).

Remark. The above proof works with any metric space instead of R^n.

End of the proof of proposition 15. Now we prove that $(31) \Rightarrow (32)$. If A is not upper semicontinuous at y_0, then there exist a closed set F and a sequence $z_k \to y_0$ such that

$$A(z_k) \cap F \neq \emptyset \text{ for every } k, \text{ while } A(y_0) \cap F = \emptyset.$$

Then

$$A(z_k) \cap F = [A(z_k) \setminus A(y_0)] \cap F.$$

If $x_k \in [A(z_k) \setminus A(y_0)] \cap F$ then by (31) x_k has a cluster point $\bar{x}$, since $A(z_k) \subset A(y_1)$ for every sufficiently large k. Continuity of g_j yields $\bar{x} \in A(y_0)$. Since $\bar{x} \in F$ we get a contradiction. □

Theorem 18. Without loss of generality, let $J_A = \{1, 2, ..., l\}$. Denote by g_A the subvector of g indexed by J_A, thus $g_A(x_0) = 0$, and let g_I be indexed by the complement of J_A, thus $g_I(x_0) < 0$. Let R_+^m be the nonnegative orthant of R^m. Define the maps

$$N : R^n \rightrightarrows R^{l+q} \text{ and } M : R^n \rightrightarrows R^{m+q}$$

as

$$N(x) = \begin{pmatrix} \nabla g_A(x_0) \\ \nabla h(x_0) \end{pmatrix} (x - x_0) + \begin{pmatrix} R_+^l \\ \{0\} \end{pmatrix},$$

$$M(x) = \begin{pmatrix} g(x_0) \\ h(x_0) \end{pmatrix} + \begin{pmatrix} \nabla g(x_0) \\ \nabla h(x_0) \end{pmatrix} (x - x_0) + \begin{pmatrix} R_+^m \\ \{0\} \end{pmatrix}$$

We need the following

19 Lemma Each of the following three conditions is equivalent to (33):
(36) $N(R^n) = R^l \times R^q$;
(37) $0 \in$ int $M[B(x_0, r)]$ for every $r > 0$;
(38) for every $r > 0$ there exists $\sigma > 0$ such that the map

$$y \rightrightarrows S(y) = M^{-1}(y) \cap B(x_0, r)$$

is Lipschitz on $B(0, \sigma)$, i.e.

$$\text{haus} [S(y_1), S(y_2)] \leq \text{const.} |y_1 - y_2|$$

whenever $y_1, y_2 \in B(0, \sigma)$.

Taking the lemma for granted, the key step of the proof of theorem 18 will be based on an abstract result concerning generalized equations of the form

(39) \qquad\qquad find $z \in \Omega$ such that $T_p(z) \in F(z)$,

where Ω is a closed set of a Banach space X, p is a parameter from a metric space P with fixed element $0, (z, p) \to T_p(z)$ maps $X \times P$ into a Banach space Y and $F : X \rightrightarrows Y$. We considered this equation in section V. 5, see V. (117).

Let $z_0 \in \Omega$, $y_0 = T_0(z_0) \in F(z_0)$ and let $T_p(z_0)$ be continuous at $p = 0$. Suppose that, for every p near zero, T_p is Fréchet differentiable near z_0 and the derivative $T_p'(z)$ is continuous in (p, z) at $(0, z_0)$. In section V. 5 (proof of theorem 23) we showed that, if

$$D_\rho(p) = \sup\{\frac{\|T_p(x) - T_p(z) - T_0'(z_0)(x - z)\|}{\|x - z\|} : (x, z) \in B(z_0, \rho) \cap \Omega\},$$

then under the above assumptions $D_\rho(p) \to 0$ as $p \to 0$ and $\rho \to 0$. Denote by $\Sigma(p)$ the set of solutions to (39). As a consequence of lemma V. 24 we obtain the following

20 Lemma Under the above assumptions let $H(y)$ be the set of solutions to the problem

(40)   find $z \in \Omega$ such that $T_0'(z_0)(z - z_0) + y \in F(z)$.

Suppose that for some R and $\sigma > 0$ the multifunction

(41) $y \rightrightarrows H(y) \cap B(z_0, R)$

is closed-valued and Lipschitz continuous on $B(y_0, \sigma)$. Then there exist positive constants k and S such that for every $p \in B(0, S)$ the set $\Sigma(p)$ is nonempty and

$$\text{dist } [z_0, \Sigma(p)] \le k \text{ dist } [T_p(z_0), \ F(z_0)].$$

Proof. Put

$$M_\rho = \sup\{D_\rho(p) : p \in B(0, \rho)\},$$

and choose ρ, $R \ge \rho > 0$, so small that

$$M_\rho \gamma < 1,$$

where γ is the Lipschitz constant of (41). Put

$$\Delta_S = \sup\{\|T_p(z_0) - y_0\| : p \in B(0, S)\}.$$

Let $l > \gamma/(1 - \gamma M_\rho)$. Choose ε and $S > 0$ such that

(42) $\frac{\sigma}{2} > \rho M_\rho + 2\Delta_S + \varepsilon$ and $\rho \ge l(\Delta_S + \varepsilon)$.

Take any $p \in B(0, S)$. There exists $y_\varepsilon \in F(z_0)$ such that

(43) $\|T_p(z_0) - y_\varepsilon \| \le \varepsilon + \text{ dist } [T_p(z_0), \ F(z_0)] \le \varepsilon + \Delta_S$.

By (42) and the triangle inequality

$$\|y_\varepsilon - y_0\| \le \varepsilon + \Delta_S + \|T_p(z_0) - y_0\| < \frac{\sigma}{2},$$

hence $B(y_\varepsilon, \sigma/2) \subset B(y_0, \sigma)$. Thus, the map (41) is Lipschitz on $B(y_\varepsilon, \sigma/2)$. Then, by (42) and (43), the conditions of lemma V. 24 are satisfied with y_0, δ, D_ρ and r there replaced by $y_\varepsilon, \|T_p(z_0) - y_\varepsilon\|, M_\rho$ and $l\|T_p(z_0) - y_\varepsilon\|$. Applying the lemma we conclude that there exists $z \in \sum(p)$ such that $\|z - z_0\| \le r$, that is

$$\text{dist } [z_0, \Sigma(p)] \le l(\text{dist } [T_p(z_0), F(z_0)] + \varepsilon).$$

Since the left-hand side of this inequality does not depend on ε, one may take $\varepsilon = 0$, and the proof is complete. $\square$

Let us proceed with the *proof of theorem 18*. First we show that
$(33) \Rightarrow (34)$. By lemma 19 it suffices to show that $(38) \Rightarrow (34)$. We apply lemma 20
with $X = \Omega = R^n$, $Y = R^{m+q}$ equipped with the norm $|\cdot|_\infty$, $z_0 = x_0$, $p = (u,t)$,

$$\Sigma(p) = E(u,t), \; T_p(x) = -\begin{pmatrix} g(x) - u \\ h(x) - t \end{pmatrix}, F(x) = \begin{pmatrix} R^m_+ \\ \{0\} \end{pmatrix}.$$

The linearized problem (40) is

$$0 \in \begin{pmatrix} \nabla g(x_0) \\ \nabla h(x_0) \end{pmatrix} (x - x_0) + y + \begin{pmatrix} R^m_+ \\ \{0\} \end{pmatrix}$$

and the requirement for Lipschitz continuity of its solution set (intersected with some
ball) around

$$y_0 = \begin{pmatrix} g(x_0) \\ h(x_0) \end{pmatrix}$$

is exactly (38). Then there exist $K > 0, H > K$ and $S > 0$ such that if

$$\max\{|u|_\infty, \; |t|_\infty\} < S$$

then $E(u,t) \neq \emptyset$ and

$$\text{dist}\,[x_0, E(u,t)] \leq K \; \text{dist} \left\{ \begin{pmatrix} -g(x_0) + u \\ -h(x_0) + t \end{pmatrix}, \begin{pmatrix} R^m_+ \\ \{0\} \end{pmatrix} \right\}$$
$$\leq \; H\max\{|[u - g(x_0)]_-|_\infty, \; |t|_\infty\}.$$

Since E is closed-valued, there exists $x_1 \in E(u,t)$ with

$$|x_1 - x_0| = \text{dist}\,[x_0, \; E(u,t)].$$

Now given (r,t) with $\max\{|r_-|_\infty, \; |t|_\infty\} < S$, put $u = r_-$ and apply the above argument,
obtaining that there exists $x_1 \in E(r_-, t)$ with

$$|x_1 - x_0| \leq H \; \max\{|[r_- - g(x_0)]_-|_\infty, \; |t|_\infty\}.$$

Since $E(r_-, t) \subset E(r,t)$, then $x_1 \in E(r,t)$. Furthermore, since $g(x_0) \leq 0$ then

$$[r_- - g(x_0)]_- = [r - g(x_0)]_-.$$

This completes the proof of (34).

352

$(34) \Rightarrow (33)$. Let $\max \{|u|_\infty, |t|_\infty\} < \alpha$. Then $\max \{|u_-|_\infty, |t|_\infty\} < \alpha$, as well. Then by (34) there exists $x_1 \in E(u,t)$ with

$$(44) \qquad |x_1 - x_0| \leq c \max \{|[u - g(x_0)]_-|_\infty, |t|_\infty\}$$
$$\leq c \text{ dist} \left\{ \begin{pmatrix} u \\ t \end{pmatrix}, \begin{pmatrix} g(x_0) \\ h(x_0) \end{pmatrix} + \begin{pmatrix} R_+^m \\ \{0\} \end{pmatrix} \right\}.$$

We will prove that (34) implies (36). Then, by lemma 19, (33) will hold, as well.

Assume that (36) is not fulfilled. Then there exists $z_0 \in R^{l+q}$ such that $z_0 \notin N(R^n)$. Put

$$z_j = ((1/j)z_0, 0) \in R^{m+q}, \ j = 1, 2, \dots.$$

Then for sufficiently large j one has $|z_j|_\infty < \alpha$. By (44), for all sufficiently large j there exists $x_j \in E(z_j)$ with

$$(45) \qquad |x_j - x_0| \leq c \text{ dist} \left[z_j, \begin{pmatrix} g(x_0) \\ h(x_0) \end{pmatrix} + \begin{pmatrix} R_+^m \\ \{0\} \end{pmatrix}\right] \leq c|z_j|_\infty = (c/j)|z_0|_\infty.$$

Put $f(x) = (g_A(x), h(x))'$. Since $x_j \in E(z_j)$ then

$$(1/j)z_0 - f(x_j) \in \begin{pmatrix} R_+^l \\ \{0\} \end{pmatrix}.$$

Let

$$w_j = \nabla f(x_0)(x_j - x_0) + (1/j)z_0 - f(x_j).$$

Then $w_j \in N(R^n)$ and

$$(46) \qquad \text{dist} ((1/j)z_0, N(R^n)) \leq |(1/j)z_0 - w_j|$$
$$= |f(x_j) - \nabla f(x_0)(x_j - x_0)|$$
$$= |f(x_j) - f(x_0) - \nabla f(x_0)(x_j - x_0)|,$$

since $f(x_0) = 0$. If $x_j = x_0$ for infinitely many j then $z_0 \in N(R^n)$, a contradiction. Let $|x_j - x_0| > 0$ for all large j. Since $N(R^n)$ is a cone we have

$$\text{dist} \left[\frac{1}{j}z_0, N(R^n)\right] = \frac{1}{j} \text{ dist} [z_0, N(R^n)].$$

Using (45) and (46) we obtain

$$\text{dist} [z_0, N(R^n)] = j \text{ dist} [(1/j)z_0, N(R^n)]$$
$$= \frac{j}{|x_j - x_0|}|x_j - x_0| \text{ dist} [(1/j)z_0, N(R^n)]$$
$$\leq \frac{j}{|x_j - x_0|}(c/j) |z_0|_\infty \text{ dist} [(1/j)z_0, N(R^n)]$$
$$\leq \frac{c|z_0|_\infty}{|x_j - x_0|}|f(x_j) - f(x_0) - \nabla f(x_0)(x_j - x_0)|.$$

Since f has a continuous derivative at x_0, the right hand-side goes to zero as $j \to +\infty$. Hence $z_0 \in N(R^n)$, a contradiction. Thus, (36) holds. □

Proof of lemma 19.

$(33) \Rightarrow (36)$. Since $N(R^n)$ is a cone with vertex at zero, it is sufficient to show that $0 \in \text{int } N(R^n)$. If $0 \notin \text{int } N(R^n)$ there exists a supporting hyperplane to $N(R^n)$ at 0, that is there exists $(u, v) \neq 0$ with

$$u'y_1 + v'y_2 \geq 0 \text{ for all } (y_1, y_2) \in N(R^n).$$

This means that

$$(47) \qquad u'(\nabla g_A(x_0)(z - x_0) + y) + v' \nabla h(x_0)(z - x_0) \geq 0$$

for all $z \in R^n$ and $y \in R_+^l$. For $z = x_0$ we obtain

$$u'y \geq 0 \text{ for all } y \in R_+^l,$$

hence $u \geq 0$. Taking $z = \bar{z} + x_0$, where $\bar{z}$ is as in (33), and $y = 0$ in (47), we conclude that

$$u' \nabla g_A(x_0)\bar{z} \geq 0.$$

On the other hand $u \geq 0$ and $\nabla g_A(x_0)\bar{z} < 0$ by (33), hence $u = 0$. From (47) we obtain

$$v' \nabla h(x_0)z = 0 \text{ for all } z \in R^n$$

which yields $v = 0$, because of (33). Thus $(u, v) = 0$, a contradiction. Hence (36) holds.

$(36) \Rightarrow (37)$. Take any $y = (y_0, y_1, y_2) \in R^l \times R^{m-l} \times R^q$. By (36) there exists $z \in R^n$ such that

$$(48) \qquad \nabla g_A(x_0) \, (z - x_0) \leq_{\varepsilon} y_0, \; \nabla h(x_0)(z - x_0) = y_2.$$

Choose $r > 0$ and let $\lambda > 0$ be so small that $z_\lambda = \lambda z + (1 - \lambda)z_0 \in \text{int } B(x_0, r)$. Then, by (48),

$$(49) \qquad \nabla g_A(x_0)(z_\lambda - x_0) \leq \lambda y_0, \; \nabla h(x_0)(z_\lambda - x_0) = \lambda y_2.$$

For every sufficiently small $\mu \in (0, 1)$ we have

$$(50) \qquad g_I(x_0) + \mu \nabla g_I(x_0)(z_\lambda - z_0) \leq \mu \lambda y_1.$$

Take $q_\mu = \mu z_\lambda + (1 - \mu)x_0$, and let μ be so small that $q_\mu \in B(x_0, r)$. Then by (49) and (50)

$$g_A(x_0) + \nabla g_A(x_0)(q_\mu - x_0) \leq \lambda \mu y_0,$$

$$g_I(x_0) + \nabla g_I(x_0)(q_\mu - x_0) \leq \lambda \mu y_1,$$

$$h(x_0) + \nabla h(x_0)(q_\mu - x_0) = \lambda \mu y_2.$$

This means that $\lambda \mu y \in M[(B(x_0, r)]$ for all sufficiently small $\lambda \geq 0$. Since y is arbitrary, the proof is complete.

(37) ⇒ *(38)*. Fix $r > 0$ and denote

$$F(z) = M(z) \text{ for } z \in B(x_0, r), \; = \emptyset \text{ otherwise.}$$

The multifunction F has closed and convex graph, $0 \in$ int im F and dom F is bounded. Then we apply Robinson-Ursescu's theorem as stated (for example) in Aubin-Ekeland [1, cor. 2, p. 132], obtaining (38).

(38) ⇒ *(37)*. By (38), $B(0, \sigma) \subset$ dom S, hence $B(0, \sigma) \subset M[B(x_0, r)]$.

(37) ⇒ *(33)*. Since $h(x_0) = 0$ we have from (37) that $0 \in$ int $\nabla h(x_0)[B(x_0, r)]$, which means that $\nabla h(x_0)$ has full rank. Furthermore, from (37) there exists $\bar{z} \in R^n$ such that

$$g(x_0) + \nabla g(x_0)(\bar{z} - x_0) < 0, \; h(x_0) + \nabla h(x_0)(\bar{z} - x_0) = 0.$$

Taking into account that $g_A(x_0) = 0$ we obtain (33). □

Section 3. Solution and Multiplier Well-Posedness in Convex Programming.

Given the real Banach space X, for $u \in R^k$ we shall consider sequences of mathematical programming problems $Q_n(u)$, $n = 0, 1, 2, \ldots$ of the form:

minimize $f_n(x)$ subject to k inequality constraints

(51)
$$g_{nj}(x) \leq u_j, \; j = 1, \ldots, k.$$

Here

$$f_n : X \to (-\infty, +\infty], \; g_{nj} : X \to (-\infty, +\infty)$$

are given functions.

Under convexity assumptions, we are interested in Hadamard well - posedness of $Q_0(0)$, as far as solutions and Kuhn-Tucker multipliers are concerned. We shall consider the stable behaviour of multipliers which corresponds to small changes of the right - hand term u in (51) near $u = 0$, and to changes f_n, g_{nj} of f_0, g_{0j}.

We denote by

$$F_n(u) = \{x \in X : g_{nj}(x) \leq u_j, \; j = 1, \ldots, k\}$$

the feasible region of $Q_n(u)$, and by

$$p_n(u) = \text{val } (F_n(u), \; f_n), \; u \in R^k$$

its value (or perturbation, or marginal) function. A (Kuhn-Tucker) *multiplier* for $Q_n(u)$ is defined as any $y \in R^k$ such that

$$y \geq 0, \; p_n(u) \text{ is finite and } \inf\{f_n(x) + y'[g_n(x) - u] : x \in X\} = p_n(u).$$

Under convexity conditions, y is a multiplier for $Q_n(u)$ iff

$$-y \in \partial p_n(u)$$

(see Rockafellar [4, th. 16 p. 40]). Then corollary IV.45 shows that the stable behaviour of multipliers is related to epi - convergence of p_n to p_0, since then

$$G - \lim \partial p_n = \partial p_0.$$

This yields convergence of the multipliers of $Q_n(u_n)$ to those of $Q_0(0)$, as $u_n \to 0$.

We shall denote by $\mapsto$ either the strong or the weak convergence on X. All sequential terms (like compactness, semicontinuity, lim sup) will correspond to $\mapsto$.

If $u \in R^k$, $u < 0$ means that every $u_j < 0$; $g_0(x) \le u$ means that $g_{0j}(x) \le u_j$ for every j.

21 Theorem *Assume that*
(52) *f_n, g_{nj} are convex and sequentially lower semicontinuous;*
(53) *for every $c \in R$ the set*

$$\cup_{n=0}^{\infty}\{x \in X : f_n(x) \le c, \ g_{nj}(x) \le c, \ j = 1, ..., k\}$$

is sequentially relatively compact;
(54) *if $x_n \mapsto x$ then $\liminf f_n(x_n) \ge f_0(x)$, $\liminf g_{nj}(x_n) \ge g_{0j}(x)$, $j = 1, ..., k$;*
(55) *for some $u^* < 0$, $F_0(u^*) \ne \emptyset$ and f_0 is proper there;*
(56) *for every x and u such that $g_0(x) \le u$, $f_0(x) < +\infty$, there exists $a_n \to 0$, $a_n > 0$, such that for n sufficiently large the inequalities*

$$f_n(z) \le f_0(x) + a_n, \ g_{nj}(z) \le u_j + a_n, \ j = 1, ..., k$$

are compatible.
Then the following conclusions hold:
(57) *$u_n \to 0$, y_n is a multiplier for $Q_n(u_n)$, $y_n \to y$ for some subsequence $\Rightarrow y$ is a multiplier for $Q_0(0)$;*
(58) *for every multiplier y of $Q_0(0)$ there exist $u_n \to 0$ in R^k and multipliers y_n for $Q_n(u_n)$ such that $y_n \to y$;*
(59) *if $u_n \to 0$ and $u_{nj} > 0$ then val $(F_n(u_n), f_n) \to$ val $(F_0(0), f_0)$; if moreover $\varepsilon_n > 0, \varepsilon_n \to 0$ then*

$$\limsup [\varepsilon_n - \arg\min (F(u_n), f_n)] \subset \arg\min (F_0(0), f_0).$$

When a unique multiplier for $Q_0(0)$ exists, then of course we get

22 Corollary *Let the assumptions of theorem 21 hold. If $Q_0(0)$ has exactly one multiplier y, then if $u_n \to 0$, for every bounded sequence y_n of multipliers for $Q_n(u_n)$ we have $y_n \to y$.*

In the finite dimensional setting, by strengthening convergences and imposing coercivity of f_0 we get

23 Corollary *Let X be finite-dimensional. Then the conclusions of theorem 21 hold if (52) is assumed together with*

$$(60) \qquad epi - \lim f_n = f_0; \ g_n(x) \to g_0(x) \ for \ every \ x;$$

$$(61) \qquad g_{0j}(x) < 0, \ j = 1, ..., k, \ and \ f_0(x) < +\infty \ for \ some \ x;$$

$$(62) \qquad f_0(x) \to +\infty \ if \ x \to \infty.$$

A discussion of assumption (56).

(i) If epi-lim $f_n = f_0$, $g_n \to g_0$ continuously (both with respect to $\mapsto$) then (56) holds. For, fix x, u with $g_0(x) \le u$. By epi- convergence, there exists $z_n \mapsto x$ such that $f_n(z_n) \to f_0(x)$. Then $g_n(z_n) \to g_0(x)$, hence

$$f_n(z_n) \le f_0(x) + |f_n(z_n) - f_0(x)|,$$
$$g_{nj}(z_n) \le u_j + |g_{nj}(z_n) - g_{0j}(x)|$$

which yield (56).

(ii) Assumptions (54) and (56) plainly imply that

$$var - \lim f_n = f_0, \ var - \lim g_{nj} = g_{0j}$$

(both with respect to $\mapsto$).

(iii) Let epi-lim $p_n = p_0$ (a key step in proving theorem 21). Let $g_0(x) \le u$ and (for sake of simplicity) assume $p_n(u)$ finite. Then for some $u_n \to u$ we have $p_n(u_n) \to p_0(u)$. Thus for some suitable $x_n \in X$

$$f_n(x_n) \le p_n(u_n) + \frac{1}{n} \le p_0(u) + |p_n(u_n) - p_0(u)| + \frac{1}{n},$$
$$g_{nj}(x_n) \le u_j + |u_{nj} - u_j|$$

implying (56).

PROOFS.

Theorem 21. It suffices to prove that

$$G - \lim \partial p_n = \partial p_0$$

to get (57) and (58), as already noticed. From corollary IV.45 we need to prove that

$$(63) \qquad \text{epi-lim } p_n = p_0;$$

(64) p_n is proper, convex and lower semicontinuous on R^k for every n sufficiently large.

Proof of (63). Firstly we show that $p_n(u)$ is finite for $|u|$ sufficiently small and n sufficiently large. By (53) and semicontinuity, f_n attains its global minimum on $F_n(u)$ whenever this set is nonempty. Since f_n never takes the value $-\infty$, it follows that

$$(65) \qquad p_n(u) > -\infty \text{ for all } n \text{ and } u.$$

Arguing by contradiction, suppose that for some subsequence n_k

$$(66) \qquad p_{n_k}(u_k) = +\infty \text{ for all } k,$$

for some $u_k \to 0$. Let $x \in F_0(u^*)$ and $f_0(x) < +\infty$. Then by (56) we have $a_k \le u_{kj} - u_j^*$ for all large k and for all j. Thus by (55) and (56) and suitable points z_k

$$g_{n_k j}(z_k) \le u_{kj}, \ f_{n_k}(z_k) \le f_0(x) + u_{kj} - u_j^*.$$

So $z_k \in F_{n_k}(u_k)$ and $p_{n_k}(u_k) < +\infty$, contradicting (66). Therefore

$$(67) \qquad p_n(u) \text{ is finite for all large } n \text{ and } |u| \text{ sufficiently small.}$$

Now we show that

$$(68) \qquad u_n \to u \Rightarrow L = \liminf p_n(u_n) \ge p_0(u).$$

Notice that $p_n(u_n) > -\infty$ by (48). Let $p_0(u) <^{\cdot}+\infty$. Only the case $L < +\infty$ needs proof. For some subsequence, $p_n(u_n) \to L$. Then given $\varepsilon > 0$, for every n there is some $x_n \in X$ with

$$(69) \qquad g_{nj}(x_n) \le u_{nj}, \ f_n(x_n) \le p_n(u_n) + \varepsilon.$$

A further subsequence verifies $x_n \mapsto x$ by (53). Then by (69) and (54)

$$f_0(x) \le \liminf f_n(x_n) \le \liminf p_n(u_n) + \varepsilon,$$

$$g_{0j}(x) \le \liminf g_{nj}(x_n) \le u_j,$$

yielding $x \in F_0(u)$. Thus

$$p_0(u) \le f_0(x) \le L$$

thereby proving (68) in this case. Now let $p_0(u) = +\infty$. If for some subsequence $\sup p_n(u_n) < +\infty$, then for every n there would exist $x_n, \ c \in R$ such that

$$f_n(x_n) \le c, \ g_{nj}(x_n) \le u_{nj}.$$

Then by (53) some further subsequence satisfies

$$x_n \mapsto x.$$

Then by (54) $x \in F_0(u)$ and

$$p_0(u) \le f_0(x) \le \liminf f_n(x_n) \le c$$

contradicting $p_0(u) = +\infty$. Thus for the original sequence $p_n(u_n) \to +\infty$, and (68) is proved.

358

Now let us show that

(70)   for every $u \in R^k$ there exists $u_n \to u$ with lim sup $p_n(u_n) \leq p_0(u)$.

Let $p_0(u) < +\infty$, $x \in F_0(u)$ and $f_0(x) < +\infty$. By (56), for every large n there exists z_n such that

$$f_n(z_n) \leq f_0(x) + a_n, \ g_{nj}(z_n) \leq u_j + a_n, \ j = 1, ..., k.$$

Then $u_n \to u$, lim sup $p_n(u_n) \leq f_0(x)$, where $u_{nj} = u_j + a_n$. Thus we obtain (70) by arbitrariness of x.

Proof of (64). Since $p_0(u^*)$ is finite, by (70) for every large n we have $p_n(u_n) < +\infty$ for some u_n. Properness of p_n follows by (65) for large n. For such n let u, v, x_1, x_2 be such that (by (53))

$$f_n(x_1) = p_n(u), \ f_n(x_2) = p_n(v),$$

with $x_1 \in F_n(u)$, $x_2 \in F_n(v)$. By convexity, if $0 \leq t \leq 1$ then

$$p_n(tu + (1-t)v) \leq t \ f_n(x_1) + (1-t)f_n(x_2) = t \ p_n(u) + (1-t)p_n(v)$$

giving convexity of p_n. Fix n and let $u_h \to u$ in R^k, such that, as $h \to +\infty$,

$$L = \liminf p_n(u_h) < +\infty.$$

We can assume that $p_n(u_h) \to L$. For every $\varepsilon > 0$ and h we can find x_h such that

$$g_{nj}(x_h) \leq u_{hj}, \ j = 1, ..., k; \ f_n(x_h) \leq p_n(u_h) + \varepsilon.$$

By (53) $x_h \mapsto x$ for some subsequence, so that

$$g_{nj}(x) \leq \liminf g_{nj}(x_h) \leq u_j, j = 1, ..., k,$$

$$f_n(x) \leq \liminf f_n(x_h) \leq \liminf p_n(u_h) + \varepsilon,$$

hence

$$p_n(u) \leq f_n(x) \leq \liminf p_n(u_h).$$

This shows lower semicontinuity of p_n. So (64) is proved. By (67) and Barbu - Precupanu [1, cor. 2. 1 p. 105] we see that $\partial p_n(u_n) \neq \emptyset$ for all large n whenever $u_n \to 0$. So (57), (58) (make sense and) are proved.

Proof of (59). Let us show that

(71) $$\text{var-lim } [f_n + \text{ind } F_n(u_n)] = f_0 + \text{ind } F_0(0)$$

(with respect to $\mapsto$). If $x_n \mapsto x$ and $x \in F_0(0)$ then by (54)

(72) $$\liminf [f_n(x_n) + \text{ind } (F_n(u_n), x_n)] \geq f_0(x) + \text{ind } (F_0(0), x).$$

If $x \notin F_0(0)$ then for some j

$$g_{0j}(x) > 0.$$

Then for all sufficiently large n, by (54)

$$g_{nj}(x_n) \geq c > 0$$

hence $x_n \notin F_n(u_n)$, proving (72). Now let $x \in F_0(0)$ be such that $f_0(x) < +\infty$. Fix any subsequence of the positive integers. By (56), for every large n of a further subsequence, there exists z_n such that

$$f_n(z_n) \leq f_0(x) + a_n, \ g_{nj}(z_n) \leq a_n \leq u_{nj}, \ j = 1, ..., k.$$

Therefore

$$\limsup [f_n(z_n) + \text{ind } (F_n(u_n), z_\ast)] \leq f_0(x) + \text{ind } (F_0(0), x),$$

which entails (71) by arbitrariness of the subsequence. Now let x_n be asymptotically minimizing for $(F_n(u_n), f_n)$. Since $p_n(u_n)$ is finite by (67), we have

$$f_n(x_n) - p_n(u_n) \to 0.$$

By IV.(12) we get $\limsup p_n(u_n) \leq p_0(0)$. So

$$f_n(x_n) \leq c, \ g_{nj}(x_n) \leq c$$

for some real constant c. Then by (53) we can apply theorem IV.5, and this yields (59).

□

Remark. From the above proof we see that (59) holds true without convexity of f_n, g_{nj}.

Corollary 23. We check the assumptions of theorem 21. Of course (55) holds. From theorem 10.8 of Rockafellar [2], $g_n \to g_0$ continuously by (60), therefore implying (54), and (56) as noticed above. Finally (53) comes from

24 Lemma If X is finite - dimensional, f_n are proper, convex and lower semicontinuous, and

$$\text{epi-lim } f_n = f_0, \ f_0(x) \to +\infty \text{ as } x \to \infty,$$

then there exist constants $a > 0$, b such that

$$f_n(x) \geq a|x| - b$$

for every large n and every x.

Proof of lemma 24. The conjugate f_0^* is continuous at 0 since f_0 is coercive. Then there exist real constants $a > 0$, b such that

$$(73) \qquad\qquad |x| \leq 2a \;\Rightarrow\; f_0^*(x) < b.$$

Let $x_1, ..., x_p$ in the finite-dimensional space X be such that

$$S = \{x \in X : \; |x| < a\} \subset \text{co } \{x_1, ..., \; x_p\} \subset Z = \{x \in X : \; |x| < 2a\}.$$

Since epi-lim $f_n^* = f_0^*$ by Attouch [3, th.3.18 p.295], there exist sequences $z_{nj} \to x_j$, $j = 1, ..., p$, such that

$$f_n^*(z_{nj}) \to f_0^*(x_j).$$

Thus by (73)

$$f_n^*(z_{nj}) < b \text{ for all large } n.$$

Since

$$S \subset \text{co } \{z_{n1}, ..., \; z_{np}\} \subset Z$$

for all large n, by convexity

$$f_n^*(z) < b \text{ if } z \in S$$

whence

$$f_n^*(z) < b + \text{ ind } (S, z) \text{ for every } z.$$

By taking conjugates we get the conclusion. $\qquad\qquad\qquad\qquad\qquad\qquad\qquad$ □

Section 4. Solution and Multiplier Well-Posedness in (non) smooth Problems.

We consider sequences $Q_n(0)$ of (not necessarily convex) mathematical programming problems, as defined in section 3, with objective function f_n and inequality constraints (51) with $u = 0$, in the finite-dimensional setting $X = R^m$ (for a given m).

Assuming local Lipschitz continuity of f_n and g_{nj}, we shall see that the multiplier rule of Clarke [2] (for global solutions) is stable under epi- convergent data perturbations. Such a well - posedness result is a corollary to theorem IV.36 about the stable behaviour of generalized gradients.

Given $f_n, g_{n1}, ..., g_{nk} : R^m \to (-\infty, +\infty)$ we consider the feasible set F_n defined by

$$x \in F_n \text{ iff } g_n(x) \leq 0.$$

Standing assumption: f_n, g_{nj} are locally Lipschitz continuous on R^m.

Let x_n be any global optimal solution to $Q_n(0)$. Then by cor. 1 of Clarke [2], there exist multipliers r_n, s_{ni} such that

(74) $r_n \geq 0$, $s_{ni} \geq 0$, not all zero,

(75) $s_{nj} \, g_{nj}(x_n) = 0, j = 1, ..., k,$

(76) $0 \in r_n \, \partial f_n(x_n) + \sum_{j=1}^{k} s_{nj} \, \partial g_{nj}(x_n).$

We denote by $M_n(x_n)$ the set of all vectors $(r_n, s_{n1}, ..., s_{nk}) \in R^{k+1}$ satisfying (74), (75), (76) (i.e. multipliers for $Q_n(0)$) corresponding to the global optimal solution x_n). The next theorem deals with nonsmooth problems (and uses the notion of equi - lower semidifferentiability, as introduced in section IV.7, see p. 158).

25 Theorem *Assume that every sequence f_n, g_{nj} is equi - lower semidifferentiable and locally equibounded;*

(77) $\partial^- g_{nj}(x) \neq \emptyset \ for \ every \ x \in F_n;$

$epi - lim \, g_{nj} = g_{0j}, \ epi - lim f_n = f_0;$

for every $y \in F_0$ there exist $y_n \in F_n$ such that $lim \ sup \ f_n(y_n) \leq f_0(y)$.

Then if x_n is a global solution to $Q_n(0)$ and $x_n \to x_0$ for some subsequence, we have that x_0 is a global solution to $Q_0(0)$; $f_n(x_n) \to f_0(x_0)$;

(78) $\emptyset \neq lim \ sup \ M_n(x_n) \subset M_0(x_0).$

Example. Uniform convergence of g_n to g_0 does not imply neither value convergence, nor semicontinuity of arg min. Consider $k = m = 1$ and the problems $Q_n(0)$ with $f_n(x) = x^2$, and

$$g_0(x) = -x - 1 \ if \ x \leq -1, = 0 \ if \ |x| \leq 1, = x - 1 \ if \ x \geq 1;$$

$$g_n(x) = g_0(x) \ if \ |x| \geq 1, = -\frac{1}{n}(x - 1) \ if \ 0 \leq x \leq 1, = \frac{1}{n}(x + 1) \ if \ -1 \leq x \leq 0.$$

Now we come to the important particular case of Hadamard well - posedness in nonlinear parametric programming with smooth data.

Let the parameter space T be a first - countable Hausdorff topological space. Given the smooth functions

$$f, \, g_1, ..., \, g_k : R^m \times T \to (-\infty, +\infty)$$

for every $p \in T$ we consider the problem of minimizing $f(x, p)$ subject to

$$g_j(x, p) \leq 0, \ j = 1, ..., k.$$

Let $F(p)$ be the corresponding feasible set. Given any x in

$$\text{arg min } [F(p), f(\cdot, p)] = \text{ arg min } (p)$$

let us denote by $M(x,p)$ the corresponding set of F. John multipliers. Therefore a vector $(r, s_1, ..., s_k)' \in R^{k+1}$ belongs to $M(x,p)$ iff it is not zero and

$$r \geq 0; \ s_j \geq 0; \ s_j \ g_j(x,p) = 0, \ j = 1, ..., k;$$

$$0 = r \bigtriangledown f(x,p) + \sum_{j=1}^{k} s_j \bigtriangledown g_j(x,p).$$

(Of course the gradient operator acts here on the x - variable).

A multivalued map

$$S : T \rightrightarrows R^k$$

will be called *lower (upper) semicontinuous by inclusion* iff for every $p \in T$ and every sequence $p_n \rightarrow p$ we have

$$S(p) \subset \liminf S(p_n) \ (\limsup S(p_n) \subset S(p)).$$

The multipliers map M will be called here *upper semicontinuous by inclusion* iff $p \in T, p_n \rightarrow p, \ x_n \in \ \arg \ \min (p_n), \ w_n \in M(x_n, p_n), \ x_n \rightarrow x, \ w_n \rightarrow w$ for some subsequence imply $w \in M(x,p)$.

This condition will be used under assumptions forcing $x \in \ \arg \ \min (p)$.

26 Corollary *Let f, g be continuous and locally equibounded on $R^m \times T$.*

Let $f(\cdot, p), \ g_j(\cdot, p)$ belong to $C^2(R^m)$ with locally equi - bounded second partial derivatives, for every $p \in T$. Suppose that F is lower semicontinuous by inclusion. Then both arg min $(\cdot)$ and M are upper semicontinuous by inclusion. Moreover

$$p \rightarrow \ val \ [F(p), \ f(\cdot, p)]$$

is continuous at any $p_0 \in T$ such that there exists a bounded selection of arg min $(\cdot)$, defined on some neighborhood of p_0.

PROOFS.

Theorem 25. It is easily seen that

$$\text{var} - \lim (f_n + \ \text{ind} \ F_n) = f_0 + \ \text{ind} \ F_0.$$

Then the conclusions about x_0 and the global optimal values follow from theorem IV.5. Let us prove (78). Given

$$w_n = (r_n, \ s_{n1}, ... s_{nk}) \in M_n(x_n)$$

without loss of generality we assume

$$0 \leq r_n \leq 1, \ 0 \leq s_{nj} \leq 1, \ |w_n| = 1,$$

and by (76) we get

$$(79) \qquad 0 = r_n \, u_n + \sum_{j=1}^{k} s_{nj} \, v_{nj}$$

for some $u_n \in \partial f_n(x_n)$ and $v_{nj} \in \partial g_{nj}(x_n)$. For some subsequences

$$r_n \to r_0 \geq 0, \; s_{nj} \to s_{0j} \geq 0.$$

By lemma IV. 40 and IV. (97), u_n and v_{nj} are bounded sequences. By theorem IV.36, for some subsequences

$$u_n \to u_0 \in \partial f_0(x_0), \; v_{nj} \to v_{0j} \in \partial g_{0j}(x_0).$$

Thus by (79)

$$0 = r_0 \, u_0 + \sum_{j=1}^{k} s_{0j} \, v_{0j}.$$

By (75) and epi - convergence

$$(80) \qquad 0 = \liminf s_{nj} \, g_{nj}(x_n) \geq s_{0j} \, g_{0j}(x_0), j = 1, ..., k.$$

Given j there exists $y_n \to x_0$ such that $g_{nj}(y_n) \to g_{0j}(x_0)$. By (75) and equi - lower semidifferentiability (see IV.(89)), remembering (77),

$$s_{nj} \, g_{nj}(y_n) \geq s_{nj} \, g_{nj}(x_n) + s_{nj} z'_n(y_n - x_n) + s_{nj} \, k(x_n, y_n)$$
$$= s_{nj} \, z'_n(y_n - x_n) + s_{nj} \, k(x_n, y_n)$$

for every $z_n \in \partial^- g_{nj}(x_n)$. Since z_n is bounded (lemma IV.40) and k is continuous, it follows that $s_{0j} \, g_{0j}(x_0) \geq 0$. Hence the complementarity condition (75) is true with $n = 0$ by (80), yielding (78). □

Remark. Assumption (77) can be omitted if continuous convergence of g_{nj} to g_{0j} is assumed, thereby obtaining directly (75) with $n = 0$, the only point where (77) was used.

Corollary 26. Given $p \in T$ and $p_n \to p$, put

$$f_n(x) = f(x, p_n), \; g_{nj}(x) = g_j(x, p_n).$$

Everything follows from theorems 25 and IV.5, as soon as we show equi - lower semidifferentiability of f_n, g_{nj}. To prove this (for f_n) we remark that $\partial^- f_n = \nabla f_n$ (Ambrosetti-Sbordone [1, lemma 3.2, (iii)]). Given any ball $T \subset R^m$, x, y in T and $u_n \in \partial^- f_n(x)$, by Taylor's formula

$$f_n(y) \geq f_n(x) + u'_n(y - x) + \frac{1}{2}(y - x)' \, Hf_n(w_n)(y - x)$$

where Hf_n is the Hessian matrix of f_n and w_n is some point between x and y. By local equiboundedness of the second partial derivatives we obtain

$$f_n(y) \geq f_n(x) + u'_n(y - x) + C|y - x|^2$$

for a suitable constant C. This shows equi-lower semidifferentiability. □

Section 5.   Estimates of Value and $\varepsilon -$ Arg min for Convex Functions.

Let f_n be a sequence of convex functions on a given normed space X. Then Mosco convergence of f_n is equivalent to Mosco convergence of their epigraphs (see Attouch [3, th. 1.39 p. 98]). On the other hand, Mosco convergence is related to convergence of optimal values and $\varepsilon -$ arg min in a very significant way, as we know from chapter IV. Moreover, Mosco convergence of closed convex sets is implied by their local Hausdorff convergence, and it is equivalent to it in the finite-dimensional setting.

As we saw in section II.3, some form of local Hausdorff convergence of the epigraph of f_n to that of f_0 is related to Hadamard and Tykhonov well-posedness. The main point, we develop here, is that local Hausdorff metrics allow a quantitative approach.

More precisely, denote by $B(r)$ the closed ball of radius r around the origin (in any normed space). For given functions

$$f, g : X \to (-\infty, +\infty]$$

define $F = $ epi $f, G = $ epi g, and put

$$d(r, f, g) = \text{ maximum between sup } \{ \text{ dist } (x, G) : x \in F \cap B(r)\} \text{ and}$$
$$\sup \{ \text{ dist } (y, F) : y \in G \cap B(r)\}.$$

Here the distance between (a, t) and (b, s) in $(X \times R)$ is given by $\max\{\|a - b\|, |s - t|\}$.

27 Proposition *Let f_n be a sequence of proper, convex, lower semicontinuous functions on the real normed space X. Then $d(r, f_n, f_0) \to 0$ as $n \to +\infty$ for all sufficiently large r implies $M - \lim f_n = f_0$. The converse holds if X is finite-dimensional.*

The family of "distances" $d(r, \cdot, \cdot)$ may now be used to get a quantitive estimate of the dependence of the $\varepsilon -$ arg min and the optimal value on the data.

Notation: $\varepsilon -$ arg min f instead of $\varepsilon -$ arg min (X, f); inf f instead of inf $f(X)$; X a given real normed space.

28 Proposition *Let $f, g : X \to (-\infty, +\infty]$ be such that inf f, inf g are finite.*
(i) If there exists $r > 0$ such that for every $\varepsilon > 0$

$$B(r) \cap \varepsilon - \text{ arg min } f \neq \emptyset,$$

$$B(r) \cap \varepsilon - \text{ arg min } g \neq \emptyset$$

then

(81) $$|inf\ f - inf\ g| \leq d(s, f, g), \text{ where}$$

$$s = max\ \{r, 1 + |inf\ f|, 1 + |inf\ g|\}.$$

(ii) If f, g are convex and there exists $r > 0$ such that for every $\varepsilon > 0$

$$\varepsilon - arg\ min\ f \cup \varepsilon - arg\ min\ g \subset B(r)$$

then with s as in (i)

$$haus\ (\varepsilon - arg\ min\ f, \varepsilon - arg\ min\ g) \le cd(s, f, g)$$

where $c = 1 + (4r + 2d)/(\varepsilon + 2d), d = d(s, f, g)$.

As an important particular case, let X be a real Hilbert space. Given $u \in X$ we may estimate the Hölder continuous behaviour of the projections $p(u, C)$ as far as the dependence on C is involved, as follows.

Given $C, D \subset X$ and $r > 0$ sufficiently large, we shall consider the local Hausdorff distance as already defined in section II.3, i.e.

$$h(r, C, D) = \text{ maximum between sup } \{ \text{ dist } (x, D) : x \in C \cap B(r)\} \text{ and}$$
$$\text{sup } \{ \text{ dist } (y, C) : y \in D \cap B(r)\}.$$

Let C, D be nonempty closed convex subsets of X, and consider, for a given $u \in X$,

$$r = \|u\| + \text{ dist } (u, C) + \text{ dist } (u, D),$$
$$c = p(u, C), d = p(u, D).$$

Then by the optimality conditions characterizing c and d,

(82) $< u - c, y - c > \le 0$ if $y \in C, < u - d, z - d > \le 0$ if $z \in D$.

Since

$$\|c\| \le \|u\| + \text{ dist } (u, C) \le r, \|d\| \le r,$$

there exist some $\bar{y} \in C, \bar{z} \in D$ such that

$$\|d - \bar{y}\| = \text{ dist } (d, C), \|c - \bar{z}\| = \text{ dist } (c, D),$$

and max $\{\|c - \bar{z}\|, \|d - \bar{y}\|\} \le h(r, C, D)$. Then adding the two inequalities in (82) with $y = \bar{y}, z = \bar{z}$ we obtain

$$< u - c, \bar{y} - d + d - c > + < u - d, \bar{z} - c + c - d > \le 0.$$

Hence

$$\|c - d\|^2 \le\ < u - c, d - \bar{y} > + < u - d, c - \bar{z} >$$
$$\le \|d - \bar{y}\| \text{ dist } (u, C) + \|c - \bar{z}\| \text{ dist } (u, D)$$
$$\le [\text{ dist } (u, C) + \text{ dist } (u, D)]h(r, C, D) \le rh(r, C, D).$$

Therefore we get

29 Proposition Let C, D be nonempty closed convex subsets of the real Hilbert space X. Then for every $u \in X$

$$\|p(u, C) - p(u, D)\| \le [rh(r, C, D)]^{1/2}$$

where

$$r = \|u\| + \text{ dist } (u, C) + \text{ dist } (u, D).$$

PROOFS.

Proposition 27. Put $F_n = $ epi f_n. As already recalled, it suffices to show that

(83) $h(r, F_n, F_0) \to 0$ for all sufficiently large r,

implies $M - \lim F_n = F_0$, and conversely if X is finite-dimensional.

Assume (83) and let $x_n \in F_n$ for some subsequence be such that $x_n \to x_0$ in X. Since x_n is bounded, $x_n \in F_n \cap B(r)$ for all sufficiently large r. Then

$$\text{dist } (x_n, F_0) \to 0$$

by definition of h. Thus there exists a sequence $u_n \in F_0$ such that $\|x_n - u_n\| \to 0$, so that $x_0 \in F_0$, being a closed convex set. Now let $y \in F_0$. Then by (83) there exists a sequence $y_n \in F_n$ such that $y_n \to y$. Therefore

(84) $M - \lim F_n = F_0.$

Now assume (84) and let X be finite-dimensional. Fix $r > 0$ and let $\varepsilon > 0$. By compactness, one can find points $x_1, ..., x_p$ in $F_0 \cap B(r)$ such that

$$F_0 \cap B(r) \subset \cup_{i=1}^{p} B_i$$

where B_i is the ball of radius $\varepsilon > 0$ centered at x_i. By (84), dist $(x_i, F_n) \leq \varepsilon$ for every i and sufficiently large n. It follows that (for such n) every point of $F_0 \cap B(r)$ is at distance $\leq 2\varepsilon$ from some point of F_n, thus showing that

$$\sup\{ \text{ dist } (x, F_n) : x \in F_0 \cap B(r)\} \to 0.$$

To end the proof, we need to show that

$$\sup \{\text{dist } (y, F_0) : y \in F_n \cap B(r)\} \to 0.$$

If not, there would exist $\varepsilon > 0, r > 0$ and points y_n such that for some subsequence

$$y_n \in F_n \cap B(r), \text{ dist } (y_n, F_0) \geq \varepsilon.$$

But this contradicts (84), since some cluster point of y_n belongs to F_0.   □

Proposition 28. **Proof of (i).** Given $\varepsilon \in (0,1)$, let $x \in B(r)$ be such that $f(x) \leq$ inf $f + \varepsilon$, thus

$$z = (x, \text{ inf } f + \varepsilon) \in \text{ epi } f \cap B(s).$$

By definition of d, for every $a > 0$ there exists $(y, t) \in$ epi g such that

$$\max\{\|x - y\|, |\text{inf } f + \varepsilon - t|\} \leq \text{ dist } (z, \text{ epi } g) + a \leq d(s, f, g) + a.$$

Hence

$$\text{inf } g \leq g(y) \leq t \leq \text{ inf } f + \varepsilon + d(s, f, g) + a$$

giving

$$\text{inf } g - \text{ inf } f \leq d(s, f, g),$$

since a and ε are arbitrary. By exchanging the role of f and g we obtain the conclusion in (i).

Proof of (ii) . Let $x \in \varepsilon - \arg\min f$ and write

$$d = d(s, f, g).$$

Then by (81)

(85) $$f(x) \leq \inf\ g + d + \varepsilon$$

and by assumption $(x, f(x)) \in$ epi $f \cap B(s)$. Then, given $a > 0$, there exists (as before) some $(y, t) \in$ epi g such that

(86) $$\max\{\|x - y\|, |f(x) - t|\} \leq a + d.$$

Hence

(87) $$g(y) \leq t \leq f(x) + a + d.$$

Adding (85) and (87) we obtain

$$g(y) \leq a + 2d + \inf\ g + \varepsilon.$$

Convexity of g will imply that

(88) $$\text{dist}\ (y, \varepsilon - \arg\min g) \leq (d + a) \frac{4r + 2d + 2a}{\varepsilon + 2d}.$$

Assuming (88), we get by (86) (letting $a \to 0$)

$$\text{dist}\ (x, \varepsilon - \arg\min g) \leq d(1 + \frac{4r + 2d}{\varepsilon + 2d}).$$

Since x is arbitrary in $\varepsilon - \arg\min f$ and the role of f, g can be exchanged, we get the conclusion of (ii). So we end by proving (88). Write

$$L(b) = \{x \in X : g(x) \leq b\}.$$

It suffices to show the following property. For every $\theta > \inf\ g, h > 0$ and $y \in L(\theta + h)$, we have

(89) $$\text{dist}\ [y, L(\theta)] \leq \frac{h(\|y\| + r)}{h + \theta - \inf g}.$$

Indeed, we get (88) from (89) with

$$h = 2d + a,\ \theta = \varepsilon + \inf\ g$$

since by (86), $\|y\| \leq r + d + a$.

Proof of (89) . Given $b \in (0, 1)$, let u be such that

$$g(u) \leq \inf g + b(\theta - \inf g), \|u\| \leq r,$$

and consider

$$z = ty + (1-t)u, 0 \leq t \leq 1.$$

Then, remembering that $g(y) \leq \theta + h$, we find

$$g(z) \leq t(\theta + h) + (1-t)(b\theta + (1-b) \inf g)$$
$$= (t + (1-t)b)\theta + th + (1-t)(1-b) \inf g.$$

Hence $g(z) \leq \theta$ provided that

$$t = \frac{(1-b)(\theta - \inf g)}{h + (1-b)(\theta - \inf g)}.$$

With such a choice of t

$$\|y - z\| = (1-t)\|y - u\| \leq \frac{h(\|y\| + r)}{h + (1-b)(\theta - \inf g)}.$$

Letting $b \to 0$ we obtain (89) (since $z \in L(\theta)$). $\qquad\qquad\qquad$ □

Section 6. Lipschitz Continuity of Arg min.

We consider the following nonlinear program $P(t)$

(90) $\qquad\qquad\qquad$ minimize $f_0(x, t)$ subject to

$$f_i(x,t) = 0, \; i = 1, 2, ..., s,$$

$$f_i(x,t) \leq 0, \; i = s+1, ..., m,$$

where $x \in R^n, t \in R^p$, and

$$f_i : R^n \times R^p \to (-\infty, +\infty), i = 0, 1, ..., m,$$

are C^2 functions.

Let $\bar{t} \in R^p$ be fixed and let $\bar{x}$ be a local minimum point of $P(\bar{t})$. We are interested to find conditions under which for t in some neighbourhood of $\bar{t}$ the set of local minimizers of (90) around $\bar{x}$ can be parameterized as a Lipschitz continuous single-valued function of t.

Here we purposely detache our well-posedness investigations from global optimal solutions (as in the rest of this book), since the local theory (so to speak) is more natural in the present setting.

A Lipschitz behavior of the whole arg min cannot hold in general.

Counterexample. Let $n = 2$. Minimizing x_2 subject to $x_1^2 \leq x_2$ and $x_2 \geq t$, we obtain, for $t \geq 0$,

$$\arg\min (t) = \{(x_1, t) \in R^2 : |x_1| \leq \sqrt{t}\},$$

hence there is no constant $L > 0$ such that

$$\text{dist} \, [x, \arg\min (0)] \leq Lt \text{ for every } x \in \arg\min (t).$$

Let $\bar{z} = (\bar{x}, \bar{t})$ and let $\bar{J}$ be the set of active constraints there,

$$\bar{J} = \{i \in \{1, ..., m\} : f_i(\bar{z}) = 0\}.$$

In the sequel we introduce certain conditions used in the main theorem. Our first condition is:

(91)   the vectors $\{\nabla_x f_i(\bar{z}), i \in \bar{J}\}$ are linearly independent.

It is classically known that (91) implies existence of a unique *Lagrange multiplier* $\bar{u} \in R^m$ with $\bar{u}_i \geq 0$ if $i = s+1, ..., m$, such that the pair $(\bar{x}, \bar{u})$ satisfies the Karush-Kuhn-Tucker conditions

$$(92) \qquad \nabla_x f_0(\bar{z}) + \sum_{i=1}^{m} \bar{u}_i \nabla_x f_i(\bar{z}) = 0,$$

$$(93) \qquad \sum_{i=1}^{m} \bar{u}_i f_i(\bar{z}) = 0,$$

(see McCormick [1, th. 18 p. 216]).

Denote

$$\bar{J}_+ = \{j \in \bar{J} : \bar{u}_j > 0\}.$$

The feasible point $\bar{x}$ of $P(\bar{t})$ satisfies the strong second-order sufficient condition iff

(94)   there exists $\bar{u} \in R^m, \bar{u}_i \geq 0, i = s+1, ..., m$, such that (92), (93) hold and

$$w'Hw > 0$$

for all nonzero w such that

$$w' \nabla_x f_i(\bar{z}) = 0 \text{ for every } i \in \bar{J}_+,$$

where

$$(95) \qquad H = \nabla_{xx} f_0(\bar{z}) + \sum_{i=1}^{m} \bar{u}_i \nabla_{xx} f_i(\bar{z})$$

and $\nabla_{xx} f_i(\bar{z})$ denotes the corresponding (Hessian) matrix of second-order partial derivatives with respect to the x - variables.

30 Theorem *Let (91) and (94) hold at $(\bar{x}, \bar{t})$ and let $\bar{u}$ be the corresponding Lagrange multiplier. Then there exist neighbourhoods U of $\bar{t}$ and V of $\bar{x}$ respectively, and a (single-valued) function $(\bar{x}(\cdot), \bar{u}(\cdot)) : U \to V \times R^m$ such that for every $t \in U, \bar{x}(t)$ is the unique* *(local and global) solution to (90) on V and $\bar{u}(t)$ is the corresponding unique Lagrange multiplier.*

Moreover, the function $t \rightarrow (\bar{x}(t), \bar{u}(t)) \in V \times R^m$ is Lipschitz continuous on U.

Proof. Let

$$\bar{y}_i = \begin{cases} \bar{u}_i, & i \in \bar{J}, \\ f_i(\bar{z}), & i \in \{1, ..., m\} \setminus \bar{J}. \end{cases}$$

Define the map

$$F : R^{n+p+m} \rightarrow R^{n+m}$$

as
(96)

$$F(x,t,y) = \begin{pmatrix} \nabla_x f_0(x,t) + \sum_{i=1}^{s} y_i \nabla_x f_i(x,t) + \sum_{i=s+1}^{m} \max\{0, y_i\} \nabla_x f_i(x,t) \\ -f_1(x,t) \\ \vdots \\ -f_s(x,t) \\ \min\{0, y_{s+1}\} - f_{s+1}(x,t) \\ \vdots \\ \min\{0, y_m\} - f_m(x,t) \end{pmatrix}$$

Then conditions (92), (93) and feasibility of $\bar{x}$ can be written as

$$F(\bar{x}, \bar{t}, \bar{y}) = 0.$$

Then, by (94), $\bar{x}$ is a strict local solution to $P(\bar{t})$ (with an unique Lagrange multiplier $\bar{u}$ by (91)): see Fiacco [2, lemma 3.2.1 p. 69].

The map F is locally Lipschitzian everywhere but not necessarily differentiable at those points for which $\bar{y}_i = 0$ if $i \in \bar{J} \setminus \bar{J}_+$. If

$$\bar{J} = \bar{J}_+,$$

i.e. the strict complementarity slackness condition holds, then, since F is smooth in a neighbourhood of $(\bar{x}, \bar{t}, \bar{y})$, we can use the standard implicit function theorem obtaining that the local minimizer and the multiplier depend smoothly on the parameter. In the general case we apply Clarke's implicit function theorem (see Clarke [3, cor. p.256]) and obtain Lipschitz dependence of the parameter.

In order to apply Clarke's implicit function theorem, one has to show that the projection $\pi_{(x,y)} \partial F(\bar{x}, \bar{t}, \bar{y})$ of the generalized Jacobian ∂F on R^{n+m} at the point $(\bar{x}, \bar{t}, \bar{y})$ is nonsingular (see Clarke [3] for the definitions).

Denote by Ω the set of points in R^{n+p+m} at which F fails to be differentiable. By definition, the generalized Jacobian of F at $(\bar{x}, \bar{t}, \bar{y})$, denoted by $\partial F(\bar{x}, \bar{t}, \bar{y})$, is the convex hull of all matrices obtained as limits of sequences of Jacobian matrices $JF(x^k, t^k, y^k)$, where $(x^k, t^k, y^k) \rightarrow (\bar{x}, \bar{t}, \bar{y})$ and $(x^k, t^k, y^k) \notin \Omega$.

Observe that F can be not differentiable with respect to y only, and Ω consists of those points (x, t, y) (in a neighbourhood of $(\bar{x}, \bar{t}, \bar{y})$) for which there is at least one $y_i = 0, i \in \bar{J} \setminus \bar{J}_+$. Then the generalized Jacobian can be obtained in the following way. Take a subset $Z \subset \bar{J} \setminus \bar{J}_+$ and compute $JF(x^k, t^k, y^k)$ supposing that $y_i^k > 0$ for $i \in Z$, and $y_i^k < 0$ for $i \notin Z$. Passing to the limit with $(x^k, t^k, y^k) \to (\bar{x}, \bar{t}, \bar{y})$ we get a matrix, corresponding to the chosen Z. The convex hull of all such matrices will form the generalized Jacobian.

Taking into account that F is smooth with respect to t, we conclude that the projection $\pi_{(x,y)} \partial F(\bar{x}, \bar{t}, \bar{y})$ will be obtained in the same way as the generalized Jacobian, replacing the Jacobian JF by the matrix of the partial derivatives of F with respect to (x, y).

Without loss of generality, let

$$\bar{J} = \{1, ..., l\} \text{ and } \bar{J}_+ = \{1, ..., k\}, k < l.$$

Denote

$$B_+ = (\nabla_x f_1(\bar{z}), ..., \nabla_x f_k(\bar{z})),$$

$$B_1 = (\nabla_x f_1(\bar{z}), ..., \nabla_x f_m(\bar{z})).$$

Applying the procedure described above we present the projection of the generalized Jacobian in the form

$$\pi_{(x,y)} \partial F(\bar{x}, \bar{t}, \bar{y}) = \text{ co } \{N(J) : J \subset \bar{J} \setminus \bar{J}_+\},$$

where the matrices $N(J)$ are determined as follows:
the first n columns of $N(J)$ are the derivatives of F with respect to x;
the next k columns are the derivatives of F with respect to y_j when $j \in \bar{J}_+$;
the $j-$ th column of $N(J)$ for $j > n + k$ is either

$$(\nabla_x f_{j-n}(z)', 0, ..., 0)' \text{ if } j - n \in J,$$

or

$$(\underbrace{0, ..., 0}_{j-n}, 1, 0, ..., 0)' \text{ otherwise;}$$

the $j-$ th column for $j > n + l$ is

$$(\underbrace{0, ..., 0}_{j-n}, 1, 0, ..., 0)'.$$

Take $J \subset \bar{J} \setminus \bar{J}_+$ and let $B_0(J)$ be a matrix whose $(i - k)-$ th column is either $\nabla_x f_i(\bar{z})$ if $i \in J$, or $0 \in R^n$ otherwise. Let the diagonal matrix $I_0(J)$ of dimension $l - k$ have 0 as $(i - k)-$ th element at the diagonal if $i \in J$ and 1 otherwise. Using this notation, we have

$$N(J) = \begin{pmatrix} H & \vdots & B_+ & B_0(J) & \vdots & 0 \\ \cdots\cdots\cdots\cdots\cdots\cdots\cdots\cdots\cdots\cdots \\ & \vdots & 0 & 0 & \vdots & 0 \\ -B_1' & \vdots & 0 & I_0(J) & \vdots & 0 \\ & \vdots & 0 & 0 & \vdots & I \end{pmatrix},$$

where I is always the identity matrix and H is defined by (95).

Let the matrix $B(J)$ have as columns the vectors $\nabla_x f_i(\bar{z}), i \in J$, and let

$$M(J) = \begin{pmatrix} H & B_+ & B(J) \\ -B'_+ & 0 & 0 \\ -B'(J) & 0 & 0 \end{pmatrix}.$$

Reordering the last $m - k$ columns of $N(J)$, if necessary, we have

$$\det N(J) = \det \begin{pmatrix} H & B_+ & B(J) & 0 \\ -B'_1 & 0 & 0 & I \end{pmatrix} = \det M(J).$$

Now we need the following

31 Lemma The matrix $M(J)$ is nonsingular and $\det M(J)$ has constant sign for all nonempty $J \subset \bar{J} \setminus \bar{J}_+$.

Taking the lemma for granted we have

(97) $\qquad \det N(J) \neq 0$ and sign $\det N(J) = $ const. for all $J \subset \bar{J} \setminus \bar{J}_+$.

If

$$N \in \text{co}\ \{N(J) : J \subset \bar{J} \setminus \bar{J}_+\},$$

then N can be represented as

$$N = \sum_J \eta_J N(J),$$

where $\eta_J \geq 0, \sum_J \eta_J = 1$, taking the sums over all $J \subset \bar{J} \setminus \bar{J}_+$.

Fix some $i \in \bar{J} \setminus \bar{J}_+$. Let λ_i be the sum of those η_J for which $i \in J$. Then the $(n+i)$-th column of N is

$$\begin{pmatrix} \lambda_i \nabla_x f_i(\bar{z}) \\ 0 \\ \vdots \\ 1 - \lambda_i \\ \vdots \\ 0 \end{pmatrix}$$

Let

$$\Lambda = \text{diag}\ \{\lambda_i : i = 1, ..., l - k\}.$$

Then

$$N = N(\Lambda) = \begin{pmatrix} H & B_+ & B_0\Lambda & 0 \\ 0 & 0 & 0 & 0 \\ -B'_1 & 0 & I_0 - \Lambda & 0 \\ 0 & 0 & 0 & I \end{pmatrix}$$

where $B_0 = (\nabla_x f_{k+1}(\bar{z}), ..., \nabla_x f_l(\bar{z}))$ and I_0 is the $(l - k)$-th identity matrix.

Each column of $N(\Lambda)$ depends on at most one λ_i, hence there are numbers

$$a_1, b_1, a_2, b_2, .., a_{l-k}, b_{l-k}$$

such that

$$\det N(\Lambda) = (a_1 + \lambda_1 b_1)...(a_{l-k} + \lambda_{l-k} b_{l-k}).$$

Taking each λ_i either 0 or 1, from $N(\Lambda)$ we obtain all matrices $N(J)$. From (97) we conclude that for $i = 1, ..., l - k$,

$$a_i \neq 0, a_i + b_i \neq 0, \text{ and sign } a_i = \text{ sign } (a_i + b_i).$$

Therefore

$$\det N(\Lambda) \neq 0.$$

Thus we can apply Clarke's implicit function theorem according to which there exist a neighbourhood U of $\bar{t}$ and a Lipschitz function $(\bar{x}(\cdot), \bar{y}(\cdot)) : U \to R^{n+m}$ such that for every $t \in U, F(\bar{x}(t), t, \bar{y}(t)) = 0$. Therefore (taking U smaller if necessary), the linear independent constraints qualification (91) holds at $(\bar{x}(t), t)$, moreover

$$\bar{u}_i(t) = \bar{y}_i(t), i = 1, ..., s, \bar{u}_i(t) = \max \{0, \bar{y}_i(t)\}, i = s + 1, ..., m$$

will be the corresponding unique Lagrange multiplier and the strong second- order sufficient condition (94) will hold at $(\bar{x}(t), t, \bar{u}(t))$. The proof of theorem 30 is complete. $\square$

Proof of lemma 31 . Take some nonempty $J \subset \bar{J} \setminus \bar{J}_+$, and for simplicity let $J = \{k + 1, ..., p\}$.
Step 1: $\det M(J) \neq 0$.
   Assume the opposite. Then there exists a nonzero vector $(v', w', q')' \in R^{n+p}$ such that

(98) $$Hv + B_+ w + B(J)q = 0,$$
(99) $$- B'_+ v = 0,$$
(100) $$- B'(J)v = 0.$$

Multiplying (98) by v' from the left and using (99) and (100) we get $v'Hv = 0$. Hence, by (94), $v = 0$. Then, from (98)

$$B_+ w + B(J)q = 0$$

i.e. $(w, q) \in \ker (B_+, B(J))$. However by (91),

$$\ker (B_+, B(J)) \subset \ker (B_+, B_0) = \{0\},$$

which contradicts $(v', w', q')' \neq 0$. We obtain in particular that the matrix

$$M(\emptyset) = M_+ = \begin{pmatrix} H & B_+ \\ -B'_+ & 0 \end{pmatrix}$$

is nonsingular. Define

$$V(J) = (B(J)' \ 0) : R^{n+k} \to R^{p-k}.$$

Step 2 . The matrix

$$S(J) = V(J)M_+^{-1}V(J)'$$

is positive definite.

It is sufficient to present a proof for $J = \{k + 1, ..., l\}$, i.e. for $p = l$. Since M_+ is nonsingular there exist matrices P and Q such that

$$M_+ \begin{pmatrix} P \\ Q \end{pmatrix} = \begin{pmatrix} B_0 \\ 0, \end{pmatrix}$$

that is

(101) $$HP + B_+Q = B_0,$$
(102) $$B'_+P = 0.$$

Then

$$S(\{k + 1, ..., l\}) = S = (B'_0 \ 0) \begin{pmatrix} P \\ Q \end{pmatrix} = B'_0 P.$$

Multiplying (101) by P' from the left and using (102) we obtain

$$S = P'HP.$$

Suppose that there exists $z \neq 0$ such that $z'Sz \leq 0$. Then, denoting $y = Pz$, one has $y'Hy \leq 0$. Multiplying (102) by z from the right we have $B'_+y = 0$. Thus, the strong second-order sufficient condition (94) implies $y = 0$. Therefore $z \in \ker P$ and then (101) yields

$$B_+Qz = B_0z,$$

i.e.

$$(B_+ \ B_0) \begin{pmatrix} Qz \\ -z \end{pmatrix} = 0,$$

hence $z = 0$ by (91), a contradiction. This completes the proof of Step 2.

End of the proof. From the equality

$$\begin{pmatrix} M_+^{-1} & 0 \\ V(J)M_+^{-1} & I \end{pmatrix} \begin{pmatrix} M_+ & V'(J) \\ -V(J) & 0 \end{pmatrix} = \begin{pmatrix} I & M_+^{-1}V'(J) \\ 0 & S(J) \end{pmatrix},$$

taking into account that

$$M(J) = \begin{pmatrix} M_+ & V'(J) \\ -V(J) & 0 \end{pmatrix},$$

we conclude that

$$\det M_+^{-1} \det M(J) = \det S(J).$$

By step 2, $\det S(J) > 0$ for any $J \subset \bar{J} \setminus \bar{J}_+$, and this completes the proof. $\square$

Section 7.   Linear Programming.

We consider the linear programming problem

(103)   minimize $c'x$ subject to $Ax \leq b, x \in R^n$.

Here $c \in R^n$, the matrix $A \in R^{m \times n}$ and the vector $b \in R^m$ are the problem's data. We are interested in the Lipschitz behaviour of both

$$\text{val } (b) = \inf \{c'x : Ax \leq b\},$$

$$\text{arg min } (b) = \{x \in R^n : Ax \leq b, c'x \leq \text{ val } (b)\},$$

having fixed c and A (a strong form of Hadamard well-posedness with respect to b).

We shall denote the multifunction of the feasible points in (103) by

(104) $F(b) = \{x \in R^n : Ax \leq b\}$.

A subset $S \subset R^n$ is a *convex polyhedral set* iff it is an intersection of finitely many halfspaces, i.e. there exist an integer p, a matrix $D \in R^{p \times n}$ and a vector $q \in R^p$ such that

(105) $x \in S$ iff $x \in R^n$ and $Dx \leq q$.

A multifunction acting between Euclidean spaces is called *convex polyhedral* iff its graph is a convex polyhedral set. This property is shared by the multifunction F defined by (104).

It turns out that convex polyhedral multifunctions have a Lipschitz property. For $T \subset R^m$, a multifunction $G : T \rightrightarrows R^n$ is called *Lipschitz* (of rank L) iff there exists $L > 0$ such that for every $t', t'' \in \text{dom } G$ (i.e. $G(t'), G(t'')$ nonempty)

$$G(t') \subset G(t'') + L|t' - t''|B.$$

Here $|\cdot|$ is the Euclidean norm and B is the Euclidean unit ball.

32 Lemma If $G : R^m \rightrightarrows R^n$ is a convex polyhedral multifunction, then it is Lipschitz.

By lemma 32, the multifunction (104) is Lipschitz. Now let $b_1, b_2 \in \text{dom } F$ be such that

$$\text{val } (b_i) > -\infty, i = 1, 2.$$

Then there exist optimal solutions x_i to (103) corresponding to $b_i, i = 1, 2$. Since F is Lipschitz, there exist $L > 0$ and $u_i \in F(b_i)$ such that

$$|u_1 - x_2| \leq L|b_1 - b_2|, \ |u_2 - x_1| \leq L|b_1 - b_2|.$$

Hence

$$\text{val } (b_1) - \text{ val } (b_2) = c'x_1 - c'x_2 \leq c'u_1 - c'x_2 \leq L|c||b_1 - b_2|.$$

Exchanging the roles of b_1, b_2 we conclude that $b \to \text{ val } (b)$ is Lipschitz whenever finite-valued.

Consider now the multifunction $G : R^{m+1} \rightrightarrows R^n$ defined by

$$G(b,t) = \{x \in R^n : Ax \leq b, c'x \leq t\}.$$

Since G is a convex polyhedral multifunction, by lemma 32 G is Lipschitz (on its domain). Observe that if $b \in$ dom F and val $(b) > -\infty$

$$\arg \min (b) = G[b, \text{val } (b)].$$

Since val $(\cdot)$ is Lipschitz, arg min $(\cdot)$ is Lipschitz as well. So we get

33 Theorem *The value and arg min of the linear program (103) are Lipschitz continuous with respect to b on their (effective) domains.*

Theorem 33 cannot be generalized (without adding regularity assumptions) to well-posedness under changes of A : consider

$$\text{val } (a) = \inf\{x_1 : x_1 + ax_2 \geq 1, x_1 \geq 0, x_2 \geq 0\}.$$

Then val $(a) = 1$ if $a \leq 0$ but val $(a) = 0$ if $a > 0$.

PROOFS.

The proof of lemma 32 is based on the following fundamental result concerning approximate solutions to systems of linear inequalities. Given $a \in R^n$ put

$$a_+ = (\max (0, a_1), ..., \max (0, a_n))'.$$

34 Lemma Let A be a nonzero $n \times m$ matrix. Define $F(b)$ as in (104), $b \in R^m$. Then there exists a constant $L > 0$ (depending on A only) such that for every $b \in$ dom F and every $y \in R^n$

$$\text{dist } [y, F(b)] \leq L|(Ay - b)_+|.$$

Proof of lemma 34 Let $b \in$ dom F and $y \notin F(b)$. Denote by $a_1, a_2, ..., a_m$ the rows of A. The problem of minimizing $|x - y|^2$ subject to $x \in F(b)$ has an unique solution u (the best approximation to y from the closed convex set $F(b)$). By Kuhn-Tucker's theorem (see Mc Cormick [1, th. 2 p. 210]) there exist numbers $\lambda_1, \lambda_2, ..., \lambda_m \geq 0$ such that

$$(106) \qquad y - u = \sum_{i=1}^{m} \lambda_i a_i, \ \lambda_i(a_i'u - b_i) = 0 \text{ for all } i.$$

Now consider the family Z of all subset $J \subset \{1, ..., m\}$ such that $J \neq \emptyset$ and there exist numbers $v_i > 0$ for each $i \in J$ fulfilling

$$y - u = \sum\{v_i a_i : i \in J\}.$$

Since $u \neq y$, some $\lambda_i > 0$ in (106), hence Z is nonempty. We claim that there exists $J \in Z$ with

$$(107) \qquad 0 \notin \text{ co } \{a_i : i \in J\}.$$

Arguing by contradiction, suppose that

(108) $$0 \in \text{co}\,\{a_i : i \in J\}\,\forall J \in Z.$$

Let $J' \in Z$ be such that

(109) $$\text{card}\,J' = \min\{\,\text{card}\,J : J \in Z\}.$$

Notice that card $J' \geq 2$ since $u \neq y$ and by (108). Since $J' \in Z$ and by (108), there exist numbers $v_i > 0$ and $w_i \geq 0$ for each $i \in J'$, $\sum\{w_i : i \in J'\} = 1$, such that

$$y - u = \sum\{v_i a_i : i \in J'\}, 0 = \sum\{w_i a_i : i \in J'\},$$

hence

$$y - u = \sum\{(v_i - tw_i)a_i : i \in J'\}\,\forall\,t \in R.$$

Put

$$t_0 = \min\{\frac{v_i}{w_i} : i \in J'\ \text{and}\ w_i > 0\},$$

then

$$v_k - t_0 w_k = 0\ \text{for some}\ k \in J', v_i - t_0 w_i \geq 0\ \text{for every}\ i \in J'.$$

Thus there exists an index set $J'' \subset J' \setminus \{k\}$ with $J'' \in Z$, card $J'' <$ card J' ($J'' \neq \emptyset$ since $y \neq u$). This contradicts (109) and establishes the existence of $J \in Z$ fulfilling (106) with $\sum_{i=1}^{m}$ replaced by $\sum_{i \in J}$ and every $\lambda_i > 0$. Put now

(110) $$\lambda = \sum\{\lambda_i : i \in J\} > 0.$$

Then from (106) and (110) we have (all sums extend over $i \in J$)

(111) $$|y - u|\ \text{dist}\,(0,\ \text{co}\,\{a_i : i \in J\}) \leq |y - u||\sum \frac{\lambda_i a_i}{\lambda}|$$

$$= \frac{1}{\lambda}(y - u)'(y - u)$$

$$= \sum \frac{\lambda_i}{\lambda}(a_i'y - b_i) \leq \max\,\{(a_i'y - b_i)_+ : i \in J\}.$$

Now dist $(0,\ \text{co}\,\{a_i : i \in J\})$ is bounded away from zero by the least distance of 0 from co $\{a_i : i \in I\}$ as I ranges over all subsets of $\{1, ..., m\}$ obeying $0 \notin$ co $\{a_i : i \in I\}$. Therefore, by (111) and (107), for some constant $C > 0$ we get

$$\text{dist}\,[y, F(b)] = |y - u| \leq C \max_i[(Ay - b)_i]_+.$$

$\square$

Proof of lemma 32. Take $t_1, t_2 \in$ dom G. By definition of convex polyhedral multifunction, there exist a positive integer p, matrices $D \in R^{p \times m}, E \in R^{p \times n}$ and a vector $q \in R^p$ such that for every $t \in R^m$

$$G(t) = \{x \in R^n : Ex + Dt \leq q\}.$$

Let $z \in G(t_2)$. Applying lemma 34 we get

$$\text{dist}\,[z, G(t_1)] \leq L|(Ez + Dt_1 - q)_+|$$
$$\leq L|(Dt_1 - Dt_2)_+| \leq (\text{const.})\,|t_1 - t_2|.$$

This yields the required Lipschitz property for G.

$\square$

Section 8.   Notes and Bibliographical Remarks.

To section 1. Propositions 1, 2 and theorem 3 are extensions of the classical Berge theorem, see Berge [1, ch.VI,3] (see also Aubin [1, sec.2.5]). Results related to propositions 1 and 2 (in a sequential setting) are in Dolecki [2], Lignola-Morgan [1] and (in a topological setting) in Penot [1] and Dolecki [1]. For a local version of theorem 3 (assuming epi-upper semicontinuity of f) see Robinson [6].

Theorem 5 is due to Berdišev [4], and theorem 8 is extracted from Bednarczuk [1]. For generalizations, see Bednarczuk [2]. Theorem 9 is obtained partly on the basis of Dolecki- Rolewicz [1], Bednarczuk [1] and Lucchetti [3]. Theorems 10 and 12 are due to Lucchetti [4]. For a survey (with many extensions and further results) see Lucchetti [7].

Upper semicontinuity of arg min is obtained when $f = f(x, y)$ and $A(y) = X$ (the opposite as assumed in theorem 10) by Furi-Martelli-Vignoli [1] if $f(\cdot, y) \to f(\cdot, z)$ uniformly on X whenever $y \to z$, and every problem $[X, f(\cdot, y)]$ is well-posed in the generalized sense. Proposition 13 is due to Cavazzuti-Pacchiarotti [1] (in the setting of uniform spaces). Necessary or sufficient well- posedness conditions were obtained by Dantzig-Folkman-Shapiro [1].

A survey on parametric programming is presented in Bank-Guddat-Klatte-Tammer [1]. For semi-infinite programming see Brosowski [1]. For two-level optimization problems see Loridan-Morgan [2].

To section 2. Propositions 14 and 15 are essentially from Evans-Gould [1], who however assume that $I(y_0) \neq \emptyset$ and $A(y_0)$ is compact (in this case condition (31) reduces to the existence of $y_1 > y_0$ such that $A(y_1)$ is compact).

Extensions of the Evans-Gould results (in a compact setting) are given in Greenberg-Pierskalla [1] and Hogan [1].

Theorem 18 is due to Robinson [1] (with a somewhat simplified proof). Condition (33) is introduced by Mangasarian-Fromovitz [1]. Condition (34) (as well as lemma 20) is related to the metric regularity of the multifunction representing the feasible set, a basic notion in the set-valued analysis. For related results see Cominetti [1], Borwein [2], Borwein-Zhuang [1], Penot [2] and the references therein.

For a qualitative definition of stability of the feasible set see Guddat- Jongen-Rueckmann [1]. Conditions for local Lipschitz continuity of the value function are given in Aubin [3].

A different approach to sensitivity and stability has been developed based on non-smooth analysis and differential properties of multifunctions, see Aubin [4], Rockafellar [12] and [13], where Lipschitz properties of multifunctions appearing in mathematical programming are investigated (see also Rockafellar [11]).

To section 3. Theorem 21 and corollaries 22, 23 are generalizations from Zolezzi [11], where further results are given about solution well-posedness. For related results see Azé [1] and [2], Azé - Attouch - Wets [1], Back [1], and Lucchetti-Patrone [5] in the finite-dimensional setting. For a survey see Zolezzi [14]. Well-posedness in the duality framework of Ekeland-Temam [1] is obtained by Azé [2, ch. 1].

For parametric convex problems in finite dimension, regions of the parameter space in which value and solution Hadamard well-posedness hold (through sufficient conditions for lower semicontinuity of the feasible set) are studied in Zlobec-Gardner-Ben-Israel [1], see also Zlobec [1].

Classical notions of stability in convex programming deal with value Hadamard well-posedness, and are deeply related to duality theory, see Laurent [1, ch. 7], Barbu-Precupanu [1, ch. 3]. The stable behavior implied by lack of the duality gap can be obtained by suitable modifications of the Lagrangean, see Duffin-Jeroslow [1]. For vector optimization problems see Tanino [1].

To section 4. Theorem 25 and corollary 26 are due to Zolezzi [12], where related results may be found. A survey is contained in Zolezzi [14]. Continuity of the optimal value function, the optimal solutions and multipliers (under the Mangasarian-Fromovitz constraint qualification condition) is obtained in Gauvin-Tolle [1], see also Gauvin [1] and Gauvin-Dubeau [1]. A notion of stability gap in nonconvex programming was introduced by Rockafellar [5, sec. 4].

For smooth problems, an interesting definition of *structural stability* has been introduced by Guddat-Jongen [1], and characterized in Jongen-Weber [1]. It is based on an equivalence relation which requires homeomorphic sublevel sets, and imposes equivalence of all nearby problems (in the Withney C^2- topology); some form of stable behavior of the Kuhn-Tucker points ensues.

To section 5. Proposition 28 is due to Attouch-Wets [4] where further Lipschitz continuity results involving the epigraphical distances $d(r, \cdot, \cdot)$ may be found. Proposition 29 comes from Attouch-Wets [5], which contains quantitative estimates for local optimal solutions and optimal values via forcing functions (using a local version of theorem I.12). See also Attouch-Wets [6] for more results. Results related to proposition 29 are in Daniel [2].

Lipschitz estimates for the Hausdorff distance of arg min (A, I) to arg min (B, I) (where A, B and I are convex) in terms of haus (A, B) and $|$ val $(A, I) - $ val $(B, I)|$ are proved in Berdišev [5]. Estimates of the modulus of uniform continuity of $M \rightrightarrows$ arg min (M, I) when I is uniformly convex are given in Berdišev [3]. For saddle points problems see Azé [3].

To section 6. Theorem 30 was proved originally by Robinson [3] on the basis of a theory concerning generalized equations. The proof presented here uses some ideas from Kojima [1] and the approach in Jongen-Möbert-Rueckmann- Tammer [1]. Another proof can be obtained on the basis of theorem V. 23. The relation between Robinson's theory and Kojima's strong stability concept is analyzed in Jongen-Klatte-Tammer [1]. For a discussion of second-order sufficient conditions see e.g. Fiacco [2], and more recent results in Klatte-Tammer [1] and Rockafellar [14].

Results related to theorem 30 are given in Hager [1], Aubin [4], Cornet-Laroque [1], Dontchev-Jongen [1], Rockafellar [13], Spingarn [1], Robinson [5] and Shapiro [1], [2] . Extensions to infinite - dimensional problems are given by Ito-Kunisch [1] and Dontchev-Hager [1]. For a survey about stability characterizations see Klatte-Tammer [2]. The counterexample before theorem 30 is due to B. Schwartz (see Klatte-Kummer [1, p. 59]).

As noted in the proof of theorem 30, under the constraint qualification (91) and the second-order sufficient condition (94), together with strict complementarity slackness, the local minimizers are (locally) C^1 functions of the parameter. This fact was observed in Fiacco [1]. It turns out that conditions (91) and (94) are generically satisfied in a sense given by Spingarn-Rockafellar [1]. Hence one obtains generic smoothness of arg min.

Criteria for, and characterization of, local Lipschitz continuity of the optimal value function in terms of the behavior of suitable defined Lagrange multipliers, or based on generalized second-order conditions, are given by Rockafellar [10] and [9]. For the local Lipschitz behavior of local minimizers see Auslender [1] and Alt [1].

To section 7. Hadamard ill-posedness in linear programming was pointed out by Tykhonov [3], where a regularization method can be found. The proof of Hoffman's lemma 34 is taken from Klatte [1]. Estimates for the Lipschitz constant in theorem 33 are presented in Mangasarian-Shiau [1]. Robinson [4] extended lemma 32 showing that every polyhedral multifunction is upper Lipschitz. A survey concerning Lipschitz properties of optimal solutions of mathematical programs with polyhedral structure can be found in Klatte [2].

Results on value Hadamard well-posedness for linear programs are presented in Wets [2]. See also Ashmanov [1]. A characterization of local Lipschitz continuity under changes of all data A, b, c (generalizing earlier results of Williams [1]) is given by Robinson [2] (see also Cojocaru-Dragomirescu [1]).

Few results seem to be known about *Tykhonov well-posedness* in mathematical programming. Characterizations in semi-infinite optimization are obtained by Blatt [1] and Nürnberger [1], [2]. A notion of weak sharp minima (which is somewhat related to Tykhonov well-posedness) has been introduced in Ferris [1].

REFERENCES

Example: Attouch - Wets [ch. IX] ..., means that this work is (more or less directly) related to chapter IX of this book, however it is not mentioned explicitly in the text.

Remark. Polyak and Tykhonov are transliterated as Poliak, Tikhonov, Tyhonov etc. in some references below.

ABDYRAKHMANOV, O. - KRYAZHIMSKI, A.V. [1] Correctness of an optimal control problem. *Differential Equations* 20 (1984), 1179-1184.

ACERBI, E. - BUTTAZZO, G. [1] On the limits of periodic Riemannian metrics. *J. Analyse Math.* 43 (1983/84), 183-201.

ADAMS, R.A. [1] **Sobolev spaces.** Academic Press, 1975.

ALART, P. - LEMAIRE, B. [ch. IV] Penalization in non classical convex programming via variational convergence. Preprint, 1989.

ALT, W. [1] Stability of solutions for a class of nonlinear cone constrained optimization problems, part I: basic theory. *Numer. Funct? Anal. Optim.* 10 (1989), 1053-1064. [2] Stability of solutions to control constrained nonlinear optimal control problems. *Appl. Math. Optim.* 21 (1990), 53-68.

AMBROSETTI, A. - SBORDONE, C. [1] Γ^- convergenza e G convergenza per problemi non lineari di tipo ellittico. *Boll. Un. Mat. Ital.* 13 A (1976), 352-362.

ANDERSON, B.D.O. - MOORE, J.B. [1] **Linear optimal control.** Prentice - Hall, 1971.

ANGELOS, J.R. - SCHMIDT, D. [1] The prevalence of strong uniqueness in L^1. *Acta Math. Hungar.* 52 (1988), 83-90.

ANSELONE, P.M. [1] **Collectively compact operators, approximation theory and applications to integral equations.** Prentice Hall, 1971.

ANTIPIN, A.S. [ch.I] A unified approach to the methods of solving ill-posed extremum problems. *Moscow Univ. Math. Bull.* 28 (1973), n.2, 90-96.

ARNOLD, V.I. [1] **Ordinary differential equations.** The MIT Press, 1978.

ARTSTEIN, Z. [1] Weak convergence of set-valued functions and control. *SIAM J. Control* 13 (1976), 865-877. [ch.V] Rapid oscillations, chattering systems and relaxed controls. *SIAM J. Control Optim.* 27 (1989), 940-948. [ch.V] A variational convergence that yields chattering systems. *Ann. Inst. H. Poincaré, Anal. Non Linéaire*, suppl. vol. 6 (1989), 49-71.

ASCOLI, G. [1] Sugli spazi lineari metrici e le loro varietá lineari. *Ann. Mat. Pura Appl.* 10 (1932), 33-81.

ASHMANOV, S.A. [1] Stability conditions for linear programming problems. *USSR Comput. Math. and Math. Phys.* 21 (1981), n.6, 40-49.

ASPLUND, E. [1] Fréchet differentiability of convex functions. *Acta Math.* 121 (1968), 31-47. [2] Čebišev sets in Hilbert space. *Trans. Amer. Math. Soc.* 144 (1969), 235-240. [3] Topics in the theory of convex functions. *Theory and applications of monotone operators*, edited by A. Ghizzetti. Oderisi, 1969, 1-33.

ASPLUND, E. - ROCKAFELLAR, R.T. [1] Gradients of convex functions. *Trans. Amer. Math. Soc.* 139 (1969), 443-467.

ATTOUCH, H. [1] Convergence des solutions d'inequations variationelles avec obstacle. *Proceedings international meeting on recent methods in nonlinear analysis* (Rome 1978). Pitagora, 1979, 101-113. [2] Familles d'operateurs maximaux monotones et mesurabilité. *Ann. Mat. Pura Appl.* 120 (1979), 35-111. [3] **Variational convergence for functions and operators.** Pitman, 1984. [ch.VIII, IX] Epi-convergence and duality. Convergence of sequences of marginal and lagrangian functions. Applications to homogenization problems in mechanics. *Lecture Notes in Math.* 1190, Springer 1986, 21-56.

ATTOUCH, H. - AZÉ, D. - WETS, R. [1] On continuity properties of the partial Legendre - Fenchel transform: convergence of sequences of augmented Lagrangian functions, Moreau - Yosida approximates and subdifferential operators. *North-Holland Math. Stud.* 129 (1986), 1-42.

ATTOUCH, H. - RIAHI, H. [1] Stability results for Ekeland's variational principle. *Sém. Anal. Convexe* (1988), exp.6.

ATTOUCH, H. - SBORDONE , C. [1] Asymptotic limits for perturbed functionals of calculus of variations. *Ricerche Mat.* 29 (1980), 85-124.

ATTOUCH, H. - WETS, R. J-B. [1] Approximation and convergence in nonlinear optimization. *Nonlinear Programming* 4, edited by O. Mangasarian - R. Meyer - S. Robinson, Academic Press, 1981, 367-394. [2] A convergence theory for saddle functions. *Trans. Amer. Math. Soc.* 280 (1983), 1-41. [3] A convergence for bivariate functions aimed at the convergence of saddle values. *Lecture Notes in Math.* 979, Springer 1983, 1-42. [4] Lipschitzian stability of ε-approximate solutions in convex optimization. *IIASA working paper* 87-25 (1987). [5] A quantitative approach via epigraphic distance to stability of strong local minimizers. *Publications AVAMAC*, Univ. de Perpignan 1-1987 (1987). [ch.IX] Quantitative stability of variational systems: I. The epigraphical distance. *IIASA working paper* 88-8 (1988). To appear in Trans. Amer. Math. Soc. [6] Quantitative stability of variational systems : II. A framework for nonlinear conditioning. *IIASA working paper* 88 - 9 (1988). [ch.IX] Epigraphical analysis. *Ann.Inst. H. Poincaré Anal. Non Linéaire*, suppl. vol. 6(1989), 73-100.

AUBIN, J.-P. [1] **Mathematical methods of game and economic theory.** North Holland, 1979. [2] **Applied functional analysis.** Wiley-Interscience, 1979. [3] Further properties of Lagrange multipliers in nonsmooth optimization. *Appl. Math. Optim.* 6 (1980), 79-90. [4] Lipschitz behavior of solutions to convex minimization problems. *Math. Oper. Res.* 9 (1984), 87-111.

AUBIN, J.-P. - CELLINA, A. [1] **Differential inclusions.** Springer, 1984.

AUBIN, J.- P. - EKELAND, I. [1] **Applied nonlinear analysis**. Wiley, 1984.

AUMANN, R. [1] Integrals of set-valued functions. *J. Math. Anal. Appl.* 12 (1965), 1-12.

AUSLENDER, A. [1] Stability in mathematical programming with nondifferentiable data. *SIAM J. Control Optim.* 22 (1984), 239-254.

AVGERINOS, E.P. - PAPAGEORGIOU, N.S. [ch. V] Infinite horizon optimal control problems for semilinear evolution equations. *Appl. Anal.* 31 (1989), 267-268. [1] Variational stability of infinite-dimensional optimal control problems. *Int. J. Systems Sc.* 21 (1990), 1473-1488. [2] On the sensitivity and relaxability of optimal control problems governed by nonlinear evolution equations with state constraints. *Monatsh. Math.* 109 (1990), 1-23.

AZE', D.[1] Stability results in convex programming. *Publications AVAMAC*, Univ. de Perpignan 85-04 (1985). [2] Convergences varationelles et dualité. Applications en calcul des variations et en programmation mathematique. *Thése de doctorat d'etat*, Univ. de Perpignan, 1986. [3] Rate of convergence for the saddle points of convex-concave functions. *Internat. Series Numer. Math. 84*, Birkhäuser 1988, 9-23.

AZE', D. - ATTOUCH, H. - WETS, R. [1] Convergence of convex-concave saddle functions: applications to convex programming and mechanics. *Ann. Inst. H. Poincaré, Anal. Non Linéaire* 5 (1988), 537-572.

BACCIOTTI, A. - BIANCHINI, R.M. [1] The linear time optimal problem revisited: regularization and approximation. *Boll. Un. Mat. Ital.* 4 B (1990), 411-422.

BACK, K. [1] Convergence of Lagrange multipliers and dual variables for convex problems. *Math. Oper. Res.* 13 (1988), 74-79.

BAIOCCHI, C. - CAPELO, A. [1] **Variational and quasi variational inequalities.** Wiley, 1984.

BALL, J. - KNOWLES, G. [1] A numerical method for detecting singular minimizers. *Numer. Math.* 51 (1987), 181-197.

BALL, J. - MARSDEN, J.E. [1] Quasiconvexity at the boundary, positivity of the second variation and elastic stability. *Arch. Rational Mech. Anal.* 86 (1984), 251-277.

BALL, J. - MIZEL, V. [1] One-dimensional variational problems whose minimizers do not satisfy the Euler-Lagrange equation. *Arch. Rational Mech. Anal.* 90 (1985), 325-388.

BANK, B. - GUDDAT, J. - KLATTE, D. - KUMMER, B. - TAMMER, K. [1] **Nonlinear parametric optimization.** Akademie Verlag, 1982.

BARANGER, J. [1] Existence des solutions pour des problémes d'optimisation non convexe. *J. Math Pures Appl.* 52 (1973), 377-405.

BARANGER, J. - TEMAM, R. [1] Nonconvex optimization problems depending on a parameter. *SIAM J. Control* 13 (1975), 146-152.

BARBU, V. - PRECUPANU, T. 1] **Convexity and optimization in Banach spaces.** Academiei - D. Reidel, 1986.

BARDI, M. - SARTORI, C. [1] Approximation and regular perturbation of optimal control problems via Hamilton - Jacobi theory. *Appl. Math. Optim.* 24 (1991), 113-128.

BARTELT, M.W. - MC LAUGHLIN, H.W. [1] Characterization of strong unicity in approximation theory. *J. Approx. Theory* 9 (1973), 255-266.

BEDNARCZUK, E. [1] On upper semicontinuity of global minima in constrained optimization problems. *J. Math. Anal. Appl.* 86 (1982), 309-318. [2] Characterization of semicontinuity of solutions to abstract optimization problems. *Numer. Funct. Anal. Optim.* 9 (1987), 685-708.

BEDNARCZUK, E. - PENOT, J.P. [ch.1] On the positions of the notions of well-posed minimization problems. To appear on Boll.Un. Mat. Ital. [1] Metrically well-set minimization problems. To appear on Appl. Math. Optim.

BEER, G. [1] On Mosco convergence of convex sets. *Bull. Austral. Math. Soc.* 38 (1988), 239-253. [2] On a generic optimization theorem of Petar Kenderov. *Nonlin. Anal.* 12 (1988), 647-655. [3] On the Young-Fenchel transform for convex functions. *Proc. Amer. Math. Soc.* 104 (1988), 1115-1123. [4] Support and distance functionals for convex sets. *Numer. Funct. Anal. Optim.* 10 (1989), 15-36 [ch.IV] Infima of convex functions. *Trans. Amer. Math. Soc.* 315 (1989), 849-859. [5] Three characterizations of the Mosco topology for convex functions. *Arch. Math.* 55 (1990), 285-292.

BEER, G. - BORWEIN, J.M. [ch. IV] Mosco convergence and reflexivity. *Proc. Amer. Math. Soc.* 109 (1990), 427-436.

BEER, G. - KENDEROV, P. [1] On the argmin multifunction for lower semicontinuous functions. *Proc. Amer. Math. Soc.* 102 (1988), 107-113. [2] Epiconvergence and Baire category. *Boll. Un. Mat. Ital.* 3 B (1989), 41-56.

BEER, G. - LUCCHETTI, R. [1] Minima of quasi-convex functions. *Optimization* 20 (1989), 581-596. [2] The epi-distance topology: continuity and stability results with applications to convex optimization problems. To appear in Math. Oper. Res. [3] Convex optimization and the epi-distance topology. *Trans. Amer. Math. Soc.* 327 (1991), 795-813. [4] Solvability for constrained problems. *Quaderno* 3/1991, *Dipart. Matem. Univ. Milano*, 1991. [ch. I, II] Well-posed optimization problems and a new topology for the closed subsets of a metric space. *Quaderno* 24/1991, *Dipart. Matem. Univ. Milano*, 1991.

BENNATI, M.L. [1] Perturbazioni variazionali di problemi di controllo. *Boll. Un. Mat. Ital.* 16 B (1979), 910-922. [2] On the convergence of dual variables in optimal control problems. *J. Optim. Theory Appl.* 34(1981), 263-278.

BENSOUSSAN, A. [1] **Perturbation methods in optimal control.** Wiley and Gauthier-Villars, 1988.

BERDYŠEV, V. [1] Operator of best approximation of finite-dimensional subspaces. *Math. Notes* 16 (1974), 888-893. [2] On the modulus of continuity of an operator of best approximation. *Math. Notes* 15 (1974), 478-484. [3] Continuous dependence of an element realizing the minimum of a convex functional on the set of admissible elements. *Math. Notes* 19 (1976), 307-313. [4] Stability of a minimization problem under perturbation of the set of admissible elements. *Math. USSR Sb.* 32 (1977), 401-412. [5] Continuity of a multivalued mapping connected with the problem of minimizing a functional. *Math. USSR Izv.* 16(1981), 431-456. [6] Variations of the norm in the problem of best approximation. *Math. Notes* 29 (1981), 95-103.

BERGE, C. [1] **Espaces topologiques, fonctions multivoques.** Dunod, 1959.

BERLIOCCHI, H. - LASRY, J.M. [1] Integrandes normales et mesures paramétrées en calcul des variations. *Bull. Soc. Math. France* 101 (1973), 129-184.

BERTERO, M. [1] Problemi lineari non ben posti e metodi di regolarizzazione. *Pubbl. IAGA* n.4, CNR, 1982. [2] Regularization methods for linear inverse problems. *Lecture Notes in Math.* 1225, Springer 1986, 52-112.

BIDAUT, M.F. [1] Existence theorems for usual and approximate solutions of optimal control problems. *J. Optim. Theory Appl.* 15 (1975), 393-411.

BLATT, H.P. [1] Characterization of strong unicity in semi-infinite optimization by chain of references. *Internat. Series Numer. Math.* 72, Birkhäuser 1985, 36-46.

BOBYLEV, N.A. [1] Stability of classical solutions of variational problems of mathematical physics. *Siberian Math. J.* 26 (1985), 485-493.

BOCCARDO, L. [1] Perturbazioni di funzionali del calcolo delle variazioni. *Ricerche Mat.* 29 (1980), 213-242. [2] L^∞ and L^1 variations on a theme of Γ^- convergence. *Partial differential equations and the calculus of variations*, vol.I, Birkhäuser 1989, 135-147.

BOCCARDO, L. - CAPUZZO DOLCETTA, I. [1] Stabilita' delle soluzioni di dise-quazioni varazionali ellittiche e paraboliche quasi lineari. *Ann. Univ. Ferrara* 24 (1978), 99-111.

BOCCARDO, L. - MARCELLINI, P. [1] Sulla convergenza delle soluzioni di disequazioni variazionali. *Ann. Mat. Pura Appl.* 110(1976), 137-159.

BOCCARDO, L. - MURAT, F. [1] Nouveaux résultats de convergence dans des problémes unilatéraux. *Research Notes in Math.* 60, Pitman 1982, 64-85.

BONNANS, J.F. - LAUNAY, G. [ch. IX] On the stability of sets defined by a finite number of equalities and inequalities. *J. Optim. Theory Appl.* 70 (1991), 417-428.

BORWEIN, J.M. [1] Weak local supportability and applications to approximation. *Pacific J. Math.* 82 (1979), 323-338. [2] Stability and regular points of inequality systems. *J. Optim. Theory Appl.* 48 (1986), 9-52.

BORWEIN, J.M. - FITZPATRICK, S. [1] Mosco convergence and the Kadec property. *Proc. Amer. Math. Soc.* 106 (1989), 843-851. [2] Existence of nearest points in Banach spaces. *Canad. J. Math.* 41 (1989), 702-720.

BORWEIN, J.M. - STROJWAS, H.M. [1] Proximal analysis and boundaries of closed sets in Banach space, part II: applications. *Canad. J. Math.* 39(1987), 428-472.

BORWEIN, J.M. - ZHUANG, D.M. [1] Verifiable necessary and sufficient conditions for opennes and regularity of set-valued and single-valued maps. *J. Math. Anal. Appl.* 134 (1988), 441-459.

BOURGIN, R.D. [1] **Geometric aspects of convex sets with the Radon- Nikodým property.** Lecture Notes in Math. 993, Springer 1983.

BRAESS, D. **Nonlinear approximation theory.** Springer, 1986.

BREZIS, H. [1] **Operateurs maximaux monotones et semigroupes de contrac-tions dans les espaces de Hilbert.** North Holland, 1973. [2] **Analyse fonctionelle. Théorie et applications.** Masson, 1983.

BROSOWSKI, B. [1] On the continuity of the optimum set in parametric semiinfinite programming. *Lecture Notes Pure Appl. Math.* 85, M. Dekker 1983, 23-48. [ch.IX] A refinement of an optimality criterion and its application to parametric programming. *J. Optim. Theory Appl.* 42 (1984), 367-382.

BROSOWSKI, B. - DEUTSCH, F. - NÜRNBERGER, G. [ch.II] Parametric approxi-mation. *J. Approx. Theory* 29 (1980), 261-277.

BROWDER, F.E. [1] Nonlinear operators and nonlinear equations of evolution in Banach spaces. *Proc. Sympos. Pure Math.* 18, part 2 (1976).

BROWN, A.L. [1] A rotund reflexive space having a subspace of codimension two with a discontinuous metric projection.*Michigan Math.* J. 21(1974), 145-151.

BRUNOVSKY, P. - KOMORNIK, J. [1] The matrix Riccati equation and the noncontrollable linear-quadratic problem with terminal constraints. *SIAM J. Control Optim.* 21 (1983), 280-288.

BUCY, R.S. [1] New results in asymptotic control theory. *SIAM J. Control* 4 (1966), 397-402.

BUDAK, B.M. - BERKOVICH, E.M. [ch. I,IV] On the approximation of extremal problems I. *USSR Comput. Math. and Math. Phys.* 11(1971), n.3, 45-66; II, same journal 11(1971), n.4, 71-88.

BUDAK, B.M. - BERKOVICH, E.M. - GAPONENKO, Y.L., [1] The construction of strongly convergent minimizing sequences for a continuous convex functional. *Zh. Vychisl. Mat. Mat. Fiz.* 9 (1969), 286-299 (russian).

BUTTAZZO, G. [1] **Semicontinuity, relaxation and integral representation in the calculus of variations.** Res. Notes in Math. 207, Pitman, 1989.

BUTTAZZO, G. - CAVAZZUTI, E.[ch.V] Limit problems in optimal control theory. *Ann. Inst. H. Poincaré Anal. Non Linéaire*, suppl. vol. 6 (1989), 151-160.

BUTTAZZO, G. - DAL MASO, G. [ch.VIII] $\Gamma-$ limits of integral functionals. *J. Analyse Math.* 37 (1980), 145-185. [1] $\Gamma-$ convergence and optimal control problems . *J. Optim. Theory Appl.* 38 (1982), 385-407. [2] Singular perturbations problems in the calculus of variations. *Ann. Scuola Normale Sup. Pisa* 11 (1984), 395-430.

BUTTAZZO, G. - FREDDI, L. [ch. V] Sequences of optimal control problems with measures as controls. Preprint, Dip. Matematica Univ. Pisa 2.82 (606), 1991.

CARBONE, L. [1] Sur un problème d'homogeneisation avec des contraintes sur le gradient. *J. Math. Pures Appl.* 58 (1979), 275-297.

CARBONE, L. - COLOMBINI, F. [1] On convergence of functionals with unilateral constraints. *J. Math. Pures Appl.* 59 (1980), 465-500.

CÂRJĂ, O. [1] On variational perturbations of control problems: minimum time problem and minimum effort problem. *J. Optim. Theory Appl.* 44 (1984), 407-433. [2] On continuity of the minimal time function for distributed control systems. *Boll. Un. Mat. Ital.* 4 A (1985), 293-302. [3] A note on admissible null controllability and on variational perturbations of the minimum time problem. *Ann. Stiint. Univ. Al. I Cuza Iasi* 32 (1986), 13-19.

CARLSON, D.A. [1] Existence of finitely optimal solutions for infinite - horizon optimal control problems. *J. Optim. Theory Appl.* 51 (1986), 41-62.

CARR, J. - GURTIN, M.E. - SLEMROD, M. [ch.VIII] One-dimensional structured phase transformations under prescribed loads. *J. Elasticity* 15 (1985), 133-142.

CAVAZZUTI, E. [1] $\Gamma-$ convergenze multiple, convergenze di punti sella e di max-min. *Boll. Un. Mat. Ital.* 1 B (1982), 251-274. [2] Convergence of equilibria in the theory of games. *Lecture Notes in Math.* 1190, Springer 1986, 95-130.

CAVAZZUTI, E. - MORGAN, J. [1] Problèmes d'optimisation uniformément bien posés et mèthodes de penalisation. *Boll. Un. Mat. Ital.* 1 B (1982), 423-450. [2] Well posed saddle point problems. *Lecture Notes Pure Appl. Math.* 86, M. Dekker 1983, p. 61-76.

CAVAZZUTI, E. - PACCHIAROTTI, N. [1] Condizioni sufficienti per la dipendenza continua dei minimi da un parametro. *Boll. Un. Mat. Ital.* 13 B (1976), 261-279. [2] Convergence of Nash equilibria. *Boll. Un. Mat. Ital.* B 5 (1986), 247-266.

CESARI, L. [1] Semicontinuita' e convessita' nel calcolo delle variazioni. *Ann. Scuola Norm. Sup. Pisa* 18 (1964), 389-423. [2] **Optimization. Theory and Applications: problems with ordinary differential equations.** Springer, 1983.

CHAVENT, G. [ch.VIII] A new sufficient condition for the well-posedness of non-linear least square problems arising in identification and control. *Lecture Notes Control Inf. Sci.* 144, Springer 1990, 452-463.

CHEN GUANG-YA. [1] On the stability of perturbations of optimal control problems. *J. Math. Anal. Appl.* 97 (1983), 46-55.

CHENEY, E.W. [1] **Introduction to approximation theory.** Mc Graw - Hill, 1966.

CHERNOUSKO, F.L. - KOLMANOVSKY, V.B. [1] Computational and approximate methods for optimal control. *Soviet Math.* 12 (1979), 310-353.

CHIADO' PIAT V. [1] Convergence of minima for non equicoercive functionals and related problems. Preprint SISSA 87 M, 1988.

CLARKE, F.H. [1] Admissible relaxation in variational and control problems. *J. Math. Anal. Appl.* 51 (1975), 557-576. [2] A new approach to Lagrange multipliers. *Math. Oper. Res.* 1 (1976), 165-174. [3] **Optimization and nonsmooth analysis.** Wiley, 1983.

CLARKE, F.H. - LOEWEN, P.D. [ch.V] The value function in optimal control: sensitivity, controllability and time - optimality. *SIAM J. Control Optim.* 24 (1986), 243-263.

ČOBAN, M.M. - KENDEROV, P.S. - REVALSKI, J.P. [1] Generic well-posedness of optimization problems in topological spaces. *Mathematika* 36 (1989), 301-324.

COJOCARU, J. - DRAGOMIRESCU, M. [1] The continuity of the optimal value of a general linear program. *Rev. Roumaine Math. Pures Appl.* 27(1982), 663-676.

COLONIUS, F. [ch.V] Asymptotic behaviour of optimal control systems with low discount rates. *Math. Oper. Res.* 14 (1989), 309-316.

COLONIUS, F. - KUNISH, K. [ch.IV] Stability of perturbed optimization problems with applications to parameter estimation. *Numer. Funct. Anal. Optim.* 11 (1990/91), 873-915.

COMINETTI, R. [1] Metric regularity, tangent sets and second-order optimality conditions. *Appl. Math. Optim.* 21 (1990), 265-287.

CORNET, B. - LAROQUE, G. [1] Lipschitz properties of solutions in mathematical programming. *J. Optim. Theory Appl.* 53 (1987), 407-427.

COURANT, R. - HILBERT, D. [1] **Methods of mathematical physics**, Vol.II. Interscience, 1962.

COYETTE, M. [1] Differentiability of distance functions. *Functional analysis and approximation*, edited by P.L. Papini, Pitagora 1988, 164-182.

CUDIA, D. [1] The geometry of Banach spaces. Smoothness. *Trans. Amer. Math. Soc.* 110 (1964), 284-314.

CULLUM, J. [1] Perturbations of optimal control problems. *J. SIAM Control* 4 (1966), 473-487. [2] Perturbations and approximations of continuous optimal control problems. *Mathematical theory of control*, edited by A.V. Balakrishnan - L.W. Neustadt, Academic Press 1967, 156-169.

DACOROGNA, B. [1] **Direct methods in the calculus of variations.** Springer, 1989.

DAL MASO, G. [1] Limits of minimum problems for general integral functionals with unilateral obstacles. *Atti Accad. Naz. Lincei Rend. Cl. Sci. Fis. Mat. Natur.* 74 (1983), 55-61. [2] Convergence of unilateral convex sets. *Lecture Notes in Math.* 1190, Springer 1986, 181-190. [3] An introduction to $\Gamma-$ convergence. *Pubbl. SISSA*, 1987.

DAL MASO, G. - DEFRANCESCHI, A. [1] Limits of nonlinear Dirichlet problems in varying domains. *Manuscripta Math.* 61 (1988), 251-278. [ch.VIII] Convergence of unilateral problems for monotone operators. *J. Analyse Math.* 53 (1989), 269-289.

DAL MASO, G. - MODICA, L. [1] Nonlinear stochastic homogenization. *Ann. Mat. Pura Appl.* 144 (1986), 347-389. [2] Integral functionals determined by their minima. *Rend. Sem. Mat. Univ. Padova* 76 (1986), 255-267.

DAL MASO, G. - MOSCO, U. [1] Wiener's criterion and $\Gamma-$ convergence. *Appl. Math. Optim.* 15 (1987), 15-63.

DANIEL, J.W. **[1] The approximate minimization of functionals.** Prentice-Hall, 1971. **[ch.IX]** Stability of the solution of definite quadratic problems. *Math. Programming* 5 (1973), 41-53. **[2]** The continuity of metric projections as functions of the data. *J. Approx. Theory* 12 (1974), 234-239.

DANTZIG, G.B. - FOLKMAN, J. - SHAPIRO, N. **[1]** On the continuity of the minimum set of a continuous function. *J. Math. Anal. Appl.* 17 (1967), 519-548.

DE BLASI, F.S. - MYJAK J. **[1]** On the minimum distance theorem to a closed convex set in a Banach space. *Bull. Acad. Polon. Sci. Ser. Sci. Math.* 29 (1981), 373-376. **[2]** Some generic properties in convex and nonconvex optimization theory. *Ann. Soc. Math. Polon. Comment. Math. Prace Math.* 24 (1984), 1-14. **[3]** On almost well posed problems in the theory of best approximation. *Bull. Math. Soc. Sci. Math. R.S. Roumanie* 28 (1984), 109-117. **[4]** Ensembles poreux dans la théorie de la meilleure approximation. *C.R. Acad. Sci. Paris* 308 (1989), 353-356. **[ch. III]** Ambiguous loci in best approximation theory. Preprint, 1991.

DE BLASI, F.S. - MYJAK, J. - PAPINI, P.L. **[ch.II]** Starshaped sets and best approximation. *Arch. Math. (Basel)* 56 (1991), 41-48. **[ch.III]** Best approximation in spaces of convex sets. Preprint 48, Centro Mat. Volterra, Univ. di Roma II, 1990. **[ch. III]** Porous sets in best approximation theory. *J. London Math. Soc.* 44 (1991), 135-142.

DEFRANCESCHI, A. **[1]** G - convergence of ciclically monotone operators. Preprint SISSA 100M, 1988.

DE GIORGI, E. **[ch.IV, VIII]** Sulla convergenza di alcune successioni d'integrali del tipo dell'area. *Rend. Mat.* 8 (1975), 277-294. **[ch.IV]** Generalized limits in calculus of variations. *Topics in functional analysis* 1980-81, edited by Strocchi - Zarantonello - De Giorgi - Dal Maso - Modica. Scuola Norm. Sup. Pisa 1981, 117-148.

DE GIORGI, E. - FRANZONI, T. **[1]** Su un tipo di convergenza variazionale. *Atti Accad. Naz. Lincei Rend. Cl. Sci. Fis. Mat. Natur.* 58 (1975), 842-850. **[2]** Su un tipo di convergenza variazionale. *Rend. Sem. Mat. Brescia* 3 (1979), 63-101.

DE GIORGI, E. - SPAGNOLO, S. **[1]** Sulla convergenza degli integrali dell'energia per operatori ellittici del secondo ordine. *Boll. Un. Mat. Ital.* 8 (1973), 391-411.

DEL PRETE, I. - LIGNOLA, M.B. **[1]** On the variational properties of $\Gamma^-(d)-$ convergence. *Ricerche Mat.* 31 (1982), 179-190.

DEUTSCH, F. **[1]** Existence of best approximations. *J. Approx. Theory* 28 (1980), 132-154.

DEUTSCH, F. - KENDEROV, P. **[ch.II]** Continuous selections and approximate selection for set-valued mappings and applications to metric projections. *SIAM J. Math. Anal.* 14 (1983), 185-194.

DEUTSCH, F. - POLLUL, W. - SINGER, I. [ch.II] On set-valued metric projections, Hahn - Banach extension maps and spherical image maps. *Duke Math. J.* 40 (1973), 355-370.

DIESTEL, J. [1] **Geometry of Banach spaces. Selected topics.** Lecture Notes in Math. 485, Springer 1975.

DIESTEL, J. - UHL, J. [1] **Vector measures.** Math. Surveys 15, Amer. Math. Soc. 1977.

DMITRIEV, M.G. [1] On the continuity of the solution of Mayer's problem under singular perturbations. *USSR Comput. Math. and Math. Phys.* 12 (1972), 788-791 (russian).

DOLECKI, S. [1] Convergence of global minima and infima. *Séminaire d'analyse numerique.* Publ. Univ. P. Sabatier, Toulouse III, 1982-83, VI-1 to VI-35. [ch.IX] Lower semicontinuity of marginal functions. *Lecture Notes in Econom. and Math. Systems* 226, Springer 1984, 30-41. [2] Convergence of minima in convergence spaces. *Optimization* 17 (1986), 553-572.

DOLECKI, S. - ROLEWICZ, S. [1] A characterization of semicontinuity preserving multifunctions. *J. Math. Anal. Appl.* 65 (1978), 26-31.

DONTCHEV, A. [1] **Perturbations, approximations and sensitivity analysis of optimal control systems.** Lecture Notes in Control and Inf. Sci. 52, Springer 1983. [2] Continuity and asymptotic behaviour of the marginal function in optimal control. *Lecture Notes in Econom. and Math. Systems* 259, Springer 1985, 185-193. [3] Equivalent perturbations and approximations in optimal control. *Internat. Series Numer. Math.* 84, Birkhäuser 1988, 43-54.

DONTCHEV, A. - FARCHI, E.M. [1] Error estimates for discretized differential inclusions. *Computing* 41 (1989), 349-358.

DONTCHEV, A. - GIČEV, T. [1] Convergence of the solutions of singularly perturbed time-optimal problems. *Appl. Math. Mech.* 43 (1976), 466-474 (russian).

DONTCHEV, A. - HAGER, W. [1] Lipschitzian stability in nonlinear control and optimization. To appear in *SIAM J. Control Optim.*

DONTCHEV, A. - JONGEN, H. [1] On the regularity of Kuhn-Tucker curve. *SIAM J. Control Optim.* 24 (1986), 169-176.

DONTCHEV, A. - LEMPIO, F. [ch. VI] Difference methods for differential inclusions: a survey. *SIAM Review* 34 (1992), n.2.

DONTCHEV, A. - MORDUHOVIČ, B.S. [1] Relaxation and well-posedness of nonlinear optimal processes. *Systems Control Lett.* 3 (1983), 177-179.

DONTCHEV, A. - SLAVOV, J. [1] Lipschitz properties of the attainable set of singularly perturbed linear systems. *Systems Control Lett.* 11 (1988), 385-392. [ch.VII] Singular perturbations in a class of nonlinear differential inclusions. *Proc. IFIP Conference*, Leipzig 1989, to appear.

DONTCHEV, A. - VELIOV, V.M. [1] Singular perturbations in Mayer's problem for linear systems. *SIAM J. Control Optim.* 21 (1983), 566-581. [2] On the behaviour of solutions of linear autonomous differential inclusions at infinity. *C.R. Acad. Bulgare Sci.* 36 (1983), 1021-1024. [3] On the order reduction of linear optimal control systems in critical cases. *Lecture Notes in Control and Inform. Sci* 66, Springer 1985, 61-73. [4] Singular perturbations in linear control systems with weakly coupled stable and unstable fast subsystems. *J. Math. Anal. Appl.* 110 (1985), 1-30. [5] Singular perturbations in linear control systems with constraints. *Proceedings 25-th IEEE Conference on Decision and Control*, IEEE Control Syst. Soc. 1986, 1781-1783.

DUFFIN, R.J. - JEROSLOW, R.G. [1] Lagrangian functions and affine minorants. *Math. Programming Study* 14 (1981), 48-60.

DUNFORD, N. - SCHWARTZ, J. [1] **Linear operators , part. I.** Interscience, 1958.

DUNN, J.C. - SACHS, E. The effect of perturbations on the convergence rates of optimization algorithms. *Appl. Math. Optim.* 10 (1983), 143-157.

EDELSTEIN, M. [1] On nearest points of sets in uniformly convex Banach spaces. *J. London Math. Soc.* 43 (1968), 375-377. [2] A note on nearest points. *Quart. J. Math. Oxford* 21 (1970), 403-405.

EFIMOV, N.V. - STEČKIN, S.B. [1] Approximative compactness and Čebišev sets. *Soviet Math. Dokl.* 2 (1961), 1226-1228.

EKELAND, I. [1] Nonconvex minimization problems. *Bull. Amer. Math. Soc.* 1 (1979), 443-474.

EKELAND, I. - LEBOURG, G. [1] Generic Fréchet differentiability and perturbed optimization problems in Banach spaces. *Trans. Amer. Math. Soc.* 224 (1976), 193-216.

EKELAND, I. - TEMAM, R. [1] **Analyse convexe and problèmes variationelles.** Dunod and Gauthier-Villars, 1974.

ENGL, H.W. [ch.I] Methods for approximating solutions of ill-posed linear operator equations based on numerical functional analysis. *Internat. Series Numer. Math.* 76, Birkhäuser, 1986.

ESTEBAN, M.J. - LIONS, P.L. [1] $\Gamma-$ convergence and the concentration-compactness method for some variational problems with lack of compactness. *Ricerche Mat.* 36 (1987), 73-101.

EVANS, J.P. - GOULD , F.G. **[1]** Stability in nonlinear pogramming. *Oper. Res.* 18 (1970), 107-118.

EWING, G.M. **[1]** Sufficient conditions for asymptotic optimal control. *J. Optim. Theory Appl.* 32 (1980), 307-325.

FAN, K. - GLICKSBERG, I. **[1]** Some geometric properties of the spheres in a normed linear space. *Duke Math. J.* 25 (1958), 553-568.

FATTORINI, H.O. - FRANKOWSKA, H. **[ch.V]** Explicit convergence estimates for suboptimal controls. *Problems of control and information theory* 19 (1990), part I: 3-29; part II: 69-93.

FERRIS, M.C. **[1]** Weak sharp minima and exact penalty functions. *Computer Sciences Technical Report* 779, Univ. of Wisconsin - Madison, 1988.

FIACCO, A. **[1]** Sensitivity analysis for nonlinear programming using penalty methods. *Math. Programming* 10 (1976), 287-311. **[2] Introduction to sensitivity and stability analysis in nonlinear programming.** Academic Press, 1983.

FIACCO, A.V. - KYPARISIS, J. **[ch.IX]** Sensitivity analysis in nonlinear programming under second order assumptions. *Lect. Notes in Control and Inform. Sci.* 66, Springer 1985, 74-97.

FIORENZA, R. **[1]** Sull'esistenza di elementi uniti per una classe di trasformazioni funzionali. *Ricerche Mat.* 15 (1966), 127-153.

FITZPATRICK, S. **[1]** Metric projections and the differentiability of distance functions. *Bull. Austral. Math. Soc.* 22 (1980), 291-312.

FLAM, S.D. - WETS, R. **[1]** Existence results and finite horizon approximates for infinite horizon optimization problems. *Econometrica* 55 (1987), 1187-1209.

FLATTO, L. - LEVINSON, N. **[1]** Periodic solutions of singularly perturbed systems. *Arch. Rational Mech. Anal.* 4 (1955), 943-950.

FLEMING, W.H. **[1]** The Cauchy problem for a nonlinear first-order partial differential equation *J. Differential Equations* 5 (1969), 515-530. **[2]** Stochastic control for small noise intensities. *SIAM J. Control* 9 (1971), 473-517.

FLEMING, W.H. - RISHEL, R. **[1] Deterministic and stochastic optimal control.** Springer, 1975.

FORT, M.K., Jr. **[1]** Points of continuity of semicontinuous functions. *Publ. Math. Debrecen* 2 (1951), 100-102.

FUJII, T. - MIZUSHIMA, N. [ch.V] Robustness of the optimality property of an optimal regulator: multi-input case. *Internat. J. Control* 39 (1984), 441-453.

FURI, M. - MARTELLI, M. - VIGNOLI, A. [1] On minimum problems for families of functionals. *Ann. Mat. Pura Appl.* 86 (1970), 181-187.

FURI, M. - VIGNOLI, A. [1] On the regularization of a nonlinear ill-posed problem in Banach spaces. *J. Optim. Theory Appl.* 4 (1969), 206-209. [2] A characterization of well-posed minimum problems in a complete metric space. *J. Optim. Theory Appl.* 5 (1970), 452-461. [3] About well-posed optimization problems for functionals in metric spaces. *J. Optim. Theory Appl.* 5 (1970), 225-229.

FUSCO, N. [1] Γ convergenza unidimensionale. *Boll. Un. Mat. Ital.* 16 B (1979), 74-86.

GABASOV, R. - KIRILLOVA, F. [1] **The qualitative theory of optimal processes.** Dekker, 1976.

GAUVIN, J. [1] The generalized gradient of a marginal function in mathematical programming. *Math. Oper. Res.* 4 (1979), 458-463.

GAUVIN, J. - DUBEAU, F. [1] Differential properties of the marginal function in mathematical programming. *Math. Programming Study* 19 (1982), 101-119. [ch.IX] Some examples and counterexamples for the stability analysis of nonlinear programming problems. *Math. Programming Study* 21 (1983), 69-78.

GAUVIN, J. - TOLLE. J.W. [1] Differential stability in nonlinear programming. *SIAM J. Control Optim.* 15 (1977), 294-311.

GEORGIEV, P. [1] The strong Ekeland variational principle, the strong drop theorem and applications. *J. Math. Anal. Appl.* 131 (1988), 1-21.

GIACHETTI, D. [1] Controllo ottimale in problemi vincolati. *Boll. Un. Mat. Ital.* 2B (1983), 445-468.

GIBSON, J.S. [1] The Riccati integral equations for optimal control problems on Hilbert spaces. *SIAM J. Control Optim.* 17 (1979), 537-565.

GIČEV, T. [1] Some questions on well-posedness of a linear optimal control problem with minimal impulse. *Differential Equations* 9 (1973), part I: 1383-1393; part II: 1561-1571 (russian). [2] Singular perturbations in a class of optimal control problems with integral convex performance index. *Appl. Math. Mech.* 48 (1984), 898-903 (russian).

GILBARG, D. - TRUDINGER, N.S. [1] **Elliptic partial differential equations of second order.** Springer, 1983.

GILES, J. [1] **Convex analysis with application in differentiation of convex functions.** Res. Notes in Math. 58, Pitman 1982.

GREENBERG, H.J. - PIERSKALLA, W.P. [1] Extension of the Evans-Gould stability theorems for mathematical programming. *Oper. Res.* 20 (1972), 143-153.

GRIGORIEFF, R.D. - REEMTSEN, R.M. [ch.IV]. Discrete approximations of minimization problems. I Theory; II Applications. *Numer. Funct. Anal. Optim.* 11 (1990), 701-719; 721-761.

GROETSCH, C.W. [1] **The theory of Tikhonov regularization for Fredholm equations of the first kind.** Res. Notes in Math. 105. Pitman 1984.

GUDDAT, J. - JONGEN, H. Th. [1] Structural stability in nonlinear optimization. *Optimization* 18 (1987), 617-631.

GUDDAT, J. - JONGEN, H. Th. - RUECKMANN, J. [1] On stability and stationary points in nonlinear optimization. *J. Austral. Math. Soc.* Ser. B 28(1986), 36-56.

GUILLERME, J. [1] Convergence of approximate saddle points. *J. Math. Anal. Appl.* 137 (1989), 297-311.

HADAMARD, J. [1] Sur les problèmes aux dérivees partielles et leur signification physique. *Bull. Univ. Princeton* 13 (1902), 49-52. [2] **Lectures on Cauchy's problem in linear partial differential equations.** Dover, 1953.

HAGER, W.W. [1] Lipschitz continuity for constrained processes. *SIAM J. Control Optim.* 17 (1979), 321-338. [2] Multiplier methods for nonlinear optimal control. *SIAM J. Numer. Anal.* 27 (1990), 1061-1080.

HAGER, W.W. - MITTER, S.K. [1] Lagrange duality theory for convex control problems. *SIAM J. Control Optim.* 14 (1976), 843-856.

HÁJEK, O. [1] Geometric theory of time-optimal control. *SIAM J. Control* 9 (1971), 339-350.

HARAUX, A. - MURAT, F. [1] Influence of a singular perturbation on the infimum of some functionals. *J. Differential Equations* 58 (1985), 43-75.

HARTMAN, P. [1] **Ordinary differential equations.** Wiley, 1964.

HAUSDORFF, F. [1] **Set theory.** Chelsea, 1962.

HENRY, D. [1] Nonuniqueness of solutions in the calculus of variations: a geometric approach. *SIAM J. Control Optim.* 18 (1980), 627-639.

HERMES, H. [ch.V] The equivalence and approximation of optimal control problems. *J. Differential Equations* 1 (1965), 409-426.

HERMES, H. - LA SALLE, J.P. [1] **Functional analysis and time optimal control.** Academic Press, 1969.

HYERS, D.H. [1] On the stability of minimum points. *J. Math. Anal. Appl.* 62 (1978), 530-537. [2] Stability of minimum points for problems with constraints. *Lecture Notes Pure Appl. Math.* 100, Dekker 1985, 283-289.

HOFFMAN, A.J. [ch. IX] On approximate solutions of systems of linear inequalities. *J. Res. Nat. Bur. Standards* 49 (1952), 263-265.

HOGAN, W.W. [1] Point-to-set maps in mathematical programming. *SIAM Rev.* 15 (1973), 591-603.

HOLMES, R. [1] **A course on optimization and best approximation.** Lecture Notes in Math. 257, Springer 1972.

HOPPE, R. [1] Constructive aspects in time optimal control. *Lecture Notes in Math.* 1190, Springer 1986, 243-272.

HOPPENSTEADT, F. [1] Properties of solutions of ordinary differential equations with small parameters. *Comm. Pure Appl. Math.* 34 (1971), 807-840.

IOFFE, A.D. - TIKHOMIROV, V.M. [1] Extension of variational problems. *Trans. Moscow Math. Soc.* 18 (1968), 207-273. [2] **Theory of extremal problems.** North Holland, 1979.

ISAACSON, E. - KELLER, H. B. [1] **Analysis of numerical methods.** Wiley, 1966.

ITO, K. - KUNISCH, K. [1] Sensitivity analysis of solutions to optimization problems in Hilbert spaces with applications to optimal control and estimation. To appear.

JOLY, J. L. - DE THELIN, F. [1]. Convergence of convex integrals in L^p spaces. *J. Math. Anal. Appl.* 54 (1976), 230 - 244.

JONGEN, H. - JONKER, P. - TWILT, F. [ch. IX] **Nonlinear optimization in R^n.** P. Lang. I, 1983; II, 1986.

JONGEN , H. Th. - KLATTE, D. - TAMMER, K. [1] Implicit functions and sensitivity of stationary points. *Math. Programming* 49 (1990), 123-138.

JONGEN, H. - MÖBERT, T. - RÜCKMANN, J. - TAMMER, K. [1] On inertia and Schur complement in optimization. *Linear Algebra Appl.* 95 (1987), 97 - 109.

JONGEN, H. Th. - TWILT, F. - WEBER, G. W. [ch. IX] Semi-infinite optimization: structure and stability of the feasible set. *Memorandum 838, Univ. of Twente,* Fac. Appl. Math., 1989.

JONGEN, H. Th. - WEBER, G. W. [1] Nonlinear optimization: characterization of structural stability. *J. Global Optim.* 1 (1991), 47-64.

KADEC, M.I. [1] On weak and norm convergence. *Dokl. Akad. Nauk. SSSR* 122 (1958), 13-16 (russian).

KALL, P. [ch.IV] Approximation to optimization problems: an elementary review. *Math. Oper. Res.* 11 (1986), 9-18.

KANNIAPPAN, P. - SASTRY, S. [1] Uniform convergence of convex optimization problems. *J. Math. Anal. Appl.* 96 (1983), 1-12.

KELLEY, J. [1] **General topology.** Van Nostrand, 1955.

KENDEROV, P. [1] Continuity - like properties of set - valued mappings. *Serdica* 9 (1983), 149-160. [2] Most of the optimization problems have unique solution. *C.R. Acad. Bulgare Sci.* 37 (1984), 297-300.

KINDERLEHRER, D. - STAMPACCHIA, G. [1] **An introduction to variational inequalities and their applications.** Academic Press, 1980.

KIRILLOVA, F.M. [1] On the continuous dependence on the initial data and parameters of the solution of an optimal control problem. *Uspehi Mat. Nauk.* 17 (1962), 141-146 (russian). [2] On the correctness of the formulation of an optimal control problem. *SIAM J. Control* 1 (1962/63), 224-239.

KLATTE, D. [1] Eine Bemerkung zur parametrischen quadratischen Optimierung. *Seminarbericht* 50, *Humboldt Universität zu Berlin*, 50-tl Jahrestagung "Math. Optimierung", 1983, 174-184. [2] Lipschitz continuity of infima and optimal solutions in parametric optimization: the polyhedral case. *Parametric optimization and related topics* (Plaue, 1985), Math. Res. 35, Akademie Verlag 1987, 229-248. [ch.IX] On the Lipschitz behavior of optimal solutions in parametric problems of quadratic optimization and linear complementarity. *Optimization* 16 (1985), 819-831. [ch.IX] On the stability of local and global optimal solutions in parametric problems of nonlinear programming. Part I: basic results. *Seminarbericht* 75, *Sektion Mathematik, Humboldt Universität zu Berlin*, 1985.

KLATTE, D. - TAMMER, K. [1] On second order sufficient optimality conditions for $C^{1,1}$ - optimization problems. *Optimization* 19 (1988), 169-179. [2] Strong stability of stationary solutions and Karush - Kuhn - Tucker points in nonlinear optimization. *Ann. Oper. Res.* 27 (1990), 285-307.

KLATTE, D. - KUMMER, B. [1] On the (Lipschitz-) continuity of solutions of parametric optimization problems. *Seminarbericht* 64, *Sektion Mathematik, Humboldt-Universität zu Berlin*, 1984. [ch.IX] Stability property of infima and optimal solutions of parametric optimization problems. *Lecture Notes in Econom. and Math. Systems* 255, Springer 1985, 215-229.

KLEIN, E. - THOMPSON, A.C. [1] **Theory of correspondences.** Wiley, 1984.

KOJIMA, M. [1] Strongly stable stationary solutions in nonlinear programming. *Analysis and computation of fixed points*, edited by S.M. Robinson, Academic Press 1980, 93-138.

KOJIMA, M. - HIRABAYASHI, R. **[ch.IX]** Continuous deformation of nonlinear programs. *Math. Programming Study* 21 (1984), 150-198.

KOHN, R.V. - STERNBERG, P. [1] Local minimisers and singular perturbations. *Proc. Roy. Soc. Edinburgh* 111 (1989), 69-84.

KOKOTOVIC, P. [1] Applications of singular perturbation techniques to control problems. *SIAM. Rev.* 26 (1984), 501-550.

KOKOTOVIC, P. - KHALIL, H. - O'REILLY, J. [1] **Singular perturbation method in control : analysis and design.** Academic Press, 1988.

KOKOTOVIC, P. - O'MALLEY, R.E. Jr. - SANNUTI P. [1] Singular perturbations and order reduction in control theory. *Automatica* 12 (1976), 123-132.

KOLMANOVSKY, V.B. [1] Optimal control of certain nonlinear systems with a small parameter. *Differential Uravnenija* 11 (1975), 1584-1594 (russian).

KONYAGIN, S.V. [1] On approximation properties of closed sets in Banach spaces and the characterization of strongly convex spaces. *Soviet Math. Dokl.* 21 (1980), 418-422.

KONYAGIN, S.V. - TSAR'KOV, I.G. [1] Efimov - Stechkin spaces. *Moscow Univ. Math. Bull.* 41 (1986), 20-28.

KORSAK, A.J. [1] On the continuous dependence of controls upon parameters in fixed-time free-terminal- state optimal control. *Automatica* 6 (1970), 289-295.

KOSMOL, P. - WRIEDT, M. [1] Starke Lösbarkeit von Optimierungsaufgaben. *Math. Nachr.* 83 (1978), 191-195.

KRASNOSELSKIJ, M.A. [1] **Topological methods in the theory of nonlinear integral equations.** Pergamon Press, 1964.

KUMMER, B. **[ch. IX]** Global stability of optimization problems. *Optimization* 8 (1977), 367-383. **[ch. IX]** Linearly and nonlinearly perturbed optimization problems in finite dimension. Parametric optimization and related topics (Plaue 1985), 249-267. Akademie Verlag, 1987.

KURATOWSKI, C. [1] **Topologie**, vol. 1. Panstwowe Wydawnictwo Naukowe, Warszawa, 1958.

KUZNETZOV, N.N. - ŠIŠKIN, A.A.[1] On a many dimensional problem in the theory of quasilinear equations. *Z. Vyčisl. Mat. i Mat. Fiz.* 4 (1964), 4, suppl., 192-205 (russian).

KYPARISIS, J [ch. IX] Sensitivity analysis for variational inequalities and nonlinear complementarity problems. *Ann. Oper. Res.* 27 (1990), 143-174.

LAURENT, P.J. [1] **Approximation et optimization.** Hermann, 1972.

LAVRENTIEV, M.M. [ch.II] **Some improperly posed problems of mathematical physics.** Springer, 1967.

LEBOURG, G. [1] Problémes d'optimisation dependant d'un paramétre a valeurs dans un espace de Banach. *Séminaire d'Anal. Fonct.* 1978-79, exposé XI (1978), Ecole Polytchnique, Centre de Math. Palaiseau, 1-15. [2] Perturbed optimization problems in Banach spaces. *Bull. Soc. Math. France*, Mém. 60 (1979), 95-111.

LEE, E.B. - MARKUS, L. [1] **Foundations of optimal control theory.** Wiley, 1968.

LEIZAROWITZ, A. [1] Infinite horizon autonomous systems with unbounded cost. *Appl. Math. Optim.* 13 (1985), 19-43.

LEMAIRE, B. [ch.IV] Coupling optimization methods and variational convergence. *Internat. Series Numer. Math.* 84, Birkhäuser 1988, 163-179. [1] Approximation in multi-objective optimization. Preprint, 1989.

LEONOV, A.S. [1] On some algorithms for solving ill-posed extremal problems. *Math. USSR Sb.* 57 (1987), 229-242.

LEVIN, A.M. [1] The stability of constructing minimizing sequences of functionals. *USSR Comput. Math. and Math. Phys.* 27 (1987), 2, 106-110.

LEVITIN, E.S. - POLYAK, B.T. [1] Convergence of minimizing sequences in conditional extremum problems. *Soviet Math. Dokl.* 7 (1966), 764-767.

LIGNOLA, M.B. - MORGAN, J. [1] On the continuity of marginal functions with dependent constraints. *Rapport CRM* - 1604, Montreal 1989. [2] Convergences of marginal functions with dependent constraints. *Pubbl. Dip. Matem. Appl. "R. Caccioppoli"*, Napoli 1990. [3] Existence and approximation results for min sup problems. Preprint, 1990. [4] Topological existence and stability for min sup problems. *J. Math. Anal. Appl.* 151 (1990), 164-180.

LIONS P.L. [1] **Generalized solutions of Hamilton-Jacobi equations.** Res. Notes in Math. 69, Pitman 1982. [2] The concentration-compactness principle in the calculus of variations. The locally compact case. *Ann. Inst. H. Poincaré Anal. non Linéaire* 1 (1984); part 1, 109-145; part 2, 223-283.

LISKOVITZ, O.A. [1] Theory and methods for solving ill-posed problems. *Itogi Nauki i Techn.* 20 (1982), 116-178 (russian).

LOEWEN, P.D. [1] On the Lavrentiev phenomenon. *Canad. Math. Bull.* 30 (1987), 102-108.

LOONEY, C. [1] Convergence of minimizing sequences. *J. Math. Anal. Appl.* 61 (1977), 835-840.

LORIDAN, P. - MORGAN, J. [1] New results on approximate solutions in two level optimization. *Optimization* 20 (1989), 819-836. [ch.IV] A sequential stability result for constrained Stackelberg problems. *Ricerche Mat.* 38 (1989), 19-32. [2] On strict $\varepsilon-$ solutions for a two-level optimization problem. Preprint 33 (1990), *Dip. di Matematica ed Appl.*, Univ. di Napoli.

LUCCHETTI, R. [1] Some aspects of the connections between Hadamard and Tykhonov well - posedness of convex programs. *Boll. Un. Mat. Ital.* C 1 (1982), 337-345. [2] Hadamard and Tykhonov well-posedness in optimal control. *Methods Oper. Res.* 45 (1983), 113-126. [3] On the continuity of optimal value and optimal set in minimum problems. *Pubbl. Ist. Mat. Univ. Genova* 133 (1983). [4] On the continuity of the minima for a family of constrained minimization problems. *Numer. Funct. Anal. Optim.* 7 (1984-85), 349-362. [5] Convergence of sets and of projections. *Boll. Un. Mat. Ital.* 4 C (1985), 477-483. [6] Well-posedness, towards vector optimization. *Lecture Notes Econom. and Math. Systems* 294, Springer 1986, 194-207. [7] Stability in optimization. *Serdica* 15 (1989), 34-48. [ch.I, II, III] Topologies on hyperspaces and well posed problems. Preprint, 1990.

LUCCHETTI, R. - MALIVERT, C. [1] Variational convergences and level sets multifunctions. *Ricerche Mat.* 38 (1989), 223-237.

LUCCHETTI, R. - MIGNANEGO, F. [1] Variational perturbations of the minimum effort problem. *J. Optim. Theory Appl.* 30 (1980), 485-499. [2] Continuous dependence on the data in abstract control problems. *J. Optim. Theory Appl.* 34 (1981), 425-444.

LUCCHETTI, R. - PATRONE, F. [1] Sulla densita' e genericita' di alcuni problemi di minimo ben posti. *Boll. Un. Mat. Ital.* 15 B (1978), 225-240. [2] A characterization of Tykhonov well-posedness for minimum problems, with applications to variational inequalities. *Numer. Funct. Anal. Optim.* 3 (1981), 461-476. [3] Hadamard and Tyhonov well-posedness of a certain class of convex functions. *J. Math. Anal. Appl.* 88 (1982), 204-215. [4] Some properties of "well-posed" variational inequalities governed by linear operators. *Numer. Funct. Anal. Optim.* 5 (1982-83), 349-361. [5] Closure and upper semicontinuity results in mathematical programming, Nash and economic equilibria. *Optimization* 17 (1986), 619-628.

LUCCHETTI, R. - PATRONE, F. - TIJS, S.H. - TORRE, A. Continuity properties of solution concepts for cooperative games. *OR Spektrum* 9 (1987), 101-107.

MACKI, J. - STRAUSS, A. [1] **Introduction to optimal control theory.** Springer, 1982.

MAEHLY, H. - WITZGALL, C. [1] Tschebyscheff-Approximationen in kleinen Intervallen II. *Numer. Math.* 2 (1960), 293-307.

MALANOWSKI, K. [1] Stability and sensitivity of solutions to optimal control problems for systems with control appearing linearly. *Appl. Math. Optim.* 16 (1987), 73-91. **[ch.V,IX] Stability of solutions to convex problems of optimization.** Lecture Notes in Control and Inform. Sci. 93, Springer 1987. [2] On stability of solutions to constrained optimal control problems for systems with control appearing linearly. *Arch. Autom. Telem.* 33 (1988), 483-497.

MANGASARIAN, O.L. - FROMOVITZ, S. [1] The Fritz John necessary optimality conditions in the presence of equality and inequality constraints. *J. Math. Anal. Appl.* 17 (1967), 37-47.

MANGASARIAN, O.L. - SHIAU, T.H. [1] Lipschitz continuity of solutions of linear inequalities, programs and complementarity problems. *SIAM.J. Control Optim.* 25 (1987), 583-595.

MANIA', B. [1] Sopra un esempio di Lavrentieff. *Boll. Un. Mat. Ital.* 13 (1934), 147-153.

MARCELLINI, P. [1] Su una convergenza di funzioni convesse. *Boll. Un. Mat. Ital.* 8 (1973), 137-158. [2] Un teorema di passaggio al limite per la somma di funzioni convesse. *Boll. Un. Mat. Ital.* 11 (1975), 107-124. [3] Periodic solutions and homogenization of non linear variational problems. *Ann. Mat. Pura Appl.* 117 (1978), 139-152.

MARCELLINI, P. - SBORDONE, C. [1] Dualita' e perturbazione di funzionali integrali. *Ricerche Mat.* 26 (1977), 383-421. [2] Homogenization of non-uniformly elliptic operators. *Applicable Anal.* 8 (1978), 101-113.

MARINO, A. - SPAGNOLO, S. [1] Un tipo di approssimazione dell'operatore $\sum_{i,j} D_j(a_{i,j}(x)D_j)$ con operatori $\sum_j D_j(\beta(x)D_j)$. *Ann. Scuola Norm. Sup. Pisa* 23 (1969), 657-673.

MARTINET, B. [ch.I] Pertubation des méthodes d'optimisation. Applications. *R.A.I.R.O. Anal. Numér.* 12 (1978), 153-171.

MATZEU, M. [1] Su un tipo di continuita' dell'operatore subdifferenziale. *Boll. Un. Mat. Ital.* 14 B (1977), 480-490.

MC CORMICK, G. [1] **Nonlinear programming.** Wiley, 1983.

MC LINDEN, L. [1] Successive approximation and linear stability involving convergent sequences of optimization problems. *J. Approx. Theory* 35 (1982), 311-354.

MECCARIELLO, E. [1] Approximation results for the subdifferential of a convex functions via the ε - subdifferential. Preprint 18 (1990), *Dipartimento di Matematica ed Applicazioni*, Universita' di Napoli.

MIGNANEGO, F. [1] Asymptotic convergences of optimal controls. *Boll. Un. Mat. Ital.* 10 (1974), 202-213. [2] Optimal control of abstract evolution equations over unbounded time domains. *Boll. Un. Mat. Ital.* 15 B (1978), 844-858.

MONTESINOS, V. [ch.II] Drop property equals reflexivity. *Studia Math.* 87 (1987), 93-100.

MORDUKHOVICH, B. Sh. [1] On finite-difference approximations to optimal control systems. *Appl. Math. Mech.* 42 (1978), 431-440 (russian). [2] On the theory of difference approximations in optimal control. Dokl. Akad. Nauk. BSSR 30 (1986), 1064 - 1067 (russian). [3] **Approximation methods in problems of optimization and control**. Nauka 1988 (russian).

MORGAN, J. [1] Constrained well-posed two-level optimization problems. *Nonsmooth optimization and related topics*, edited by Clarke, Demyanov, Giannessi. Plenum Press 1989, 307-325.

MORTOLA, S. - PROFETI, A. [1] On the convergence of the minimum points of non equicoercive quadratic functionals. *Comm. Partial Differential Equations* 7 (1982), 645-673.

MOSCARIELLO, G. [1] $\Gamma-$ convergenza negli spazi sequenziali. *Rend. Accad. Sci. Fis. Mat. Napoli* 43 (1976), 333-350.

MOSCO, U. [1] Approximation of the solutions of some variational inequalities. *Ann. Sc. Norm. Sup. Pisa* 21 (1967), 373-394; same journal, 765. [2] Convergence of convex sets and of solutions of variational inequalities. *Adv. in Math.* 3 (1969), 510-585. [3] An introduction to the approximate solution of variational inequalities. *Constructive aspects of functional analysis*. CIME, Erice 1971; Cremonese 1973.

MOSOLOV, P.P. - MJASNIKOV, V.P. [ch.VIII] On the correctness of boundary value problems in the mechanics of continuous media. *Math. USSR Sb.* 17 (1972), 257-268.

NASHED, M.Z. [ch.I] On well-posed and ill-posed extremal problems. *Nonlinear phenomena in mathematical sciences*, ed. by Lakshmikantham. Academic Press 1982, 737-746. [ch. I] On nonlinear ill-posed problems I: classes of operator equations and minimization of functionals. *Lecture Notes Pure Appl. Math.* 109, Dekker, 1987, 351-373.

NIKOLSKII, M.S. [1] On the well-posedness of an optimal control problem for a linear control system with integral quadratic performance index. *Proc. Steklov Inst. Math.* 1 (1986), 197-205.

NÜRNBERGER, G. [1] Unicity in semi-infinite optimization. Parametric optimization and approximation, edited by Brosowski - Deutsch. *Internat. Series Numer. Math.* 72, Birkhäuser 1985, 231-247. [2] Global unicity in semi-infinite optimization. *Numer. Funct. Anal. Optim.* 8 (1985), 173-191.

NÜRNBERG, G. - SINGER, I. [1] Uniqueness and strong uniqueness of best approximations by spline subspaces and other subspaces. *J. Math. Anal. Appl.* 90 (1982), 171-184.

OLECH, C. [1] A characterization of L_1-weak semicontinuity of integral functionals. *Bull. Polish Acad. Sci. Math.* 25 (1977), 135-142. [2] The Lyapunov theorem: its extensions and applications. *Lecture Notes in Math.* 1446, Springer 1990, 84-103.

O'MALLEY, R.E., Jr. [1] Singular perturbations and optimal control. *Lecture Notes in Math.* 680, Springer 1978, 171-218.

OSHMAN, E.V. [1] A continuity criterion for metric projections in Banach spaces. *Math. Notes* 10 (1971), 697-701. [2] On the continuity of metric projection onto convex closed sets. *Soviet Math. Dokl.* 27 (1983), 330-332.

PAPAGEORGIOU, N.S. [1] Relaxation of infinite dimensional variational and control problems with state constraints. *Kodai Math. J.* 12(1989), 392-419. [2] Sensitivity analysis of evolution inclusions and its applications to the variational stability of optimal control problems. *Houston J. Math.* 16 (1990), 509-522. [3] Remark on the well-posedness of the problem of minimization of integral functionals. *J. Optim. Theory Appl.* 65 (1990), 171-178, [4] A convergence result for a sequence of distributed parameter optimal control problems. *J. Optim. Theory Appl.* 68 (1991), 305-320. [ch.VI] Relation between relaxability and performance stability for optimal control problems governed by nonlinear evolution equations. *Internat. J. Systems Sci.* 22 (1991), 237-259.

PAPAGEORGIOU, N.S. - KANDILAKIS, D.A. [ch.II] Convergence in approximation and nonsmooth analysis. *J. Approx. Theory* 49 (1987), 41-54.

PARK, S.H. [1] Uniform Hausdorff strong uniqueness. *J. Approx. Theory* 58 (1989), 78-89.

PATRONE, F. [1] On the optimal control for variational inequalities. *J. Optim. Theory Appl.* 22 (1977), 373-388. [2] Nearly-orthogonal sequences: applications to not well-posed minimum problems and linear equations. *Boll. Un. Mat. Ital.* 3 C (1984), 143-156. [3] Well posedness as an ordinal property. *Riv. Mat. Pura Appl.* 1 (1987), 95-104. [4] Most convex functions are nice. *Numer. Funct. Anal. Optim.* 9 (1987), 359-369. [5] Well-posed minimum problems for preorders. *Rend. Sem. Mat. Univ. Padova* 84 (1990), 109-121.

PATRONE, F. - REVALSKI, J. [1] Constrained minimum problems for preorders: Tikhonov and Hadamard well-posedness. *Boll. Un. Mat. Ital.* 5B (1991), 639-652. [2] Characterization of Tykhonov well-posedness for preorders. *Math. Balkanica* 5 (1991), 146-155.

PATRONE, F. - TORRE, A. [1] Characterization of existence and uniqueness for saddle point problems and related topics. *Boll. Un. Mat. Ital.* 5 C (1986), 175-184.

PEDEMONTE, O. [1] On the convergence of constrained minima. *Numer. Funct. Anal. Optim.* 5 (1982-83), 249-265. [2] On perturbations of quadratic functionals with constraints. *Control Cybernet.* 13 (1984), 15-26.

PENOT, J-P. [1] Continuity properties of performance functions. *Lecture Notes Pure Appl. Math.* 86, Dekker 1983, 77-90. [2] Metric regularity, openness and Lipschitzian behavior of multifunctions. *Nonlinear Anal.* 13 (1989), 629-643.

PENOT, J-P. - STERNA-KARWAT, A. [1] Parametrized multicriteria optimization; order continuity of the marginal multifunctions. *J. Math. Anal. Appl.* 144 (1989), 1-15.

PERVOZVANSKY, A.A. - GAITSGORY, V.G. [1] **Decomposition, aggregation and approximate optimization**. Nauka, 1979 (russian).

PETROV, N.N. [1] On the continuity of the Bellman function with respect to a parameter. *Vestnik Leningrad Univ. Math.* 7 (1979), 169-176.

PHELPS, R.R. [ch.I] Convex sets and nearest points. *Proc. Amer. Math. Soc.* 8 (1957), 790-797. [1] **Convex functions, monotone operators and differentiability.** Lecture Notes in Math. 1364, Springer 1989.

PIERI, G. [1] Variational perturbations of the linear-quadratic problem. *J. Optim. Theory Appl.* 22 (1977), 63-77. [2] Sulla stabilita' variazionale del problema del minimo tempo. *Boll. Un. Mat. Ital.* 15 B (1978), 108-118. [3] Variational characterization of the uniqueness of the optimal state for the minimal-time problem. *J. Optim. Theory Appl.* 30 (1980), 635-642. [4] Well-posedness and bang-bang principle. *Numer. Funct. Anal. Optim.* 5 (1982), 77-84.

PIONTKOWSKI, O. [1] Über die Minimierung nichtlinearer Funktionale in normierten Räumen. *Uspehi Mat. Nauk.* 29 (1974), 225-226 (russian).

POLLOCK, J. - PRITCHARD, A.J. [1] The infinite time quadratic cost control problem for distributed systems with unbounded control action. *J. Inst. Maths. Applics.* 25 (1980), 287-309.

POLYAK, B.T. [1] Existence theorems and convergence of minimizing sequences in extremum problems with restrictions. *Soviet Math. Dokl.* 7 (1966), 72-75. [2] **Introduction to optimization.** Optimization Software, 1987.

PORACKÁ - DIVIŠ, Z. [1] Existence theorems and convergence of minimizing sequences in extremum problems. *SIAM J. Math. Anal.* 2 (1971), 505-510.

PSHENICHNYI, B.N. [1] **Convex analysis and extremal problems.** Nauka, 1980 (russian)

REVALSKI, J. [1] Generic properties concerning well posed optimization problems. *C.R. Acad. Bulgare Sci.* 38 (1985), 1431-1434. [2] Variational inequalities with unique solution. Mathematics and Education in Mathematics. *Proc. 14th Spring Conference of the Union of Bulgarian Mathematicians*; Sofia 1985, 534-541. [3] Generic well posedness in some classes of optimization problems. *Acta Univ. Carolin. - Math. Phys.* 28 (1987), 117-125. [4] An equivalence relation between optimization problems connected with the well-posedness. *C.R. Acad. Bulgare Sci.* 41 (1988), 11-14. [5] Well-posedness almost everywhere in a class of constrained convex optimization problems. *Proc. 17th Spring Conference of the Union of Bulgarian Mathematicians*, 1988, BAS, 348-353. [ch.I, III] Well-posedness of optimization problems: a survey. *Functional analysis and approximation*, ed. by P.L. Papini, Pitagora 1988, 238-255.

REVALSKI, J. - ZHIVKOV, N. [1] Well-posed constrained optimization problems in metric spaces. To appear on J. Optim. Theory Appl.

RIBIERE, G. [1] Regularization d'operateurs. *RAIRO* 1, 5(1967), 57-79.

ROBINSON, S.M. [1] Stability theory for systems of inequalities, part II: differentiable nonlinear systems. *SIAM J. Numer. Anal.* 13 (1976), 497-513. [2] A characterization of stability in linear programming. *Oper. Res.* 25 (1977), 435-447. [3] Strongly regular generalized equations. *Math. Oper. Res.* 5 (1980), 43-62. [4] Some continuity properties of polyhedral multifunctions. *Math. Programming Study* 14 (1981), 206-214. [5] Generalized equations and their solutions, part II: applications to nonlinear programming. *Math. Programming Study* 19 (1982), 200-221. [6] Local epi-continuity and local optimization. *Math. Programming* 37 (1987), 208-222. [ch.IX] Local structure of feasible sets in nonlinear programming, part III: stability and sensitivity. *Math. Programming Study* 30 (1987), 45-66. Corrigenda. *Math. Programming* 49 (1990), 143.

ROCKAFELLAR, R.T. [1] Level sets and continuity of conjugate convex functions. *Trans. Amer. Math. Soc.* 123 (1966), 46-63. [2] **Convex analysis.** Princeton University Press, 1970. [3] Optimal arcs and the minimum value function in problems of Lagrange. *Trans. Amer. Math. Soc.* 180 (1973), 53-83. [4] **Conjugate duality and optimization.** Regional conference series in appl. math. 16. SIAM 1974. [5] Augmented Lagrange multiplier functions and duality in nonconvex programming. *SIAM J. Control* 12 (1974), 268-285. [6] Integral functionals, normal integrands and measurable selections. *Lecture Notes in Math.* 543, Springer 1976, 157-207. [7] Clarke's tangent cones and the boundaries of closed sets in R^n. *Nonlinear Anal.* 3 (1979), 145-154. [8] **The theory of subgradients and its applications to problems of optimization. Convex and nonconvex functions.** Heldermann 1981. [9] Lagrange multipliers and subderivatives of optimal value function in nonlinear programming. *Math. Programming Study* 17 (1982), 28-66.

[10] Marginal values and second-order necessary conditions for optimality. *Math. Programming* 26 (1983), 245-286. [11] Extensions of subgradient calculus with application to optimization. *Nonlinear Anal.* 9 (1985), 665-698. [12] Lipschitzian stability in optimization: the role of nonsmooth analysis. *Lecture Notes in Econom. and Math. Systems* 255, Springer 1985, 55-73. [13] Lipschitzian properties of multifunctions. *Nonlinear Anal.* 9 (1985), 867-885. [14] Second order optimality conditions in nonlinear programming obtained by way of pseudoderivatives. *Math. Oper. Res.* 14 (1989), 462-484.

ROLEWICZ, S. [1] On Δ - uniform convexity and drop property. *Studia Math.* 87 (1987), 181-191.

ROUBICEK,T. [ch.IX] Stable extensions of constrained optimization problems. *J. Math. Anal. Appl.* 141 (1989), 120-135.

SAKSENA, V.R. - O'REILLY, J. - KOKOTOVIC, P. [1] Singular perturbations and time-scale methods in control theory: survey 1976-1983. *Automatica* 20 (1984), 273-293.

SALINETTI, G. - WETS, R. J-B. [1] On the relations between two types of convergence for convex functions. *J. Math. Anal. Appl.* 60 (1977), 211-226.

SALVADORI, A. [1] On the $M-$ convergence for integral functional on L^p_X. *Atti Sem. Mat. Fis. Univ. Modena* 33 (1984), 137-154.

SBORDONE, C. [ch.IV, VIII] La $\Gamma-$ convergenza e la G- convergenza. *Rend. Sem. Mat. Univ. Politec. Torino* 40 (1982), 25-51.

SEIDMAN, T.I. [1] Nonconvergence results for the application of least-squares estimation to ill-posed problems. *J. Optim. Theory Appl.* 30 (1980), 535-547. [2] Convergent approximation methods for ill-posed problems. *Control Cybernet.* 10 (1981), 31-49 and 51-71.

SENDOV, B. - POPOV, V. [1] **The averaged moduli of smoothness.** Wiley, 1988.

SHAPIRO, A. [1] Sensitivity analysis of nonlinear programs and differentiability properties of metric projections. *SIAM J. Control Optim.* 26 (1988), 628-645. [2] Perturbation theory of nonlinear programs when the set of optimal solutions is not a singleton. *Appl. Math. Optim.* 18 (1988), 215-229.

SHUNMUGARAJ, P. - PAI, D.V. [1] On approximate minima of a convex functional and lower semicontinuity of metric projections. *J. Approx. Theory* 64 (1991), 25-37. [2] On stability of approximate solutions of minimization problems. Preprint, 1990.

SINGER, I. [1] Some remarks on approximative compactness. *Rev. Roumaine Math. Pures Appl.* 9 (1964), 167-177. [2] **Best approximation in normed linear spaces by elements of linear subspaces.** Springer, 1970.

SMARZEWSKI, R. [1] Strongly unique minimization of functionals in Banach spaces with applications to theory of approximation and fixed points. *J. Math. Anal. Appl.* 115 (1986), 155-172. [2] Strongly unique best approximation in Banach spaces. II. *J. Approx. Theory* 51 (1987), 202-217.

ŠOLOHOVIČ, V.F. [1] Unstable extremal problems and geometric properties of Banach spaces. *Soviet Math. Dokl.* 11 (1970), 1470-1472. [2] On stability of extremal problems in Banach spaces. *Math. USSR - Sb.* 14 (1971), 417-427.

SONNEBORN, L.M. - VAN VLECK, F.S. [1] The bang-bang principle for linear control systems. *SIAM J. Control* 2 (1965), 151-159.

SONNTAG, Y. [1] Interprétation géométrique de la convergence d'une suite de convexes. *C.R. Acad. Sci. Paris* 282 (1976), 1099-1100. [2] Convergence au sens de U. Mosco: théorie et applications à l'approximation de solutions d'inequations. *Thèse, Université de Provence*, 1982. [3] Approximation des problèmes en optimization. Applications. *Publ. Math. Appl. Marseille-Toulon*, 83/1(1983). [ch.II] Approximation et pénalisation en optimisation. *Publ. Math. Appl. Marseille-Toulon*, 83/2 (1983). [4] Convergence des suites d'ensembles. Université de Provence, 1987.

SPAGNOLO, S. [1] Sul limite delle soluzioni di problemi di Cauchy relativi all'equazione del calore. *Ann. Scuola Norm. Sup. Pisa* 21 (1967), 657-699. [2] Sulla convergenza di soluzioni di equazioni paraboliche ed ellittiche. *Ann. Scuola Norm. Sup. Pisa* 22 (1968), 571-597. [3] Convergence in energy for elliptic operators. Numerical solutions of partial diff. equations - *III Synspade* 1975. Academic Press, 1976, 469-498.

SPINGARN, J.E. [1] Fixed and variable constraints in sensitivity analysis. *SIAM J. Control Optim.* 18 (1980), 297-310.

SPINGARN, J.E. - ROCKAFELLAR, R.T. [1] The generic nature of optimality conditions in nonlinear programming. *Math. Oper. Res.* 4 (1979), 425-430.

STASSINOPOULOS, G.I. [1] Continuous dependence of constrained solutions of linear differential inclusions. *J. Inst. Maths. Applics.* 25 (1980), 161-175.

STASSINOPOULOS, G.I. - VINTER, R.B. [1] Continuous dependence of solutions of a differential inclusion on the right hand side with applications to stability of optimal control problems. *SIAM J. Control Optim.* 17 (1979), 432-449.

STEČKIN, S.B. [1] Approximative properties of subsets of Banach spaces. *Rev. Roumaine Math. Pures Appl.* 8 (1963), 5-18 (russian).

STEGALL, C. [1] Optimization of functions on certain subsets of Banach spaces. *Math. Ann.* 236 (1978), 171-176. [2] Optimization and differentiation in Banach spaces. *Linear Algebra Appl.* 84 (1986). 191-211.

STERNA - KARWAT, A. [1] Continuous dependence of solutions on a parameter in a scalarization method. *J. Optim. Theory Appl.* 55 (1987), 417-434.

STERNBERG, P. [ch.VIII] The effect of a singular perturbation on nonconvex variational problems. *Arch. Rat. Mech. Anal.* 101 (1988), 207-260.

STUDNIARSKI, M. [ch.IX] Sufficient conditions for the stability of local minimum points in nonsmooth optimization. *Optimization* 20 (1989), 27-35.

SUSSMANN, H. [1] Analytic stratifications and control theory. *Proceedings Internat. Congress of Math.* Helsinki 1978, 865-871. [2] Small-time local controllability and continuity of the optimal time function for linear systems. *J. Optim. Theory Appl.* 53 (1987), 281-296.

TADUMADZE, T.A. [1] Continuity of the minimum of an integral functional in a nonlinear optimum control problem. *Differential Equations* 20 (1984), 716-720. [ch.V] **Some topics on the qualitative theory of the optimal control.** Tbilissi Univ. Press, 1983 (russian).

TALENTI, G. [1] Sui problemi mal posti. *Boll. Un. Mat. Ital.* 15 A (1978), 1-29.

TANINO, T. [1] Stability and sensitivity analysis in convex vector optimization. *SIAM J. Control Optim.* 26 (1988), 521-536. [ch. IX] Stability and sensitivity analysis in multiobjective nonlinear programming. *Ann. Oper. Res.* 27 (1990), 97-114.

TAUBERT, K. [1] Converging multistep methods for initial value problems involving multivalued maps. *Computing* 27 (1981), 123-136.

TYKHONOV, A.N. [ch.1] Stability of inverse problems. *Dokl. Akad. Nauk. SSSR* 39 (1943), 176-179 (russian). [1] Systems of differential equations containing a small parameter in the derivative. *Mat. Sbornik* 31 (1952), 575-586 (russian). [2] Solution of incorrectly formulated problems and the regularization method. *Soviet Math. Dokl.* 4 (1963), 1035-1038. [3] Improper problems of optimal planning and stable methods for their solution. *Soviet Math. Dokl.* 6 (1965), 1264-1267. [4] Methods for the regularization of optimal control problems. *Soviet Math. Dokl.* 6 (1965), 761-763. [5] On the stability of the functional optimization problem. *USSR Comp. Math. Math. Phys.* 6 (1966), 4, 28-33.

TIKHONOV, A. - ARSÉNINE, V. [1] **Méthodes de résolution de problèmes mal posés.** Traduction francaise, MIR 1976. [2] **Methods for solving ill-posed problems.** Nauka, 1986 (russian).

TYKHONOV, A.N. - GALKIN, V.Y. - ZAIKIN, P.N. [ch.V] Direct methods for the solution of optimal control problems. *Zh. Vychisl. Matem. i Matem. Fiz.* 7 (1967), 416-423 (russian).

TJUHTIN, V.B. [1] An error estimate for approximate solutions in one-side variational problems. *Vestnik Leningrad Univ. Math.* 14 (1982), 247-254.

TODOROV, M.I. [1] Generic existence and uniqueness of the solution to linear semi-infinite optimization problems. *Numer. Funct. Anal. Optim.* 8 (1985-86), 541-556. [ch.III] Properties resembling well-posedness in the linear vector semi-infinite optimization *C.R. Acad. 'Bulgare Sci.* 42 (1989), 23-25.

TRUDZIK, L.I. [ch.IX] Perturbed convex programming in reflexive Banach spaces. *Nonlinear Anal.* 9 (1985), 61-78.

TSUKADA, M. [1] Convergence of best approximations in a smooth Banach space. *J. Approx. Theory* 40 (1984), 301-309.

VAINBERG, M. [1] Le problème de la minimisation des fonctionelles non linéaires. *CIME IV ciclo*, 1970. Cremonese, 1971.

VALADIER, M. [1] Sur un théorème de relaxation d'Ekeland-Temam. *Sèm. Anal. Convexe* 11 (1981), exposé 5.

VASIL'EV, F.P. [1] **Methods for solving extremum problems.** Nauka, 1981 (russian).

VASIL'EVA, A.B. [1] Asymptotic behaviour of solutions to certain problems involving nonlinear differential equations containing a small parameter multiplying the highest derivatives. *Russian Math. Surveys* 18 (1963), 13-81.

VASIL'EVA, A.B. - DIMITRIEV, M.G. [1] Singular perturbations in optimal control problems. *Itogi Nauki i Techn. Math. Analysis* 20 (1982), 3-77 (russian).

VASIN, V.V. [1] Discrete approximation and stability in extremal problems. *USSR Comput. Math. and Math. Phys.* 22, 4 (1982), 57-74.

VEL'OV [1] On the local properties of Bellman's function for nonlinear time-optimal control problems. *Serdica* 10 (1984), 68-77.

VEL'OV, V.M. - DONCHEV, A.L. [1] Continuity of the family of trajectories of linear control systems with respect to singular perturbations. *Soviet Math. Dokl.* 35 (1987), 283-286.

VERVAAT, W. [1] Une compactification des espaces fonctionelles C et D; une alternative pour la démontration de theorèmes limites fonctionelles. *C.R. Acad. Sci. Paris* 292 (1981), 441-444.

VINTER, R.B. - PAPPAS, G. [1] A maximum principle for nonsmooth optimal control problems with state constraints. *J. Math. Anal. Appl.* 89 (1982), 212-232.

VISINTIN, A. [1] On the well-posedness of some optimal control problems. *Atti Accad. Naz. Lincei Rend. Cl. Sci. Fis. Mat. Natur.* 75 (1983), 34-41. [2] Strong convergence results related to strict convexity. *Comm. Partial Differential Equations* 9 (1984), 439-466.

VLASOV, L.P. [1] Almost convex and Chebychev sets. *Math. Notes* 8 (1970), 776-779. [2] Approximative properties of sets in normed linear spaces. *Russian Math. Surveys* 28 (1973), 6, 1-66. [3] Continuity of the metric projection. *Math. Notes* 30 (1981), 906-909.

WARGA, J. [1] Relaxed variational problems. *J. Math. Anal. Appl.* 4 (1962), 111-128.

WETS. R.J. - B. [1] A formula for the level sets of epi-limits and some applications. *Lecture Notes in Math.* 979, Springer 1983, 256-268. [ch.IV] On a compactness theorem for epi-convergent sequences of functions. Mathematical Programming, Cottle - Kelmanson - Korte eds., 347-355. Elsevier (North-Holland), 1984.[2] On the continuity of the value of a linear program and of related polyhedral-valued multifunctions. *Math. Programming Study* 24 (1985), 14-29.

WIJSMAN, R.A. [1] Convergence of sequences of convex sets, cones and functions. *Bull. Amer. Math. Soc.* 70 (1964), 186-188. Convergence of sequences of convex sets, cones and functions II. *Trans. Amer. Math. Soc.* 123 (1966), 32-45.

WILLIAMS, A.C. [1] Marginal values in linear programming. *J. Soc. Indust. Appl. Math.* 11 (1963), 82-94.

WITOMSKI, P. [1] Théorèmes d'existence en densité pour une classe de problèmes d'optimisation non convexes. *Bull. Soc. Roy. Sci. Liége* 46 (1977), 5-7.

WRIEDT, M. [ch.II] Stetigkeit von Optimierungsoperatoren. *Arch. Math.* 28 (1977), 652-656.

WU LI. [1] Strong uniqueness and Lipschitz continuity of metric projections: a generalization of the classical Haar theory. *J. Approx. Theory* 56 (1989), 164-184.

ZALINESCU, C. [1] Continuous dependence on data in abstract control problems. *J. Optim. Theory Appl.* 43 (1984), 277-306.

ZHIKOV, V.V. [ch.VIII] Averaging of functionals of the calculus of variations and elesticity theory. *Math. USSR Izv.* 29 (1987), 33-66.

ZLOBEC, S. [1] Survey on input optimization. *Optimization* 18 (1987), 309-348.

ZLOBEC, S. - GARDNER, R. - BEN-ISRAEL, A. [1] Regions of stability for arbitrarily perturbed convex programs. *Mathematical Programming with Data Perturbations* I, ed. by A.V. Fiacco, Dekker 1982, 69-89.

ZOLEZZI, T. [1] Su alcuni problemi debolmente ben posti di controllo ottimo. *Ricerche Mat.* 21 (1972), 184-203. [2] Necessary conditions for optimal controls of elliptic or parabolic problems. *SIAM J. Control* 10 (1972), 594-607. [3] Condizioni necessarie di stabilita' variazionale per il problema lineare del minimo stato finale. *Boll. Un. Mat. Ital.* 7 (1973), 142-150. [4] On convergence of minima. *Boll. Un. Mat. Ital.* 8 (1973),

246-257. [5] Su alcuni problemi fortemente ben posti di controllo ottimo. *Ann. Mat. Pura Appl.* 45 (1973), 147-160. [6] On some asymptotic minimum problems. *Rend. Sem. Mat. Univ. Padova* 52 (1974), 93-116. [7] Characterizations of some variational perturbations of the abstract linear-quadratic problem. *SIAM J. Control Optim.* 16 (1978), 106-121. [8] On equiwellset minimum problems. *Appl. Math. Optim.* 4 (1978), 209-223. [9] Well posed optimization problems for integral functionals.*J. Optim. Theory Appl.* 31 (1980), 417-430. [10] A characterization of well-posed optimal control systems. *SIAM J. Control Optim.* 19 (1981), 604-616. [11] On stability analysis in mathematical programming. *Math. Programming Study* 21 (1984), 227-242. [12] Continuity of generalized gradients and multipliers under perturbations. *Math. Oper. Res.* 10 (1985), 664-673. [13] Well posedness and stability analysis in optimization. *North Holland Math. Studies* 129 (1986), 305-320. [14] Stability analysis in optimization. *Lecture Notes in Math.* 1190, Springer 1986, 397-419. [15] Well-posedness and the Lavrentiev phenomenon. To appear on SIAM J. Control Optim.

NOTATIONS.

$AC([0,T])$: space of all absolutely continuous functions on $[0,T]$.

a.e. : almost everywhere.

arg min (X,I) : set of all global minimizers of (X,I).

arg min (y) : page 335.

A' : transpose of A.

$A \subset\subset \Omega$: cl $A \subset \Omega$.

A^ε : set of all points x such that dist $(x,A) < \varepsilon$.

$^\varepsilon A$: set of all points x such that dist $(x,A) \le \varepsilon$.

A^+ : pseudoinverse of A, page 31.

$\alpha(A)$: Kuratowski number of A, page 25.

B : closed unit ball.

$BC^0(X)$: space of all real-valued, continuous functions on X that are bounded from below.

$BLS(X)$: space of all real-valued, lower semicontinuous functions that are bounded from below on X.

$B(r)$: closed ball of center 0 and radius r.

$B(x,a)$: closed ball of center x and radius a.

card J : number of the elements of J.

$C^0(A)$: space of all continuous functions on A.

$C^0([a,b])$: normed space of all continuous functions on $[a,b]$, equipped with the uniform norm.

$CC(X)$: space of all real-valued, continuous, convex functions on X.

cl (A) : closure of A.

Cl (X) : space of all nonempty closed subsets of X.

co A : convex hull of A.

Conv (X) : space of all nonempty closed convex subset of X.

(Conv $(X), I)$: page 48

$C(X)$: space of all extended real-valued, convex, lower semicontinuous functions on X.

$C_0(X)$: space of all extended real-valued, proper, convex and lower semicontinuous functions on X.

$C^1(X)$: space of all real-valued functions that are continuously Fréchet differentiable on X.

$C(X, a)$: set of all extended real-valued, convex, lower semicontinuous functions on X.

diam A : diameter of A.

dist (x, A) : distance between x and A.

dom F : effective domain of the multifunction F = set of the points x such that $F(x)$ is nonempty.

dom I : effective domain of the extended real-valued function I = set of the points x such that $-\infty < I(x) < +\infty$.

$d(r, f, g)$: page 364.

$d(x, y)$: distance between x and y.

∇I : gradient of I.

$\nabla_x I$: gradient of I with respect to the $x-$ variablès.

$\nabla_{xx} f(z)$: Hessian matrix, with respect to the $x-$ variables, of f at z.

$\partial^- f(x)$: page 158.

$\partial I(x)$: subdifferential, or generalized gradient, or generalized Jacobian of I at x.

∂R : boundary of R.

$e(A, B)$: page 55.

epi f : epigraph of f.

epi - lim : page 140.

extr U : set of all extremal points of U.

$\varepsilon - \arg \min (X, I)$: set of all $x \in X$ such that $I(x) \le \varepsilon + \inf I(X)$ if $\inf I(X) > -\infty$; such that $I(x) \le -1/\varepsilon$ if $\inf I(X) = -\infty$.

$F^{-1}(V)$: set of all points x such that $F(x) \cap V$ is nonempty.

gph T : graph of T.

$G(s, w) - \lim$: page 157.

$G(w, s) - \lim$: page 156.

G_δ : countable intersection of open sets.

haus (A, B) : Hausdorff distance between A and B, page 55.

$H - \lim$: Hausdorff limit.

$h(r, C, D)$: page 57.

I^* : conjugate function of I, page 15.

ind K : indicator function of K; $= 0$ on K, $= +\infty$ outside K.

int A : interior of A.

$I(X)$: image of X by $I = \{I(u) : u \in X\}$.

$I_n \to I$ uniformly on bounded sets : page 56.

$JF(x)$: Jacobian matrix of F at x.

kA : set of all points $kx, x \in A$.

(K, A, u) : page 71.

$K - \lim$: Kuratowski limit.

lev (I, t) : set of all points x such that $I(x) = t$.

$\liminf A_n$: set of all points $x = \lim x_n$, for some sequence $x_n \in A_n$.

$\limsup A_n$: set of all points $x = \lim x_j$, for some sequence $x_j \in A_{n_j}$ and some subsequence n_j.

loc. $H - \lim$: page 57.

$LS(X)$: space of all proper, extended real-valued, lower semicontinuous functions on X.

$LSC(X)$: space of all real-valued, lower semicontinuous, convex functions on X.

meas A : Lebesgue measure of A.

$M - \lim$: Mosco limit; page 48 (sets), page 163 (functions).

nbh (x) : set of all open neighborhoods of x.

$0(\varepsilon)$: page 254.

0_k^δ : page 262.

$o(t)$: any function such that $o(t)/t \to 0$ as $t \to 0$.

$p(x, K)$: best approximation to x from K.

$q(s-), q(s+)$: limit of $q(t)$ as $t \to s$ from the left, from the right.

$r_- = $ negative part of the vector r : page 347.

$|r|_\infty$: page 347.

Re $\sigma(A)$: page 250.

$SC(X)$: space of all extended real-valued lower semicontinuous functions on X.

seq. epi-lim : page 126.

sp A : linear subspace spanned by A.

strong lim inf K_n : set of all points x such that $x_n \to x$ for some sequences $x_n \in K_n$.

sub lev (I, t) : set of all points x such that $I(x) \le t$.

supp (D, q) : value at q of the support function of D.

$t_k \downarrow t$: t_k decreases and converges to t.

$t_k \uparrow t$: t_k increases and converges to t.

$u|_A$: restriction of u to A.

$u'v$: euclidean scalar product of u, v in R^n.

$u < 0 \ (u \leq 0)$: every component of the vector u is $< 0 \ (\leq 0)$.

val (X, I) : optimal value of $(X, I) = \inf I(X)$.

val (y) : page 335.

var - lim : page 122.

$V(f)$: total variation of f.

weak seq. lim sup K_n : set of all points x such that $y_p \rightharpoonup x$ for some sequences $y_p \in K_{n_p}$ and some subsequence n_p.

$W^{1,1}(0, T)$: space of all absolutely continuous functions on $[0, T]$.

$W^{1,p}(\Omega)$: Sobolev space of the real-valued functions on Ω, whose first partial derivatives belong to $L^p(\Omega)$.

$W_0^{1,p}(\Omega)$: space of the functions in $W^{1,p}(\Omega)$ that vanish on the boundary of Ω.

$W^{1,\infty}(0, T)$: space of the Lipschitz continuous functions on $[0, T]$.

X^* : dual space of X.

$X-$ convergence in X^* : weak-star convergence in X^*.

(X, I) : problem of minimizing $I(u)$ subject to $u \in X$.

$(X, \text{strong })$: the Banach space X equipped with the strong convergence.

$(X, \text{weak })$: the Banach space X equipped with the weak convergence.

$|x|$: norm of the point $x \in R^n$.

$Y_\varepsilon(t, s)$: page 251.

$\rightharpoonup$: weak convergence.

$\rightharpoonup$: weak - star convergence.

$< \cdot, \cdot >$: duality pairing between X^* and X.

INDEX.

420

Printing: Druckhaus Beltz, Hemsbach
Binding: Buchbinderei Schäffer, Grünstadt